The Essential Naturalist

The Essential Naturalist

Timeless Readings in Natural History

EDITED BY
MICHAEL H. GRAHAM,
JOAN PARKER,
AND
PAUL K. DAYTON

Terri,
This is better
than the cave paintings!
Joan Parker

Terry
Keep Natural History Alive!
Mike Graham

THE UNIVERSITY OF CHICAGO PRESS
Chicago and London

Michael H. Graham is associate professor at the Moss Landing Marine Laboratories at San Jose State University. Joan Parker is the head librarian at Moss Landing Marine Laboratories. Paul K. Dayton is professor at the Scripps Institute of Oceanography at the University of California, San Diego.

The University of Chicago Press, Chicago 60637
The University of Chicago Press, Ltd., London

Printed in the United States of America

20 19 18 17 16 15 14 13 12 11   1 2 3 4 5

ISBN-13: 978-0-226-30569-1 (cloth)
ISBN-13: 978-0-226-30570-7 (paper)
ISBN-10: 0-226-30569-4 (cloth)
ISBN-10: 0-226-30570-8 (paper)

Library of Congress Cataloging-in-Publication Data

The essential naturalist : timeless readings in natural history / edited by Michael H. Graham, Joan Parker, and Paul K. Dayton.
p. cm.
Includes bibliographical references.
ISBN-13: 978-0-226-30569-1 (cloth : alk. paper)
ISBN-10: 0-226-30569-4 (cloth : alk. paper)
ISBN-13: 978-0-226-30570-7 (pbk. : alk. paper)
ISBN-10: 0-226-30570-8 (pbk. : alk. paper)
1. Natural history. I. Graham, Michael H. (Michael Hall) II. Parker, Joan, 1955– III. Dayton, Paul Kuykendall, 1941–
QH9.E87 2011
508—dc22
2011000183

∞ The paper used in this publication meets the minimum requirements of the American National Standard for Information Sciences—Permanence of Paper for Printed Library Materials, ANSI Z39.48-1992.

Contents

Acknowledgments

There was a time, not too long ago, when MHG and PKD (student and mentor) would spend hours discussing "natural history gems," lost papers penned by adventurous souls who spent their lives wandering, observing, thinking, and writing, only to have their words filed away out of the reach of the casual natural history student . . . presumably their intended audience. It excited us when we came across a prescient paper that was clearly in line with much of the current ecological thinking of our day, but was overlooked because it was not written in English; had no statistics; appeared in a memoir, symposium proceeding, or bulletin; was found only through scouring the literature cited sections of other papers (what to do for papers published before such sections became common!); or was written by an unknown author. We became convinced that the resurrection of these papers would be of great value to contemporary ecologists and evolutionary biologists. Therefore, MHG organized a symposium for the Western Society of Naturalists (WSN) annual meeting in Monterey, California, December 1999, when PKD was the WSN president. The title of the symposium was "Paradigms in Ecology: Past, Present, and Future." The goal was to bring together leaders in ecology and evolutionary biology to debate the greatest developments in their fields and to honor the contributions of their natural history heroes. The symposium led to the publication of a special feature on "Paradigms in Ecology" in the journal *Ecology* in 2002. The success of that symposium and the genesis of this project were due mostly to the enthusiasm and curiosity of the WSN students and their mentors, to the invited speakers (Shahid Naeem, Carlos Robles, Jeremy Jackson, Peter Grant, Mark Hixon, David Reznick, and Bob Paine), and to the generous financial support of the Hall Family Foundation. This project would not have come together without their contributions.

Upon publication of the special feature, MHG and PKD began researching papers that might serve as a basis for a natural history compendium. We surveyed hundreds of scientists for their natural history gems and spent countless hours researching the hundreds of leads. We thank all of these colleagues for their wonderful suggestions. It was a pleasure to spend so many hours in the library, and we are sorry that space limitations hindered our ability to resurrect everyone's favorites. We also used graduate student seminar courses at the Moss Landing Marine Laboratories to winnow down the candidate papers, and again we thank the participating students and faculty for their enthusiastic efforts. But we still found ourselves overwhelmed by the literature and in need of professional help. In 2004, JP joined our crusade, bringing with her the invaluable skills of a librarian of science and an intrinsic desire to get original contributions into the hands of contemporary readers. Our group of section editors was assembled soon after, and a long list of contributed papers was developed. We greatly thank these colleagues (Bob Paine, Gage Dayton, Harry Greene, Nancy Knowlton, Shahid Naeem, and Peter Grant) for the *years* of dedication and patience to see this project through. The library staff at MLML made special efforts to handle innumerable special requests generated by this project; Laurie Hall and Gala Wagner Haskins deserve special thanks. We thank Lajos Szoboszlai and Richard Palmer for their generous time in translating original French writings into English. We have also received continued support and enthusiasm from our editor, Christie Henry, at the University of Chicago Press. Christie began helping us formulate our plans in 2002 and has seen us through many revisions, contributed papers, titles, and even children. Her dedication to this project was critical to our success, and we thank her, Cathy Pfister, and the rest of the UCP Board of Editors for their support.

Finally, it would have been impossible to complete a project of such magnitude without the tirelessness of our families. MHG thanks his parents for helping to keep him financially stable as a student during the beginning of the process, and especially to his father, Jon Hall, for his support of the original symposium, the random gifts of natural history gems from British rare book stores, and his willingness to sit up late at night and discuss the history of natural history. MHG thanks Lara Ferry for seeing this project through to the end. Finally, MHG thanks his children Ryan and Alison for being so curious about nature, giving us hope that the next generation will share their interest. JP also thanks her parents, who long ago established foundations for her academic career. They encouraged inquisitiveness and curiosity, taught her to read almost before she could walk, and picked up the bills for many, many years of college.

They should never forget that any past, present, and future successes are a direct product of their love and support. PKD has long worried that the age of fast electronic literature searches has truncated the understanding of the rich history of ecology and has offered a course on the "classics" of environmental sciences. We hope that this book will inspire scholars to appreciate their intellectual foundations and dig out other important but often forgotten papers.

From the Editors

The production of this book was made difficult by the various formats of the texts (e.g., books, notes, reports, and formal scientific papers), languages, vernacular, and prose. In each case, we strived to reproduce the writing in a format that maintained the author's original intent, at least as we perceived the intent to be. It was, however, inevitable that we omit certain figures, tables, references, footnotes, etc. from various writings in order to provide a final product that could be appreciated by the widest possible audience. It was also necessary to add certain text to help define words or provide context; in all cases these additions are listed in italics within brackets. We apologize if any of these omissions or modifications prove offensive to the readers, as this was not our intent. We hope that this book will provide an impetus for interested readers to explore the original texts, in their original form, and we have included explicit formal citations at the end of each reading to facilitate this pursuit.

A Foundation Built by Giants

Michael H. Graham

The attraction of humans to nature is clear and ubiquitous. Whether it is a means to find resources for survival, to promote commerce, or to stimulate the senses, the understanding of how natural organisms live within their environment is of intrinsic value to all people. At the beginning of human history, cave paintings and early writings focused largely on descriptions of nature. Over millennia, such descriptions evolved into advanced techniques for identifying organisms, complex schemes for classifying them, intricate models for studying their interrelationships, and numerous scientific societies for debating the results. In the past, all of these scientific pursuits would have been termed Natural History, and the practitioners natural historians or naturalists. Today, answers to complicated questions about the functioning of the natural world are pursued by thousands of ecologists, evolutionary biologists, taxonomists, organismal biologists, paleontologists, environmental scientists, etc. Indeed, many debate whether Natural History remains a valid scientific pursuit, what the characteristics of Natural History really are (is it a purely descriptive field?), and whether natural historians are equipped for hypothesis-oriented science. Despite such debate, the utility of understanding how nature works remains timeless and has never been denied.

We define Natural History as *the systematic study of natural organisms through observations*. We prefer this definition to others because it focuses attention on the organisms (as opposed to molecules) and observations (as opposed to models), yet it does not explicitly state the forms of study. We further assert that it follows that Natural History writing is the *description of the results of such studies* and that such writing may take different forms: from colloquial nature writing to expeditionary reports to more formal scientific pursuits. Our goal with this book has never been to exhaustively explore the characteristics of Natural History nor document its historical development; the ongoing series of Egerton (2001–present) is a sensational resource for such activities. The contributors to this book are all naturalists, and in our own ways we have defended and promoted the utility of Natural History in science through our politics, research, teachings, and writings (e.g., Grant 2000, Graham and Dayton 2002, Dayton 2003, Greene 2005). This book is also not an attempt to get readers to experience nature firsthand, although we believe that this is a likely and meaningful outcome. Instead, we advance here the simple idea that our current understanding of how the world functions comes from the collective knowledge of generations of naturalists that came before, and exist among, modern-day scientists. Borrowing from the classic metaphor of Bernard of Chartres (John of Salisbury 1159), we consider that contemporary naturalists are today viewing nature from the "shoulders of giants" whether they recognize it or not. Finally, our only tangible connection to the contributions of these giants is through their writings. Thus, our explicit goal is for this book to serve as a means of promoting the *reading* and *writing* of Natural History and as an entry point for the exploration of original literary contributions by early and contemporary naturalists. These are essential activities for scientists and nonscientists alike.

The papers selected for this book are arranged around central themes that focus on of the diversity of Natural History writing forms and the subsequent utility of Natural History literature in the modern world: inspiration, exploration, initiation, intuition, and unification. Many of the papers are old and rarely encountered through traditional literature searches, or were published in sources that are difficult to obtain; these may serve as a primary source for Natural History students. Given the breadth of Natural History, however, the selections can by no means be considered representative of the best that Natural History has to offer. Indeed, readers must acknowledge that many of the facts or interpretations within the selected papers have ultimately been revised, and that some of the concepts forwarded by the author may appear as naïve relative to modern perspectives. Goode (1886) recognized this in his presidential address to the Biological Society of Washington:

> There were then, it is certain, many men equal in capacity, in culture, in enthusiasm, to the naturalists of today, who were giving careful attention to the study of precisely the same phenomena of nature. The misfortune of men of science in the year 1785 was that they had three generations fewer of scientific predecessors than have we. Can it be doubted that the scientists of some period long distant will look back upon the work of our own times as archaic and crude, and catalogue our books among the "curiosities of scientific literature?"

Four generations later, Goode's words ring ever more true. Our goal, therefore, was simply to select interesting readings that advance our chosen themes. The sections are introduced with commentaries by contemporary naturalists that provide additional entry points into the Natural History literature. The commentaries represent the diversity of impacts that Natural History has upon its practitioners; they are inherently personal. Together, these readings and commentaries are a unique anthology of Natural History and a modern testament to the timeless nature of the field.

Most naturalists can list the seminal readings that were influential in shaping their views of nature and their desire to pursue Natural History. Such papers effectively put visions of nature into words and are often sensational, utilizing a variety of personal writing styles. Certainly the first papers read by young naturalists can tend towards the colloquial, with little interest by the authors in making direct scientific contributions. Nonetheless, by definition, these papers are inspirational. In the first section, "Inspiration," Robert Paine introduces a series of rousing papers by drawing upon a lifetime interest in Natural History and in the education and development of naturalists. The reader is confronted with feelings of adventure, disappointment, and devotion to nature, and may be left with desires to explore, investigate, conserve, and put on a jacket; this is the essence of "nature writing."

After inspiration, the next step in practicing Natural History is obviously the collection of personal observations of nature. Although such observations can occur anywhere and anytime, some of the most useful and compelling Natural History contributions have come from naturalists observing new things, in new places, at just the right time. For thousands of years, colonialism, exploration, and even missionary activities have placed learned people and keen observers in distant unknown places of the world. In many cases, subsequent writings of these naturalists remain as the sole descriptions of unique species, habitats, and human cultures. In the second section, "Exploration," Gage Dayton, Paul Dayton, and Harry Greene demonstrate the importance of expeditionary reports for understanding the natural history of remote lands, and the serendipitous discoveries that often arise during such explorations. Whether on a distant island, aboard a royal yacht, or in the outskirts of Texas, these contributions make lasting impacts on our understanding of the functioning of our world and provide a unique baseline against which to gauge the impact of contemporary human activities.

That many of the world's most famous and successful ecologists and evolutionary biologists began with naturalist roots is evidence of the modern utility of Natural History. The most popular and well-cited papers of the most influential natural scientists have been reprinted and commented on elsewhere (e.g., Real and Brown 1991, Lomolino, Sax, and Brown 2004), clearly demonstrating the authors' perception and superior contributions to their fields. Rarely, however, do these reprinted works identify Natural History as the almost universal stage upon which these giants learned their trades. In "Initiation," Nancy Knowlton highlights early contributions to Natural History by eminent scientists that have been overshadowed by the popularity of their subsequent work. In no case will these papers be recognized as the seminal works of the authors, but in every case they demonstrate good Natural History as a prerequisite for success in natural sciences.

In *The Structure of Scientific Revolutions*, Thomas Kuhn (1962) described in detail how society's inability to accept, explain, or even understand certain scientific contributions has historically repressed the contribution's impact on our understanding of how nature works. The inherent breadth of Natural History as a field, and the focus on detailed observation and exploration, often places naturalists at the frontiers of science. In the fourth section, "Intuition," Shahid Naeem introduces a series of naturalists whose observations uncovered patterns that would become fundamental ecological/evolutionary principles beyond their time. Some of the contributions were representative of distinguished careers in which a new technique, theory, or perspective continuously placed the naturalist at the forefront of discovery. For example, in addition to his codiscovery of the theory of natural selection with Charles Darwin (Darwin and Wallace 1858), Alfred Russel Wallace also unknowingly founded the field of biogeography through a series of works including *The Malay Archipelago* (Wallace 1869), *The Geographical Distribution of Animals* (Wallace 1876), and *Island Life* (Wallace 1880). *Island Life* in particular, which far predates MacArthur and Wilson's (1967) *Theory of Island Biogeography*, could still serve as a foundation for a modern biogeography course, although the following quote from the introduction shows that even the most gifted giants have flaws:

> Since, if we once admit that continents and oceans may have changed places over and over again (as many writers maintain), we lose all power of reasoning on the migrations of ancestral forms of life, and are at the mercy of every wild theorist who chooses to imagine the former existence of a now submerged continent to explain the existing distribution of a group of frogs or a genus of beetles. (Wallace 1880, 10)

Other contributions in this section reflect unique accomplishments that, though groundbreaking, never achieved much attention or recognition and need not be branded as transformational; they were simply ahead of their time. The discovery of such gems by the literature-savvy naturalist can be rewarding.

Today, some may argue that Natural History is too descriptive for naturalists to make useful contributions to the process-oriented studies of ecology and evolution. Yet again, in our view, naturalists simply pursue an understanding of how nature functions: the same simple goal of ecologists and evolutionary biologists. In its infancy, Natural History was inherently descriptive, as there were few established patterns in nature to warrant a search for explanatory processes. Everything changed with the events leading up to and including the publication of *The Origin of Species* (Darwin 1859), after which evolutionary biology and ecology emerged as theory-driven disciplines for understanding nature. Still, process is irrelevant without pattern, and thus ecological and evolutionary contributions suffer in the absence of the context provided by Natural History. In the final section, "Unification," Peter Grant develops a chronicle of readings to demonstrate the union of ecology and evolutionary biology in a contemporary Natural History framework. Thus, Grant helps to redefine Natural History in a modern perspective by casting doubt on the depiction of natural description and pattern formation as antiquated pastimes destined to oblivion.

We learned much in developing this Natural History anthology, uncovering new leads with every new citation, bibliography, or passing reference. Each day we found a new paper, a new author, and a new perspective, many of which could easily have been included in this book. It is to the endlessness, the perpetuity, the complexity, and the integrity of the Natural History literature that modern natural scientists owe their success. In the end, we owe everything to these often-faceless giants and their Natural History foundation upon which our contemporary understanding of nature has been built. We hope that the words in this book will promote interest in both the past and future study of nature . . . our Natural History.

LITERATURE CITED

Darwin, C. 1859. On the Origin of Species by Means of Natural Selection, or the Preservation of Favoured Races in the Struggle for Life. John Murray, London.

Darwin, C., and A. R. Wallace. 1858. On the tendency of species to form varieties; and on the perpetuation of varieties and species by natural means of selection. Journal of the Proceedings of the Linnean Society of London. Zoology 3: 46–50.

Dayton, P. K. 2003. The importance of the natural sciences to conservation. American Naturalist 162: 1–13.

Egerton, F. N. 2001–present. A history of the ecological sciences. Ongoing series in the Bulletin of the Ecological Society of America.

Goode, G. B. 1886. The beginnings of natural history in America. Proceedings of the Biological Society of Washington 3: 35–105.

Graham, M. H., and P. K. Dayton. 2002. On the evolution of ecological ideas: Paradigms and scientific progress. Ecology 83: 1481–1489.

Grant, P. R. 2000. What does it mean to be a naturalist at the end of the twentieth century? American Naturalist 155: 1–12.

Greene, H. W. 2005. Organisms in nature as a central focus for biology. Trends in Ecology and Evolution 20: 23–27.

John of Salisbury. 1159. The metalogicon: A twelfth-century defense of the verbal and logical arts of the trivium. Translated from Latin by D. D. McGarry (1962). University of California Press, Berkeley.

Kuhn, T. S. 1962. The Structure of Scientific Revolutions. University of Chicago Press, Chicago.

Lomolino, M. V., D. F. Sax, and J. H. Brown (eds.). 2004. Foundations of Biogeography: Classic Papers with Commentaries. University of Chicago Press, Chicago.

MacArthur, R. H., and E. O. Wilson. 1967. The Theory of Island Biogeography. Princeton University Press, Princeton, New Jersey.

Real, L. A., and J. H. Brown (eds.). 1991. Foundations of Ecology: Classic Papers with Commentaries. University of Chicago Press, Chicago.

Wallace, A. R. 1869. The Malay Archipelago; The Land of the Orang-utan and the Bird of Paradise; A Narrative of Travel With Studies of Man and Nature. Macmillan, London.

Wallace, A. R. 1876. The Geographical Distribution of Animals; With A Study of the Relations of Living and Extinct Faunas as Elucidating the Past Changes of the Earth's Surface. Macmillan, London.

Wallace, A. R. 1880. Island Life: Or, The Phenomena and Causes of Insular Faunas and Floras, Including a Revision and Attempted Solution of the Problem of Geological Climates. Macmillan, London.

Inspiration

Robert T. Paine

NATURE, NATURAL HISTORY, AND NATURALISTS

Louis Agassiz's dictum "study nature, not books" remains eternally current. Remediation and conservation based on understanding the resident species' ecological quirks and needs are likely to be less successful and effective without the deeply rooted knowledge let alone appreciation of how nature functions. But how does one become a student of nature? Access to the living world, or *nature*, that astonishing tapestry of living things, their patterns, activities, smells, sounds, and interactions is essential. Yet the accumulated intimate knowledge of organisms, their *Natural History*, is becoming more difficult to acquire; Natural History requires access to nature. Here I focus on a third and related term, *naturalist* or *natural historian.* Becoming and being a naturalist requires more than mere enthusiasm—it requires the development of many talents. For instance, many people, young or old, lay or scientific, are committed to conservation or addicted to some favored taxon (e.g., falcons, whales, sea otters). These people are not necessarily naturalists. The paragraphs that follow provide my perceptions on factors contributing to a naturalist's ontogeny. That development begins with ample opportunity to observe nature, and surely adult encouragement is an essential ingredient. But it also seems to require intense curiosity about her workings, a freedom to roam, and a capacity to survive, even enjoy, being a loner. Bates (1950) stressed that the necessary observational skills develop early; they are difficult if not impossible to teach later on. Bates wrote (1950, 254), "The commonest first sign of a developing naturalist is the collecting habit."

Fundamental assumptions of this viewpoint are that a child can read and has access to books, and that a deep fascination with the organic world is subsequently nurtured or at least not suppressed. Thus the intuition, observational skills, and interpretative abilities will develop early. Superb naturalists of course existed well before the printing press—witness stone age cave art—suggesting that there is no requirement that a naturalist be literate. But in the presence of today's mounting environmental crises, an interest in the science underlying biological patterns, in addition to advocacy and adoration, is required. I suspect that intuitive naturalists are increasingly uncommon in academic and regulatory circles because of decreasing access to natural environments, as suggested by Balmford et al. (2002). They asked British children aged four to eleven to identify flash cards of both common native wildlife and Pokemon "species." At age four, identification success of wildlife was substantially greater than that of Pokemon species, 32% vs. 7%. By age eight, children were "typically identifying Pokemon 'species' better than organisms such as oak trees and badgers" (Balmford et al. 2002, 2367). There can be little room for optimism in these results.

For these reasons, the writings in this section do not embody hard, data-rich science. Their general hallmarks are some complex of biology and adventure. They are never tainted by detailed statistical tables. The writing, usually, is entirely comprehensible to the six- to thirteen-year-old reader. Humor and those anthropomorphisms forbidden in more "scientific" literature are conspicuous. The authors were observers, popularizers, even adventurers, but by catching the excitement and enthusiasm for their experiences, they were able to inspire young, impressionable minds. Scientists know that animals communicate: owls hoot, crickets chirp, whales sing and even have dialects. Humans routinely talk to their pets—"here kitty, kitty," "bad dog"; some even teach their parrots to swear. Receptive children are left with the perception that animals communicate, a perception enhanced by a wealth of children's books. The Brothers Grimm (e.g., in their story of Little Red Riding Hood) mastered the technique. I learned from Thornton W. Burgess that porcupines don't throw their quills, and ineffectively challenged my fourth-grade teacher with this factoid. Rudyard Kipling's Jungle Books have mongooses talking to cobras and birds (Rikki-Tikki-Tavi), or Mowgli discussing life with a bear, wolves and even a tiger. Such communications between humans and animals, or at least intimate associations, extend to some of the earliest human writings, from Sumaria, in the third millennium BC (Sanders 1972). Aesop's fables and the story of Romulus and Remus continued that tradition. Thus, at an early age the boundary between fantasy and reality is blurred, with many of the stories presenting the natural world as an interesting place. I believe this contributes to both a fear and loathing of nature but also an abiding fascination.

This part of the book introduces a hierarchy of inspirational readings for aspiring naturalists. First, keen observers describe the intimate details of their chosen taxon. Anthropomorphisms may abound, but they never diminish the basic message. Second, the adventure of biological exploration is illustrated with three works from the human perspective and one from that of the organisms. These writings were never intended to be propaganda or advertisement for the importance of Natural History. Nonetheless, they achieved that goal. And finally, we identify some synthetic analyses by a renowned naturalist. How does one go from single species to a biologically complex universe? An observant child can't wander outdoors without seeing robins yanking earthworms from the soil, caterpillars chewing holes in leaves, or starfish eating mussels; children may enjoy catching fish on natural lures. The sections that follow thus represent both personal favorites—writings that triggered my youthful interests—and, more importantly, communicate the challenge and adventure that knowing something special about the fabric of nature endows.

FORMATIVE READINGS

My first nature love was birds, though butterflies and salamanders were close seconds. I had ready access to all three volumes of Edward Forbush's (1858–1929) *Birds of Massachusetts* (1927), which I read with enthusiasm, skipping the prescient parts dealing with hurricanes and other environmental issues while focusing on the adventures of his naturalist correspondents. The imagery was captivating:

> A gang of men were startled one winter's day a few years ago in Medfield woods by the sight of what seemed like a huge bird "with four wings" whirring past. The "bird" dropped into the snow not far away. It turned out to be a Goshawk and a Barred Owl locked in deadly clasp. Both were dead when picked up. (Forbush 1927, 2: 122)

Such accounts can excite a youthful imagination. Nature, red in tooth and claw, is an often-violent venue, but it provides a stimulation that must be healthier than the brutality permeating much of today's TV and computer games. I spent many happy hours searching the local woodlands, finding at best only a few blue jay feathers. Forbush's writing style was hardly dry or boring. More importantly, the language employed was understandable, and clearly conveyed the importance, for example, of knowing the Natural History of great horned owls or black-throated blue warblers.

Another example of inspirational writings is Jim Corbett's (1875–1955) *Man-eaters of Kumoan* (1944). Many of his stories are packed with drama, suspense, and even personal danger, as he tracked down man-eating tigers and leopards in rural India. His understanding of the basic ecology and Natural History of many of the surrounding species form the environmental context. The thrill of the chase was a central theme, but that was broadly supplemented by behavioral observations, his interpretations of sundry alarm calls, and equally, a very humanistic regard for the local human population. I learned much later that Corbett was given a very rare award, "Freedom of the Forests," for his services to rural India. Its title conveys his compassion for and understanding of all the forest's residents. I also recognized later that both Forbush and Corbett were avid hunters, and their observations of nature were of central understanding to their success. However, both evolved into ardent conservationists; tigers, as Corbett feared, are now nearly extinct on the Indian subcontinent (Check 2006). The two brief examples reprinted here should serve both as an appetizer to read more of Corbett's writings (Hawkins 1978) but also, taken together, as a convincing statement about the continuing satisfaction of knowing a little about our natural world. Hawkins, Corbett's editor, indicates that the original tiger footage shot by Corbett is preserved in the British Film Archive. Remarkably, doubts have been expressed whether Corbett wrote his own stories. He was sixty-nine when his first book was published; Hawkins dismisses this, even while suggesting that help from Corbett's sister enhanced the enticing style.

Three examples round out our choices for formative readings, though many others could be found. Emmett Dunn's (1894–1956) description of the lure of salamanders expresses the retrospective thoughts of a mature scientist. By his account (Dunn 1926), at eighteen he had written to an authority for advice; the frank response, essentially that *salamanders need study*, launched a lifetime of research and adventure minimally glimpsed in the paragraphs reprinted here. Jean Fabre (1823–1915) was a superb self-taught naturalist who spent an isolated life, until near its end, observing insects in rural France. The sphecid wasps are large, solitary, deadly-looking, but passive creatures. They catch, subdue, lay their eggs in, and then bury large prey: grasshoppers, locusts, and spiders, for instance. Fabre's account in *The Hunting Wasps* (in Teale 1949) is captivating, both for the detail and his seemingly boundless curiosity about how to substantiate his preliminary interpretations with experimental manipulations. Teale's (1949) accounting of Fabre's work provides numerous other examples of his observational genius. John Crompton's (1893–1972) descriptions of the intimate details of spider life history reek with enthusiasm for the taxon. Spiders were sacred objects in the house I grew up in. One could easily observe orb weavers at work, usually on house flies deprived of their wings, or see wolf spiders skulking around the basement. Crompton's book (1954) provided sometimes-lurid details, often in a language not permitted in more scientific literature. But it seemed biologically accurate (one could repeat his observations), and it added greatly to a youth's interest.

The art of seductive writings like those of Fabre and Crompton has not been lost, and numerous modern examples exist. I identify four as a sampler. Greene (1997) introduced a book on snakes with a story about fer de lances. It's an enticing beginning, and a reader is never disappointed in further reading. Pianka and Vitt (2003), describing the biology of lizards, are equally successful. The authors' autobiographical statements attest to the potent attractions of nature; the subsequent text holds one's attention. Brodo, Sharnoff, and Sharnoff (2000) present lichens as "the essence of wildness." As in the other taxon-centric examples, stunning photographs effectively portray the beauty of our organic world while the text develops the underlying Natural History and ecology. Finally, Holldobler and Wilson's (1990) treatise, while more scientific, and purposely intended for a more mature audience, presents ants with an enthusiasm and deep knowledge that would be contagious for readers of any age. These four examples reflect the face of modern Natural History, unavoidably more quantitative and taxonomically explicit than their predecessors. They delve into the biological nuances of specific taxa, and are written with an infectious enthusiasm that cannot be mistaken by a literate audience of any age.

BIOLOGICALLY MOTIVATED ADVENTURE

A second category of readings may captivate young minds with the pure adventure of biological endeavor. Adventure dominates these writings, but the heroics are often biologically motivated. Dinosaurs are the initial sparks of inspiration for many small children, including Stephen Jay Gould. Gallenkamp (2001) has recounted Roy Chapman Andrews' explorations of the Mongolian Gobi desert in the 1920s, which brought dinosaur diversity and ecology to many. William Beebe's (1877–1962) exploration of the deep oceans while submerged in a bathysphere shouldn't be read by the claustrophobic. The sample reprinted here (Beebe 1934) catches the sense of the excitement of observing midwater fishes, in their natural medium, for the first time.

The account by Apsley Cherry-Garrard (1886–1959) of a south polar, midwinter, man-hauled sledding expedition to collect emperor penguin eggs makes one cold just in the reading. Biology mattered to the participants; they were willing to die for it (Cherry-Garrard 1930). Difficult choices had to be made in choosing a sampler from the seventy-one pages of unadulterated hardship, commitment, and camaraderie. *The Winter Journey* presents an astonishing melding of penguin biology and human endurance against formidable odds.

Another reading to be avoided by the claustrophobic is a brief account by Peter Nicholson (1945–) of his love affair with wombats (Nicholson 1963). He crawled into their spacious subterranean tunnels, became close friends with one, and offers fascinating observations on their "cultural" diversity. These Australian marsupials come alive in his writing; he was about years old at the time of these observations and adventures.

There are equally captivating yet vicarious accounts of species life histories. Rachel Carson's (1907–1964) description of the American eel (*Anguilla rostrata*) is disturbingly anthropomorphic to an adult reader, but fully captivating to a child insensitive to such imperfections. Her account in the *Odyssey of the Eel* (1941) of an individual's life history builds on the risks of migrating from fresh to salt water and back. The tale conveys the life cycle complexity in dramatic yet simple prose. The message is one of mystery, danger, and ultimate survival.

Biological adventures are appealing, and when told effectively, as by Beebe, Cherry-Garrard, and Nicholson, can be infectious. But so are the explorations of the American west by Lewis and Clark (Jones 2002) and the adventures of Bering and Steller along the Aleutian chain (Frost 2003); some of Steller's detailed observations of the Commander Islands are reprinted here in the "Exploration" section. Our increasingly impaired natural environment diminishes the chances to experience biological exploration. Biological science will be the poorer when most real-time hands-on experiences, at whatever spatial scale, are replaced and must be obtained vicariously.

SYNTHETIC NATURAL HISTORY

This part of the book concludes with an example of an outstanding naturalist generalizing his observations by placing them into a broader context. Several such works were available. An early work, Forbes's essay *The Lake as a Microcosm* (1925) remains a remarkably modern synthesis of habitat and trophic interactions. The lake assemblage is portrayed as interactive and dynamic, with black bass playing a very significant role. Though its accessibility to young readers, and thus its motivational potential, are questionable, nonetheless it describes a complexity certain to appeal to a mind curious about how nature works. Because it has been reprinted elsewhere (Real and Brown 1991), we have not duplicated it here. Our chosen example (MacGinitie 1938) was clearly aimed at a scientifically educated audience but reflected a lifetime of careful observation and contemplation. George MacGinitie (1889–1989) continued to expand and develop his talents as a natural historian throughout a long and productive career. His basic credo was:

> Individual life histories of members of the community are absolutely necessary to a fuller understanding of the sociology of marine animal communities. (MacGinitie 1938, 52)

The excerpt reprinted here was chosen to illustrate not only his taxonomic breadth and keen eye for detail, but also his willingness to tinker with specimens in the laboratory and speculate on such currently popular issues as dispersal and disturbance.

I must mention two more recent examples of collating years of accumulated Natural History into a synthetic view of community structure and function. Charles Elton, arguably the founding father of community ecology, spent more than twenty-five years studying Wytham Woods close to Oxford, U.K. He hoped (Elton 1966, 17) to produce a generalized framework for all community studies, because "the ecological enquirer is, more than most scientific people, apt to find himself lost in a large labyrinth of interactions and variables." One of Elton's early papers is reprinted here in "Initiation." A more recent book written in the same spirit confronts a vastly more diverse community, a tropical rain forest. Leigh (2002) focuses on diversity, patterns and interactions, often accompanied by stunning photographs. His book catches the eye, and should encourage interest, exploration, and synthesis. Further examples of what could be termed "applied Natural History" exist. Rosenfeld with Paine (2002) illustrate the hallmarks of biological interactions on marine rocky shores, and discuss their consequences. Wessels's *Reading the Forested Landscape* (1997), permits one to reconstruct the history of New England woodlands by knowing how to interpret subtle signs, much like a detective at a crime scene.

A FUTURE GENERATION OF NATURALISTS?

Becoming a naturalist is a road increasingly less traveled for a host of obvious reasons: dwindling habitats, species extinctions, social pressures on the young, and the dangers associated with just roaming around observing and poking into the dark corners of nature. For our crowded planet, even the concept of "freedom of the forest" may be vanishing. On the positive side, however, lie the stunning array of field guides, educational TV programs, modern zoos that often realistically mimic a species environment, community gardens, and hosts of taxon-focused societies, many of which sponsor field trips. But acquiring the intuition of a naturalist requires at least two further components. The first is a curiosity extending beyond simple attraction and enthusiasm, often expressed as handling, poking, observing, and collecting fragments of nature. When my daughters were young I took them and friends on skunk cabbage–smelling expeditions. A great time was had by all; only their clothes suffered. The second requires a step beyond curiosity, dealing with the broader implications of whatever was observed. For instance, last night a coyote ate a neighbor's cat in my backyard. Where is the child who will wonder what this means for the resident songbirds, as in Crooks and Soule (1999)? Nature is dramatic at every scale, packed with beauty and action, as the readings in this section demonstrate. Youthful curiosity will always be present. The challenge is how to expand it to the level of that intuition essential to interpreting Darwin's "tangled bank."

LITERATURE CITED

Balmford, A., L. Clegg, T. Coulson, and J. Taylor. 2002. Why conservationists should heed Pokémon. Science 295: 2367.

Bates, M. 1950. The Nature of Natural History. Scribner, New York.

Beebe, W. 1934. Half Mile Down. Harcourt Brace and Company, New York.

Brodo, I. M., S. D. Sharnoff, and S. Sharnoff. 2000. Lichens of North America. Yale University Press, New Haven.

Carson, R. L. 1941. Under the Sea Wind. Simon and Schuster, New York.

Check, E. 2006. The tiger's retreat. Nature 441: 927–930.

Cherry-Garrard, A. 1930. The Worst Journey in the World. Dial Press, New York.

Crompton, J. 1954. The Life of Spiders. New American Library of World Literature, New York.

Crooks, K. R., and M. E. Soule. 1999. Mesopredator release and avifaunal extinctions in a fragmented system. Nature 400: 563–566.

Dunn, E. R. 1926. The Salamanders of the Family Plethodontidae. Smith College, Northampton, Massachusetts.

Elton, C. E. 1970. The Pattern of Animal Communities. Butler and Tanner, Ltd., Frome.

Forbush, E. H. 1927. Birds of Massachusetts and Other New England States. Vol. 2. Norwood Press, Norwood, Massachusetts.

Frost, O. 2003. Bering. Yale University Press, New Haven.

Gallenkamp, C. 2001. Dragon Hunter. Penguin Books, New York.

Greene, H. W. 1997. Snakes: The Evolution of Mystery in Nature. University of California Press, Berkeley.

Hawkins, R. E. 1978. Jim Corbett's India. Oxford University Press, Oxford.
Holldobler, B., and E. O. Wilson. 1990. The Ants. Harvard University Press, Cambridge.
Jones, L. Y. 2002. The Essential Lewis and Clark. Harper Collins, New York.
Leigh, E. G. 2002. A Magic Web: The Forest of Barro Colorado Island. Oxford University Press, Oxford.
MacGinitie, G. E. 1938. Notes on the natural history of some marine animals. American Midland Naturalist 19: 207–219.
MacGinitie, G. E. 1939. Littoral marine communities. American Midland Naturalist 21: 28–55.
Nicholson, P. J. 1963. Wombats. Timbertop Magazine, Geelong Grammar School 8: 32–38.
Pianka, E. R., and L. J. Vitt. 2003. Lizards. University of California Press, Berkeley.
Real, L. A., and J. H. Brown (eds.). 1991. Foundations of Ecology: Classic Papers with Commentaries. University of Chicago Press, Chicago.
Rosenfeld, A. W., with R. T. Paine. 2002. The Intertidal Wilderness. University of California Press, Berkeley.
Sanders, N. K. 1972. The Epic of Gilgamesh. Penguin Books, London.
Teale, E. W. 1949. The Insect World of J. Henri Fabre. A. Teixeira de Mattos, trans. Dodd, Mead, New York.
Wessels, T. 1997. Reading the Forested Landscape. Countryman Press, Woodstock.

The Great Horned Owl (1927)

Edward H. Forbush

The Great Horned Owl is the most morose, savage and saturnine of all New England birds. We can hardly wonder that certain Indian tribes regarded this fowl as the very personification of the Evil One, or that they feared its influence and regarded its visits to their dwellings as portentous of disaster or death. Brewster writes as follows of the Great Horned Owl:

> Despite his dignified bearing and handsome plumage, the Great Horned Owl does not make a pleasing addition to the aviary, for of all New England birds he is perhaps the most irredeemably morose and untamable. Even a Rattlesnake is scarce less amenable to kindly human advances, but whereas it may resent them with active and perhaps fatal aggression, the

> Owl will be no more than coldly unresponsive provided he is not touched. Much of what is doubtless his true nature may be read in his staring, yellow-girdled eyes. Soulless at all times, they express either vacuous or sullen indifference when he is resting undisturbed; malignant resentment, should he be annoyed or threatened; bloodthirsty lust, when his gaze is fixed on coveted prey. Yet he is not addicted to quarrelling with other large birds, even when closely confined with them, or to killing beyond his need for food. Such forbearance might redound more to his credit were it less obviously the outcome of mere selfish distaste for needless exertion of every kind. If he be ever moved to affection for any living creature, except, perhaps, for his mate, with whom he is accustomed to pair for life, the existence of such emotion is certainly not betrayed by any outward sign.

The Great Horned Owl is not only the most formidable in appearance of all our owls, but it is the most powerful. The Great Gray Owl and the Snowy Owl may appear larger but the Great Horned Owl exceeds them in courage, weight and strength. Indeed it little regards the size of its victim for it strikes down geese and turkeys many times its weight, and has even been said at times to drive the Bald Eagle away from its airy and domicile its own family therein. In one case the owl laid its eggs in a cavity in the side of an eagle's nest and both families occupied the same habitation.

Recently a correspondent wrote to me saying that a Horned Owl had struck the claws of both feet into the back of his large collie dog. This bird may have been misled by a white patch on the dog, as the white on the back of a skunk is its favorite mark. Any white moving object is likely to attract the attention of this owl at night and bring on an attack, and it may be that old Bubo strikes at the white head of the eagle during the hours of darkness when the larger bird is at a disadvantage. Dr. William Wood wrote to Dr. Merriam the story of a man who was riding on horseback at night through a large tract of woods and wearing a white beaver hat, when something struck his head with such force that it took his hat, and for a time he feared that the top of his head had gone with it. The attack was noiseless and the poor man, thinking that he was pursued by Satan himself, fled at top speed for about three miles. I came very near to a similar experience in the autumn of 1918. One night before retiring, while wearing a white canvas hat, I stepped out behind my cabin in the Wareham woods for a little gymnastic exercise. I had barely begun the calisthenics when just above my right ear there sounded a startling *waugh* followed by a string of vituperative owl language, which drifted swiftly away toward a pine tree on the edge of the woods where some guinea fowls were want [*sic*] to roost. The wings of that owl must have almost brushed my ear, yet no sound of them was audible. Probably the great bird, watching from the top of some tall pine, had seen my moving hat in the moonlight, and as he shot over the cabin roof had realized his mistake just in time.

Sometimes when the nest of the Horned Owl is disturbed, the bird attacks the intruder and inflicts painful wounds, though usually it is extremely careful not to trust itself by daylight within gunshot of a human being. Mr. M. A. Carriker, Jr. relates that when he climbed to a nest containing well grown young, the female attacked him so suddenly that he had barely time to throw up his gloved hand before his face when she struck it with terrific force tearing the glove, severely lacerating the hand and almost causing him to lose his hold on the tree! Another intruder at the nest of one of these birds was struck a stunning blow on the back of the neck by the female with such force that he was dazed for an instant as well as wounded, having received six slashes from her powerful claws. Mr. F. N. Whitman is still another climber who has suffered terrific blows from a ferocious pair of these dangerous birds, which nearly broke his grip on the branch besides leaving him cut and bleeding.

My lifelong friend, Mr. John A. Farley, who is responsible for the correction of some errors, which otherwise might have appeared in this volume, has studied closely the nesting of this owl but only once was he molested by the bird. While climbing to a nest he was struck a hard blow from behind. The powerful talons of the owl penetrated his overalls and heavy winter clothing and drew blood. Mr. John F. Carleton also informs me that he was struck from behind under similar conditions by one of these birds. Mr. H. H. Pitman and Mr. Howard Jones also record similar attacks. Mr. Pitman was struck and wounded three times. Dr. Frank Overton recounts the continual attacks of a especially vindictive pair of Horned Owls, and gives a striking series of close-range pictures taken on these occasions. In my experiences at the nests of these birds no attack was ever made. The mother bird usually disappeared at once and did not return; but one dull day in early March I had climbed to a nest on a great oak near Lake Quinsigamond and had stowed safely in my pocket the two newly-hatched young, when the powerful mother, who had been circling the tree, suddenly swooped down, alighted on a limb within a few feet of my face, and, leaning forward until she hung almost head downward, stared directly into my eyes with an expression of such fiendish ferocity that I prepared to defend myself. She soon swept away again, however, and continued her restless circling. Mr. William F. Lisk, who has found many nests of the Horned Owl tells of a night watchman in a wooden mill who was accustomed to pass through a grove of pine and other trees on the banks of Swift River near his home. He was attacked there one day in May by a large bird, which struck him on the head and shoulders repeatedly. He picked up a stick to defend himself, but this proved to be rotten and useless, and the bird fairly drove him out, so that he ran toward home, beating the bird off with his hands. Mr. Lisk says that the bird was undoubtedly a Great Horned Owl that was heard hooting there. Probably it had fledged young somewhere near. Mr. Joseph B. Underhill tells of a female Great Horned Owl (whose mate he had caught and confined) that attacked him and struck him a stunning blow on the forehead, inflicting wounds and drawing much blood!

A Horned Owl has been known to flutter over the ground like a Ruffed Grouse uttering short "wailing" notes and beating the ground with one wing and then the other, as if wounded, as a protest against the climbing by an intruder to its nest with young.

If taken early enough, young Horned Owls sometimes make interesting and tractable pets. A young bird of this species, confined at the rooms of the Worcester Natural History Society, was so heavy that his mere weight made his sharp claws sink into one's wrist. Yet he was the gentlest of pets, uttering continually his peeping cries.

The Horned Owl is swift and skillful on the wing and threads its way with ease among the trees of the forest. Only one case has come to my notice where it has struck an object inadvertently. Dr. E. H. Perkins, writing from Unity, Maine, says that he heard a sudden noise outside the house, and the next morning he went out and picked up a dead owl of this species. The bird appeared to have been confused by a lighted window and had struck the building.

The courtship of the Horned Owl is a curious performance. The male goes through peculiar contortions, nodding, bowing, flapping its wings and using, meanwhile, the choicest and most persuasive owl language. One motion which seems common to all owls is a rotary movement of the head which is raised to the full length of the neck, then swung to one side and dropped as low at least as the feet, and then swung to the other side and raised again, giving the owl a ludicrous appearance. Whatever may be said about the fierceness and ferocity of owls, no one can accuse them of being unfaithful to their young. Mr. Farley tells me that more than once he has found the nest well lined with a handsome, yellow mass of warm feathers from the mother's breast, which seemed analogous to the downy lining of a duck's nest. She sits closely on her eggs during the cold days and long nights of late February and early March. Often the snow covers both her and the nest and then if she is driven away by an intruder, the nest will be found covered with snow surrounding the imprint of her body, showing where she has faithfully outstayed the storm. The young remain in the nest and continue to grow for one long month, and during all that time they are well cared for and provided with a quantity of food. Mr. Ned Dearborn tells a tale which shows both parental finesse and devotion. Two young birds were taken from the nest and placed on the ground below, in the hope that their parents would come near enough to be shot. Soon an owl was heard hooting near by, and one of the watchers started in pursuit of the bird. The hooting gradually receded as the man with the ready gun advanced. Presently the man who remained behind was surprised to see one of the parents swoop to the ground and bear off one of the young in its talons, regardless of the unready gun, which roared far behind her!

In January and February these owls become vocal. Later in the year the birds are less vocal, but now and then, especially in autumn, their hooting may be heard. The following is taken from my notes of September 5, 1904, when I was staying at one of the Brewster cabins on the Concord River:

> Tonight just after dark as I sat in the cabin a sound came from down river like someone trying to imitate a dog, which was baying on the other shore; then it changed to higher, gabbling tones, gradually coming nearer, and sounding more like the talk and laughter of a boatload of women coming up the river. Soon it was followed by the unmistakable hoo-hoo, hoo, hoo-hoo of the Great Horned Owl. I stepped out of doors and answered it and the bird soon came quite near. I then went indoors, and it alighted on one of the trees above the cabin where it remained for some time, conversing with me in owl language. That bird made more peculiar and diverse sounds than I have heard from any other owl; for awhile its vocalizing was confined to soft cooing tones like wu woo and wu wa. Then, becoming bolder it launched forth a volley of interrogative wahoos and double hoots that were startling when coming from a distance of less than twenty yards. I answered occasionally and the bird kept up its hooting for half an hour. The upward slide of its loud waugh-hoo was inimitable; beginning at C the note waugh ran up to F at its ending followed by the emphatic hoo which fell again to C. Also it called wau-hoo hoo ooo, oo-oo without the rising inflection, or wu waugh hoo-hoo-hoo, hoo-hoo, hoo, hoo-hoo, etc. To me this was the principal event of the day.

One moonlit night the owl began hooting away off in the swamp nearly a mile away and every hoot was answered by the note of a male Black

Duck across the river. The hooting came nearer and nearer and the drake answered, until suddenly the call of the owl boomed out directly over the cabin roof. From then on that drake kept very quiet.

In August 1920, as I lay on the ground beside a small lake west of Waquoit Bay listening to numerous mosquitoes in conclave just outside my netting, a pair of owls began their hooting in the darkness. Answering their calls I soon enticed them to the trees above me. One was a particularly deep-toned hooter. The other, probably the male, seemed to be a lightweight. I have noticed in other cases that one of a pair, presumably the female, has a much deeper voice than the other.

In the forests of northern New England we usually hear the Horned Owl hooting in the swamps but sometimes they are on the mountains. A case of this kind occurred in western Massachusetts during the winter of 1919–20. European Hares had been introduced by some one in eastern New York near the Massachusetts line and had increased rapidly in numbers and spread into western Massachusetts where they had become a serious pest to the farmer and orchardist. In November and December of 1919 a flight of owls came into that region and Mr. Walter Pritchard Eaton told me that numbers were heard about the mountain in Sheffield. There they preyed upon the hares. He said that the mournful hoot of these great owls began about 3:30 P.M. in December and continued until late at night all along the wooded mountainside. He had heard as many as four at once near his house. These owls overran the region and many returned the next winter. The following spring European Hares were very scarce in that country.

The Horned Owl is a nocturnal bird, most active in the dusk of the evening and on moonlit nights, but it may be heard hooting at times at midnight in the "dark of the moon." It hunts at night, yet it can see perfectly in the daytime. It will hardly allow a man to come within gunshot in daylight, though under the protection of darkness it may come very near him. The best opportunity to see one by day is to advance upon it when it is mobbed by the noisy angry Crows. Then by keeping behind trees and moving up quickly while the Crows are calling, one may rarely get a glimpse of the owl. But it always starts before the Crows, which, no sooner do they see the owl go, than they follow in their usual uproarious pack. The bird often hunts just before dusk on dull days, and on very dark days it may be out before the middle of the afternoon. I remember once looking down from a cliff on the side of a mountain at Mount Desert upon an owl hunting over the meadows below. This was about 4 P.M. and the view of the owl's back and wings as it quartered over the land like a Marsh Hawk hunting for mice was all that could be desired.

Horned Owls are not regarded as migratory, but in winter they often leave regions where suitable food is scarce and go where it can be found. They are not always successful in this search, however, and now and then one is found starved and frozen in the snow. Such specimens when discovered are very much emaciated. Mr. H. A. P. Smith writing from Digby, Nova Scotia, March 14, 1923, says that a Great Horned Owl was found there sitting upright in an apple tree frozen stiff. Probably his tightly clinched talons froze to the limb and held him there in death; but the bird would not have been frozen had it had food enough to keep up the animal heat in its body. In this species the extensor (fused) tendons of the leg, which retract the toes and clinch the claws pass through a groove in the tarso-metatarsus and then into and through a hole in the bone, as a rope passes through a pulley block. The tendons are thus held firmly in place as they pass over and through what in the human being would be the heel. When the leg is bent and drawn up the tendons also are drawn tight, clenching the toes and driving in the talons. Thus when the owl settles down on a limb, its toes and talons lock round it and hold the bird firmly in sleep or as in the above case even in death. This arrangement of the tendons and their connection with the machinery of the toes forms a powerful equipment with which to strike its prey. When the bird strikes a victim, the force of the blow alone tends to bend the legs, contract the tendons and drive the claws into a vital part. In addition to this the powerful muscles of the legs are brought into play to force the talons home. Thus its speed, weight and muscular power all combine to give the bird the force to overcome animals many times its own weight.

Horned Owls migrate usually from regions where food is scarce in winter and thus escape starvation. Probably there is some migration of these birds every year from the northern parts of their range toward more southern regions. In some seasons of scarcity this migration increases to considerable proportions. Says Mr. E. O. Grant writing on November 12, 1924, "There were lots of Great Horned Owls coming in before I left township 11, range 16, Aroostook County, Maine." Major Mark Robinson wrote November 20, 1921 from Algonquin National Park in the highlands of Ontario. "We observed quite a migration of owls November 2 and 3. The woods were ringing with their hooting. They were gone on the fourth." The spring migration northward is not so evident as the fall exodus. Many of these northern birds are shot by gunners in autumn and winter and many more are taken in traps, especially around game farms. As many as 50 or 75 of these owls are taken in some years about a single game farm. When pressed by hunger they will eat any small mammal that they find in a trap and often they are caught at such times in baited traps.

Miss Florence Pease in a letter to me in 1918 said that in the previous winter a large owl was reported in Conway, Massachusetts with a steel trap on one leg and that in 1919 one alighted on a factory in Connecticut in a similar plight, causing so much excitement in the factory that the works were shut down.

Every living thing above ground in the woods on winter nights pays tribute to the Great Horned Owl except the larger mammals and man. Ordinarily when there is good hunting this owl has a plentiful supply of food, and when there is game enough it slaughters an abundance and eats only the brains; but in winter when house rats and mice keep mostly within the buildings, when woodchucks and skunks have "holed up" and when field mice are protected by deep snow—then if rabbits are scarce and starvation is imminent, the owl will attack even the domestic cat, and usually with success. A farmer brought me a Great Horned Owl one winter day that had killed his pet tomcat on the evening of the previous day. The cat was out walking in the moonlight on one of his usual expeditions in search of unattended females,—when the farmer heard a wail of mortal agony and opening the door saw Mr. Cat in the grasp of the owl. Before he could get his gun and shoot the bird the cat was no more. Its vitals had been torn out. Usually the noiseless flight of the owl enables it to take the cat by surprise and seize it by the back of the neck and the small of the back, when all is soon over; but if the cat is not taken by surprise and is quick enough to turn on the owl, the episode is likely to have a different ending. Sometimes the owl "wakes up the wrong customer" as the following incident related to me by Mr. J. A. Farley clearly shows:

> Mr. Zenas Langford, for many years superintendent of streets in the southern part of Plymouth, tells me that a few years ago in the Pine Hills, he came upon a Horned Owl in trouble with a black snake. As he went along the cart road, the two suddenly 'fell into it.' Plainly the owl had caught the snake, but the reptile had twisted itself around the bird so that it was unable to fly, and fell to the ground with its prey. Mr. Langford says that the owl had grasped the snake about six inches below its head, but the part of the snake below the owl's talons had twisted itself around the bird tightly. There was at least one light turn around the owl's neck. Mr. Langford could not see that the snake had bitten the owl, which however, was nearly exhausted. The owl nevertheless had not relaxed its hold on the snake. Neither of the creatures had given up. Mr. Langford killed the snake, which measured 4 feet. The owl was so weakened and helpless that it could not fly; it seemed to have been choked. Mr. Langford wrapped it in a blanket, took it home, kept it for a week and then let it go. The

porcupine, also, does not yield up his life, without doing all the harm possible. Witness the Horned Owl, liberally besprinkled with porcupine quills, which was shot in the Province of Quebec in December 1907, and reported by Rev. G. Eifrig.

Horned Owls kill and eat many skunks, and seem to care little for the disagreeable consequences of attacking these pungent animals. Many of the owls that I have handled give olfactory evidence of the habit. They kill both wild and domesticated ducks, picking them up skillfully out of the water at night, and no goose is too large for them to tackle. Occasionally, however, the owl comes to grief in its attack on some waterfowl. At Lanesboro, Massachusetts, in February, 1918, one was found floating dead in a pool with a duck that it had killed. No one knew the details of this midnight tragedy. It may be that the owl was weakened by hunger and was unable to rise from the water. Where hens, chickens and turkeys are allowed to roost at night in the trees, many fall victims to this owl, which is said to alight on the branch beside its chosen prey, crowd it off the limb and then strike it in the air. Some guinea hens roosting in trees on my farm disappeared in some such manner. Usually the owl does not stop to eat his victim on the spot but bears it away.

Mr. A. A. Cross of Huntington informed me that in December 1918, one of his neighbors left a live guinea hen in a sack and it disappeared. The next day he found the remains of the fowl in the orchard. The sack had been torn open and about half of the bird had been eaten. Two traps were set near it and the next night a splendid specimen of the Great Horned Owl was taken.

In eating its prey this owl usually begins at the head and eats backward. The birds are plucked, but small mammals such as mice and rats are swallowed whole, headfirst. Mr. E. O. Grant says that he saw an owl strike a Ruffed Grouse in a bush and bear it away without even checking his flight. "The owl flew up a stream with a stream of feathers trailing in his wake. I think," says Mr. Grant, "that he must have had that bird pretty well plucked before he had gone 60 rods." The Horned Owl is no respecter of persons. It kills weaker owls from the Barred Owl down, most of the hawks and such nocturnal animals as weasels and minks. It is the most deadly enemy of the Crow, taking old and young from their nests at night and killing many at their winter roosts. Game birds of all kinds, poultry, a few small birds, rabbits (especially bush rabbits), hares, squirrels, gophers, mice, rats, woodchucks, opossums, fish, crawfish and insects are all eaten by this rapacious bird. It is particularly destructive to rats. Mr. E. O. Niles tells of a nest of this owl on his farm containing two young owls and several dead rats. On the ground below the nest were the bodies of 113 rats, recently killed, with their skulls opened and the brains removed. When we find young in the nest of one of these owls, they are well supplied with game. Mr. H. O. Green writes that he found remains of a skunk, a Crow and a Pheasant at one nest and Mr. Joseph Peters says that the parent birds that supplied another nest brought black ducks, rabbits, rats, snakes, a Red Phalarope, a Virginia Rail, two Woodcocks, a Bob-white, a Northern Flicker, a Pheasant and some small birds. In eating their food the fierce young birds grasped the game firmly in the talons and tore at it with their beaks. They began their feast on the larger birds and mammals by consuming the head and then working backward through the carcass. Of 127 stomachs of Great Horned Owls examined by Dr. A. K. Fisher, 31 contained poultry or game birds; 8, other birds; 13, mice; 65, other mammals; 1, a scorpion; 1, fish; 10, insects and 17 were empty.

In the wilderness the Great Horned Owl exerts a restraining influence on both the game and the enemies of game, for it destroys both and thus does not disturb the balance of nature. But on the farm or the game preserve, it cannot be tolerated.

Forbush, E. H. 1927. Birds of Massachusetts and Other New England States. *Vol. 2. Norwood Press, Norwood; "The Great Horned Owl," pp. 221–228.*

Just Tigers (1944)

Jim Corbett

I think that all sportsmen who have had the opportunity of indulging in the twin sports of shooting tigers with a camera and shooting them with a rifle will agree with me that the difference between these two forms of sport is as great, if not greater, than the taking of a trout on a light tackle in a snow-fed mountain stream, and the killing of a fish on a fixed rod on the sun-baked bank of a tank.

Apart from the difference in cost between shooting with a camera and shooting with a rifle, and the beneficial effect it has on our rapidly decreasing stock of tigers, the taking of a good photograph gives far more pleasure to the sportsman than the acquisition of a trophy; and further, while the photograph is of interest to all lovers of wild life, the trophy is only of interest to the individual who acquired it. As an illustration I would instance Fred Champion. Had Champion shot his tigers with a rifle instead of with a camera his trophies would long since have lost their hair and been consigned to the dustbin, whereas the records made by his camera are a constant source of pleasure to him, and are of interest to sportsmen in all parts of the world.

It was looking at the photographs in Champion's book, *With a Camera in Tiger-Land*, that first gave me the idea of taking photographs of tigers. Champion's photographs were taken with a still camera by flashlight and I decided to go one better and try to take tiger pictures with a cine camera by daylight. The gift by a very generous friend of a Bell & Howell 6-mm camera put just the weapon I needed into my hands, and the 'Freedom of the Forests', which I enjoy, enabled me to roam at large over a very wide field. For ten years I stalked through many hundreds of miles of tiger country, at times being seen off by tigers that resented my approaching their kills, and at other times being shooed out of the jungle by tigresses that objected to my going near their cubs. During this period I learnt a little about the habits and ways of tigers, and though I saw tigers on, possibly, two hundred occasions I did not succeed in getting one satisfactory picture. I exposed films on many occasions, but the results were disappointing owing either to over-exposure, under-exposure, obstruction of grass or leaves, or cobwebs on the lens; and in one case owing to the emulsion on the film having melted while being processed.

Finally, in 1938, I decided to devote the whole winter to making one last effort to get a good picture. Having learnt by experience that it was not possible to get a haphazard picture of a tiger, my first consideration was to find a suitable site, and I eventually selected an open ravine fifty yards wide, with a tiny stream flowing down the centre of it, and flanked on either side by dense tree and scrub jungle. To deaden the sound of my camera when taking pictures at close range I blocked the stream in several places, making miniature waterfalls a few inches high. I then cast round for my tigers, and having located seven, in three widely separated areas, started to draw them a few yards at a time to my jungle studio. This was a long and difficult job, with many setbacks and disappointments, for the area in which I was operating is heavily shot over, and it was only by keeping my tigers out of sight that I eventually got them to the exact spot where I wanted them. One of the tigers for some reason unknown to me left the day after her arrival, but not before I had taken a picture of her; the other six I kept together and I exposed a thousand feet of film on them. Unfortunately it was one of the wettest winters we have ever had and several hundred feet of the film were ruined through moisture on the lens, under-exposure, and packing of the film inside the camera due to hurried and careless threading. But, even so, I have got approximately six hundred feet of film of which I am inordinately proud, for they are a living record of six full-grown tigers—four males, two of which are over ten feet, and two females, one of which is a white tigress—filmed in daylight, at ranges varying from ten to sixty feet.

The whole proceeding from start to finish took four and a half months, and during the countless hours I lay near the tiny stream and my miniature waterfalls, not one of the tigers ever saw me.

The stalking to within a few feet of six tigers in daylight would have been an impossible feat, so they were stalked in the very early hours of the morning, before night had gone and daylight come—the heavy winter dew making this possible—and were filmed as light, and opportunity, offered.

Hawkins, R. E. 1978. Jim Corbett's India. *Oxford University Press, Oxford; "Just Tigers," pp. 241–243. Reprinted with permission from Oxford University Press.*

Looking Back (1944)

Jim Corbett

Between the catapult and the muzzle-loader periods there was a bow-and-arrow interlude which I look back on with very great pleasure, for though I never succeeded in impaling bird or beast with an arrow, I opened my credit account—with my small savings—with the Bank of Nature during that period, and the jungle lore I absorbed during the interlude, and later, has been a never-ending source of pleasure to me.

I have used the word 'absorbed', in preference to 'learnt', for jungle lore is not a science that can be learnt from textbooks; it can, however, be absorbed, a little at a time, and the absorption process can go on indefinitely, for the book of nature has no beginning, as it has no end. Open the book where you will, and at any period of your life, and if you have the desire to acquire knowledge you will find it of intense interest, and no matter how long or how intently you study the pages your interest will not flag, for in nature there is no finality.

Today it is spring, and the tree before you is bedecked with gay bloom. Attracted by this bloom a multitude of birds of many colours are flitting from branch to branch, some drinking the nectar from the flowers, others eating the petals, and others again feeding on the bees that are busily collecting honey. Tomorrow the bloom will have given place to fruit and a different multitude of birds will be in possession of the tree. And each member of the different multitudes has its allotted place in the scheme of nature. One to beautify nature's garden, another to fill it with melody, and yet another to regenerate the garden.

Season after season, year after year, the scene changes. A new generation of birds in varying numbers and species adorn the tree. The tree loses a limb—torn off in a storm—gets stackheaded and dies, and another tree takes its place; and so the cycle goes on.

On the path at your feet is the track of a snake that passed that way an hour before sunrise. The snake was going from the right-hand side of the path to the left, was three inches in girth, and you can be reasonably certain that it was of a poisonous variety. Tomorrow the track on the same path, or on another, may show that the snake that crossed it five minutes earlier was traveling from left to right, that it was five inches in girth, and that it was non-poisonous.

And so the knowledge you absorb today will be added to the knowledge you will absorb tomorrow, and on your capacity for absorption, not on any fixed standard, will depend the amount of knowledge you ultimately accumulate. And at the end of the accumulating period—be that period one year or fifty—you will find that you are only at the beginning, and that the whole field of nature lies before you waiting to be explored and to be absorbed. But be assured that if you are not interested, or if you have no desire to acquire knowledge, you will learn nothing from nature.

I walked with a companion for twelve miles through a beautiful forest from one camp to another. It was the month of April and nature was at her best. Trees, shrubs, and creepers were in full bloom. Gaily coloured butterflies flitted from flower to flower, and the air, filled with the scent of flowers, throbbed with the song of birds. At the end of the day my companion was asked if he had enjoyed the walk, and he answered, 'No. The road was very rough.'

I was traveling, shortly after World War I, from Bombay to Mombassa in the British India liner *Karagola.* There were five of us on the upper deck. I was going to Tanganyika to build a house, the other four were going to Kenya—three to shoot and one to look at a farm he had purchased. The sea was rough and I am a bad sailor, so I spent most of my time dozing in a corner of the smoke room. The others sat at a table nearby playing bridge, smoking, and talking, mostly about sport. One day, on being awakened by a cramp in my leg, I heard the youngest member of the party say, 'Oh, I know all about tigers. I spent a fortnight with a Forest Officer in the Central Provinces last year.'

Admittedly two extreme cases, but they will serve to emphasize my contention that if you are not interested you will see nothing but the road you walk on, and if you have no desire to acquire knowledge and assume you can learn in a fortnight what cannot be learnt in a lifetime, you will remain ignorant to the end.

Hawkins, R. E. 1978. Jim Corbett's India. *Oxford University Press, Oxford; "Looking Back," pp. 249–250. Reprinted with permission from Oxford University Press.*

Foreword to *The salamanders of the Family Plethodontidae* (1926)

Emmett R. Dunn

Much water has flowed under many bridges since the January of the year of our Lord one thousand nine hundred and thirteen, when a college Sophomore opened a letter and read a sentence, the results hereof are the following pages. It has been thirteen years since the eighteen-year-old boy read those few words and in those thirteen years for him the surf has whitened the shores of Caribbean islands; the slopes of the Balsams have been blue in the distance; hawks have soared a thousand feet below the naked peak of Sharp Rock; the iridescent wings of morpho have fluttered through glades in the rain forest and in the mist of the cataract at Xico; from the mooring at Vera Cruz tall Orizaba has stood against the western sky; the low sun has shone on the ice of Ixtaccihuatl, most unforgettable of mountains; step after step, for him, far out at sea, Chirripo Grande, "nie von Menschenfuss betreten," has climbed aloft; and from shaken

Irazu, while the ash cloud of the eruption rose above his head and floated, a black pall beaten by the fierce wind, he has seen through a break in the clouds, far to the Southwards, the sheer and menacing shaft of rock which is Cerro de la Muerte.

And in that letter and in that sentence there was no hint of high peaks or of distant places, of mountains seen and unclimbed and forever remembered, of the coughing of tigers by night in the black forest, there were only these words in that letter from Leonhard Stejneger "I would like to point out to you, however, that Herpetology offers fields more in need of investigation than the snakes, viz., the salamanders."

Bald and prosaic words these, true though they were. Bald and prosaic has much of the work been, and, for those who venture further in what lies hereinafter, there is aridity and to spare. This is deplorable enough, but unavoidable, and while the written accounts of genera and of species unquestionably make dull reading, and were, it must be confessed, dull writing, yet the taking of specimens and the gathering of data and all that went therewith, and the remembrance thereof, has more than outweighed the dullness.

And it must have been so with the many others whose work has here been synthesized. It cannot have been all drudgery, the work of the men who have accumulated the specimens and the information in the 125 years since first Palisot de Beauvais collected around Philadelphia, and since Sonnini described the *Salamandra rubra* which be brought back. For every specimen of the 12,000 and odd about which this book has been written was taken in the field by collectors who found them on their own land and close to their own homes, and discovered strange creatures in familiar places, or by collectors who half a world away from their own country met places and animals and men that they had not seen. They were dull, and not the work, if there was not for them some thrill therein.

One hundred and fifty-three times some one of these creatures has been described as new, and, whether in field or in museum, in the recognition of a novelty as it was first divulged to intelligent and appreciative eyes, there has come a peculiar and high satisfaction. Highest of course when the collector realizes that he has caught and holds in his hands an unknown animal, only less high when home from the hunt he finds a new beast among his spoils, and high enough when, envisaging the spoils of another, he sees among them one strange to him and to science.

Nor is a novelty necessary. Many a one has dreamed of some rare creature, and has traveled far, and known much discomfort in search thereof, and the final meeting is overpayment. Conrad, in the twentieth chapter of *Lord Jim*, has depicted the appropriate emotions, and Stein felt there what many another has felt elsewhere.

It is not in the collecting of specimens, in the observing of habits, in the examination of museum material, or in the search through the literature that one encounters ennui. Rather it is after all the material is at hand, and all the information is gathered that the unalloyed drudgery begins.

Now that drudgery is over, and I can remember how puzzling the first few specimens of the commonest species were to me, and how I foolishly imagined that the first grown males of *Desmognathus fuscus* that I caught and examined were the desiderate *Leurognathus*, because they too lacked teeth on the prevomers.

And I can remember the first gigantic specimen of *quadra-maculatus* that I took at Brevard; and how I saw for the first time *gutto-lineata* in the light of an electric torch, and it was climbing on a bush, and how it leapt violently down into a deep ditch and was lost to me, and how many days went by before I ever saw another. And I remember *Plethodon metcalfi* at See-Off, when half the ridges of the Southern Appalachians rode in dark blue waves across the French Broad valley, and how few of them there were on that bare and wind swept crest. And how, on the slopes of the Grandfather, some chance turned log disclosed the red band on the back of *yonahlossee*, and I knew that an unknown species was before me.

I can remember climbing White Rock on a special trip for *jordani* which I had never seen; I climbed a thousand feet before I saw the first, but I had thirty-seven when I came down.

At Navarro there was the big, gray, *robustus* that lived in the fallen tree in the forest, and, in a pasture near Jalapa, among stones at the foot of a tree, the pair of black-and-golden *bellii*.

Or I remember how the snake I saw at La Loma, when I turned the leaf and it slid into the water, changed into *collaris* when I caught it; or how I came by night upon the type of *Oedipus colonneus*.

Yet it is the places that stick in my mind, which I had not seen but for *Desmognathus* or for *Oedipus*.

Cedar Rock in the rain, when the face of it was a waterfall a thousand feet high and a half mile wide, water pouring out from the spruce forest on top, and spreading out over the great dome and down with a mighty roaring to its hidden base.

The cliffs of Lost Cove and the Escarpment of the Blue Ridge down which we went all one long day, with our senses of direction clean gone, and dark came on, and we followed trails in the woods which ended blindly, until across three ridges we heard a cow bell and the bark of a dog.

The silence of the spruce forest on White Top, and the thick moss on the ground there.

The rocky pool on White Rock, with the Tennessee valley three thousand feet below, the Cumberland Front a dim line to the west, Guyot and Clingman's Dome looming up through the mist that hung along the line of the Smokies, when all there was between me and the Richland Balsams was miles of air and a hawk soaring.

Coming into port at Limon, with the pelicans and frigate birds rattling by on their coarse wings, and a look to them as if they had been drawn by Dare.

The oak forest on Irazu, when our horse hooves made no sound in the soft ash, and the leaves of the oaks were gray from the ash, and it was very silent as we rode: until about dusk the wind swooped from behind the shoulder of the volcano, very cold with a cold rain, and we stopped to put on our ponchos, looking back and far below, where the lights were coming on in Cartago and Paraiso, cities of the plain.

Riding by night from Alajuela to San Pedro de Poas over bridges across torrents, which roared and were white in the moonlight.

On top of a train with Mexican soldiers riding from Xico to Jalapa, when Cofre de Perote towered above us like some misshapen lion, and the white cone of Orizaba rose above hill after hill as the train went by.

And events not wholly connected with salamanders come into my mind and need no effort of recollection.

The heave of the earth and the glimpse of the under fires in the aforementioned eruption of Irazu; the fer-de-lance at the ford of the Bananito, when I picked up the bridle of a mule and a six-foot snake with it; the first time I heard the red howlers, alone in the woods above Navarro; night at La Loma, and something coughing, and Celso who said "*aqui viene el tigre*"; the peons, drunk and fighting, on pay day at Guapiles; the sudden appearance of forty horsemen with rifles and banderillos one night at Cerro del Gallo; and the terrible weariness of the path to Tlamacas.

Places and happenings, people and animals, winds at sea and on mountains, and all that I remember, it was no unpleasant path I trod when I followed this "questing beast" or that, nor have all of them been unattainable, and with this foreword I commend the remainder to the reader, who will, I doubt not, turn elsewhere before many pages, *Bueno, vaya usted con Dios.*

Dunn, E. R. 1926. The salamanders of the Family Plethodontidae. *Smith College, Northampton, Massachusetts; foreword, pp. v–viii. Reprinted with permission of Smith College.*

The Wisdom of Instinct (1918)

Jean Henri Fabre

To paralyze her prey, the Languedocian Sphex [*a hunting wasp*], I have no doubt, pursues the method of the Cricket-huntress and drives her lancet repeatedly into the Ephippiger's [*the European bush cricket or green locust*] breast in order to strike the ganglia of the thorax. The process of wounding the nerve-centres must be familiar to her; and I am convinced beforehand of her consummate skill in that scientific operation. This is an art thoroughly known to all the Hunting Wasps, who carry a poisoned dart that has not been given them in vain. At the same time, I must confess that I have never yet succeeded in witnessing the deadly performance. This omission is due to the solitary life led by the Languedocian Sphex.

When a number of burrows are dug on a common site and then provisioned, one has but to wait on the spot to see how one huntress and now another arrive with the game which they have caught. It is easy in these circumstances to try upon the new arrivals the substitution of a live prey for the doomed victim and to repeat the experiment as often as we wish. Besides, the certainty that we shall not lack subjects of observation, as and when wanted, enables us to arrange everything in advance. With the Languedocian Sphex, these conditions of success do not exist. To set out expressly to look for her, with one's material prepared, is almost useless, as the solitary insect is scattered one by one over vast expanses of ground. Moreover, if you do come upon her, it will most often be in an idle hour and you will get nothing out of her. As I said before, it is nearly always unexpectedly, when your thoughts are elsewhere engaged, that the Sphex appears, dragging her Ephippiger after her.

This is the moment, the only propitious moment to attempt a substitution of prey and invite the huntress to let you witness her lancet thrusts. Quick, let us procure an alternative morsel, a live Ephippiger! Hurry, time presses: in a few minutes, the burrow will have received the victuals and the glorious occasion will be lost! Must I speak of my mortification at these moments of good fortune, the mocking bait held out by chance? Here, before my eyes, is matter for interesting observations; and I cannot profit by it! I cannot surprise the Sphex' secret for the lack of something to offer her in the place of her prize! Try it for yourself, try setting out in quest of an alternative piece with only a few minutes at your disposal, when it took me three days of wild running about before I found weevils for my Cerceres! And yet I made the desperate experiment twice over. Ah, if the keeper had caught me this time, tearing like mad through the vineyards, what a good opportunity it would have been for crediting me with robbery and having me up before the magistrate! Vine-branches and clusters of grapes: not a thing did I respect in my mad rush, hampered by the trailing shoots. I must have an Ephippiger at all costs; I must have him that moment. And once I did get my Ephippiger during one of these frenzied expeditions. I was radiant with joy, never suspecting the bitter disappointment in store for me.

If only I arrive in time, if only the Sphex be still engaged in transport work! Thank heaven, everything is in my favour! The Wasp is still some distance away from her burrow and still dragging her prize along. With my forceps I pull gently at it from behind. The huntress resists, stubbornly clutches the antennae of her victim and refuses to let go. I pull harder, even drawing the carter back as well; it makes no difference: the Sphex does not loose her hold. I have with me a pair of sharp scissors, belonging to my little entomological case. I use them and promptly cut the harness-ropes, the Ephippiger's long antenna. The Sphex continues to move ahead, but soon stops, astonished at the sudden decrease in the weight of the burden which she is trailing, for this burden is now reduced merely to the two antenna, snipped off by my mischievous wiles. The real load, the heavy, pot-bellied insect, remains behind and is instantly replaced by my live specimen. The Wasp turns round, lets go the ropes that now draw nothing after them and retraces her steps. She comes face to face with the prey substituted for her own. She examines it, walks round it gingerly, then stops, moistens her foot with saliva and begins to wash her eyes. In this attitude of meditation, can some such thought as the following pass through her mind:

"Come now! Am I awake or am I asleep? Do I know what I am about or do I not? That thing's not mine. Who or what is trying to humbug me?"

At any rate, the Sphex shows no great hurry to attack my prey with her mandibles. She keeps away from it and shows not the smallest wish to seize it. To excite her, I offer the insect to her in my fingers; I almost thrust the antennæ under her teeth. I know that she does not suffer from shyness; I know that she will come and take from your fingers without hesitation, the prey that you have snatched from her and afterwards present to her. But what is this? Scorning my offers, the Sphex retreats instead of snapping up what I place within her reach. I put down the Ephippiger, who, obeying a thoughtless impulse, unconscious of danger, goes straight to his assassin. Now we shall see! Alas, no: the Sphex continues to recoil, like a regular coward, and ends by flying away. I never saw her again. Thus ended, to my confusion, an experiment that had filled me with such enthusiasm.

Later and by degrees, as I inspected an increasing number of burrows, I came to understand my failure and the obstinate refusal of the Sphex. I always found the provisions to consist, without a single exception, of a female Ephippiger, harbouring in her belly a copious and succulent cluster of eggs. This appears to be the favourite food of the grubs. Well, in my hurried rush through the vines, I had laid my hands on an Ephippiger of the other sex. I was offering the Sphex a male. More farseeing than I in this important question of provender, the Wasp would have nothing to say to my game: "A male, indeed, is that a dinner for my larvæ? What do you take them for?"

What nice discrimination they have, these dainty epicures, who are able to differentiate between the tender flesh of the female and the comparatively dry flesh of the males! What an unerring glance, which can distinguish at once between the two sexes, so much alike in shape and colour! The female carries a sword at the up of her abdomen, the ovipositor wherewith the eggs are buried in the ground; and that is about the only external difference between her and the male. This distinguishing feature never escapes the perspicacious Sphex; and that is why, in my experiment, the Wasp rubbed her eyes, hugely puzzled at beholding swordless a prey, which she well knew carried a sword when she caught it. What must not have passed through her little Sphex brain at the sight of this transformation?

Let us now watch the Wasp when, having prepared the burrow, she goes back for her victim, which, after its capture and the operation that paralyzed it, she has left at no great distance. The Ephippiger is in a condition similar to that of the Cricket sacrificed by the Yellow-winged Sphex, a condition proving for certain that stings have been driven into her thoracic ganglia. Nevertheless, a good many movements still continue; but they are disconnected, though endowed with a certain vigour. Incapable of standing on its legs, the insect lies on its side or on its back. It flutters its long antennæ and also its palpi; it opens and closes its mandibles and bites as hard as in the normal state. The abdomen heaves rapidly and deeply. The ovipositor is brought back sharply under the belly, against which it almost lies flat. The legs stir, but languidly and irregularly; the middle legs seem more torpid than the others. If pricked with a needle, the whole body shudders convulsively; efforts are made to get up and walk, but without success. In short, the insect would be full of life, but for its inability to move about or even to stand upon its legs. We have here therefore a wholly local paralysis, a paralysis of the legs, or rather a partial abolition and ataxy of their movements. Can this very incomplete inertia be caused by some special arrangement of the victim's nervous system, or does it come from this, that the Wasp perhaps administers only a single prick, instead of stinging each ganglion of the thorax, as the Cricket-huntress does? I cannot tell.

Still, for all it's shivering, its convulsions, its disconnected movements, the victim is nonetheless incapable of hurting the larva that is meant to devour it. I have taken from the burrow of the Sphex Ephippigers struggling just as lustily as when they were first half paralyzed; and nevertheless the feeble grub, hatched but a few hours since, was digging its teeth into the gigantic victim in all security; the dwarf was biting into the colossus without danger to itself. This striking result is due to the spot selected by the mother for laying her egg. I have already said how the Yellow-winged Sphex glues her egg to the Cricket's breast, a little to one side, between the first and second pair of legs. Exactly the same place is chosen by the White-edged Sphex; and a similar place, a little farther back, towards the root of one of the large hind-thighs, is adopted by the Languedocian Sphex, all three thus giving proof, by this uniformity, of wonderful discernment in picking out the spot where the egg is bound to be safe.

Consider the Ephippiger pent in the burrow. She lies stretched upon her back, absolutely incapable of turning. In vain, she struggles, in vain she writhes: the disordered movements of her legs are lost in space, the room being too wide to afford them the support of its walls. The grub cares nothing for the victim's convulsions: it is at a spot where naught can reach it, not tarsi, nor mandibles, nor ovipositor, nor antennæ, a spot absolutely stationary, devoid of so much as a surface tremor. It is in perfect safety, on the sole condition that the Ephippiger cannot shift her position, turn over, get upon her feet; and this one condition is admirably fulfilled.

But, with several heads of game, all in the same stage of paralysis, the larva's danger would be great. Though it would have nothing to fear from the insect first attacked, because of its position out of the reach of its victim, it would have every occasion to dread the proximity of the others which, stretching their legs at random, might strike it and rip it open with their spurs. This is perhaps the reason why the Yellow-winged Sphex, who heaps up three or four Crickets in the same cell, practically annihilates all movement in its victims, whereas the Languedocian Sphex, victualling each burrow with a single piece of game, leaves her Ephippigers the best part of their power of motion and contents herself with making it impossible for them to change their position or stand upon their legs. She may thus, though I cannot say so positively, economize her dagger thrusts.

While the only half-paralyzed Ephippiger cannot imperil the larvae, fixed on a part of the body where resistance is impossible, the case is different with the Sphex who has to cart her prize home. First, having still, to a great extent, preserved the use of its tarsi, the victim clutches with these at any blade of grass encountered on the road along which it is being dragged; and this produces an obstacle to the hauling-process which is difficult to overcome. The Sphex, already heavily burdened by the weight of her load is liable to exhaust herself with her efforts to make the other insect relax its desperate grip in grassy paces. But this is the least serious drawback. The Ephippiger preserves the complete use of her mandibles, which snap and bite with their customary vigour. Now what these terrible nippers have in front of them is just the slender body of the enemy at a time when she is in her hauling-attitude. The antennæ, in fact, are grasped not far from their roots, so that the mouth of the victim dragged along on its back faces either the thorax or the abdomen of the Sphex, who, standing high on her long legs, takes good care, I am convinced not to be caught in the mandibles yawning underneath her. At all events, a moment of forgetfulness, a slip, the merest trifle can bring her within the reach of two powerful nippers, which would not neglect the opportunity of taking a pitiless vengeance. In the more difficult cases at any rate, if not always, the action of those formidable pincers must be done away with; and the fish-hooks of the legs must be rendered incapable of increasing their resistance to the process of transport.

How will the Sphex go to work to obtain this result? Here man, even the man of science, would hesitate, would waste his time in barren efforts and would perhaps abandon all hope of success. He can come and take one lesson from the Sphex. She, without ever being taught it, without ever seeing it practiced by others, understands her surgery through and through. She knows the most delicate mysteries of the physiology of the nerves, or rather she behaves as if she did. She knows that under her victim's skull there is a circlet of nervous nuclei, something similar to the brain of the higher animals. She knows that this main centre of innervation controls the action of the mouth-parts and moreover is the seat of the will, without whose orders not a single muscle acts; lastly, she knows that, by injuring this sort of brain, she will cause all resistance to cease, the insect no longer possessing any will to resist. As for the mode of operating, this is the easiest matter in the world to her; and, when we have been taught in her school, we are free to try her process in our turn. The instrument employed is no longer the sting; the insect, in its wisdom, has deemed compression preferable to a poisoned thrust. Let us accept its decision, for we shall see presently how prudent it is to be convinced of our own ignorance in the presence of the animal's knowledge. Lest by editing my account I should fail to give a true impression of the sublime talent of this masterly operator, I here copy out my note as I penciled it on the spot, immediately after the stirring spectacle.

The Sphex finds that her victim is offering too much resistance, hooking itself here and there to blades of grass. She then stops to perform upon it the following curious operation, a sort of *coup de grâce*. The Wasp, still astride her prey, forces open the articulation of the neck, high up, at the nape. Then she seizes the neck with her mandibles and, without making any external wound, probes as far forward as possible under the skull, so as to seize and chew up the ganglia of the head. When this operation is done, the victim is utterly motionless, incapable of the least resistance, whereas previously the legs, though deprived of the power of connected movement needed for walking, vigorously opposed the process of traction.

There is the fact in all its eloquence. With the points of its mandibles, the insect, while leaving uninjured the thin and supple membrane of the neck, goes rummaging into the skull and munching the brain. There is no effusion of blood, no wound, but simply an external pressure. Of course, I kept for my own purposes the Ephippiger paralyzed before my eyes, in order to ascertain the effects of the operation at my leisure; also of course, I hastened to repeat in my turn, upon live Ephippigers, what the Sphex had just taught me. I will here compare my results with the Wasp's.

Two Ephippigers whose cervical ganglia I squeeze and compress with a forceps fall rapidly into a state resembling that of the victims of the Sphex. Only, they grate their cymbals if I tease them with a needle; and the legs still retain a few disordered and languid movements. The difference no doubt is due to the fact that my patients were not previously injured in their thoracic ganglia, as were those of the Sphex, who were first stung on the breast. Allowing for this important condition, we see that I was none too bad a pupil and that I imitated pretty closely my teacher of physiology, the Sphex. I confess, it was not without a certain satisfaction that I succeeded in doing almost as well as the insect.

As well? What am I talking about? Wait a bit and you shall see that I still have much to learn from the Sphex. For what happens is that my two patients very soon die: I mean, they really die; and, in four or five days, I have nothing but putrid corpses before my eyes. And the Wasp's Ephippiger? I need hardly say that the Wasp's Ephippiger, even ten days after the operation, is perfectly fresh, just as she will be required by the larva for which she has been destined. Nay, more: only a few hours after the operation under the skull, there reappeared, as though nothing had occurred, the disorderly movements of the legs, antennæ, palpi, ovipositor and mandibles; in a word, the insect returned to the condition wherein it was before the Sphex bit its brain. And these movements were kept up after, though they became feebler every day. The Sphex had merely reduced her victim to a passing state of torpor, lasting amply long enough to enable her to bring it home without resistance; and I, who thought myself her rival, was but a clumsy and barbarous butcher: I killed my prize. She, with her inimitable dexterity, shrewdly compressed the brain to produce a lethargy of a few hours; I, brutal through ignorance, perhaps crushed under my forceps that delicate organ, the main seat of life. If anything could prevent me from blushing at my defeat, it would be the conviction that very few, if any, could vie with these clever ones in cleverness.

Ah, I now understand why the Sphex does not use her sting to injure the cervical ganglia! A drop of poison injected here, at the centre of vital force, would destroy the whole nervous system; and death would follow soon after. But it is not death that the huntress wishes to obtain; the larvae have not the least use for dead game, for a corpse, in short, smelling of corruption; and all that she wants to bring about is a lethargy, a passing torpor, which will put a stop to the victim's resistance during the carting process, this resistance being difficult to overcome and moreover dangerous for the Sphex. The torpor is obtained by a method known in laboratories of experimental physiology: compression of the brain. The Sphex acts like a Flourens, who, laying bare an animals brain and bearing upon its cerebral mass, forthwith suppresses intelligence, will, sensibility and movement. The pressure is removed; and everything reappears. Even so do the remains of the Ephippiger's life reappear, as the lethargic effects of a skillfully directed pressure pass off. The ganglia of the skull, squeezed between the mandibles but without fatal contusions, gradually recover their activity and put an end to the general torpor. Admit that it is all alarmingly scientific.

Fortune has her entomological whims: you run after her and catch no glimpse of her; you forget about her and behold, she comes tapping at your door! How vainly I watched and waited, how many useless journeys I made to see the Languedocian Sphex sacrifice her Ephippigers! Twenty years pass; these pages are in the printer's hands; and one day early this month, on the 8th of August 1878, my son Emile comes rushing into my study: "Quick!" he shouts. "Come quick: there's a Sphex dragging her prey under the plane-trees outside the door of the yard!"

Emile knew all about the business, from what I had told him to amuse him when we used to sit up late, and better still from similar incidents which he had witnessed in our life out doors. He is right. I run out and see a magnificent Languedocian Sphex dragging a paralyzed Ephippiger by the antennæ. She is making for the hen-house close by and seems anxious to scale the wall, with the object of fixing her burrow under some tile on the roof; for, a few years ago, in the same place, I saw a Sphex of the same species accomplish the ascent with her game and make her home under the arch of a badly-joined tile. Perhaps the present Wasp is descended from the one who performed that arduous climb.

A like feat seems about to be repeated; and this time before numerous witnesses, for all the family, working under the shade of the plane-trees, come and form a circle around the Sphex. They wonder at the unceremonious boldness of the insect, which is not diverted from its work by a gallery of onlookers; all are struck by its proud and lusty bearing, as, with raised head and the victim's antennæ firmly gripped in its mandibles, it drags the enormous burden after it. I, alone among the spectators, feel a twinge of regret at the sight:

"Ah, if only I had some live Ephippigers!" I cannot help saying, with not the least hope of seeing my wish realized.

"Live Ephippigers?" replies Emile. "Why, I have some perfectly fresh ones, caught this morning!"

He dashes upstairs, four steps at a time, and runs to his little den, where a fence of dictionaries encloses a park for the rearing of some fine caterpillars of the Spurge Hawkmoth. He brings me three Ephippigers, the best that I could wish for, two females and a male.

How did these insects come to be at hand, at the moment when they were wanted, for an experiment tried in vain twenty years ago? That is another story. A Lesser Grey Shrike had nested in one of the tall plane trees of the avenue. Now a few days earlier, the mistral, the brutal northwest wind of our parts, blew with such violence as to bend the branches as well as the reeds; and the nest, turned upside down by the swaying of its support, had dropped its contents, four small birds. Next morning, I found the brood upon the ground; three were killed by the fall, the fourth was still alive. The survivor was entrusted to the cares of Emile, who went Cricket-hunting twice a day on the neighbouring grass-plots for the benefit of his young charge. But Crickets are small and the nursling's appetite called for many of them. Another dish was preferred, the Ephippiger, of whom a stock was collected from time to time among the stalks and prickly leaves of the eryngo. The three insects, which Emile brought me, came from the Shrike's larder. My pity for the fallen nestling had procured me this unhoped-for success.

After making the circle of spectators stand back so as to leave the field clear for the Sphex, I take away her prey with a pair of pincers and at once give her in exchange one of my Ephippigers, carrying a sword at the end of her belly, like the game which I have abstracted. The dispossessed Wasp stamps her feet two or three times; and that is the only sign of impatience that she gives. She goes for her new prey, which is too stout too obese even to try to avoid pursuit, grip it with her mandibles by the saddle-shaped corselet, gets astride and, curving her abdomen, slips the end of it under the Ephippiger's thorax. Here, no doubt, some stings are administered, though I am unable to state the number exactly, because of the difficulty of observation. The Ephippiger, a peaceable victim, suffers herself to be operated on without resistance; she is like the silly sheep of our slaughterhouses. The Sphex takes her time and wields her lancet with a deliberation, which favours accuracy of aim. So far, the observer has nothing to complain of; but the prey touches the ground with its breast and belly and exactly what happens underneath escapes his eye. As for interfering and lifting the Ephippiger a little, so as to see better, that must not be thought of: the murderess would resheathe her weapon and retire. The act that follows is easy to observe. After stabbing the thorax, the tip of the abdomen appears under the victim's neck, which the operator forces open by pressing the nape. At this point, the sting probes with marked persistency, as if the prick administered here were more effective than elsewhere. One would be inclined to think that the nerve-centre attacked is the lower part of the esophageal chain; but the continuance of movement in the mouthparts—the mandibles, jaws and palpi—controlled by this seat of innervation shows that such is not the case. Through the neck, the Sphex reaches. Simply the ganglia of the thorax, or at any rate the first of them, which is more easily accessible through the thin skin of the neck than through the integuments of the chest.

And in a moment it is all over. Without the least shiver denoting pain, the Ephippiger becomes henceforth an inert mass. I remove the Sphex' patient for the second time and replace it by the other female at my disposal. The same proceedings are repeated, followed by the same result. The Sphex has performed her skilful surgery thrice over, almost in immediate succession, first with her own prey and then with my substitutes. Will she do so a fourth time with the male Ephippiger whom I still have left? I have my doubts, not because the Wasp is tired, but because the game does not suit her. I have never seen her with any prey but females, who, crammed with eggs, are the food which the larvæ appreciate above all others. My suspicion is well founded; deprived of her capture, the Sphex stubbornly refuses the male whom I offer to her. She runs hither and thither, with hurried steps, in search of the vanished game; three or four times, she goes up to the Ephippiger, walks round him, casts a scornful glance at him; and at last she flies away. He is not what her larvæ want; experiment demonstrates this once again after an interval of twenty years.

The three females stabbed, two of them before my eyes, remain in my possession. In each case, all the legs are completely paralyzed. Whether lying naturally, on its belly or on its back or side, the insect retains indefinitely whatever position we give it. A continued fluttering of the antennæ, a few intermittent pulsations of the belly and the play of the mouthparts are the only signs of life. Movement is destroyed but not susceptibility; for, at the least prick administered to a thin-skinned spot, the whole body gives a slight shudder. Perhaps, some day, physiology will find in such victims the material for valuable work on the functions of the nervous system.

Teale, E. W. 1949. The Insect World of J. Henri Fabre *(translated by A. Teixeira de Mattos). Dodd, Mead, New York; "The Wisdom of Instinct," pp. 42–52.*

The Wolf Spiders (1954)

John Crompton [pseudonym of John Battersby Crompton Lamburn]

One cannot make rules about spiders nor lump them together. Nor can one go into their lives species by species even if these lives were known. We shall study the wolf spider's life by taking again *Lycosa narbonnensis* or the Black-bellied Tarantula as our model, and again Fabre will be the class master during most of the lesson.

Lycosa, when mature, lives in a burrow above which she has erected a parapet, and from this parapet she surveys the territory over which she holds sway. Should any wayfarer approach she crouches down and gets ready for a quick takeoff. Hers is the method of the cheetah, an animal that is a famous sprinter, but no stayer. A gazelle or buck that can keep in front of a cheetah for a hundred yards will get away, but the odds are it will not keep in front, for the cheetah beyond a doubt, holds the world's record for the hundred yards.

So the wayfarer, plodding along, awakens to the fact that something is coming down on him like an express train. One backward look is sufficient; the wayfarer accelerates frantically. But usually it is hopeless; the Tarantula has had a flying start and is on him almost before he has started. The wayfarer is then either eaten on the spot or dragged to the cellar and eaten there.

Lycosa, the Wolf Spider, is not the sadist that Aranea is. If the prey is large she kills it immediately, for she lacks ropes to tie it and has no web to embarrass it. This means of course that she must often lunch in our own manner, crunching her meat. But she does not really eat meat; no spider does. She chews to extract the fluid. Not a nice meal to watch.

In the neighborhood of the castle of the Wolf Spider tales must circulate. The inhabitants must talk together and shake their heads. Doubtless many a traveler inquiring the way meets with a significant pause and is advised to take a roundabout route. If the traveler asks why, he hears vague mutterings about an ogre and about people disappearing. Anyway he will be well advised to take the route the locals suggest.

But many pass by the castle and the ogress waxes fat. Later on she kills less. For one thing she is bulging with food, and for another a desire for other things than blood stirs in her. With the heat of August she feels the need of a mate. It is no use her waiting where she is. The eager lover she envisages will not come into her bedroom—not if he can help it. Sheep do not rush into slaughterhouses. The prospective bridegroom is eager, too, for her but he wishes to meet his bride in a place where he has a chance of getting away after the nuptials are over.

So Lycosa leaves her burrow and goes in search of a lover, and in due course, in obedience to that strict disciplinarian, Nature, a small creature appears and stands trembling before her. The bridegroom has arrived.

I shall deal in another chapter with the mating of spiders: what I propose to do now is to study Lycosa as a mother.

We start with this same Lycosa observed by Fabre's children under a bush. The children called their father and he came running. The nuptials were over and the bridegroom had not been quick enough. He had lingered longer than he should in the arms of his bride and she now stood over him eating him. The morbid circle composed of Fabre and his children waited until the obese female had munched and drained her husband dry, then they captured her and put her in a cage, which stood on sand in Fabre's study.

In this cage in due course Fabre was able to examine the making of the cocoon of the Wolf Spider. A brief summary will be sufficient for, as I have said, the making of a spider's cocoon is a very complicated process. The cocoons of the web weavers were stationary affairs, edifices, little houses, placed in certain fixed spots; the cocoon of the Wolf Spider is a large pill that she straps to her person and carries about. It is not so complicated as the cocoons of most of the web weavers, but since it has to be made detached and portable it presents certain difficulties. It requires space for the making, a cleared patch, so she must leave her burrow to do it. She left her burrow in any case in order to meet the unfortunate whom for a brief period she called husband. Now, as a widow—grass or otherwise according to the alertness and dodging power of her spouse—she probably remains away from home for the ten days or so that must elapse before the laying of her eggs. She takes refuge under stones when necessary and is just as great a menace to passers by—not quite, perhaps; never svelte, her belly is now enormous and must take something off her speed.

Away from her burrow the ogress herself is not of course immune from danger. Other pregnant mothers are abroad. It is the hunting wasp season and among the hunting wasps are those who feed their children on nothing but spiders. A smaller, shapelier, deadlier mother may descend on her, overpower her, drug her, and drag her to the cruelest fate that callous Nature ever devised. There are birds, too, and other enemies, but assuming she escapes these dangers she selects a pitch and lays down a carpet. Though made of silk it is a rough affair, more a mat than a carpet. When one lacks a table and has delicate work to do one does lay down a mat. The Chinese do it; so do the Indians and many others. It acts as a floor and a protection from dirt and sand. On this floor Lycosa weaves a disc. It is a different affair from the mat, beautifully made from closely woven white silk, and forms a kind of bowl. In the center of this bowl Lycosa lays a circular heap of eggs, which are then covered with a napkin, which she weaves over them. The most laborious part of the affair is to come: the silk disc is carefully detached from the mat to which it has been tied and then, legs and jaws working hard, is folded over the bag of eggs, after which the sides are sewn together. The result is a white silk globe nearly as large as the mother, with one side thick and the other side thinner.

The mother is tired. The making of the cocoon has been exhausting work. Moreover, in addition to manufacturing eggs her inside has been drained of what, for a wolf spider, is a great deal of silk. She clasps the globe to her and rests for the remainder of the day.

Mother love is a strange thing. We marvel at it in our own kind, for a newborn child is not really enticing. Still less enticing, one would think, would be a mass of jellied eggs. Yet upon that mass, even though enclosed in a case, Lycosa now lavishes an affection greater than that displayed by the most obsessed of human mothers. And what makes it stranger is the character of the mother herself. We have seen the way she treated the father, the part creator of those precious globules. And we have seen her, morose, solitary, dour, dangerous, thinking only of her stomach. That mother love, transcendent, self-denying, prepared always for the supreme sacrifice, should spring from such a source is strange. Human mothers have something that needs them, yet in spite of their infatuation one cannot imagine them crooning over a number of billiard balls in a box, which, in effect, is all the wolf spider mother has to cherish. And, or so one supposes, she knows nothing of the future. The human mother thinks her child will turn into something even more beautiful. Lycosa does not even know that her eggs will turn into anything at all. And when they do they jettison quite half her affection.

Next day Lycosa will be found to have tied the globe to her with silk tapes and to be moving about with it. She carries it slung beneath her. It is almost as large as herself and at first her movements are greatly hampered. But she gets used to it and by raising her legs as she goes can clear most obstructions. Later on she learns to run quite quickly with it. In a manner of speaking she is a mother that is never seen without her perambulator, and mother and perambulator often go flying along at more speed than one would think right for such a combination. In these chases, which are generally after game, the cocoon is often knocked off, but it is a matter of seconds for Lycosa to strap it on again.

So attached is she, in the figurative sense, to her globular burden that she will fight to the death to retain it. As Savory says, "The possession of a cocoon changes the mother's entire outlook on life." If it is taken away from her she goes frantic. She searches everywhere. She is incredulous, horrified.

Bonnet once found a Wolf Spider carrying her cocoon and threw her into the pit of an ant lion. Aware of her danger the spider tried to run away. Too late; the ant lion seized the underslung cocoon between its jaws. The mother struggled and in these struggles the cocoon became detached. Now was the mother's chance to save her life, but instead of running away she turned around in the pit and seized the cocoon, which was being drawn under the sand. Her strength of course could not prevail against that of the ant lion, but she retained her hold and was drawn under the sand together with her beloved bag. And she would have been buried alive and that would have been the end of her, but for Bonnet, who dug her out. He got no thanks, for her first act was to hurl herself into the pit again. Bonnet fished her out a second time and for some time kept her from suicide by warding her away with a twig from the fatal chasm. So obviously the Wolf Spider prefers death to life without her cocoon.

Equally striking is her delight when it is given back to her and she straps the beloved possession to her person once more and goes on her way rejoicing.

Some observers say that if the burden is not restored the mother dies of grief in a few days. This is to be doubted. Those careful observers, the Peckhams, say that Lycosa, having been deprived of her capsule will still eagerly seize it if it is restored to her after 16 or 17 hours. Sometimes she will respond after 24 hours. But after two days she never has any further interest in it.

Dearly as she loves her pill it is just a pill and loud has been the laughter of naturalists at the fact that after it has been taken away she cannot recognize it from others similar. But before we vertebrates laugh too loudly let us make quite sure that we are in a position to do so. Any bird will brood over misshapen stones substituted for her eggs. Hedge sparrows do not throw out the cuckoo's egg, so different from their own. Fowls will sit on potatoes or nothing at all for four weeks or more. A cat, agonized by the loss of her young will suckle baby rats. A sheep dog in similar circumstances will suckle young foxes. A cow whose calf has been taken away will make known its bereavement to all the neighborhood. Yet its lugubrious bellows will cease if the calf's skin stuffed with straw is placed by its side. The story of Romulus and Remus was considered credible by the ancient Romans, and substitution of babies has been made without even human mothers realizing it.

Therefore I am not going to comment too scathingly on Lycosa in the experiments that Fabre is about to conduct.

He took a Lycosa's cocoon away from her. He did it with his famous pair of forceps, which she attacked with such fury, that he could hear her fangs grating on the steel. Immediately he gave her a cocoon taken from another Lycosa. She took it, strapped it to her, and walked contentedly away.

Well, even Fabre did not jeer very much at this. He merely said contemptuously, "Her own or another's: it is all one to the spider."

He next tried the same experiment but substituted a cocoon of a different shape, made by another species of spider. This too was taken with as much relief as the other.

"Let us," goes on Fabre, "penetrate yet further into the wallet bearer's stupidity."

So he made a ball of cork of the same size as Lycosa's cocoon. He made it carefully and polished it with a file. It was accepted.

"The silly creature . . . embraces the cork ball, fondles it . . . fastens it to her spinnerets and drags it after her."

Worse is to come. Fabre made several of these cork balls—four or five—and arranged them with the real cocoon among them. The bereaved Lycosa simply rushed up and took the first.

"This obtuseness," concludes Fabre, "baffles me. We will leave her alone; we know all that we want to know about her poverty of intellect."

Fabre thought he had sounded bottom in the abyss of Lycosa's stupidity. But he had not, although he never knew it. Locket records that he saw a *Lycosa palustris* running about with a small snail shell tied to her. And this was not an experiment. The creature must have lost her cocoon somehow and after hours of frantic search decided that a snail shell was better than nothing.

But when dealing with spiders, experiments are not of as much value as those made upon many other animals. Spiders are individualists (a horrid but useful word). The experiments have now gone further. Savory tells us that a species in Africa, *Palystes natalius*, invariably selects her own cocoon from that of another spider even of her own species. And yet this same discriminating spider after having bungled in the making of her cocoon and produced a very misshapen object, selected a properly shaped one made by another in preference to her own. So scent has no bearing on the subject. Sight may have; for another of this species selected another's cocoon after her own had been stained with aniline dye.

Other genera also, of the family Theridiidae, will have no dealings at all with any substitutes.

By the middle of September in the case of our particular Lycosa, the pill bursts and two hundred young come pouring out. When the last has emerged the mother unstraps her bag and throws it away. But if she thinks she is now a free female she is very much mistaken. These two hundred know where they are going; they are going with mother. You almost need a lens to see them, but they are perfect spiders in every respect and shaped like their parents. They have their wits about them too and do not delay. The bursting of the pill has scattered them around a bit but they pull themselves together and all come running up to the mountainous form that is their parent. They find her long legs and stream up them and soon are perched on her back like so many passengers on a bus. Their bodies are packed in rows and their legs intermingled together. Fortunately no conductor is present or there would be wholesale ejections and long ago the "No room on top" would have sounded. Once aboard and seated they are very quiet and good.

It is a long journey on which these diminutive passengers have embarked. It will last about six months, and they have brought no provisions. It is a hazardous journey too; the driver is careless and sometimes makes insufficient allowance for obstructions so that the passengers are swept from their seats. On these occasions the bus usually stops and waits and the fares run up and climb back to their places.

The mother has changed. She is still a good mother (to carry two hundred children about for six months speaks for itself) but she does not dote on them now as she doted on them when they were inanimate pellets in a box. Her attitude is that of a good-natured but indifferent riding donkey.

These spider babies pass the whole of the six or seven months without food; at least so far as observers keeping them under scrutiny in cages can ascertain. How they do it is a mystery. It is not as if they were comatose and expending no energy. They expend quite a lot one way and another. Even when seated on top they often have to cling together to retain their positions on the swaying vehicle, especially when the vehicle sees some prey and accelerates violently. And when they fall off they expend still more energy, for the mother does not always stop for them. Yet this energy is never replaced by food. The mother likes to "sun" her babies whenever she can, turning herself from time to time so that all get the rays and Fabre suggests that they get their energy direct from the sun. Scientists however say this is nonsense; only vegetables can do that.

The mother still likes her food. Babies or no babies, she runs down what game her burdened state allows. And when she is eating it the young sit above viewing the orgy with complete indifference. They are no more interested than human passengers are interested when their coach takes in petrol.

Fabre, of course, has not failed to experiment with these riders. He brushed a complete party off in such a way that they fell among the legs of another Lycosa mother who was already carrying more than her full complement of excursionists. The mother of the first party he whisked away. Party No. 1 immediately climbed on to the top of the new vehicle, without protest either from the driver or the legitimate fares.

He then repeated the experiment, but instead of brushing the youngsters into the vicinity of another Lycosa, he selected an Aranea, a type that does not cater for passenger traffic at all. The urchins with one accord ran up Aranea's legs and Aranea shook her legs in horror and disgust. Several juveniles were knocked spinning but some reached the top, and most of those that had been ejected ran back and again attempted the ascent. Again many were flung away, but again many reached the target and took their seats sedately with the others.

Their complacency was short-lived: Aranea rolled over on her back, killing some and maiming others. And she had to roll several times before she could get rid of them all.

The young, unless brushed off, sit where they are, but in dangerous situations they alight of their own accord. One such dangerous situation is when two wolf mothers come face to face. In such an event there will be no friendly discussion about the care of children: there will simply be a fight. The children themselves realize this and lose no time in dismounting and making themselves scarce. Peering from cover, they then watch a fight that is no kid-gloved affair.

Eventually one panting matron stands over the other. Fabre has recorded such a fight and shown us the victorious one holding down her opponent, gripping her with her legs and pressing down on her belly. Among female spiders no quarter is given or expected. The prone one spits viciously and the other waits her time to avoid those snapping jaws and bite into the brain.

Fabre also relates how, after the *coup de grâce* had been given and the conqueress was eating her opponent, all the children came forth from hiding and climbed on to her back. And there they sat until mother—for she was "mother" now to all of them—had finished her meal.

It follows therefore that a matron, if a good fighter, may go out with her own children in the morning and come back in the evening carrying three lots packed several tiers deep.

Three is the limit. Fabre experimented and managed to get three loads on to one mother's back but could not manage a fourth. The mother herself raised no objection, but the load became top-heavy and swayed when the mother moved so that those perched on the summit fell off.

In winter all live in the burrow and the mother eats little, if anything. But in spring she is up and about again and taking her toll of passersby. The little ones are still on her back and they are exactly the same size as when they left the cocoon—one could hardly expect growth on a diet of air.

Some people, of the class who only give a spider one vague look, have asked me, why, if wolf spiders are so common, they have never seen a wolf spider running about the garden carrying young on her back? The answer is that they have. To them a wolf spider even with a three-tier deep lot of passengers would seem merely a spider with a large grayish colored back.

In late spring new urges stir in the bosoms of the passengers. The long, long trip is over so far as this particular vehicle is concerned. They alight in batches and are soon seen running away as if they had an urgent appointment somewhere else. They have, for their journey is not over.

The mother shows no emotion at their departure. And they on their part show as much concern as we show when we alight at a junction and go from one train to another. They are well advised. For all her apparent indifference the mother's patience is wearing a bit thin. In a few months she will marry again and have more children and before that time she wants some liberty. She wants to start another reign of terror in the countryside. So if any children stay too long they disappear. Passengers are no longer allowed on top, but there is plenty of room inside.

The little spiders hurry on. They run up short grasses and down again and try others. Already some have reached the tops of railings or tall herbage. It is as if each one carried a banner with a strange device—Excelsior.

And shortly they go higher still. One by one, like prophets snatched to heaven, they are ascending into the blue without apparent agency.

This ballooning, this rising upwards of young spiders, is not confined to wolf spiders: it is the preliminary launching of most species of all our groups and we shall have more to say about it later. Those that escape the many dangers will descend. It may be only a few yards from the starting place or it may be many miles.

There is always something sad in the breaking up of family life and this case seems sadder than most. For six months none of them have quarreled. Can such harmony continue? No, it cannot; hence the trip by air to widely separated parts. The brother that sat next to a sister on the mother's back and must have got to know her very well, had better not get near to her again.

Descending to land, each casts off its harness. The discarded silken trail floats away and the young spider looks around. It is feeling a completely new sensation—hunger. There are forms around smaller even than itself and one of these it catches and eats—much, no doubt, to the surprise of a stomach that has not yet functioned. It is the most accommodating of all stomachs, being able to cope with much or nothing at all. Had we in these days of indigestion and gastric ulcers the stomachs of—but we have not the stomachs of spiders so let us forget it.

Crompton, J. 1954. The Life of Spiders. *The New American Library of World Literature, New York; "The Wolf Spiders," pp. 66–74.*

Contour Diving (1934)

William Beebe

The four final bathysphere [*spherical, unpowered, and tethered deep-sea submersible devised by Otis Barton in 1928*] dives of 1930 were devoted to what I might describe as contour diving, to steal part of an aviation term. This is decidedly more risky than deep dives in the open sea, but is of equal scientific importance. It opens up an entirely new field of possibilities, the opportunity of tracing the change from shallow-water fauna, corals, fish, etc., to those of mid-water, with the hope finally of observing the disappearance of the latter, and the change, gradual or abrupt, into the benthic, or deep-sea, forms of life. We know absolutely nothing of this at present, as the transition zone is so rough and untrawlable that there is no method known of learning anything about it. Nets are torn to shreds and dredges catch almost at once, the wire breaks and not a single organism comes up.

I worked out the simplest method during the last dives of 1930. I brought the *Gladisfen* and *Ready* [*Beebe's tugboat and barge*] as close to shore as I dared on a day of perfect calm with a slight offshore wind and there began diving, with the bottom, nine or ten fathoms down, actually in sight from the deck. We were lowered to within two fathoms of the reef while the *Ready* drifted slowly seaward. As it turned out, the first two dives were probably the most foolhardy things we could have done, for the sphere drifted backward, and from my window I could see only the bottom over which we had passed. If our projecting wooden landing gear had caught in a sudden rise of reef it would have gone very hard with us. As it was, I could only flatten my eyes against the quartz and try to adapt my elevation orders to what the contour promised. Once a crag passed two feet beneath us and I had a most unpleasant moment while we were rushed up 30 feet.

Early on the following day, June twentieth, Barton affixed a double wooden rudder of boards to orient the sphere, so that in our subsequent contour dives we swung around and faced forward. Another improvement was the shifting of the shackle to the posterior hole, so that the whole apparatus tilted slightly downward in front. I had the lead heaved constantly and telephoned to me, and so rapidly were my orders transmitted to the man at the winch that we rose and fell swiftly as we progressed slowly seaward, now ordering a fathom or two of elevation to escape a projecting coral crag, then dropping down into a submarine valley until the bottom again became visible. In spite of a constant watch ahead, accidents were on several occasions barely avoided.

Two years later, in 1932, when we were diving from the deck of a large tug, one of the contour dives was marked by the narrowest escape which we ever experienced. We had already hurdled two low coral reefs, twenty and thirty feet high. A group of large fish held our attention directly below. I could not quite identify them and they were whirling around some focus of attraction when a dark shadow fell across the window. I looked up and saw that we were drifting rapidly toward an enormous crag or part of a coral reef, towering fifty feet or more above us, and covered on its almost perpendicular slope with great outreaching crags and sharp, water-worn hooks and snags. I sent the most urgent S.O.S. on the wire to haul us up as rapidly as possible, and we could almost hear the hissing steam as the winch began to turn at full speed. Fortunately there were no clamps to cut free. As we ascended we swung nearer and nearer the cliff, and the waving sea fans and great anemones grew larger, and we were so close that every detail, every small fish became visible.

I fully expected to strike and had already formulated the next order, which would have been to let us out as rapidly as we had been drawn up and to go astern full speed. In this way we might have slipped down the reef without becoming entangled and when the tug had backed over the reef we would have swung clear.

But again the clarity of the fused quartz windows deceived us and we just cleared the summit, passing so close to it that I am sure our wooden base must have brushed the finger-like plumes on the reef top.

Even if we had struck, no harm might have resulted. We might have bumped and scraped up and over the top. But a straight blow on one of the quartz windows, or getting badly tangled in one of those steelhard, outreaching crags would not have been so good.

On the other side I had us lowered 65 feet as the bottom sloped rather steeply, and when we were again within a few feet of the bottom I saw below us a wide beach of white sand, mixed with water-worn pebbles and shells and sloping against the outer base of the great reef. I could even see ripple marks and could distinguish the various kinds of shells. This was to me one of the most interesting discoveries of all my dives. It was undoubtedly the old foreshore of Bermuda, the ancient beach, which was above water at the last glacial period, say twenty-five thousand years ago. At that time there was so much ice locked up on the continents that the oceans were 250 feet lower than they are at present, and the dry land area of Bermuda was then doubtless measured in hundreds instead of tens of miles. Not far beyond this beach of olden time an abrupt and awful drop led down into a bluish-black abyss, where the bottom was lost and could not be recovered without too much risk.

Visibility was usually excellent, except close in-shore and for a few days after a severe storm or hurricane, when cloudiness put an end to the work. When the water was clear I could make out and identify all coral and algal growths, and fish down to two inches in length. After several years work in the diving helmet, I was familiar with the Bermuda fauna down to six and eight fathoms, and now the thing which impressed me most as I went deeper was the increased size of the fish: snappers, grunts, angelfish, and chubs, trumpetfish, surgeons, parrots, and jacks—all were as large or larger than I had ever seen them when diving in shallower water near shore. Now and then a fish was seen larger than any of its kind ever taken in Bermuda, and this in spite of the fact that angling is carried on down to ninety fathoms.

Certain species of mid-water fish offered unexpected problems. The two most abundant were the blue chromis, *Demoisellea cyaneus*, and the smooth sardine, *Sardinella anchovia.* The former holds a place on the Bermuda list solely on the basis of a single doubtful record of seventy years ago, while there are few published records of this sardine. Yet on these shallow dives I saw school after school of each, hundreds of chromis swimming loosely, and tens of thousands of sardines in dense formations. When the latter sighted the bathysphere they turned downward as one fish, and poured past like elongated, silvery raindrops. The chromis usually passed on a horizontal plane. In the West Indies recently I saw these two species in vast numbers about the shallowest reefs near shore.

Once I saw an interesting exchange of courtesy, one which I have observed many times when diving near shore. The giant cærulean parrotfish browse on hard coral as a horse tears off mouthfuls of grass. After an interval of feeding, when the teeth and jaws and scales of the head are covered with debris, the fish upends in mid-water and holds itself motionless while a school of passing wrasse, all tiny in comparison with the big fish, rush from all sides and begin a systematic cleaning of the large fish's head. As in most relationships between different species of animals, this is founded on mutual benefit, the parrotfish getting a free cleaning, and the wrasse finding a supply of particles of food ready at hand.

On the very last dive of 1930, we were 30 feet down with the bottom at least a hundred feet beneath, when, without the slightest warning, the green water rained blue parrotfish. They were all deep cærulean blue (*Scarus cæruleus*) almost unmarked, and they varied from about six inches to four feet. Hundreds and hundreds streamed obliquely past and downward, unending lines of vivid blue, and they extended far beyond my vision in every direction. Some were the merest shadow ghosts of parrotfish in the distance, others almost brushed the glass, and the downpour did not cease—we merely passed through it. It seemed as if all the parrotfish of Bermuda had suddenly decided to leave for the depths of the open sea.

Once before, a few miles to the westward, when I descended on a particularly rough day in my helmet to a depth of 30 feet I saw a similar migration or gathering of the blue parrotfish clans. On this occasion they were filled with curiosity about me and milled about for five minutes, fairly blanketing me—almost obliterating the surrounding seascape.

Well out from shore on one of these contour dives I had the thrill of suddenly seeing a thin, endless sea serpent. We were drifting slowly along, now lifting over a toothed ridge or settling down into a valley of caverns and gorges when, without warning, I saw a long black line undulating over the bottom, clearly visible when over a bed of sand, or vanishing behind a mass of giant sea-plumes. A second glance revealed it as the deep-sea transatlantic cable resting quietly on its bed and carrying innumerable messages of hope and fear, joy and death. Kipling's words took on a new significance and I shall never send a cable again without this memory, nor shall I ever forget the breathtaking belief of the first few seconds.

Another important phase of this method of observation is the physical geography of the bottom. I have been able to describe and map over a mile of bottom seen from five to twenty feet elevation, traversing steadily seaward. After passing the great loop of the cable, all visible life ceased, and we drifted over a wide expanse of desert, with no fish or plumes or living coral. I have no idea of the significance of this dead zone.

I have never succeeded in following the bottom lower than 350 feet. Increasing cloudiness of the water and greater obscurity have made it impossible to distinguish anything, and the danger of getting hung up and snapped off on some projecting cliff is too constant to progress blindly.

With a calm sea, a steady offshore breeze or current, and our searchlight in working order it will be possible sometime to make a systematic survey of the Bermudian insular shelf. Even to repeat the conditions under which we worked would mean perfect facility for recognizing the change in species of fish and such invertebrates as echinoderms and horny corals; to see and name and note the bottom limits of alga: and brain coral; and finally to watch for the end of water-worn and air-worn rocks, and the beginnings of the lava flows of old Bermuda Mountain.

Beebe, W. 1934. Half Mile Down. *Harcourt Brace and Company, New York; "Contour Diving," pp. 138–145.*

The Winter Journey (1930)

Apsley Cherry-Garrard

To me, and to everyone who has remained here the result of this effort is the appeal it makes to our imagination, as one of the most gallant stories in Polar History. That men should wander forth in the depth of a Polar night to face the most dismal cold and the fiercest gales in darkness is something new; that they should have persisted in this effort in spite of every adversity for five full weeks is heroic. It makes a tale for our generation which I hope may not be lost in the telling.

Scott's diary, at Cape Evans

The view from eight hundred feet up the mountain was magnificent and I got my spectacles out and cleared the ice away time after time to look. To the east a great field of pressure ridges below, looking in the moonlight as if giants had been ploughing with ploughs which made furrows fifty or sixty feet deep: these ran right up to the Barrier edge, and beyond was the frozen Ross Sea, lying flat, white and peaceful as though such things as blizzards were unknown. To the north and northeast the Knoll. Behind us Mount Terror on which we stood, and over all the grey limitless Barrier seemed to cast a spell of cold immensity, vague, ponderous, a breeding-place of wind and drift and darkness. God! What a place!

There was now little moonlight or daylight, but for the next forty-eight hours we used both to their utmost, being up at all times by day and night, and often working on when there was great difficulty in seeing anything; digging by the light of the hurricane lamp. By the end of two days we had the walls built, and banked up to one or two feet from the top; we were to fit the roof cloth close before banking up the rest. The great difficulty in banking was the hardness of the snow, it being impossible to fill in the cracks between the blocks, which were more like paving stones than anything else. The door was in, being a triangular tent doorway, with flaps which we built close in to the walls, cementing it with snow and rocks. The top folded over a plank and the bottom was dug into the ground.

Birdie was very disappointed that we could not finish the whole thing that day: he was nearly angry about it, but there was a lot to do yet and we were tired out. We turned out early the next morning (Tuesday 18th) to try and finish the igloo, but it was blowing too hard. When we got to the top we did some digging but it was quite impossible to get the roof on, and we had to leave it. We realized that day that it blew much harder at the top of the slope than where our tent was. It was bitterly cold up there that morning with a wind force 4–5 and a minus thirty temperature.

The oil question was worrying us quite a lot. We were now well in to the fifth of our six tins, and economizing as much as possible, often having only two hot meals a day. We had to get down to the Emperor penguins somehow and get some blubber to run the stove, which had been made for us in the hut. The 19th being a calm fine day we started at 9.3°, with an empty sledge, two ice axes, Alpine rope, harnesses and skinning tools. Wilson had made this journey through the Cape Crozier pressure ridges several times in the *Discovery* [*the ship used by Robert Scott to explore Antarctica, 1901–4*] days. But then they had daylight, and they had found a practicable way close under the cliffs, which at the present moment were between us and the ridges.

As we neared the bottom of the mountain slope, farther to the north than we had previously gone, we had to be careful about crevasses, but we soon hit off the edge of the cliff and skirted along it until it petered out on the same level as the Barrier. Turning left handed we headed towards the sea-ice knowing that there were some two miles of pressure between us and Cape Crozier itself. For about half a mile it was fair going, rounding big knobs of pressure but always managing to keep more or less on the flat and near the ice-cliff which soon rose to a very great height on our left. Bill's idea was to try and keep close under this cliff, along that same *Discovery* way which I have mentioned above. They never arrived there early enough for the eggs in those days: the chicks were hatched. Whether we should now find any Emperors, and if so whether they would have any eggs, was by no means certain.

However, we soon began to get into trouble, meeting several crevasses every few yards, and I have no doubt crossing scores of others of which we had no knowledge. Though we hugged the cliffs as close as possible we found ourselves on the top of the first pressure ridge, separated by a deep gulf from the ice-slope which we wished to reach. Then we were in a great valley between the first and second ridges: we got into huge heaps of ice pressed up in every shape on every side, crevassed in every direction: we slithered over snow-slopes and crawled along drift ridges, trying to get in towards the cliffs. And always we came up against impossible places and had to crawl back. Bill led on a length of alpine rope fastened to the toggle of the sledge; Birdie was in his harness also fastened to the toggle, and I was in my harness fastened to the rear of the sledge, which was of great use to us both as a bridge and a ladder.

Two or three times we tried to get down the ice-slopes to the comparatively level road under the cliffs, but it was always too great a drop. In that dim light every proportion was distorted; some of the places we actually did manage to negotiate with ice axes and Alpine rope looked absolute precipices, and there were always crevasses at the bottom if you slipped. On the way back I did slip into one of these and was hauled out by the other two standing on the wall above me.

We then worked our way down into the hollow between the first and second large pressure ridges, and I believe on to the top of the second. The crests here rose fifty or sixty feet. After this I don't know where we went. Our best landmarks were patches of crevasses, sometimes three or four in a few footsteps. The temperatures were lowish (−37°), it was impossible for me to wear spectacles, and this was a tremendous difficulty to me and handicap to the party: Bill would find a crevasse and point it out; Birdie would cross; and then time after time, in trying to step over or climb over on the sledge, I put my feet right into the middle of the cracks. This day I went well in at least six times; once, when we were close to the sea, rolling into and out of one and then down a steep slope until brought up by Birdie and Bill on the rope.

We blundered along until we got into a great cul-de-sac, which probably formed the end of the two ridges, where they butted on to the sea-ice. On all sides rose great walls of battered ice with steep snow-slopes in the middle, where we slithered about and blundered into crevasses. To the left rose the huge cliff of Cape Crozier, but we could not tell whether there were not two or three pressure ridges between us and it, and though we tried at least four ways, there was no possibility of getting forward.

And then we heard the Emperors calling.

Their cries came to us from the sea-ice we could not see, but which must have been a chaotic quarter of a mile away. They came echoing back from the cliffs, as we stood helpless and tantalized. We listened and realized that there was nothing for it but to return, for the little light which now came in the middle of the day was going fast, and to be caught in absolute darkness there was a horrible idea. We started back on our tracks and almost immediately I lost my footing and rolled down a slope into a crevasse. Birdie and Bill kept their balance and I clambered back to them. The tracks were very faint and we soon began to lose them. Birdie was the best man at following tracks that I have ever known, and he found them time after time. But at last even he lost them altogether and we settled we must just go ahead. As a matter of fact, we picked them up again, and by then were out of the worst: but we were glad to see the tent.

The next morning (Thursday, 20 June) we started work on the igloo at 3 a.m. and managed to get the canvas roof on in spite of a wind which harried us all that day. Little did we think what that roof had in store for us as we packed it in with snow blocks, stretching it over our second sledge, which we put athwartships across the middle of the longer walls. The windward (south) end came right down to the ground and we tied it securely to rocks before packing it in. On the other three sides we had a good two feet or more of slack all round, and in every case we tied it to rocks by lanyards at intervals of two feet. The door was the difficulty, and for the present we left the cloth arching over the stones, forming a kind of portico. The whole was well packed in and over with slabs of hard snow, but there was no soft snow with which to fill up the gaps between the blocks. However, we felt already that nothing could drag that roof out of its packing, and subsequent events proved that we were right.

It was a bleak job for three o'clock in the morning before breakfast, and we were glad to get back to the tent and a meal, for we meant to have another go at the Emperors that day. With the first glimpse of light we were off for the rookery again.

But we now knew one or two things about that pressure which we had not known twenty-four hours ago; for instance, that there was a lot of alteration since the *Discovery* days and that probably the pressure was bigger. As a matter of fact it has been since proved by photographs that the ridges now ran out three-quarters of a mile farther into the sea than they did ten years before. We knew also that if we entered the pressure at the only place where the ice-cliffs came down to the level of the Barrier, as we did yesterday, we could neither penetrate to the rookery nor get in under the cliffs where formerly a possible way had been found. There was only one other thing to do—to go over the cliff. And this was what we proposed to try and do.

Now these ice-cliffs are some two hundred feet high, and I felt uncomfortable, especially in the dark. But as we came back the day before we had noticed at one place a break in the cliffs from which there hung a snowdrift. It might be possible to get down that drift.

And so, all harnessed to the sledge, with Bill on a long lead out in front and Birdie and myself checking the sledge behind, we started down the slope, which ended in the cliff, which of course we could not see. We crossed a number of small crevasses, and soon we knew we must be nearly there. Twice we crept up to the edge of the cliff with no success, and then we found the slope: more, we got down it without great difficulty and it brought us out just where we wanted to be, between the land cliffs and the pressure.

Then began the most exciting climb among the pressure that you can imagine. At first very much as it was the day before—pulling ourselves and one another up ridges, slithering down slopes, tumbling into and out of crevasses and holes of all sorts, we made our way along under the cliffs which rose higher and higher above us as we neared the black lava precipices which form Cape Crozier itself. We straddled along the top of a snow ridge with a razor-backed edge, balancing the sledge between us as we wriggled: on our right was a drop of great depth with crevasses at the bottom, on our left was a smaller drop also crevassed. We crawled along, and I can tell you it was exciting work in the more than half darkness. At the end was a series of slopes full of crevasses, and finally we got right in under the rock on to moraine, and here we had to leave the sledge.

We roped up, and started to worry along under the cliffs, which had now changed from ice to rock, and rose 800 feet above us. The tumult of pressure which climbed against them showed no order here. Four hundred miles of moving ice behind it had just tossed and twisted those giant ridges until Job himself would have lacked words to reproach their Maker. We scrambled over and under, hanging on with our axes, and cutting steps where we could not find a foothold with our crampons. And always we got towards the Emperor penguins, and it really began to look as if we were going to do it this time, when we came up against a wall of ice which a single glance told us we could never cross. One of the largest pressure ridges had been thrown, end on, against the cliff. We seemed to be stopped, when Bill found a black hole, something like a fox's earth, disappearing into the bowels of the ice. We looked at it: 'Well, here goes!' he said, and put his head in, and disappeared. Bowers likewise. It was a longish way, but quite possible to wriggle along, and presently I found myself looking out of the other side with a deep gully below me, the rock face on one hand and the ice on the other. 'Put your back against the ice and your feet against the rock and lever yourself along,' said Bill, who was already standing on firm ice at the far end in a snow pit. We cut some fifteen steps to get out of that hole. Excited by now, and thoroughly enjoying ourselves, we found the way ahead easier, until the penguins' call reached us again and we stood, three crystallized ragamuffins, above the Emperors' home. They were there all right, and we were going to reach them, but where were all the thousands of which we had heard?

We stood on an ice foot, which was really a dwarf cliff some twelve feet high, and the sea-ice, with a good many ice-blocks strewn upon it, lay below. The cliff dropped straight, with a bit of an overhang and no snowdrift. This may have been because the sea had only frozen recently; whatever the reason may have been it meant that we should have a lot of difficulty in getting up again without help. It was decided that someone must stop on the top with the Alpine rope, and clearly that one should be I, for with short sight and fogged spectacles which I could not wear I was much the least useful of the party for the job immediately ahead. Had we had the sledge we could have used it as a ladder, but of course we had left this at the beginning of the moraine miles back.

We saw the Emperors standing all together huddled under the Barrier cliff some hundreds of yards away. The little light was going fast: we were much more excited about the approach of complete darkness and the look of wind in the south than we were about our triumph. After indescribable effort and hardship we were witnessing a marvel of the natural world, and we were the first and only men who had ever done so; we had within our grasp material which might prove of the utmost importance to science; we were turning theories into facts with every observation we made, and we had but a moment to give.

The disturbed Emperors made a tremendous row, trumpeting with their curious metallic voices. There was no doubt they had eggs, for they tried to shuffle along the ground without losing them off their feet. But when they were hustled a good many eggs were dropped and left lying on the ice, and some of these were quickly picked up by eggless Emperors who had probably been waiting a long time for the opportunity. In these poor birds the maternal side seems to have necessarily swamped the other functions of life. Such is the struggle for existence that they can only live by a glut of maternity, and it would be interesting to know whether such a life leads to happiness or satisfaction.

The men of the *Discovery* found this rookery where we now stood. They made journeys in the early spring but never arrived early enough to get eggs and only found parents and chicks. They concluded that the Emperor was an impossible kind of bird who, for some reason or other, nests in the middle of the Antarctic winter with the temperature anywhere below seventy degrees of frost, and the blizzards blowing, always blowing, against his devoted back. And they found him holding his precious chick balanced upon his big feet, and pressing it maternally, or paternally (for both sexes squabble for the privilege) against a bald patch in his breast. And when at last he simply must go and eat something in the open leads near by, he just puts the child down on the ice, and twenty chickless Emperors rush to pick it up. And they fight over it, and so tear it that sometimes it will die. And, if it can, it will crawl into any ice-crack to escape from so much kindness, and there it will freeze. Likewise many broken and addled eggs were found, and it is clear that the mortality is very great. But some survive, and summer comes; and when a big blizzard is going to blow (they know all about the weather), the parents take the children out for miles across the sea-ice, until they reach the threshold of the open sea. And there they sit until the wind comes, and the swell rises, and breaks that ice floe off; and away they go in the blinding drift to join the main pack ice, with a private yacht all to themselves.

You must agree that a bird like this is an interesting beast, and when, seven months ago, we rowed a boat under those great black cliffs and found a disconsolate Emperor chick still in the down, we knew definitely why the Emperor has to nest in mid-winter. For if a June egg was still without feathers in the beginning of January, the same egg laid in the summer would leave its produce without practical covering for the following winter. Thus the Emperor penguin is compelled to undertake all kinds of hardships because his children insist on developing so slowly, very much as we are tied in our human relationships for the same reason. It is of interest that such a primitive bird should have so long a childhood.

But interesting as the life history of these birds must be, we had not traveled for three weeks to see them sitting on their eggs. We wanted the embryos, and we wanted them as young as possible, and fresh and unfrozen, that specialists at home might cut them into microscopic sections and learn from the previous history of birds throughout the evolutionary ages. And so Bill and Birdie rapidly collected five eggs, which we hoped to carry safely in our fur mitts to our igloo upon Mount Terror, where we could pickle them in the alcohol we had brought for the purpose. We also wanted oil for our blubber stove, and they killed and skinned three birds—an Emperor weighs up to 6½ stones [*41 kg or 91 lb.*].

The Ross Sea was frozen over, and there were no seal in sight. There were only 100 Emperors as compared with 2000 in 1902 and 1903. Bill reckoned that every fourth or fifth bird had an egg, but this was only a rough estimate, for we did not want to disturb them unnecessarily. It is a mystery why there should have been so few birds, but it certainly looked as though the ice had not formed very long. Were these the first arrivals? Had a previous rookery been blown out to sea and was this the beginning of a second attempt? Is this bay of sea-ice becoming unsafe?

Those who previously discovered the Emperors with their chicks saw the penguins nursing dead and frozen chicks if they were unable to obtain a live one. They also found decomposed eggs which they must have incubated after they had been frozen. Now we found that these birds were so anxious to sit on something that some of those, which had no eggs, were sitting on ice! Several times Bill and Birdie picked up eggs to find them lumps of ice, rounded and about the right size, dirty and hard. Once a bird dropped an ice nest egg as they watched, and again a bird returned and tucked another into itself, immediately forsaking it for a real one, however, when one was offered.

Meanwhile a whole procession of Emperors came round under the cliff on which I stood. The light was already very bad and it was well that my companions were quick in returning: we had to do everything in a great hurry. I hauled up the eggs in their mitts (which we fastened together round our necks with lampwick lanyards) and then the skins, but failed to help Bill at all. 'Pull,' he cried, from the bottom: 'I am pulling,' I said. 'But the line's quite slack down here,' he shouted. And when he had reached the top by climbing up on Bowers's shoulders, and we were both pulling all we knew Birdie's end of the rope was still slack in his hands. Directly we put on a strain the rope cut into the ice edge and jammed—a very common difficulty when working among crevasses. We tried to run the rope over an ice-axe without success, and things began to look serious when Birdie, who had been running about prospecting and had meanwhile put one leg through a crack into the sea; found a place where the cliff did not overhang. He cut steps for himself, we hauled, and at last we were all together on the top—his foot being by now surrounded by a solid mass of ice.

We legged it back as hard as we could go: five eggs in our fur mitts, Birdie with two skins tied to him and trailing behind, and myself with one. We were roped up, and climbing the ridges and getting through the holes was very difficult. In one place where there was a steep rubble and snow slope down I left the ice-axe half-way up; in another it was too dark to see our former ice-axe footsteps, and I could see nothing, and so just let myself go and trusted to luck. With infinite patience Bill said: 'Cherry, you *must* learn how to use an ice-axe.' For the rest of the trip my wind-clothes were in rags.

We found the sledge, and none too soon, and now had three eggs left, more or less whole. Both mine had burst in my mitts: the first I emptied out, the second I left in my mitt to put into the cooker; it never got there, but on the return journey I had my mitts far more easily thawed out than Birdie's (Bill had none) and I believe the grease in the egg did them good. When we got into the hollows under the ridge where we had to cross, it was too dark to do anything but feel our way. We did so over many crevasses, found the ridge and crept over it. Higher up we could see more, but to follow our tracks soon became impossible, and we plugged straight ahead and luckily found the slope down which we had come. All day it had been blowing a nasty cold wind with a temperature between −20° and −30°, which we felt a good deal. Now it began to get worse. The weather was getting thick and things did not look very nice when we started up to find our tent. Soon it was blowing force 4, and soon we missed our way entirely. We got right up above the patch of rocks which marked our igloo and only found it after a good deal of search.

I have heard tell of an English officer at the Dardanelles who was left, blinded, in No Man's Land between the English and Turkish trenches. Moving only at night, and having no sense to tell him which were his own trenches, he was fired at by Turk and English alike as he groped his ghastly way to and from them. Thus he spent days and nights until, one night, he crawled towards the English trenches, to be fired at as usual. 'Oh God! What can I do!' someone heard him say, and he was brought in.

Such extremity of suffering cannot be measured: madness or death may give relief. But this I know: we on this journey were already beginning to think of death as a friend. As we groped our way back that night, sleepless, icy, and dog-tired in the dark and the wind and the drift, a crevasse seemed almost a friendly gift.

'Things must improve,' said Bill next day, 'I think we reached bedrock last night.' We hadn't, by a long way.

It was like this.

We moved into the igloo for the first time, for we had to save oil by using our blubber stove if we were to have any left to travel home with, and we did not wish to cover our tent with the oily black filth which the use of blubber necessitates. The blizzard blew all night, and we were covered with drift which came in through hundreds of leaks: in this wind-swept place we had found no soft snow with which we could pack our hard snow blocks. As we flensed some rubber from one of our penguin skins the powdery drift covered everything we had.

Though uncomfortable this was nothing to worry about overmuch. Some of the drift, which the blizzard was bringing, would collect to leeward of our hut and the rocks below which it was built, and they could be used to make our hut more weatherproof. Then with great difficulty we got the blubber stove to start, and it spouted a blob of boiling oil into Bill's eye. For the rest of the night he lay, quite unable to stifle his groans, obviously in very great pain: he told us afterwards that he thought his eye was gone. We managed to cook a meal somehow, and Birdie got the stove going afterwards, but it was quite useless to try and warm the place. I got out and cut the green canvas outside the door, so as to get the roof cloth in under the stones, and then packed it down as well as I could with snow, and so blocked most of the drift coming in.

It is extraordinary how often angels and fools do the same thing in this life, and I have never been able to settle which we were on this journey. I never heard an angry word: once only (when this same day I could not pull Bill up the cliff out of the penguin rookery) I heard an impatient one: and these groans were the nearest approach to complaint. Most men would have howled. 'I think we reached bedrock last night,' was strong language for Bill. 'I was incapacitated for a short time,' he says in his report to

Scott. Endurance was tested on this journey under unique circumstances, and always these two men with all the burden of responsibility which did not fall upon myself, displayed that quality which is perhaps the only one which may be said with certainty to make for success, self-control.

We spent the next day—it was 21 July—in collecting every scrap of soft snow we could find and packing it into the crevasses between our hard snow blocks. It was a pitifully small amount but we could see no cracks when we had finished. To counteract the lifting tendency the wind had on our roof we cut some great flat hard snow blocks and laid them on the canvas top to steady it against the sledge which formed the ridge support. We also pitched our tent outside the igloo door. Both tent and igloo were therefore eight or nine hundred feet up Terror: both were below an outcrop of rocks from which the mountain fell steeply to the Barrier behind us, and from this direction came the blizzards. In front of us the slope fell for a mile or more down to the ice-cliffs, so windswept that we had to wear crampons to walk upon it. Most of the tent was in the lee of the igloo, but the cap of it came over the igloo roof, while a segment of the tent itself jutted out beyond the igloo wall.

That night we took much of our gear into the tent and lighted the blubber stove. I always mistrusted that stove, and every moment I expected it to flare up and burn the tent: but the heat it gave, as it burned furiously, with the double lining of the tent to contain it, was considerable.

It did not matter, except for a routine which we never managed to keep, whether we started to thaw our way into our frozen sleeping bags at 4 in the morning or 4 in the afternoon. I think we must have turned in during the afternoon of that Friday, leaving the cooker, our finnesko, a deal of our foot-gear, Bowers's bag of personal gear, and many other things in the tent. I expect we left the blubber stove there too, for it was quite useless at present to try and warm the igloo. The tent floor cloth was under our sleeping bags in the igloo.

'Things must improve,' said Bill. After all there was much for which to be thankful. I don't think anybody could have made a better igloo with the hard snow blocks and rocks which were all we had: we would get it airtight by degrees. The blubber stove was working, and we had fuel for it: we had also found a way down to the penguins and had three complete, though frozen eggs: the two which had been in my mitts smashed when I fell about because I could not wear spectacles. Also the twilight given by the sun below the horizon at noon was getting longer.

But already we had been out twice as long in winter as the longest previous journeys in spring. The men who made those journeys had daylight where we had darkness, they had never had such low temperatures, generally nothing approaching them, and they had seldom worked in such difficult country. The nearest approach to healthy sleep we had had for nearly a month was when during blizzards the temperature allowed the warmth of our bodies to thaw some of the ice in our clothing and sleeping-bags into water. The wear and tear on our minds was very great. We were certainly weaker. We had a little more than a tin of oil to get back on, and we knew the conditions we had to face on that journey across the Barrier: even with fresh men and fresh gear it had been almost unendurable.

And so we spent half an hour or more getting into our bags. Cirrus cloud was moving across the face of the stars from the north, it looked rather hazy and thick to the south but it is always difficult to judge weather in the dark. There was little wind and the temperature was in the minus twenties. We felt no particular uneasiness. Our tent was well dug in, and was also held down by rocks and the heavy tank off the sledge which were placed on the skirting as additional security. We felt that no power on earth could move the thick walls of our igloo, nor drag the canvas roof from the middle of the embankment into which it was packed and lashed.

'Things must improve,' said Bill.

I do not know what time it was when I woke up. It was calm, with the absolute silence, which can be so soothing or so terrible as circumstances dictate. Then there came a sob of wind, and all was still again. Ten minutes and it was blowing as though the world was having a fit of hysterics. The earth was torn in pieces: the indescribable fury and roar of it all cannot be imagined.

Cherry-Garrard, A. 1930. The Worst Journey in the World. *Dial Press, New York; "The Winter Journey," pp. 267–281.*

Wombats (1963)

Peter J. Nicholson

The Australian marsupial called the wombat got its name from the word used by the New South Wales aborigines. When white men first came to Australia they extended from central Queensland down through Victoria to the Western Australian border. Now they are disappearing: there are a few colonies in Queensland, but the hairy-nosed wombat has disappeared from Victoria and is restricted to a coastal fringe of South Australia. The Bass Strait Wombat is verging on extinction. The wombat I have studied is the common wombat of the eastern highlands of Victoria and the Monaro district of New South Wales—*Vombatus hirsutus*.

During 1960, while I was at Timbertop [*an elementary school in Victoria, Australia*], I had a wonderful opportunity to do what I had always wanted to do—to study wombats in their natural surroundings and to explore their underground burrows.

The main study locality is between 1,500 and 4,000 feet above sea level. The country is rugged and covered with wet sclerophyll forest. Here wombat burrows were plentiful except in the deep fern gullies.

The wombat belongs to a very ancient Australian family of marsupials. Fossils of its ancestors are well known: one is Diprotodon. The wombat family differs from most other marsupials, in making burrows underground and in the arrangement of the teeth and some of its bones. Like its close relation, the koala, it has no tail.

At Timbertop I was able to assemble a complete skeleton: the teeth are more like a rodent's than a marsupial's and have evolved from its gnawing habits. It has only one pair of upper and lower incisors; none of the teeth have roots. They continue to grow from the base as they are worn down by grinding the roots and grasses which it eats.

The wombat is a buff to grayish-black, coarse-haired, thickset animal about three feet long, and it stands only about one foot three inches high. It is heavy (about seventy pounds in weight) and the weight is carried by its short thick legs. It is very strong and when seen in a hurry looks like a tank as it takes all types of hills and gullies in its stride. It leaves a footprint like a foreshortened human footprint in mud or snow.

ALBINO

In one burrow I found a full-grown wombat which had pinky-white skin under its muddy, light coloured fur. It had large red eyes and no visible pigment anywhere. I concluded it was an albino. It was very timid and stayed about sixteen feet from the entrance in a day nest. Although I always looked for it out in the daytime, it came out only well after the sun had gone down.

BURROWING

The wombat has powerful long claws and the front paws are hollowed out rather like a hand. With these it digs burrows into the hillside or on flat country. The longest burrow I have explored was about sixty feet, to counting the network of tunnels.

It has been said that a wombat always digs with the fore-paws while lying on its side, thrusting the soil out with its feet. I have often watched a wombat dig. It sits firmly on its rump and hind legs, digs the earth with its front paws and thrusts the dirt out to one side. The dirt is moved out of the burrow with the front and back paws as the wombat moves out backwards. In the burrow a good deal of time is spent scratching around, continually moving the dirt of the floor and making nests in the half light about twelve feet from the entrance.

The burrow is the personal home containing usually one, but sometimes two, adult wombats at one time. The wombat is a social animal and often goes visiting other burrows. There are well-beaten tracks between burrows. They keep their burrows clean and always defecate away from the burrow area.

I have never seen wombats actually mating. The breeding season extends from about the end of March until June. The mother wombat usually has only one young, or very rarely two. She suckles her young in a pouch in which it remains until it is fully furred, about mid-November. The mother makes a simple nursery of near-dead bracken fronds and bark at the end of a tunnel. The young wombat learns tunneling in the mother's burrow and gradually digs its own small burrow inside the mother's. About four months after it leaves the pouch it leaves the burrow and goes in search of a deserted burrow or place to dig its own. Often it meets failure when it digs in a creek bed and the burrow is flooded.

If the wombat finds a deserted burrow, which may contain the dead body of another wombat, it will either dig another tunnel or widen the burrow to by-pass the dead wombat. It will not disturb a fairly freshly dead wombat but will toss the bones around if only the skeleton is left. Eventually the skeleton will become buried in the floor dirt; here it will start to decompose. The wombat will usually dig about nine feet from the old chamber to make its own chamber, which it will fairly comfortably line with bracken fronds and bark. If the new wombat is a female the chamber will be slightly larger and lined.

Wombats are very strong—Mr. David Fleay told me that he has known one crush a dog by pressing the dog with its back and head against the walls of the burrow. Wild dogs fear their strength but some will live in old deserted burrows.

BEHAVIOUR OUTSIDE THE BURROW

The wombat's behaviour seems to be directly related to the intensity of light. On a bright sunny day a wombat will be deep down in the burrow usually sleeping; on a less bright day he may be found closer to the entrance, and on a very dull day he may be out in his sit or just outside the burrow in the half-light. The sit is a slightly excavated area about three-square feet and up to ten yards from the entrance of the burrow. It is usually in a sheltered position with a fairly good view. They spend some time sun baking in the morning and evening, and sometimes during a dull day. It is hard to catch them there as these sense your approach, but if you sit quietly or watch through binoculars from some vantage point, you can observe them each morning or evening.

I always thought wombats were completely nocturnal and very much alike, but they show a lot of divergence in their behaviors. I often saw wombats out playing and feeding at all times during the day, provided there was not much sun. In an area of over a hundred square miles I found many different wombat communities. Along the Howqua River the wombats were brown-furred and ferocious. In the King River–Mt. Buttercup area they were a dark grey and played about a lot during the day. In the Mt. Buller–Mt. Timbertop area they were a light grey, fairly inactive, friendly type. This is not taking into account dirt colouration, as some burrows penetrate thick red clay while others are in orange clay, and some even go through partly decomposed granite rock. Some communities of wombats were less friendly than others and those near a camp where there were dogs were definitely hostile. One female wombat living in the Howqua River bank, possibly with young, chased me out of the burrow by grunting and advancing at me. This is the only time that I have been chased out by a wombat. Once I was chased out by a wild dog.

It is amusing to watch wombats in the snow. They seem to have a real sense of sport; they toboggan down snow slopes at quite a speed. They keep the front paws on the snow and curve the hind legs and rump under to give a large flat area which acts as a skid. The wombat will not toboggan continually on the one slope but uses this trick as a quick conveyance down a slope while he is making a journey in search for food. I believe the wombat is not taught this by its mother but acquires it in its youthful frolicking in the snow. I have followed or tracked wombats up and down slopes for more than two miles. When a wombat is traveling with a purpose it is very hard to keep up with him and remain unseen. I used a small Yukon pack toboggan and only then could I keep up with the wombat. If you chase a wombat it is impossible to catch it as it moves up or down a slope with equal ease, and will disappear down any burrow.

Outside the burrow a good deal of their time is spent in feeding—either digging to find roots or fungi, or eating grasses. Inside the burrow they do a lot of scratching and are the host of a large blood-sucking tick, many of which I collected by finding replete ones which had fallen off in the burrow.

A WILD WOMBAT AS A FRIEND

Wombats have been studied as domesticated pets by trapping and, not so far as I have heard, by making friends with a wild one. At the beginning of my observations I thought wombats might be dangerous if I met one in a burrow. Soon I learned that they were usually friendly and very inquisitive but not until second term did I start in earnest to try to make friends with one. I picked a fairly young wombat and spent an hour or so each day with it. This time was never wasted as he usually gave me some demonstration of digging or burrow life. I always had a torch with me but never pointed it directly at him, and it always contained, on these occasions, dull batteries. For quite long periods the torch was off so that he could examine me. Occasionally he would come up to me and sniff my arms and examine my face and hair inquisitively while I imitated his friendly grunt.

After about three months of knowing me he followed me out of the burrow as I was leaving it. I sat down near his sit and he in it. The day was overcast and very dull. He then came to me and examined me closely, putting his forepaws on my legs and sniffing up and down my legs. After this he would usually follow me out on a dull day. I never attempted to feed him and it seemed that he was only inquisitive. He gave me the impression of being an intelligent, one-track-mind person. He used to love to be scratched.

I also saw a wombat spend about twenty minutes scratching itself against the bases of wattle trees. It is easy to find wombat burrows by looking for muddy-based trees or logs and following tracks from them.

CHARTING WOMBAT BURROWS

This I did by driving pegs into the floor of the burrow and stretching cord between these. The cord was then measured and the angle between the cords at the post was noted. Slopes inside the burrow were noted only by the difficulty in crawling up or down them. Sometimes it was very difficult to back up a slope in a narrow tunnel. The danger in becoming stuck was small, as I always had someone outside the burrow if it was a new deep burrow or if I was excavating and sifting dirt for bones.

In the winter when the ground was heavy with water, underground work was too dangerous. I occasionally crawled down burrows and usually found the wombat deep down, fast asleep in the nest. This had to be done carefully and no excavating was done. Twice I had very minor caveins on my back. During the winter I found four burrows blocked by dirt, and from only one of these burrows did the wombat dig free. There were often caveins where the wombat could easily get out by climbing up and over the dislodged earth.

EXCAVATING

I excavated only old burrows which had skeletons in them, and did no excavating in the burrow of the friendly wombat. Where possible I took only the floor dirt, which I sieved carefully, removing all bones. In very deep burrows about every twenty feet, at a branch, I would enlarge and deepen the burrow so that I could turn around.

In old burrows when the loose floor earth had been removed the tunnel was deep, sometimes up to four feet. Bones that were found at this depth were in various stages of decomposition. A bright torch was used in this work and usually inserted in a hole made in the wall. It would not have been worth open-cutting these burrows as most of them were into a steep hillside. Excavating was done in clay or dirt, but not in rocky soil, as it was too uncomfortable to work in and was hard to sieve.

There were usually insects in wombat burrows. I found some blowflies clinging to the wall all winter, apparently hibernating. In the spring I brought two out; both revived and buzzed off, showing no signs of weakness. No insects in the burrows ever bothered me.

My study, conducted over a period of a year, was incomplete, but it did supply me with most of the information which I had set out to find. It also taught me that you cannot say positively what such an intelligent animal will do. He will have certain habits but, as he is at least as intelligent as a dog, he is unpredictable. Each wombat had a definite personality. I can understand why they make good domestic pets. To have a wombat as an inquisitive wild bush friend is more than I could have hoped for when I started my study.

Nicholson, P. J. 1963. "Wombats," Timbertop Magazine, *Geelong Grammar School 8: 32–38. Reprinted with permission of Peter Nicholson.*

Journey to the Sea (1941)

Rachel L. Carson

There is a pond that lies under a hill, where the threading roots of many trees—mountain ash, hickory, chestnut oak, and hemlock—hold the rains in a deep sponge of humus. The pond is fed by two streams that carry the runoff of higher ground to the west, coming down over rocky beds grooved in the hill. Cattails, bur reeds, spike rushes, and pickerel-weeds stand rooted in the soft mud around its shores and, on the side under the hill, wade out halfway into its waters. Willows grow in the wet ground along the eastern shore of pond, where the overflow seeps down a grass-lined spillway, seeking its passage to the sea.

The smooth surface of the pond is often ringed spreading ripples made when shiners, dace, or other minnows push against the tough sheet between air and water, and the film is dimpled, too, by the hurrying feet of small water insects that live among the reeds and rushes. The pond is called Bittern Pond, because never a spring passes without a few of these shy herons nesting in its bordering reeds, and the strange, pumping cries of the birds that stand and sway in the cattails, hidden in the blend of lights and shadows, are thought by some who hear them to be the voice of an unseen spirit of the pond.

From Bittern Pond to the sea is two hundred miles as a fish swims. Thirty miles of the way is by narrow hill streams, seventy miles by a sluggish river crawling over the coastal plain, and a hundred miles through the brackish water of a shallow bay where the sea came in, millions of years ago, and drowned the estuary of a river.

Every spring a number of small creatures come up the grassy spillway and enter Bittern Pond, having made the two-hundred-mile journey from the sea. They are curiously formed, like pieces of slender glass rods shorter than a man's finger. They are young eels, or elvers, that were born in the deep sea. Some of the eels go higher into the hills, but a few remain in the pond, where they live on crayfish and water beetles and catch frogs and small fishes and grow to adulthood.

Now it was autumn and the end of the year. From the moon's quarter to its half, rains had fallen, and all the hill streams ran in flood. The water of the two feeder streams of the pond was deep and swift and jostled the rocks of the streambeds as it hurried to the sea. The pond was deeply stirred by the inrush of water, which swept through its weed forests and swirled through its crayfish holes and crept up six inches on the trunks of its bordering willows.

The wind had sprung up at dusk. At first it had been a gentle breeze, stroking the surface of the pond to velvet smoothness. At midnight it had grown to a half gale that set all the rushes to swaying wildly and rattled the dead seed heads of the weeds and plowed deep furrows in the surface waters of the pond. The wind roared down from the hills, over forests of oak and beech and hickory and pine. It blew toward the east, toward the sea two hundred miles away.

Anguilla, the eel, nosed into the swift water that raced toward the overflow from the pond. With her keen senses she savored the strange tastes and smells in the water. They were the bitter tastes and smells of dead and rain-soaked autumn leaves, the tastes of forest moss and lichen and root-held humus. Such was the water that hurried past the eel, on its way to the sea.

Anguilla had entered Bittern Pond as a finger-long elver ten years before. She had lived in the pond through its summers and autumns and winters and springs, hiding in its weed beds by day and prowling through its waters by night, for like all eels she was a lover of darkness. She knew every crayfish burrow that ran in honeycombing furrows through the mudbank under the hill. She knew her way among the swaying, rubbery stems of spatterdock, where frogs sat on the thick leaves; and she knew where to find the spring peepers clinging to grass blades, bubbling shrilly, where in spring the pond overflowed its grassy northern shore. She could find the banks where the water rats ran and squeaked in play or tussled in anger, so that sometimes they fell with a splash into the water-easy prey for a lurking eel. She knew the soft mud beds deep in the bottom of the pond, where in winter she could lie buried, secure against the cold—for like all eels she was a lover of warmth.

Now it was autumn again, and the water was chilling to the cold rains shed off the hard backbones of the hills. A strange restiveness was growing in Anguilla the eel. For the first time in her adult life, the food hunger was forgotten. In its place was a strange, new hunger, formless and ill defined. Its dimly perceived object was a place of warmth and darkness—darker than the blackest night over Bittern Pond. She had known such a place once in the dim beginnings of life, before memory began. She could not know that the way to it lay beyond the pond outlet over which she had clambered ten years before. But many times that night, as the wind and the rain tore at the surface film of the pond, Anguilla was drawn irresistibly toward the outlet over which the water was spilling on its journey to the sea. When the cocks were crowing in the farmyard over the hill, saluting the third hour of the new day, Anguilla slipped into the channel spilling down to the stream below and followed the moving water.

Even in flood, the hill stream was shallow, and its voice was the noisy voice of a young stream, full of gurglings and tricklings and the sound of water striking stone and of stone rubbing against stone. Anguilla followed the stream, feeling her way by the changing pressure of the swift water currents. She was a creature of night and darkness, and so the black water path neither confused nor frightened her.

In five miles the stream dropped a hundred feet over a rough and boulder-strewn bed. At the end of the fifth mile it slipped between two hills, following along a deep gap made by another and larger stream years before. The hills were clothed with oak and beech and hickory, and the stream ran under their interlacing branches.

At daybreak Anguilla came to a bright, shallow riffle where the stream chattered sharply over gravel and small rubble. The water moved with a sudden acceleration, draining swiftly toward the brink of a ten-foot fall where it spilled over a sheer rock face into a basin below. The rush of water carried Anguilla with it, down the steep, thin slant of white water and into the pool. The basin was deep and still and cool, having been rounded out of the rock by centuries of falling water. Dark water mosses grew on its sides and stoneworts were rooted in its silt, thriving on the lime which they took from the stones and incorporated in their round, brittle stems. Anguilla hid among the stoneworts of the pool, seeking a shelter from light and sun, for now the bright shallows of the stream repelled her.

Before she had lain in the pool for an hour another eel came over the falls and sought the darkness of the deep leaf beds. The second eel had come from higher up in the hills, and her body was lacerated in many places from the rocks of the thin upland streams she had descended. The newcomer was a larger and more powerful eel than Anguilla, for she had spent two more years in fresh water before coming to maturity.

Anguilla, who had been the largest eel in Bittern Pond for more than a year, dived down through the stoneworts at sight of the strange eel. Her passage swayed the stiff, limy stems of the chara and disturbed three water boatmen that were clinging to the chara stems, each holding its position by the grip of a jointed leg, set with rows of bristles. The insects were browsing on the film of desmids and diatoms that coated the stems of the stoneworts. The boatmen were clothed in glistening blankets of air which they had carried down with them when they dived through the surface film, and when the passing of the eel dislodged them from their quiet anchorage they rose like air bubbles, for they were lighter than water.

An insect with a body like a fragment of twig supported by six jointed legs was walking over the floating leaves and skating on the surface of the water, on which it moved as on strong silk. Its feet depressed the film into six dimples, but did not break it, so light was its body. The insect's name meant "a marsh treader," for its kind often lived in the deep sphagnum moss of bogs. The marsh treader was foraging, watching for creatures like mosquito larvae or small crustaceans to move up to the surface from the pool below. When one of the water boatmen suddenly broke through the film at the feet of the marsh treader, the twig like insect speared it with the sharp stilettos projecting beyond its mouth and sucked the little body dry.

When Anguilla felt the strange eel pushing into the thick mat of dead leaves on the floor of the pool, she moved back into the dark recess behind the waterfall. Above her the steep face of the rock was green with the soft fronds of mosses that grew where their leaves escaped the flow of water, yet were always wet with fine spray from the falls. In spring the midges came there to lay their eggs, spinning them in thin, white skeins on the wet rocks. Later when the eggs hatched and the gauzy-winged insects began to emerge from the falls in swarms, they were watched for by bright-eyed little birds who sat on overhanging branches and darted open-mouthed into the clouds of midges. Now the midges were gone, but other small animals lived in the green, water-soaked thickets of the moss. They were the larvae of beetles and soldier flies and crane flies. They were smooth-bodied creatures, lacking the grappling hooks and suckers and the flattened, stream-molded bodies that enabled their relatives to live in the swift currents draining to the brink of the falls overhead or a dozen feet away where the pool spilled its water into the stream bed. Although they lived only a few inches from the veil of water that dropped sheer to the pool, they knew nothing of swift water and its dangers; their peaceful world was of water seeping slow through green forests of moss.

The beginning of the great leaf fall had come with the rains of the past fortnight. Throughout the day, from the roof of the forest to its floor, there was a continuous down drift of leaves. The leaves fell so silently that the rustle of their settling to the ground was no louder than the thin scratching of the feet of mice and moles moving through their passages in the leaf mold.

All day flights of broad-winged hawks passed down along the ridges of the hills, going south. They moved with scarcely a beat of their outspread wings, for they were riding on the updrafts of air made as the west wind struck the hills and leaped upward to pass over them. The hawks were fall migrants from Canada that had followed down along the Appalachians for the sake of the air currents that made the flight easier.

At dusk, as the owls began to hoot in the woods, Anguilla left the pool and traveled downstream alone. Soon the stream flowed through rolling farm country. Twice during the night it dropped over small milldams that were white in the thin moonlight. In the stretch below the second dam, Anguilla lay for a time under an overhanging bank, where the swift currents were undercutting the heavy, grassy turf. The sharp hiss of the water over the slanting boards of the dam had frightened her. As she lay under the bank the eel that had rested with her in the pool of the waterfall came over the milldam and passed on downstream. Anguilla followed, letting the current take her bumping and jolting over the shallow riffles and gliding swiftly through the deeper stretches. Often she was aware of dark forms moving in the water near her. They were other eels, come from many of the upland feeder creeks of the main stream. Like Anguilla, the other long, slender fishes yielded to the hurrying water and let the currents speed their passage. All of the migrants were roe eels, for only the females ascend far into the fresh-water streams, beyond all reminders of the sea.

The eels were almost the only creatures that were moving in the stream that night. Once, in a copse of beech, the stream made a sharp bend and scoured out a deeper bed. As Anguilla swam into this rounded basin, several frogs dived down from the soft mud bank where they had been sitting half out of water and hid on the bottom close to the bole of a fallen tree. The frogs had been startled by the approach of a furred animal that left prints like those of human feet in the soft mud and whose small black mask and black-ringed tail showed in the faint moonlight. The raccoon lived in a hole high up in one of the beeches near by and often caught frogs and crayfish in the stream. He was not disconcerted by the series of splashes that greeted his approach, for he knew where the foolish frogs would hide. He walked out on the fallen tree and lay down flat on its trunk. He took a firm grip on its bark with the claws of his hind feet and left forepaw. The right paw he dipped into the water, reaching down as far as he could and exploring with busy, sensitive fingers the leaves and mud under the trunk. The frogs tried to burrow deeper into the litter of leaves and sticks and other stream debris. The patient fingers felt into every hole and crevice, pushed away leaves and probed the mud. Soon the coon felt a small, firm body beneath his fingers—felt the sudden movement as the frog tried to escape. The coon's grip tightened and he drew the frog quickly up onto the log. There he killed it, washed it carefully by dipping it into the stream, and ate it. As he was finishing his meal, three small black masks moved into a patch of moonlight at the edge of the stream. They belonged to the coon's mate and their two cubs, who had come down the tree to prowl for their night's food.

From force of habit, the eel thrust her snout inquisitively into the leaf litter under the log, adding to the terror of the frogs, but she did not molest them as she would have done in the pond, for hunger was forgotten in the stronger instinct that made her a part of the moving stream. When Anguilla slipped into the central current of water that swept past the end of the log, the two young coons and their mother had walked out onto the trunk and four black-masked faces were peering into the water, preparing to fish the pool for frogs.

By morning the stream had broadened and deepened. Now it fell silent and mirrored an open woods of sycamore, oak, and dogwood. Passing through the woods, it carried a freight of brightly colored leaves-bright-red, crackling leaves from the oaks, mottled green and yellow leaves from the sycamores, dull-red, leathery leaves from the dogwoods. In the great wind the dogwoods had lost their leaves, but they held their scarlet berries. Yesterday robins had gathered in flocks in the dogwoods, eating the berries; today the robins were gone south and in their place flurries of starlings swept from tree to tree, chattering and rattling and whistling to one another as they stripped the branches of berries. The starlings were in bright new fall plumage, with every breast feather spear-tipped with white.

Anguilla came to a shallow pool formed when an oak had been uprooted in a great autumn storm ten years before and had fallen across the stream. Oak dam and pool were new in the stream since Anguilla had ascended it as an elver in the spring of that year. Now a great mat of weeds, silt, sticks, dead branches, and other debris was packed around the massive trunk, plastering all the crevices, so that the water was backed up into a pool two deep. During the period of the full moon the eels lay in the oak-dam pool, fearing to travel in the moon-white water of the stream almost as much as they feared the sunlight.

In the mud of the pool were many burrowing, wormlike larvae—the young of lamprey eels. They were not true eels, but fishlike creatures whose skeleton was gristle instead of bone, with round, tooth-studded mouths that were always open because there were no jaws. Some of the young lampreys had hatched from eggs spawned in the pool as much as four years before and had spent most of their life buried in the mud flats of the shallow stream, blind and toothless. These older larvae, grown nearly twice the length of a man's finger, had this fall been transformed into the adult shape, and for the first time they had eyes to see the water world in which they lived. Now, like the true eels, they felt in the gentle flow of water to the sea something that urged them to follow, to descend to salt water for an interval of sea life. There they would prey semiparasitically on cod, haddock, mackerel, salmon, and many other fishes and in time would return to the river, like their parents, to spawn and die. A few of the young lampreys slipped away over the log dam every day, and on a cloudy night, when rain had fallen and white mist lay in the stream valley, the eels followed.

The next night the eels came to a place where the stream diverged around an island grown thickly with willows. The eels followed the south channel around the island, where there were broad mud flats. The island had been formed over centuries of time as the stream had dropped part of its silt load before it joined the main river. Grass seeds had taken root; seeds of trees had been brought by the water and by birds; willow shoots had sprung from broken twigs and branches carried down in floodwaters; an island had been born.

The water of the main river was gray with approaching day when the eels entered it. The river channel was twelve feet deep and its water was turbid because of the inpouring of many tributary streams swollen with autumn rains. The eels did not fear the gloomy channel water by day as they had feared the bright shallows of the hill streams, and so this day they did not rest but pushed on downstream. There were many other eels in the river—migrants from other tributaries. With the increase in their numbers the excitement of the eels grew, and as the days passed they rested less often, pressing on downstream with fevered haste.

As the river widened and deepened, a strange taste came into the water. It was a slightly bitter taste, and at certain hours of the day and night it grew stronger in the water that the eels drew into their mouths and passed over their gills. With the bitter taste came unfamiliar movements of the water—a period of pressure against the down flow of the river currents followed by slow release and then swift acceleration of the current.

Now groups of slender posts stood at intervals in the river, marking out funnel shapes from which straight rows of posts ran slanting toward the shore. Blackened netting, coated with slimy algae, was run from post to post and showed several feet above the water. Gulls were often sitting on the pound nets, waiting for men to come and fish the nets so that they could pick up any fish that might be thrown away or lost. The posts were coated with barnacles and with small oysters, for now there was enough salt in the water for these shellfish to grow.

Sometimes the sandspits of the river were dotted with small shore birds standing at rest or probing at the water's edge for snails, small shrimps, worms, or other food. The shore birds were of the sea's edge, and their presence in numbers hinted of the nearness of the sea.

The strange, bitter taste grew in the water and the pulse of the tides beat stronger. On one of the ebb tides a group of small eels—none more than two feet long—came out of a brackish-water marsh and joined the migrants from the hill streams. They were males, who had never ascended the rivers but had remained within the zone of tides and brackish water.

In all of the migrants striking changes in appearance were taking place. Gradually the river garb of olive brown was changing to a glistening black, with underparts of silver. These were the colors worn only by mature eels' about to undertake a far sea journey. Their bodies were firm and rounded with fat-stored energy that would be needed before the journey's end. Already in many of the migrants the snouts were becoming higher and more compressed, as though from some sharpening of the sense of smell. Their eyes were enlarged to twice their normal size, perhaps in preparation for a descent along darkening sealanes.

Where the river broadened out to its estuary, it flowed past a high clay cliff on its southern bank. Buried in the cliff were thousands of teeth of ancient sharks, vertebrae of whales, and shells of mollusks that had been dead when the first eels had come in from the sea, eons ago. The teeth, bones, and shells were relics of the time when a warm sea had overlain all the coastal plain and the hard remains of its creatures had settled down into its bottom oozes. Buried millions of years in darkness, they were washed out of the clay by every storm to lie exposed, warmed by sunshine and bathed by rain.

The eels spent a week descending the bay, hurrying through water of increasing saltiness. The currents moved with a rhythm that was of neither river nor sea, being governed by eddies at the mouths of the many rivers that emptied into the bay and by holes in the muddy bottom thirty or forty feet beneath. The ebb tides ran stronger than the floods, because the strong outflow of the rivers resisted the press of water from the sea.

At last Anguilla neared the mouth of the bay. With her were thousands of eels, come down, like the water that brought them, from all the hills and uplands of thousands of square miles, from every stream and river that drained away to the sea by the bay. The eels followed a deep channel that hugged the eastern shore of the bay and came to where the land passed into a great salt marsh. Beyond the marsh, and between it and the sea, was a vast shallow arm of the bay, studded with islands of green marsh grass. The eels gathered in the marsh, waiting for the moment when they should pass to the sea.

The next night a strong southeast wind blew in from the sea, and when the tide began to rise the wind was behind the water, pushing it into the bay and out into the marshes. That night the bitterness of brine was tasted by fish, birds, crabs, shellfish, and all the other water creatures of the marsh. The eels lay deep under water, savoring the salt that grew stronger hour by hour as the wind-driven wall of seawater advanced into the bay. The salt was of the sea. The eels were ready for the sea—for the deep sea and all it held for them. Their years of river life were ended.

The wind was stronger than the forces of moon and sun, and, when the tide turned an hour after midnight, the salt water continued to pile up in the marsh, being blown upstream in a deep surface layer while the underlying water ebbed to the sea.

Soon after the tide turned, the seaward movement of the eels began. In the large and strange rhythms of a great water which each had known in the beginning of life, but each had long since forgotten, the eels at first moved hesitantly in the ebbing tide. The water carried them through an inlet between two islands. It took them under a fleet of oyster boats riding at anchor, waiting for daybreak. When morning came, the eels would be far away. It carried them past leaning spar buoys that marked the inlet channel and past several whistle and bell buoys anchored on shoals of sand or rock. The tide took them close under the lee shore of the larger island, from which a lighthouse flashed a long beam of light toward the sea.

From a sandy spit of the island came the cries of shore birds that were feeding in darkness on the ebb tide. Cry of shore bird and crash of surf were the sounds of the edge of the land—the edge of the sea.

The eels struggled through the line of breakers, where foam seething over black water caught the gleam of the lighthouse beacon and frothed whitely. Once beyond the wind-driven breakers they found the sea gentler, and as they followed out over the shelving sand they sank into deeper water, unrocked by violence of wind and wave.

As long as the tide ebbed, eels were leaving the marshes and running out to sea. Thousands passed the lighthouse that night, on the first lap of a far sea journey—all the silver eels, in fact, that the marsh contained. And as they passed through the surf and out to sea, so also they passed from human sight and almost from human knowledge.

Carson, R. L. 1941. Under the Sea Wind. *Simon and Schuster, New York; "Journey to the Sea," pp. 211–231. "Journey to the Sea," from* Under the Sea Wind

Notes on the Natural History of Some Marine Animals (1938)

George E. MacGinitie

PORIFERA

Although sponges have but few enemies, they are subject to the predacious activities of certain other animals, especially gastropods, of which the nudibranchs are the most important. *Lamellaria stearnsii*, a yellowish gastropod with an internal shell, seems to feed almost exclusively upon sponges, mainly the yellow sponge *Lissodendoryx noxiosa*. I have also seen the cushion star, *Patiria miniata*, making a meal of this incrusting sponge.

Sponges usually remain fixed after the larvae become attached, but on the mud flats of estuaries of southern California there is a sponge, *Tetilla mutabilis*, which, when settling, fastens itself in the mud by a strong, slender stalk. This stalk, which is about two inches long, consists of a number of fine strands which are fringed at their lower ends so that they anchor the animal firmly. The sponge, which possesses but a single osculum when young, continues to grow until it reaches the size of a pigeon's egg or a little larger, whereupon it usually breaks loose from the stalk and is rolled about over the mud flats by the tides. It may move a considerable distance from the place where it settled, continuing to grow until it becomes an irregular mass, six inches or more in diameter, having several oscula. I have never observed this sponge burrowing by contraction as some other sponges do. On April 25, 1935, I found thousands of young *Tetilla*, ranging from one-fourth inch to two inches in length, anchored in the mud near Harbor Island in Newport Bay.

Also, it is usually considered that sponges are not readily grown in the laboratory, but at the Kerckhoff Marine Laboratory larvae of *Sycon coronatum* often come in with the circulating water and become fixed in great numbers over the sides of the aquaria or upon solid objects in the aquaria, where, unless removed, they continue to develop until they reach a length of about three centimeters.

COELENTERATA

The eolid nudibranch, *Antiopella aureocincta*, eats the heads of the solitary hydroid, *Corymorpha palma*. The eolid also makes another use of the hydroid, for it deposits its eggs upon the bases of the latter. The nudibranch is so nearly the color of the heads of this hydroid that only close scrutiny reveals the presence of the mollusc. In this region *Corymorpha* often occurs on the mud flats, about one to every four square inches, in areas many feet wide and fifty or a hundred feet long.

The larvae of the sea pansy, *Renilla köllikeri*, settle in the water at or below low tidemark, and many of them migrate to the portions of the flats exposed at low tides. This migration is accomplished slowly by changing position from day to day. The stem is withdrawn from the substratum and the colony moves a distance of a foot or two per day by a peristaltic movement, which progresses throughout the entire colony. At Anaheim Slough (17 miles north of here), on December 26, 1936, I found many young *Renilla*, measuring from 0.7 cm. to 5 cm., which had thus migrated to the exposed mud flats.

Stylatula gracilis, a sea pen that is very common in the mud flats of southern California, is easily reared in the laboratory. When the sea pens are brought in and allowed to lie in a dish of salt water, they will give off eggs and sperm if they are ripe. Eggs laid one day become planulae the following day, and three days later the larvae have settled to the bottom. By this time they possess tentacles and perform peristaltic movements along the body. If they are fed they will continue to grow, developing new heads every few days on alternate sides of the forming colony.

ECHIUROIDEA

Spawning in Urechis.—Eggs and sperm of *Urechis caupo* are stored in six segmental organs as they are produced, and spawning takes place during a short season, usually in the spring or at the beginning of summer as the temperature of the water rises. Since spawning is inhibited, at least for a time, by bringing the worms into the laboratory, it is thus possible to keep a supply of embryological material when the specimens in the field are spawned out. Worms are best kept in the laboratory by placing them in glass U-shaped tubes. One male *Urechis* which I have kept thus in the laboratory for a period of two or three years spawned on May 24th and 25th of this year, and I was fortunate enough to observe the process both times.

Just prior to spawning the *Urechis* came nearly to the opening of the burrow, whereupon three ridges were thrown around the body of the animal so that the creases were just anterior to each of the three pairs of gonopores, and the gonopores themselves were somewhat protruded and turned toward the anterior end of the body, and, therefore, toward the opening of the burrow. The openings of all six gonopores became quite conspicuous; this was followed by several retching movements, as though the animal were attempting to regurgitate, and then sperm issued in a stream from each gonopore. When the sperm ceased to be expelled, the animal underwent a violent peristalsis, the waves running from the posterior to the anterior end, causing the sperm to pour out of the tube. The retching, followed by the violent antiperistalsis, was performed three distinct times. On both days, after spawning the worm went back to the bottom of the tube, pumped vigorously for some time, and then resumed feeding. During spawning the body of the worm was much more elongated than normally. The spawning on May 24th took place at 4:30 p.m., that on the following day at 9:20 a.m. The storage organs were practically emptied the first day of spawning, but the movements and procedure on the second day were the same as for the first spawning, although very little sperm was discharged.

ANNELIDA

On January 27, 1935, an examination of the faecal pellets of the tube-dwelling annelid, *Chaetopterus variopedatus*, showed that they contained larval pelecypods. When the pellets were broken apart with needles, these young clams, apparently larval Teredos, opened their valves, spread their veliger wings, and swam about actively, none the worse for their trip through the digestive tract of the worm. Larval clams, which were embedded near the surface of the pellets, were able to escape unaided within a few hours. Undisturbed pellets were placed in a dish and allowed to remain for four days until they began to disintegrate, whereupon the larval clams, which were thus liberated from the central portions of the pellets, swam away. The pellets are cylindrical, with rounded ends, and average 5 mm. in length by 0.8 mm. in diameter.

DISTRIBUTION OF ESTUARINE LARVAE

The preceding note concerning *Chaetopterus* brings up another subject which has been the cause of considerable discussion by biologists who are interested in the distribution of marine animals. Distribution of animals which live along the open beach is, of course, rather easily accounted for, but animals which are indigenous only to estuaries furnish a more complex problem.

To begin with, it should be kept in mind that the invasion of larvae of a certain species from another estuary, occurring once every thousand years, would be sufficient to maintain a constancy of specific characters. This is true mainly because of the comparatively static conditions of marine estuaries over long periods of time, with the resulting lack of stimulus to evolutionary change. This condition is also indicated by the fact that investigation of recent fossil remains show that these animals have changed but little over a period of several million years. Some of the forms found in marine estuaries have a rather long larval period, as for example, the echiuroid worm, *Urechis caupo*, which has a larval period of from 40 to 60 days. An animal which has a larval period as long as this could depend entirely upon currents for its distribution from estuary to estuary. In times past other forms could easily have been carried from place to place, say on the bodies of whales with their covering of barnacles. The females of the California gray whale made use of the estuaries along the coast of California and Lower California as a place of refuge when giving birth to their young. Today, with the great numbers of "foul bottom ships" passing along the coast of all countries, a means of distribution is provided for practically all forms of larvae of estuarine animals. Since *Teredo* and other pelecypod larvae are able to withstand trips through the alimentary tracts of other animals, they may be thus conveyed long distances from their place of origin. *Chaetopterus variopedatus* is known to build its tube among the sessile animals which foul the bottoms of floats and boats which are left in this bay (Newport Bay, California) a year or more without cleaning. Because of the fact that *C. variopedatus* is a native of the Atlantic, and as late as 1922 was not known to occur on this coast, it would not be surprising if this animal migrated to the Pacific Coast on the bottom of some ship. I found *C. variopedatus* at Newport Bay on January 20, 1929. Mr. S. A. Glassell, who recently has collected in the Gulf of Lower California, reports that this species occurs abundantly there.

OPHIUROIDEA

Estuarine and ocean bottom forms subsist mainly upon the organic materials contained in the detritus of the surface of the mud. Much of this organic material is bacteria, and is contained in the top centimeter or so. Even though deeply buried, all detritus feeders that I have studied, except *Callianassa californiensis*, obtain their food directly from the surface.

Amphiodia barbarae, a serpent star, lives in the sand or mud, buried to a depth of about three or four inches, and extends the tips of its arms to the surface to obtain its food. It, therefore, obtains its food in somewhat the same manner as does the clam, *Macoma nasuta*. This habit of living necessitates long arms; the disk of one of these serpent stars measured only 0.65 cm. in diameter, whereas the arms measured 15 cm. in length.

Two other interesting ophiurans, *Amphiodia psara* and *Ophiacantha eurythra* possess the unusual ability for serpent star of being able to swim rather efficiently. The former swims by extending one arm ahead, trailing two behind it, and using the remaining two arms as wings. However, it possesses this ability to swim efficiently only when young and while the rays are short in relation to the size of the disk as compared with their length in more mature specimens. *Ophiacantha eurythra* swims by using four arms as two pairs of wings and trailing one arm behind. In this case also I think it can be said that the young specimens swim better than the older ones, but, because of the fact that this serpent star drops off the distal portion of its arms when disturbed by the handling necessary in dredging it, the arms become sufficiently short that any sized specimens I have so far found are always able to swim. This latter species also makes what may be termed hopping movements along the bottom of a dish in the laboratory, using the same motions that it does in swimming. Either species may swim for as long as two or three minutes at a time. *Ophiacantha eurythra* is more easily stimulated to perform swimming movements than is *Amphiodia psara.*

There is no tendency with either species to use the same arms for the same function during different swimming excursions. That is, *Amphiodia psara* may extend any arm ahead, or *Ophiacantha eurythra* may trail any arm behind. The tendency is for the serpent stars to start in the opposite direction from that from which the stimulus is applied, and, therefore, I think it can be said that swimming is performed as an escape reaction.

GASTROPODA

The tooth shell, *Dentalium neohexagonum*, as dredged off Newport Beach, often has its ovaries nearly filled with cercariae. I do not know to what trematode these cercariae belong.

The large gastropod snail, *Bursa californica*, nearly always uses one half of a large clamshell in which to deposit its egg cases. Occasionally other objects, such as a piece of rock, etc., may be used. The snail remains with its foot covering the circle of capsules until all the eggs are hatched. When removed from its "nest" this snail will hunt for the clamshell with its deposit of cases, and, upon finding it, will at once resume the brooding position; but as soon as the larvae escape from the cases the snail immediately leaves the shell and is no longer interested. The elongated capsules are laid on the inside of a clamshell, and are placed radially with their flat side toward the center, eventually forming a disk of capsules about two and one half inches in diameter. I have counted as many as 274 capsules in one group, and each capsule contains from 2500 to 3000 eggs. One snail spent five days laying its eggs, which began hatching twenty days later.

It is generally thought that the brooding habit or parental care is a great advantage in the rearing of young, and that, in general, the numbers of eggs laid by animals which show these traits are less, and, in some cases at least, the size of the egg, and, therefore, the amount of yolk is greater in those animals which brood their eggs. I doubt if this is true for marine gastropods in general. For example, there is another snail, *Kellettia kellettii*, which does not brood its eggs, yet is more abundant than *Bursa*, though it is about the same size as the latter and lives under practically the same conditions. *Bursa* is somewhat more abundant in rocky regions than *Kellettia*, but both live at about the same depth and their food habits are much alike. I have counted the approximate numbers of eggs laid by both snails in what I judge to be average layings. *Bursa californica* laid 822,000 eggs, and *Kellettia kellettii* laid 196,000. Since *Bursa* broods its eggs and *Kellettia* does not, it seems that the reverse should be true.

I think that the comparison of these two snails bears out my contention that the greatest mortality occurs at the time of metamorphosis and establishment of the young in their permanent habitat and the activities, which are typical of the adult animal. The two main factors which substantiate this hypothesis are: (a) metamorphosis may not take place in a suitable region for settling, and (b) the surface of the ocean bottom, whether rock, sand, or mud, is inhabited by a horde of predacious feeders which depend to a greater or less extent upon these settling larvae for their food. I am inclined to think that it is the ability of the *Kellettia* to get along better on the more open ocean bottom that accounts for their being more numerous than *Bursa*. If this is true, it greatly extends, over that of *Bursa*, the region in which *Kellettia* may settle. Predacious feeders are more numerous in rocky regions, for most of them need refuges from their enemies, and, though certain predacious worms, crustaceans, and fishes bury in the surface of the mud, many more, in comparison, secure refuge among the seaweed and rock crevices of rocky shores.

So far as I am able to determine, the eggs of marine molluscs are not used by other animals for food, and so we are confronted with the apparent fact that the brooding habit of *Bursa* is of no advantage. If not, why did this habit evolve. So far as I have been able to observe, brooding serves two purposes among marine animals: first, to keep the eggs clean that gas exchange may take place (as in certain fishes, octopi, and crustacea), and, second, to shorten the free swimming larval period that settling may take place in a suitable location near adult colonies (as in *Ostrea lurida*, *Teredo*, *Aletes squamigerus*, and *Crepidula onyx*).

PELECYPODA

It is generally the case that when any species of pelecypod spawns, sperm and eggs are given off simultaneously by all the ripe members of the species of that particular region. The spawning of one stimulates spawning among the others. The larvae derived from such a spawning, though motile, are entirely at the mercy of the conditions of the water, that is, currents, tides, storms, etc. Perhaps it is the rule that whole spawnings are entirely lost, being carried to places in which there would be no opportunity for settling where conditions would be right for starting a new colony. The brood may remain fairly well concentrated or may become widely scattered, due to the factors mentioned above, and but few of the offspring find suitable places for establishment. Again it may happen that the fairly concentrated brood may be in a location where nearly all of the larvae which have survived the predacious attacks of other planktonic animals, will, upon metamorphosing and settling, find conditions optimum for the establishment of a rich colony. The segregation of the broods depends to a great extent on the length of the free-swimming stage, i.e., the shorter this stage the better the segregation at the time of settling, other things being equal. Therefore, aside from the numbers which are lost by falling prey to other plankton feeders, it depends upon the location of the brood at the time of metamorphosis or at the time of settling, whether any or most of them will survive. As one watches the tide flats from year to year, he often finds areas where a pelecypod of a certain species has settled in great numbers. Often these young clams will do well for a time, but, because of shifting sands, spring tides or other causes, they may be entirely wiped out before reaching maturity. I have observed many such occurrences among colonies of *Macoma, Tagelus, Schizothaerus, Sanguinolaria, Barnea, Pholadidea, Zirfaea, Tellina, Donax, Mactra*, and the mussels. The establishment of a permanent colony is rare. Of course, this is not surprising when one considers that most female pelecypods lay eggs in quantities exceeding a million at a time, and may lay several times per season.

The fronds of a single plant of *Pelagophycus porra* which was picked up as it was floating by the laboratory were covered on both sides by the scallop, *Pecten latiauritus monotimeris*. There was an average of 285 of these molluscs per square foot of frond surface, making a total of 85,080 on this one plant. Undoubtedly a brood of this pecten had been carried by the current to the neighborhood of this seaweed, and had used the seaweed as a place of attachment when time for settling came. They were all of a uniform size, ranging between 13 and 15 mm. in diameter. There was also an indication that a great many had fallen off. A flatworm which was present in considerable numbers was devouring them at a great rate. It is doubtful that any of the pectens would have reached maturity on this plant. They just happened to settle there.

I think this accounts for the fact that a small clam, *Donax gouldii*, settles in patches along the shores of southern California. Often they are washed out by a single tide, and at such times hundreds of pounds of them are usually gathered by people and made into chowder.

The mussel, *Mytilus californianus*, and starfish, *Pisaster ochraceus*, follow a cycle of succession as definite as that of the rabbit and the bobcat. When the number of mussels becomes great, the starfish increases proportionately and devours the bed of mussels. The starfish then becomes scarce and the mussels reestablish themselves, only to have the cycle repeat itself. The cycle takes from five to ten years, or even longer. However, at no place are the mussels entirely exterminated, and in some places factors which are disadvantageous to the starfishes may exert themselves to such an extent that the mussel beds are not greatly harmed. A starfish may cling to a rock as strongly as a mussel when it gives its whole attention to the act, but when it goes in search of food, as it must, it is not able to withstand the heavy beating of the surf that a mussel bed can. Another factor which prevents a mussel bed from ever being exterminated is that the mussels are able to live farther above low tide mark than can starfishes. A starfish might have time to make a trip to the highest mussel when the tide is in, but it would not have time to tent up over the mussel, open it and digest it. Mussels can close their valves and withstand drying between tides, but starfishes cannot. In those places where mussels are not cleaned up by starfishes, they attach one on top of the other until the bed becomes so thick that the lower ones are smothered, and a heavy storm will often wash rocks almost entirely bare of mussels, so that, even though the starfishes do not destroy them, they destroy and reestablish themselves. In the northern part of their range, where the surf is heavy, this latter type of succession may readily be observed, while in southern California the starfish-mussel cycle is more evident.

CEPHALOPODA

In contrast with *Paroctopus bimaculatus*, which subsists mainly on shellfish, *P. apollyon* feeds chiefly on crabs. The young of *P. apollyon* feed on hermit crabs, which they pull from the shells and devour. As these octopi grow larger, they use larger crabs for food, using the web between their tentacles to fold around their prey. In an aquarium an octopus attacked a crab so large that it could not be entirely surrounded by the web; so the octopus, by spreading its tentacles out on the glass, tented up over the crab and pressed it against the glass. When it was first seized, the crab made a few vigorous struggles, one cheliped grasping an edge of the web. Within 20 seconds this cheliped, as well as the other, opened wide, in the manner assumed when the crab take a defensive attitude, then slowly began to close. As the chelae closed, the abdomen began to unbend in the manner characteristic of death, the left maxilliped opened, then the right, the appendages quivered, and from the branchial canals issued a slight brownish coloration. At the end of 45 seconds the crab was, to all outward appearances, entirely dead. During this time the crab was in no way attacked by the beak of the octopus, and it was not eaten until 20 minutes later. Crabs are no doubt killed by a secretion from the "salivary" glands, of which the octopus has two pairs, one pair being fairly large. In order to eat the crab the octopus opened it at the dorsal juncture between carapace and abdomen, the place where the break comes when a crab molts. The octopus then pulled off the back of the crab, ate the viscera first, dropped the back, and then one by one pulled off the legs, cleaned out and ate the contents, and dropped the empty shell of each as it was finished.

The southern species, *Paroctopus bimaculatus*, has a permanent secluded, dark cavity in which it lives, and from which it issues forth at dusk in search of its food. It may even build its own burrow in an old shell bed, or under a rock or other object. It has the habit of gathering loose shells with which to close the entrance of its burrow during the daytime. Because of this habit, its burrows are easily found on the mud flats when the tide is out.

When in search of food it glides smoothly over the mud, inserting a tentacle into this or that opening. Those octopi living in proximity to mussel beds live mainly on mussels, while others farther up the estuary live mainly on pectens. Both *Pecten circularis aequisulcatus* and *P. latiauritus monotimeris* are fairly abundant in such localities, the former lying loose on the mud flats, the latter attaching by a byssus thread to the eel grass, *Zostera.*

Eggs of the squid, *Loligo opalescens*, are laid in long finger-like bunches which are attached to anything on the ocean bottom that is solid, such as seaweed, rocks, shells, or debris. The bunches may contain over 100 "fingers." One hundred nineteen were counted in one cluster, but one finger contained only empty egg capsules. Such a cluster may have been laid by more than one squid. Many other molluscs will lay a number of empty capsules at the end of a spawning. The length of time for development within the egg case varies from twenty-one to twenty-eight days. This difference in time may be due to a difference in the amount of yolk in particular batches of eggs. I have observed that toward the latter part of the development time, if the embryos with a small amount of yolk sac still remaining are taken from the egg capsule, they will drop the remaining yolk sac, and will become, to all appearances, fully developed embryos ready to meet the outside environment. Hatching of a cluster of eggs continues over a period of about one week. This may indicate that a cluster of squid eggs is laid by one individual, and that it requires about one week to lay them. The average number of eggs per finger is about one hundred. When the fingers are laid, they are about one inch long and one fourth inch in diameter, but the jelly surrounding them takes in water and swells until the finger becomes about three fourths of an inch in diameter and three and one half inches long and each egg is surrounded by a considerable space.

During the spawning season, which occurs mainly during April, May, June, and July, the season being earlier in the southern and later in the northern part of the range of the squid, the eggs are so plentiful and conspicuous it would seem that few of them would escape being eaten by other animals. Such, however, is not the case, for the eggs are seldom eaten by any other animal. Perhaps one of the worst enemies is the annelid worm, *Capitella dizonata*, which develops in the jelly at the same time with the embryo squids, but by the time the worms have devoured the greater portion of the jelly surrounding the embryos, the latter hatch. The jelly which surrounds the egg capsules in the fingers not only seems to hold no attraction as food for other animals, but is actually very difficult to digest. A finger of eggs which was fed to an anemone was ingested and then disgorged after two hours, none the worse for the experience.

At another time a starfish, *Patiria miniata*, surrounded a finger of eggs with its stomach, holding the gelatinous mass against the glass wall of the aquarium for between 76 and 77 hours before maceration began to take place. The finger of eggs was then digested by the starfish during the next 24 hours. It is doubtful if this starfish would have attempted to eat these eggs had it not been in the aquarium without food.

PISCES

Because of their shape, serpents have an advantage in that they are able to enter places in search of food that otherwise would be inaccessible. The moray eel, *Gymnothorax mordax*, which lives along the rocky coasts of southern and lower California, no doubt is successful for exactly the same reason. Its muscular serpent-like body allows it to slide among the narrow crevices and into the darkest corners of rock piles in search of food. Its food consists almost exclusively of octopi. If the octopus is small enough, it is swallowed whole. When the octopus adheres firmly to the head of the eel with its tentacles, the eel throws a loop in the posterior portion of its body and pulls its head quickly backward through this loop. This loosens the octopus by sliding the sucking cups of the tentacles from the eel's head. If the pull is at right angles to the surface adhered to, the sucking cup of an octopus will hold very firmly to the eel, but the cups will slide along the surface easily, and this accounts for the ability of the eel to so quickly slide octopus tentacles from its head. This performance may be repeated several times during the time an eel is swallowing an octopus. If the octopus is too large to be swallowed whole, the eel will grasp a single tentacle in its mouth and rotate its body so rapidly that the tentacle is quickly twisted from the octopus. This spinning movement is executed rapidly and requires practically no more space for its execution than a hole large enough to accommodate the straightened body of the eel. The eel makes practically no use of its eyes in searching out its prey, but depends altogether upon scent and movement. Moray eels are attracted by movement, and I have known them to come from their hiding places to investigate my hand as it moved about in the water when taking specimens. Abalone fishermen are seldom bitten by the eels, probably because of the commotion they make while searching for the abalones. However, many people have been bitten by these eels, and only last summer a man who had one of his fingers badly lacerated by an eel came to the laboratory for first aid.

After an eel has been exposed to the ink discharged by an octopus it is unable to locate the cephalopod, even though it may come in contact with it. This inhibiting effect upon its senses usually lasts for more than an hour. Octopi are instinctively afraid of the moray eel and will often discharge their ink into the water even though the eel comes no closer than within a foot or two of their body. In this case at least, the ink of the octopus is efficient not because of its blinding effect, but because of its inhibition on the sense of smell.

After a moray eel is kept in the laboratory for some time without its natural food it will gladly accept mussels which are taken from the shell. If one holds the mussel meat firmly by means of the byssus threads the eel will throw a loop in its body and jerk ones hand loose from the material. It will do this even if lifted entirely above the water.

Too, if a fish is handled until it has discharged its mucus over its surface and then is replaced in the aquarium, the eel will begin looking for it just as he does if an octopus is placed in the aquarium, and will eat the fish if he can catch it. This is another reason why I think that the eel hunts by scent and movement rather than by sight.

The ocean sunfish, *Mola mola*, uses Scyphozoa (jellyfish) for food, perhaps exclusively. This is not an original statement, but because of the fact that recent publications have stated that the food of this fish is not known, and since one even goes so far as to suggest plankton as a possible source of food, the statement is repeated here. I have seen *Mola mola* eating jellyfish many times. The fish begins at the edge of the bell and devours the jellyfish bite by bite, then moves to another and repeats the process, eating one jellyfish after another. The extremely small size of the mouth and gill slits would preclude the possibility of this fish making use of plankton for food. Even though it possessed a larger mouth and gill openings, it would still of necessity have to be equipped with some form of gill raker for straining out the plankton, and would also have to be equipped with locomotory organs much more efficient than those it now possesses. It is a very slow swimming fish, and one can often approach within a few feet of it without making any effort to move away. Jellyfish may seem a meager diet, but when one considers that the mesogloea of these animals may be rather well supplied with food stored in amœbocyte cells, it is evident that the nutrient value of jellyfish may be much greater than one would at first suppose.

MacGinitie, G. E. 1938. "Notes on the natural history of some marine animals." American Midland Naturalist *19: 207–219. Reprinted with permission of University of Notre Dame.*

Exploration

Gage H. Dayton, Paul K. Dayton, and Harry W. Greene

> In this point of view the discovery of every new species is important, and their correct description and accurate identification absolutely necessary. The most obscure and minute species are for this purpose of equal value with the largest and most brilliant, and a correct knowledge of the distribution and variations of a beetle or a butterfly as important as those of the eagle or the elephant. It is to the elucidation of these apparent anomalies that the efforts of the philosophic naturalist are directed; and we think, that if this highest branch of our science were more frequently alluded to by writers on natural history, its connexion with geography and geology discussed, and the too prevalent idea—that Natural History is at best but an amusement, a trivial and aimless pursuit, a useless accumulating of barren facts—would give place a more correct view of a study, which presents problems as vast, as intricate, and as interesting as any to which the human mind can be directed, whose objects are as infinite as the stars of heaven and infinitely diversified, and whose field of research extends over the whole earth, not only as it now exists, but also during the countless changes it has undergone from the earliest geological epochs. (Wallace 1857, 480)

Contemporary science owes a rarely acknowledged debt to early naturalists, some well known (e.g., Charles Darwin, Henry Bates, Alfred Wallace, and Alfred Wegener), and others less so (e.g., William Dampier and Henry N. Moseley). Many modern scientific theories and entire research programs often are based on the work of these naturalists. Fundamental observations captured in their notes and journals are important contributions in their respective fields.

New discoveries are founded upon carefully recorded observations that provide a springboard for new questions and the development of sound hypotheses. We are now in an era in which students are too infrequently trained to observe and understand nature. As we make great leaps in understanding the molecular makeup of organisms, it is important to maintain a focus on the organisms themselves and their unique natural histories and to recognize the wisdom passed down by the early naturalists. Teaching Natural History and educating students in the value of descriptive science is quickly disappearing from modern curricula. This is especially disturbing at a time when species are vanishing from the earth at an alarming rate. We must recognize that it is important to integrate subdisciplines of biological sciences in order to achieve a broad understanding of the processes that give rise to and maintain species diversity (Grant 2000). Certainly a thorough knowledge of Natural History is a fundamental first step to successful conservation (Dayton 2003).

This section of the book is a compilation of early articles by natural historians and scientists, highlighting their astute descriptions of the natural world. Most of these studies represent efforts that are not well known; they were chosen here to stress the importance of making keen observations and recording them in detail. The papers include species descriptions, field notes taken during voyages and sampling trips, and details of naturalist's encounters with living organisms and new environments. Their observations illustrate the importance of maintaining detailed records of flora, fauna, climate, and physiographic conditions, and they provide us with baseline data that we can use to compare historic conditions with those of the present day. These early works showcase the authors' ability to record their findings in meticulous detail, and perhaps equally importantly, their passion for discovery.

It is important to realize that many early observations and descriptions were lost or never published due to harsh conditions or social attitudes of the time. A sad example of this is Beatrix Potter's work on the symbiotic relationship between algae and fungi. In 1897, Potter was at the forefront of her field in describing the complex relationship of fungi and algae in the formation of lichens (Linder 1989). She was the first scientist in England to advance the Swiss botanist Simon Schwendener's hypothesis that lichens were dual organisms, a hypothesis that was soundly rejected by British botanists at the time and was still being challenged nearly 100 years later (see Honegger 2000). Potter's paper "On the germination of the spores of Agaricineae" was read by George Masse (women were not allowed to be present) at the 1897 Linnean Society of London meeting. Unfortunately, her paper was never published, and when it was later destroyed in a fire, the details of her work were lost. Potter's experiences trying to "get into" the scientific community discouraged her, and she decided to focus her efforts on writing children's books, which she found much more heartening (Gilpatrick 1972).

The goal in choosing the papers in this section of the book was to expose readers to early, largely descriptive studies, to acknowledge that these were excellent scientists who were dedicated to recording the natural world, and to recognize that their work is still of great value. These outstanding ecologists left a legacy that has served as a foundation for present-day questions and modern theories. The work in this section should be read not only for its scientific content, but also for the ability of the authors to convey a lucid and dynamic view of the natural world they experienced. The "art" of describing the natural world is rarely taught in contemporary ecology, now often characterized by brief, concise reports of highly focused tests. There is much to be learned from the early naturalists about perceiving details in nature and keeping meticulous records of observations, and it is important that our educational system recover the tradition of experiencing nature firsthand so that future ecologists have a "sense of place" of the organisms and habitats they study.

One such naturalist was Henry N. Moseley (1844–1891). Moselely recorded numerous new observations the reports on his voyage around the world on the *Challenger* from 1862 to 1876 (Moseley 1879). His ability to vividly describe the events of his trip brings the voyage to life as well as provides readers with a tangible sense of the flora and fauna he encountered. His accounts of collecting rock penguins on Inaccessible Island in the late 1800s give the reader a feel for what being on the island and walking amongst the penguin rookeries smelled and felt like; you can almost hear the shrieks of the penguins and feel their relentless attacks on the explorers' legs as they pass through the immense colonies. His passion for natural history and discovery speaks loudly in his notes as he writes:

> That on a voyage like this, where there is so much uncertainty, it is always best to take the very first opportunity, and I have always landed on the places we visited with the very first boat, even if it were only for an hour in the evening. It may come on to blow, and another chance may never occur. I strongly advise any naturalist similarly situated to do the same. (Moseley 1879, 113)

He goes on to record that during his visit to Nightingale Island he observed only two fur seals; however, he also reports that only four years earlier a single ship harvested over 1400 fur seals during a single visit. Such observations are not only fun to read, they provide insight into the status of native species and the impacts humans had on populations during this time period.

The contribution of these early scientists is even more remarkable considering the working conditions they faced. Consider William Dampier (1652–1715), perhaps better known as a buccaneer than as a natural historian. Dampier's descriptions of the flora and fauna he encountered during his buccaneering days include a wealth of information for later natural historians, evolutionary ecologists, and conservation biologists (Dampier 1697). In recording his visit to the Galapagos Islands, he describes the basking behavior of green sea turtles, references the unique characteristics of the islands' cacti, and notes the impact sailors have on the unique naïve tortoises and iguana unaccustomed to land predators. The attention to detail and the meticulous recording are even more impressive when we consider that Dampier was a privateer actively involved in sea battles and the capturing of ships, competing buccaneers, and wealth. Equally remarkable is Georg Steller (1709–1746), who describes the physiographic conditions and gives detailed accounts of plants and animals on Bering Island (Steller 1925). Steller's knack for describing the flora and fauna of Bering Island is even more impressive when his plight is put in perspective. Steller and members of the ship's crew were stranded on Bering Island and had to overwinter before they could build a boat and sail to shore the following spring. During this time nearly half of the crew, including Bering, died from scurvy. His written observations of Steller's sea cow (*Hydrodamalis gigas*) and the speckled cormorant (*Phalacrocorax perspicillatus*) provide the most thorough descriptions of these species, both of which were later hunted to extinction; the sea cow was wiped out only twenty-seven years after its description by Steller (Stejneger 1887). Such accounts from seventeenth- and eighteenth-century world travelers have provided important baseline information about the distribution of various organisms (Grant 2000, Jackson et al. 2001).

In 1926, John K. Strecker (1875–1933) came to the conclusion that "*civilization in its progress is fast sounding the death knell of the majority of species of living animals*" and took it upon himself to publish his field notes on the natural history of Texas reptiles and amphibians. He did so with the understanding that his observations were important for future scientists working on these species, and went on to state "*we must endeavor to have published every fragment of data concerning their ways of living.*" The point brought up by Strecker, that it is critical to publish ones findings, is important. While detailed observations can lead to new discoveries, if scientists fail to publish their findings the information is of limited importance as it is not available for others to expand upon or to use as baseline data.

By publishing his work (Strecker, 1926), Strecker has provided important natural history data that has proven to be useful for scientists studying the natural history of these, and closely related, organisms.

One of the papers in this chapter is a short work by the eighteenth century naturalist Alexander von Humboldt (1769–1859). Von Humboldt was born in Germany in 1769 and at the age of 27 teamed up with Aime Bonpland, a botanist, to travel the seas and record the natural world. His discoveries of currents (such as the Humboldt Current) and his meticulously detailed maps of the rivers and drainages in South America helped lay the foundation for modern geography and played an important role in the development of Charles Darwin's and Alfred Wallace's later works. In his paper entitled *Account of the Electrical Eels, and the Method of catching them in South America by means of Wild Horses (1820)*, Humboldt describes how he and his colleagues ascertain the electrical action of the eel. His description shows the creative sparks of inquisitiveness and scientific methodology that are fundamental to science. Working in flooded backwaters of the Llanos the "Father of Modern Geography" labors with local inhabitants to "fish with horses" and then prods and pokes at electric eels in order to understand the electrical capabilities of these unique animals. Although it is not often referenced, this paper illustrates Humboldt's curiosity and his ability to observe, test, and conclude. It is a wonderful example of the scientific methodology in action, revealing a simple and effective approach to understanding how a theretofore utterly novel process works. Such formulations of meaningful questions, and the development of a methodology to answer these questions, as well as forming hypotheses that are grounded in the relevancy of an organisms natural history, are fundamental for discovery, understanding, and the progress of science.

Naturalists and explorers who were in the right place at the right time and realized that their collections or notes would be of value to science have collected many important observations and data. A wonderful example of this is Prince Albert's (1848–1922) *Comments on the cephalopods found in the stomach of a sperm whale* (Albert I of Monaco, 1913/1914). Prince Albert of Monaco dedicated a great deal of money and time to sampling deepwater fauna. On one trip he happened across fishing boats in pursuit of a sperm whale and followed them in order to witness the hunt and catch a glimpse of the whale. After a long fight the sperm whale died next to the boat and proceeded to vomit several cephalopods of "fantastic size." Realizing the scientific value of these specimens, which live at great depths and had not been previously described, Prince Albert stirred up the dense vomit and was able to collect several specimens that were new to science. By seizing the moment and taking advantage of the serendipitous occasion, Prince Albert provided science with the first specimens of deepwater cephalopods.

One can be in the right place at the right time and fail to seize the opportunity. Not realizing you are in a unique situation is likely the most common cause of failing to take advantage of the moment; however, not having the tools to collect desired data can also lead to failed opportunities. A great example of the ingenuity of many early natural historians is captured in James Murray's (1865–1914) paper *On collecting at Cape Royds* (1910). Murray was a biologist on Shackleton's 1908 trip to Antarctica on the *Nimrod*. Long before scuba, Murray wanted to collect marine organisms that lived along the sea floor. This was especially hard to do since the sea surface was frozen solid during the winter months. Getting through the ice was the first hurdle that they overcame as they picked and shoveled their way through the ice to maintain their holes. In order to collect organisms along the bottom they maintained two holes in the ice and pulled a tethered dredge that scooped up organisms as it passed over the ocean floor. Their efforts were rewarded with buckets full of specimens that were then carefully examined and sorted under dim light in their hut. The creativity of these early natural historians, and their drive to obtain information (often collected in extreme conditions), provided science with new information and ingenious sampling methods.

Modern ecologists study the processes regulating the distribution and abundance of species. Marine ecologists consider wave action one of the most important environmental variables, yet modern kelp ecologists focus on the interplay between depth, available light, currents, and grazing in influencing the distribution and abundance of kelps. While such research continues to be published, few recognize the early adventures of the one of the best subtidal ecologists ever. In July 1931, twenty-three-year-old Jack Kitching (1908–1996) donned a diving contraption he had constructed of a biscuit tin glued to a piece of a milk churn. This had 150 feet of garden hose attached to a bicycle tire pump as a source of air. His first dive was at Wembury Bay on the Devon coast of England, and he immediately understood all the relationships that modern kelp ecologist's study. He took his equipment to Scotland, where, in 1932, he began an underwater research project that addressed all of the modern questions. He provided the first detailed description of underwater forests and animals in which all the appropriate environmental parameters were discussed, and the year after his initial dive, he began a project in Scotland in which he used photo cells to actually measure light and hedge cutters to cut the fronds off the kelps to evaluate the importance of light on succession. In addition, he carefully evaluated the drag of the kelp on the benthic currents.

His research eventually moved to Ireland, where he continued to evaluate important ecological questions at least a decade earlier than other ecologists. The paper we include is his very first in this series (Kitching et al. 1934), a remarkable example of how exploratory science can be integrated with novel technology to study previously unknown species associations and communities.

The final work in this section is by Thomas Hariot (1560–1621). Hariot, known for his contributions to math, physics, astronomy, navigation, and cartography, was a trained observer. His acute attention to detail and his description of the natural resources and native inhabitants of America likely played an important role in shaping European colonization of America. In contrast to disparaging accounts by surviving colonists from the failed colony of Virginia, Hariot (1588) provided a detailed description of rich natural resources that were the source of prosperity for indigenous people and of potential wealth and power to colonialists. Noting the rich growth of corn and legumes in fields that required little tilling and no fertilizer, abundant fish, and diverse game species, he noted similarities and differences between the natural resources of America and those of England. His ability to provide succinct accounts of the flora and fauna and to articulate the link between indigenous people and their rich natural resources painted a favorable picture of the potential habitability of America. Prior to his book, the only knowledge of the Natural History of what would become the United States came from mostly religious accounts by Spanish explorers in Florida that included lore of elephants, tigers, and giraffes!

Although detailed observations can lead to new discoveries, being a keen observer is not the same as being a naturalist; however, it is certainly a prerequisite. The difference is that naturalists are able to synthesize their observations and make sense of patterns that they see in nature by illuminating processes; this is how ecological wisdom is cultivated. This section of the book sheds light on the importance of recording nature in detail, and offers a glimpse into the minds of several early naturalists who were dedicated to recording nature and disseminating their findings to others. The driving force behind the desire to study and conserve the natural world is a sense of wonder and curiosity about the intricacies of nature and a compelling passion for discovery. One thread linking all of the papers in this section is wonder, passion, and regard for detail. These naturalists were the pioneers of ecological science, often foraging into unknown lands and meticulously observing and recording the organisms that lived there. Like one putting together the pieces of a puzzle they astutely collected the details and arranged them to form the larger picture of Natural History. The ecological wisdom of these early natural historians speaks throughout time.

LITERATURE CITED

Albert I of Monaco. 1913/1914. Résultats des campagnes scientifiques accomplies sur son yacht par Albert 1er, Prince Souverain de Monaco. No. 17.

Dampier, W. 1697. A New Voyage Around the World. Reprint: Dover Publications, New York, 2007.

Dayton, P. K. 2003. The importance of the natural sciences to conservation. American Naturalist 162: 1–13.

Gilpatrick, N. 1972. The secret life of Beatrix Potter. Natural History 155: 1–12.

Grant, P. R. 2000. What does it mean to be a naturalist at the end of the twentieth century? American Naturalist 155: 1–12.

Greene, H. W., and R. W. McDiarmid. 2005. Wallace and Savage: Heros, theories, and venomous snake mimicry. In Ecology and Evolution in the Tropics: A Herpetological Perspective, M. A. Donnelly, B. I. Crother, C. Guyer, M. H. Wake, and M. E. White, eds., 190–208. University of Chicago Press, Chicago.

Hariot, T. 1588. A Briefe and True Report of the Newfoundland of Virginia. Reprint: Dover Publications, New York, 1972.

Honegger, R. 2000. Great discoveries in bryology and lichenology: Simon Schwendener (1829–1919) and the dual hypothesis of lichens. Bryologist 103: 307–313.

Jackson, J. B. C., M. X. Kirby, W. H. Berger, K. A. Bjorndal, L. W. Botsford, B. J. Bourque, R. H. Bradbury, R. Cooke, J. Erlandson, J. A. Estes, T. P. Hughes, S. Kidwell, C. B. Lange, H. S. Lenihan, J. M. Pandolfi, C. H. Peterson, R. S. Steneck, M. J. Tegner, and R. R. Warner. 2001. Historical overfishing and the recent collapse of coastal ecosystems. Science 293: 629–638.

Kitching, J. A., T. T. Macan, and H. C. Gilson. 1934. Studies in sublittoral ecology. I. A submarine gully in Wembury Bay, South Devon. Journal of the Marine Biological Association 19:677–705.

Linder, L. 1989. The Journal of Beatrix Potter: From 1881–1897. Penguin Group, London.

Mallet, J., and M. Joron. 1999. Evolution of diversity in warning color and mimicry: Polymorphisms, shifting balance, and speciation. Annual Review of Ecology and Systematics 30: 201–233.

Moseley, H. N. 1879. Notes by a Naturalist on the "Challenger": Being an Account of Various Observations Made During the Voyage of H.M.S. "Challenger" Round the World, in the Years 1872–1876, Under the Commands of Capt. Sir G. S. Nares and Capt. F. T. Thomson. MacMillan and Co., London.

Murray, J. 1910. On collecting at Cape Royds. British Antarctic Expedition 1907–1909 1: 1–15.

Stejneger, L. 1887. How the great northern Sea-cow (Rytina) became exterminated. American Naturalist 12: 1046–1054.

Steller, G. W. 1925. Steller's journal of the sea voyage from Kamchatka to America and return on the second expedition, 1741–1742. Trans. L. Stejneger. Vol. 2 in Bering's Voyages: An Account of the Efforts of the Russians to Determine the Relation of Asia

and America, F. A. Golder, ed. American Geographical Society Research Series no. 2. New York.
Strecker, J. K. 1926. Chapters from the Life-Histories of Texas Reptiles and Amphibians: Part 1. Contributions from Baylor University Museum, no. 8.
Von Humboldt, A. 1820. Account of the electrical eels, and the method of catching them in South American by means of wild horses. Edinburgh Philosophical Journal 2: 242–249.
Wallace, A. R. 1857. On the natural history of the Aru Islands. Annals and Magazine of Natural History 20: 355–376.

Inaccessible Island and Nightingale Island (1879)

Henry N. Moseley

Inaccessible Island, October 16th, 1873—The ship moved over to Inaccessible Island and kept close under its high cliffs all night.

Inaccessible Island lies W. by S. ½ S. of Tristan da Cunha, distant about 23 miles; i.e., from the Peak of Tristan to the centre of Inaccessible Island. The island is about 4½ miles in length, from east to west, and about 2 miles broad, 4 square miles in area. The highest point of the island is 1,840 feet in altitude. We anchored on the northeast side.

All night the penguins were to be heard screaming on shore and about the ship, and as parties of them passed by, they left vivid phosphorescent tracks behind them as they dived through the water alongside.

In the morning we had a view of the island. It presented on this side a range of abrupt cliffs, about 1,000 feet in height, of much the same structure as those of Tristan, viz., successive layers of basalt, traversed by vertical or oblique dykes, but mostly by narrow vertical ones. At the foot of the cliffs are some very steep debris slopes extending in one place a long way up the cliff, but not so as to render the ascent possible.

In front of these stretches a strip of narrow uneven ground, formed of large detached rocks and detritus from the cliffs above, which terminates seawards in a beach of black boulders and large pebbles. In one place, where the cliff is somewhat lower than elsewhere, there is a waterfall, which at the time of our visit was scantily supplied with water, but from the marks left by it on the rocks and vegetation, evidently attains much greater dimensions in rainy weather. The cascade pours right down from the high cliff above into a dark pool of peaty water on the beach below. The rocks about its course are covered with masses and green incrusting plants.

The face of the cliff generally is sprinkled over with green, the vegetation consisting principally of tussock grass (*Spartina arundinacea*), *Apiun graveolens* (a small sedge), *Sonchus olcraceus* (Sow thistle), *Rumex* (Dock), and ferns: with dark green patches of *Phylica arborea* on the debris slopes and ledges. The strip of accessible lower shore land is mostly covered with a dense growth of tall grass, called by the Tristan people "tussock" but quite different in structure from the well-known tussock of the Falklands, though in outward habit resembling it very closely.

Amongst the grass are several patches or small coppices of *Phylica arborea* trees, which keep the ground beneath them free from tussock, it being covered instead with a thick growth of sedges, ferns, and mosses, which form an elastic carpet on the dark peaty soil. Amongst the moss creeps *Nertera depressa*, with its bright red berries, and the Potentilla-like *Acaena ascendens* grows here and there together with the "tea-plant" of the islanders.

The stems and branches of the Phylica trees are covered with lichens in tufts and variously coloured crusts, and the branches of the trees meeting overhead these little islands as it were, in the seas of tall grass, afford most pleasant shady retreats, which seem a perfect paradise after the terrible struggle and fight through the penguin rookery, which it is necessary to endure in order to reach them.

In the early morning, we made out with a glass two men standing on the shore gazing at the ship. The Captain went on shore first, and brought off the men, who proved to be the two Germans we had heard of at Tristan da Cunha. They were overjoyed at the chance of escape from the island; we gave them breakfast, and heard something of their story.

They both spoke English, one of them remarkably well. They were brothers; one of them had been an officer in the German army during the war, the other one a sailor. They had got landed at Inaccessible Island by a whaling vessel, in the hopes that they would be able to make a considerable sum by killing fur seals, and taking their skins. They had been bitterly disappointed.

After breakfast, I landed with one of the Germans as guide with a large party. We passed through a broad belt of water, covered with the floating leaves of the wonderful seaweed *Macrocystis pyrifera*, which here, as at Tristan and Nightingale Island, forms a sort of zone around the greater part of the island, and of which we afterwards saw so much at Kerguelen's Land.

As we approached the shore, I was astonished at seeing a shoal of what looked like extremely active very small porpoises or dolphins. I could not imagine what the things could be, unless they were indeed some most marvelously small Cetaceans; they showed black above and white beneath, and came along in a shoal of fifty or more, from seawards towards the shore at a rapid pace, by a series of successive leaps out of the water, and splashes into it again, describing short curves in the air, taking headers out of the water and headers into it again; splash, splash, went this marvelous shoal of animals, till they went splash through the surf on to the black stony beach, and there struggled and jumped up amongst the boulders and revealed themselves as wet and dripping penguins, for such they were.

Much as I had read about the habits of penguins, I never could have believed that the creatures I saw thus progressing through the water, were birds, unless I had seen them to my astonishment thus make on shore. I had subsequently much opportunity of watching their habits.

We landed on the beach; it was bounded along its whole stretch at this point by a dense growth of tussock. The tussock (*Spartina arundinacea*), is a stout coarse reed-like grass: it grows in large clumps, which have at their base large masses of hard woody matter, formed of the bases of old stems and roots.

In penguin rookeries, the grass covers wide tracts with a dense growth like that of a field of standing corn, but denser and higher, the grass reaching high over one's head.

The Falkland Island "tussock" (*Dactylis cæspitosa*) is of a different genus, but it seems to have a similar habit. Here there is a sort of mutual-benefit-alliance between the penguins and the tussock. The millions of penguins sheltering and nesting amongst the grass, saturate the soil on which it grows, with the strongest manure, and the grass thus stimulated grows high and thick, and shelters the birds from wind and rain, and enemies, such as the predatory gulls.

On the beach were to be seen various groups of penguins, either coming from or going to the sea. There is only one species of penguin in the Tristan group; this is, *Eudyptes saltator*, or the "well diving jumper." The birds stand about a foot and a half high; they are covered, as are all penguins, with a thick coating of close-set feathers, like the grebe's feathers that muffs are made of. They are slate grey on the back and head, snow white on the whole front, and from the sides of the head projects backwards on each side a tuft of sulphur yellow plumes. The tufts lie close to the head when the bird is swimming or diving, but they are erected when it is on shore, and seem then almost by their varied posture, to be used in the expression of emotions, such as inquisitiveness and anger.

The bill of the penguin is bright red, and very strong and sharp at the point, as our legs testified before the day was over; the iris is also red. The penguin's iris is remarkably sensitive to light. When one of the birds was standing in our "work room" on board the ship with one side of its head turned towards the port, and the other away from the light, the pupil on the one side was contracted almost to a speck, whilst widely dilated on the other; Captain Carmichael observed the same fact. The birds are subject to great variations in the amount of light they use for vision, since they feed at sea at night as well as in the daytime.

It seems remarkable that there should be only one species of penguin at the Tristan da Cunha group, since in most localities several species occur together. It would have seemed probable that a species of "jackass" penguin (*Spheniscus*), should occur on the islands, since one species (*S. magellanicus*), occurs at the Falkland Islands and Fuegia, and another (*S. demersus*), at the Cape of Good Hope, intermediate between which two points Tristan da Cunha lies. The connection between these two widely separated *Sphenisci* is wanting; it perhaps once existed at Tristan, and has perished.

Most of the droves of penguins made for one landing-place, where the beach surface was covered with a coating of dirt from their feet, forming a broad tract, leading to a lane in the tall grass about a yard wide at the bottom, and quite bare, with a smoothly beaten black roadway; this was the entrance to the main street of this part of the "rookery," for so these penguin establishments are called.

Other smaller roads led at intervals into the rookery to the nests near its border, but the main street was used by the majority of birds. The birds took little notice of us, allowing us to stand close by, and even to form ourselves into a group for the photographer, in which they were included.

This kind of penguin is called by the whalers and sealers "rock-hopper," from its curious mode of progression. The birds hop from rock to rock with both feet placed together, scarcely ever missing their footing. When chased, they blunder and fall amongst the stones, struggling their best to make off.

With one of the Germans as guide, I entered the main street. As soon as one was in it, the grass being above one's head, one was as if in a maze, and could not see in the least where one was going to. Various lateral streets lead off on each side from the main road, and are often at their mouths as big as it, moreover, the road sometimes divides for a little and joins again: hence it is the easiest thing in the world to lose one's way, and one is quite certain to do so when inexperienced in penguin rookeries. The German, however, who was our guide on our first visit, accustomed to pass through the place constantly for two years, was perfectly well at home in the rookery and knew every street and turning.

It is impossible to conceive the discomfort of making one's way through a big rookery, haphazard, or "across country" as one may say. I crossed the large one here twice afterwards with the seamen carrying my basket and vasculum, and afterwards went through a larger rookery still, at Nightingale Island.

You plunge into one of the lanes in the tall grass, which at once shuts out the surroundings from your view. You tread on a slimy black damp soil composed of the birds' dung. The stench is overpowering, the yelling of the birds perfectly terrifying; I can call it nothing else. You lose the path, or perhaps are bent from the first in making direct for some spot on the other side of the rookery.

In the path only a few droves of penguins, on their way to and from the water, are encountered, and these stampede out of your way into the side alleys. Now you are, the instant you leave the road, on the actual breeding ground. The nests are placed so thickly that you cannot help treading on eggs and young birds at almost every step.

A parent bird sits on each nest, with its sharp beak erect and open ready to bite, yelling savagely "caa, caa, urr, urr," its red eye gleaming and its plumes at half-cock, and quivering with rage. No sooner are your legs within reach than they are furiously bitten, often by two or three birds at once: that is, if you have not got on strong leather gaiters, as on the first occasion of visiting a rookery you probably have not.

At first you try to avoid the nests, but soon find that impossible; then maddened almost, by the pain, stench and noise, you have recourse to brutality. Thump, thump, goes your stick, and at each blow down goes a bird. Thud, thud, you hear from the men behind as they kick the birds right and left off the nests, and so you go on for a bit, thump and smash, whack, thud, "caa, caa, urr, urr," and the path behind you is strewed with the dead and dying and bleeding.

But you make miserably slow progress, and, worried to death, at last resort to the expedient of stampeding as far as your breath will carry you. You put down your head and make a rush through the grass, treading on old and young haphazard, and rushing on before they have time to bite.

The air is close in the rookery, and the sun hot above, and out of breath, and running with perspiration, you come across a mass of rock fallen from the cliff above, and sticking up in the rookery; this you hail as "a city of refuge." You hammer off it hurriedly half a dozen penguins who are sunning themselves there, and are on the look-out, and mounting on the top take out your handkerchief to wipe away the perspiration and rest a while, and see in what direction you have been going, how far you have got, and in which direction you are to make the next plunge. Then when you are refreshed, you make another rush, and so on.

If you stand quite still, so long as your foot is not actually on the top of a nest of eggs or young, the penguins soon cease biting at you and yelling. I always adopted the stampede method in rookeries, but the men usually preferred to have their revenge and fought their way every foot.

Of course it is horribly cruel thus to kill whole families of innocent birds, but it is absolutely necessary. One must cross the rookeries in order to explore the island at all, and collect the plants, or survey the coast from the heights.

These penguins make a nest which is simply a shallow depression in the black dirt scantily lined with a few bits of grass, or not lined at all. They lay two greenish white eggs about as big as duck eggs, and both male and female incubate.

After passing through the rookery, we entered one of the small coppices I have already described. Hopping and fluttering about amongst the trees and herbage, were abundance of a small finch and a thrush; no other land birds were seen. The finch (*Neospiza acuhnæ*) looks very like a green-finch, and is about the same size.

The thrush (*Nesocichla eremita*) looks like a very dark-coloured song thrush, but it is peculiar for its remarkably strong acutely ridged bill. It is peculiar to the Tristan group. It feeds especially on the berries of the little Nertera; but also is fond of picking the bones of the victims of the predatory gull (*Stercorarius antarcticus*). The finch eats the fruit of the Phylica.

It was here that we first encountered that remarkable tameness, and ignorance of danger in birds which has been so constantly noticed by voyagers landing on little frequented islands, and notably by Darwin, who dilates on the fact in his account of the Galapagos Archipelago.

The thrush and finch hopped unconcernedly within a yard or two of us, whilst stone after stone was hurled at them, and till they were knocked over, and often sat still on a bough to be felled with a walking stick. By whistling a little as one approached them, numbers could be thus killed, and yet the Germans, with their house close by, had been constantly thus killing the thrushes for eating for two years. The birds are, however, not quite so tame in Tristan Island.

The finch seems to have become extinct in Tristan da Cunha itself. Von Willemoes Suhm was told that the Tristan da Cunha people had tried to introduce the bird into their island.

We were in search of another land bird, a kind of Water-Hen (*Galinula nesiotis*), which is found on the higher plateau at Tristan, and is described by the inhabitants as scarcely able to fly. We could not meet with a specimen. Only very few inhabit the low land under the cliffs, and we were not able to land at the only place from which the higher main plateau of the island is to be reached.

The Germans said that the Inaccessible Island bird is much smaller than *G. nesiotis*, and differs from it in having finer legs and a longer beak. This is, however, hardly probable, since the Tristan species occurs at Gough Island.

The family of *Gallinulidæ* is remarkably widely spread, and one of these birds is in several instances the inhabitant of some isolated island group; several occur thus in the Pacific. This is curious, since one would at first perhaps think these birds bad flyers, but they are not, and are not uncommonly met with on the wing at sea far from land, just as we met with Water-rails between Bermuda and Halifax.

Sitting on the treetops with the thrushes were numerous "noddies" of the same two species as those of St. Paul's Rocks. It was strange to see birds, which one had met with on the equator living in common with boobies, here mingling with Antarctic forms. The noddy however ranges far north also, even occasionally to Ireland.

The whole of the peaty ground underneath the trees in the Phylica woods is bored in all directions with the holes of smaller sea birds, called by the Germans "night birds," a Prion and a Puffinus.

The burrows that these birds make are of about the size of large rats' holes. They traverse the ground everywhere, twisting and turning, and undermining the ground, so that it gives way at almost every step.

I went along the beach, and through a second wood towards the waterfall, where was the hut of the Germans, and their potato ground. A flock of thirty or forty predatory gulls (*Stercorarius antarcticus*) were quarrelling and fighting over, the bodies of penguins, the skins of which had been taken in considerable numbers by our various parties on shore. The Skua is a gull which has acquired a sharp curved beak, and sharp claws at the tips of its webbed toes. The birds are thoroughly predaceous in their habits, quartering their ground on the look-out for carrion, and assembling in numbers where there is anything killed, in the same curious way as vultures.

They steal eggs and young birds from the penguins when they get a chance, but their principal food here appears to be the night birds, especially the Prions, which they drag from their holes, or pounce on as they come out of them. The place was strewed with the skeletons of Prions, with the meat torn off them by these gulls, which leave behind the bones and feathers.

The Antarctic Skua is very similar in appearance to the large northern Skua. The two species were at first considered by naturalists to be identical; they differ however, especially in the structure of the bill. The Skua is of a dark brown colour, not unlike that of most of the typical birds of prey. We met with the bird constantly afterwards on our southern voyage, as far down even as the Arctic Circle; and a specimen was noticed by Ross further south still, in Possession Island.

The hut of the Germans was a comfortable one of stone, thatched with tussock and with a good frame window and door, and comfortable bunks to sleep on. There used to be wild goats on the top of Inaccessible Island, and there are still plenty of wild pigs. The ferine pigs were, as the Germans told me, of various colouring, and showed no tendency to uniformity; but the goats were almost invariably black, only one or two had a few white markings about head, neck, and chest. The sows used to be seen with litters of seven or eight young, but in a few days the number dwindled to one or two; the sows probably eating their young. The young suffered often from a sort of scrofula in which the glands about the neck became much enlarged.

The pigs now remaining are mostly boars: they are very hairy and have long tusks. The hogs are fierce, and one of the Germans told me that one once regularly hunted him, as if to attempt to kill him for food. The pigs feed mainly on birds and their eggs, but eat also the roots of the tussock and wild celery; they have nearly exterminated a penguin rookery on the south side of the island, but a few penguins remain, who have learnt to build in holes under stones, where the pigs cannot, reach them.

This fact is curious, as showing how easily circumstances may arise, such, that in an island even so small as Inaccesible, one colony of birds may develop a totally new habit, whilst other colonies of the same species preserve their original customs. And yet how strong is the tendency in birds to preserve their habits! I know of no more striking instance of this than the fact that the Apteryx of New Zealand (*A. australis*) considers it necessary to put as much of its head as it can under its rudiment of a wing, when it goes to sleep.

The pigs cannot get down the cliffs to the rookeries on the north side of the island.

One penguin at the Falkland Islands (*Spheniscus magellanicus*) regularly nests in burrows, sometimes twenty feet long. Another species of the same genus (*Spheniscus minor*), breeds in neat holes burrowed in sandbanks, at New Zealand.

On the beach are large banks of seaweed, but as at Tristan the heavy surf so batters the weeds, that it is difficult to find a serviceable specimen. An *Octopus* is very common amongst the stones, about the edge of the surf. I caught several attracted by the washing of the penguins' flesh and skins in the water. A *Chiton*, *Patella* and *Buccinum* are also common about the shore, as at Tristan.

All night long the penguins on shore in the rookery kept up an incessant screaming, no doubt lamenting the terrible invasion to which they had been subjected. The sound at a distance was not unlike that which one hears from tree frogs in the south of Europe, "Caa Quark, Caa Quark, Ca Caa Ca Caa." In the morning we moved to Nightingale Island, taking the Germans with us.

Nightingale Island, Oct. 17th, 1873—Nightingale Island, the smallest of the Tristan group, lies 20½ miles S.W. of Tristan Island, and about 22 miles N.W. by W. of Inaccessible Island. The island is about 1 ½0th miles long, by less than one mile broad; its area is thus not more than one square mile. We steamed up to the northwest side in the morning.

In the northeast is a rocky peak, from which an elevated ridge runs down to the sea on the east side, whence the Peak is accessible. On the north side it is impracticable, being too precipitous. A lower ridge stretches N.E. and S.W. on the south side of the island, and a broad valley separates the western termination of this ridge from the high ground and peaks on the N.E.; the highest peak is 1,100 feet in height, and the highest point of the lower ridge, 960 feet.

The whole of the lower land, and all but the steepest slopes of the high land and its actual summits, are covered with a dense growth of tussock, which occupies also even the ledges and short slopes between the bare perpendicular rocks of the Peak. The lower ridge is covered with the grass on all except its very summit, where amongst huge irregularly piled boulders of basalt, grow the same ferns as are found in Inaccessible Island, and *Phylica arborea* trees. The summit of the highest ridge appears to have a similar vegetation, the tussock ceasing there.

In the sea of tall grass, clothing the wide main valley of the island on its south side, are patches of Phylica trees, growing in many places thickly together, as at Inaccessible Island, with a similar vegetation devoid of tussock, beneath them. The appearance of the tall grass, when seen from a distance, is most deceptive; as we viewed the island from the deck of the ship, about a quarter of a mile off, we saw a green coating of grass, coming everywhere down to the verge of the wave-wash on the rocks, and stretching up comparatively easy looking slopes towards the summit of the Peak.

The grass gave no impression of its height and impenetrability, and one of the surveyors started off jauntily to go to the top of the Peak and make a surveying station. On closer inspection, however, the real state of the case might be inferred, for there was plainly visible a dark sinuous line leading from the sea, right inland through the thickest of the tussock. This was a great penguin road, and the whole place was one vast penguin rookery, and the grass that looked like turf to walk on, was higher than a man's head.

I made out with my glass a great drove of penguins on the rocks under the termination of the road, and I went below at once to put on my thickest gaiters.

We pulled onshore through beds of kelp, and landed on shelving rocks leading up to caves, the haunt of the Fur Seals in the proper season. We met the surveyors coming back, well pecked and dead beat, having given up the Peak in despair.

The shelving rock is composed of volcanic conglomerate, full of irregular fragments and rounded lumps of hard basalt, and various scoriaceous forms; in places also of a similarly derivative rock of a reddish colour, but devoid of larger embedded fragments. In a cliff about forty feet in height, adjoining and rising from the shelves, are beds of fine-grained volcanic sandstone rock, banded yellow and black, and horizontally bedded, probably of submarine formation.

These beds constitute the whole mass of two or three small outlying rocks or islands lying to the N.E., and are there also horizontal. These beds appear about twenty feet thick in the cliff, and above them is a layer of basalt of about the same thickness, which extends east and west, capping the softer beds and conglomerates. This layer is evidently a lava flow of comparatively late date, as it seems to have run down the valley between the two ridges, and to have come from the south; its upper surface is a little rounded, higher in the centre, and thinning off at the edges, as may be seen in the section exposed in the cliff.

It is on the almost level upper surface of this flow that the great penguin rookery lies. The island has evidently, like Inaccessible Island, undergone immense denudation, and there is no trace of any centres of action remaining. In the low cliffs of the coast, numerous caves are formed by the eating out by waves of the softer strata underlying the hard cap of basalt.

The caves are so numerous as to form a striking feature in the appearance of the island as it is approached from seawards; such caves are not apparent at Inaccessible or Tristan da Cunha Islands.

The caves with the sloping ledges leading up to them are frequented as was said by fur seals. Four years before 1,400 seals had been killed on the island by one ship's crew; they are much scarcer now, but the island is visited regularly once a year by the Tristan people, as is also Inaccessible Island. The Germans only killed seven seals at Inaccessible Island, but the Tristan people killed forty there in December, 1872. Two seals were seen by us in the water about the rocks, but none on land.

The sloping rock ledges are covered with a thin coating of dark green *Ulva*, which, when dry, has a peculiar almost metallic glance. A short scramble up the rocks brought us at once face to face with the tall grass and penguins.

The party broke up into small groups, each choosing what it thought the best route for penetrating the enemy's country. I made along the rocks to the point where, as I had seen from the ship, the main street ended: here were hundreds of penguins coming from and going to the sea in droves, or hurrying along singly to catch up some drove, or lolling about on the rocks, basking; the moving ones going along hop, hop, hop, just like men in a sack race.

The hard rock was actually polished, and had its irregularities smoothed off where the feet of the birds had worn it down at the entrance to the street. No doubt the Diatom skeletons present in the food and dung of the penguins, and always in abundance in the mud of their rookeries, adhering to their dirty feet, acts as polishing powder and assists the wearing process.

The street did not open by a single definite mouth towards the sea, but split up into numerous channels leading down to a number of easy tracks through the rocks. A little way in there was a clear open track six feet wide, and in places as much as eight or ten feet in width.

On each side narrow alleys led at nearly right angles to the rows of nests with which the whole space on either side of the main street was taken up.

Amongst the penguins here were numerous nests of the yellow-billed Albatross (*Diomedea culminata*) called by the Tristan people "Mollymauk," variously spelt in books, Molly Hawk, Mollymoy, Mollymoc, Mallymoke. It is, as are most of the sealers' names in the South, a name originally given to one of the Arctic birds, the Fulmar, and then transferred to the Antarctic from some supposed or real resemblance.

In the same manner the name given by northern whalers to the Little Auk is given in the South to the Diving Petrel of Kerguelen's Land. And the term "clap match" given to the female southern fur seals by the sealers, is the name originally given by the Dutch to the hooded seal or "bladdernose" of Greenland (*Cystocephalus*), and is a corruption of the word "Klapmuts," a bonnet, "the seal with a bonnet." It is curious that in this case the term should have been thus transferred to so very different a seal, which has nothing resembling a hood, but the word is so peculiar that there can be no doubt about its origin.

Various similar corruptions are in use as terms for southern animals. The name Albatross itself is the Spanish word "alcatraz" a "gannet." The Spanish no doubt called the albatrosses they met with "gannets," their familiar sea bird, just as common sailors will call every sea bird a gull, and a foreigner's corruption of the word became adopted as a special name for the bird.

The name Penguin is another instance in point. The word was not coined, as often supposed, by the early Dutch navigators, from the Latin word "pinguis," but is, as has been shown by M. Roulin, and others, a Breton or Welsh word, "pen gwenn," "white head," the name originally given to European sea birds with white heads, probably to the Puffin (*Mormon fratercula*). The name Pingouin is applied in modern French to the Great and Little Auk. In early voyages the name is applied to various exotic sea birds. In early Dutch travels the true meaning of the word is given, and it is stated to be English.

The Mollymauk is an albatross about the size of a goose, head, throat, and under part pure white, the wings grey, and the bill black with a yellow streak on the top and with a bright yellow edge to the gape, which extends right back under the eye. The yellow shows out conspicuously on the side of the head. It is not thus shown in Gould's coloured figures. The bird is extremely handsome. They take up their abode in separate pairs anywhere about in the rookery, or under the trees, where there are no penguins, which latter situation they seem to prefer.

They make a cylindrical nest of tufts of grass, clay, and sedge, which stands up from the ground. The nest is neat and round. There is a shallow concavity on the top for the bird to sit on, and the edge overhangs somewhat, the old bird undermining it, as the Germans said, during incubation, by pecking away the turf of which it is made.

I measured one nest, which was 14 inches in diameter and 10 inches in height. The nests when deserted and grass-grown make most convenient seats. The birds lay a single egg, about the size of a goose's, or somewhat larger, but elongate, with one end larger than the other, as are all albatross eggs.

The egg is held in a sort of pouch whilst the bird is incubating. The bird has thus to be driven right off the nest before the egg is dropped out of the pouch and it can be ascertained whether there is one there or no.

The birds when approached sit quietly on their nests or stand by them, and never attempt to fly; indeed they seem, when thus bent on nesting, to have forgotten almost the use of their wings.

Captain Carmichael, in his account of Tristan da Cunha, relates how he threw one of the birds over a cliff and saw it fall like a stone without attempting to flap, and yet these birds will soar after a ship over the sea as cleverly as any other albatross; indeed, the same peculiarity occurs in the case of the large albatross when nesting.

When bullied with a stick or handled on the nests, the birds snap their bills rapidly together with a defiant air, but they may be pushed or poked off with great ease. Usually a pair is to be seen at each nest, and then by standing near a short time one may see a curious courtship going on.

The male stretches his neck out, erects his wings and feathers a bit, and utters a series of high-pitched rapidly repeated sounds, not unlike a shrill laugh. As he does this he puts his head close up against that of the female.

Then the female stretches her neck straight up, and turning up her beak utters a similar sound, and rubs bills with the male again. The same maneuver is constantly repeated.

The albatrosses make their nests sometimes right in the middle of a penguin road, but the two kinds of birds live perfectly happily together. I saw no fighting, though, small as the penguins are, I think they could easily drive out the Mollymauks if they wished it.

The ground of the rookery is bored in all directions by the holes of Prions and petrels, which thus live under the penguins. Their holes were not so numerous in the rookery at Inaccessible Island as here. The holes add immensely to the difficulties of traversing a rookery, since as one is making a rush, the ground is apt to give way, and give one a fall into the black filthy mud amongst a host of furious birds; which have then full chance at one's eyes and face.

Besides the mollymauks and petrels, one or two pairs of Skuas had nests on a few mounds of earth in the rookery. How these mounds came there I could not understand.

The Skuas' eggs are closely like those of the lesser blackbacked gull, and two in number. The birds swooped about our heads as we robbed the nests, but were not nearly so fierce as those we encountered further south. All round their nests were scattered skeletons of Prions.

I, with three sailors carrying my botanical cases, attempted to scale the Peak; we had a desperate struggle through long grass and penguins, and at last had to come back beaten, and made for the Phylica patches, where the ground was clear. Thence I fought my way through the grass up to the top of the lower ridge of the island, but though there were no penguins on this slope, I never had harder work in my life.

I had to stop every ten yards or so for breath, the growth of the grass was so dense. My men lost me and never reached the top. On the summit I found the rest of the party which had come on shore, full of the hardships they had suffered in getting through the rookery, and looking forward with no pleasure to the prospect of going back again through it.

Two spaniels had been brought on shore and were taken through the rookery, partly by being carried, partly dragged. One of these was lost on the way back; he would not face the penguins and could not be carried all the way, so got left behind, and I fear must have died and been eaten by Skuas.

Poor old "Boss," Lieutenant Channer's pet, though one-eyed and too old to be much good for shooting, was a favourite, and we were all very sorry for him. Three volunteers charged back into the rookery in search, but it was of no use. He was frightened to death and would not answer to a call.

The dogs brought to Inaccessible Island by the two Germans ran wild in the penguin rookery, notwithstanding their exertions to keep them at home, and finally the dogs had to be shot. They fed themselves on the eggs and young.

After getting through the rookery on to the rocks, it was amusing to see the party arrive singly and in twos at all sorts of points of the edge of the rookery and on the verge of the cliff, having lost their direction, and often to their disgust having to turn back through the edge of the rookery again to reach some spot where they could get down to the sea.

The penguins were having their evening bath and pluming themselves on our arrival. The number of birds here must be enormous. At least one-fourth of the surface of the island and small outlyers, for these also are rookeries, must be covered by them; taking thus a space a quarter of a mile square, and allowing two only to a square yard, there would be nearly 400,000 penguins.

The rookery has evidently once been larger than at present, since a good part of the tall grass, now not occupied by birds, had old deserted nests amongst it. Probably the number of birds varies considerably each season.

One of the most remarkable facts about the penguins is that they are migratory; they leave Inaccessible Island, as the Germans told us, in the middle of April after moulting, and return, the males in the last week of July, the females about August 12th; and I do not think it possible that the Germans could have been mistaken. Whither can they go, and by what means can they find their way back? The question with regard to birds that fly is difficult enough, but it may always be supposed that they steer their course by landmarks seen at great distances from great heights, or that they follow definite lines of land. In the present case the birds can have absolutely no landmarks, since from sea level Tristan da Cunha is not visible from any great distance; the birds cannot move through the water with anything approaching the velocity of birds of flight; they have however, the advantage of a constant presence of food. The question of the aquatic migration of penguins and seals seems a special one, and presents quite different difficulties to that of the migration of birds of flight. The penguins certainly do not go to the Cape of Good Hope nor St. Helena, and they cannot live at sea altogether.

Moseley, H. N. 1879. Notes by a Naturalist on the "Challenger." *MacMillan and Co., London; "Tristan da Cunha," pp. 114–133.*

The Islands Galapagos (1697)

William Dampier

The Islands Galapagos (May 1684)

The 19th day in the evening we sailed from the island Lobos with Captain Eaton in our company. We carried the three flour prizes with us, but our first prize laden with timber we left here at an anchor; the wind was at south by east which is the common tradewind here, and we steered away N.W. by north intending to run into the latitude of the isles Galapagos, and steer off west, because we did not know the certain distance, and therefore could not shape a direct course to them. When we came within 40 minutes of the Equator we steered west, having the wind at south, a very moderate gentle gale.

It was the 31st day of May when we first had sight of the islands Galapagos: some of them appeared on our weather bow, some on our lee bow, others right ahead. We at first sight trimmed our sails and steered as nigh the wind as we could, striving to get to the southernmost of them but, our prizes being deep laden, their sails but small and thin, and a very small gale, they could not keep up with us; therefore we likewise edged away again a point from the wind to keep near them; and in the evening the ship that I was in and Captain Eaton anchored on the east side of one of the easternmost islands, a mile from the shore, in sixteen fathom water, clean, white, hard sand.

The Galapagos Islands are a great number of uninhabited islands lying under and on both sides of the Equator. The easternmost of them are about 110 leagues [*~330 miles or 480 km*] from the Main. They are laid down in the longitude of 181, reaching to the westward as far as 176; therefore their longitude from England westward is about 68 degrees. But I believe our hydrographers do not place them far enough to the westward. The Spaniards who first discovered them, and in whose charts alone they are laid down, report them to be a great number stretching northwest from the Line, as far as 5 degrees north, but we saw not above 14 or 15. They are some of them 7 or 8 leagues long, and 3 or 4 broad. They are of a good height, most of them flat and even on the top; 4 or 5 of the easternmost are rocky, barren and hilly, producing neither tree, herb, nor grass, but a few dildoe-trees, except by the seaside. The dildoe-tree is a green prickly shrub that grows about 10 or 12 foot high, without either leaf or fruit. It is as big as a man's leg, from the root to the top, and it is full of sharp prickles growing in thick rows from top to bottom; this shrub is fit for no use, not so much as to burn. Close by the sea there grows in some places bushes of burton-wood, which is very good firing. This sort of wood grows in many places in the West Indies, especially in the Bay of Campeachy and the Samballoes. I did never see any in these seas but here. There is water on these barren islands in ponds and holes among the rocks. Some other of these islands are mostly plain and low, and the land more fertile, producing trees of divers sorts unknown to us. Some of the westernmost of these islands are nine or ten leagues long and six or seven broad; the mould deep and black. These produce trees of great and tall bodies, especially mammee-trees, which grow here in great groves. In these large islands there are some pretty big rivers; and in many of the other lesser islands there are brooks of good water. The Spaniards when they first discovered these islands found multitudes of Guanos, and land-turtle or tortoise, and named them the Galapagos Islands. I do believe there is no place in the world that is so plentifully stored with those animals. The Guanos here are fat and large as any that I ever saw; they are so tame that a man may knock down twenty in an hour's time with a club. The land-turtle are here so numerous that 5 or 600 men might subsist on them alone for several months without any other sort of provision: they are extraordinary large and fat; and so sweet that no pullet eats more pleasantly. One of the largest of these creatures will weigh 150 or 200 weight, and some of them are 2 foot, or 2 foot 6 inches over the challapee or belly. I did never see any but at this place that will weigh above 30-pound weight. I have heard that at the isle of St. Lawrence or Madagascar, and at the English Forest, an island near it called also Don Mascarin and now possessed by the French, there are very large ones, but whether so big, fat, and sweet as these, I know not. There are 3 or 4 sorts of these creatures in the West Indies. One is called by the Spaniards hecatee; these live most in fresh-water ponds, and seldom come on land. They weigh about 10 or 15 pounds; they have small legs and flat feet, and small long necks. Another sort is called terapen; these are a great deal less than the hecatee; the shell on their backs is all carved naturally, finely wrought, and well clouded: the backs of these are rounder than those before mentioned; they are otherwise much of the same form: these delight to live in wet swampy places, or on the land near such places. Both these sorts are very good meat. They are in great plenty on the isles of Pines near Cuba: there the Spanish hunters when they meet them in the woods bring them home to their huts, and mark them by notching their shells, then let them go; this they do to have them at hand, for they never ramble far from thence. When these hunters return to Cuba, after about a month or six weeks' stay, they carry with them 3 or 400 or more of these creatures to sell; for they are very good meat, and every man knows his own by their marks. These tortoise in the Galapagos are more like the hecatee except that, as I said before, they are much bigger; and they have very long small necks and little heads. There are some green snakes on these islands, but no other land animal that I did ever see. There are great plenty of turtledoves so tame that a man may kill 5 or 6 dozen in a forenoon with a stick. They are somewhat less than a pigeon, and are very good meat, and commonly fat.

There are good wide channels between these islands fit for ships to pass, and in some places shoal water where there grows plenty of turtle-grass; therefore these islands are plentifully stored with sea-turtle of that sort which is called the green turtle. I have hitherto deferred the description of these creatures therefore I shall give it here.

There are 4 sorts of sea turtle, namely, the trunk-turtle, the loggerhead, the hawksbill, and the green turtle. The trunk-turtle is commonly bigger than the other, their backs are higher and rounder, and their flesh rank and not wholesome. The loggerhead is so called because it has a great head, much bigger than the other sorts; their flesh is likewise very rank, and seldom eaten but in case of necessity: they feed on moss that grows about rocks. The hawksbill-turtle is the least kind, they are so called because their mouths are long and small, somewhat resembling the bill of a hawk: on the backs of these hawksbill turtle grows that shell which is so much esteemed for making cabinets, combs, and other things. The largest of them may have 3 pound and a half of shell; I have taken some that have had 3 pound 10 ounces: but they commonly have a pound and a half or two pound; some not so much. These are but ordinary food, but generally sweeter than the loggerhead: yet these hawksbills in some places are unwholesome, causing them that eat them to purge and vomit excessively, especially those between the Samballoes and Portobello. We meet with other fish in the West Indies of the same malignant nature. These hawksbill-turtles are better or worse according to their feeding. In some places they feed on grass, as the green tortoise also does; in other places they keep among rocks and feed on moss or seaweeds; but these are not so sweet as those that eat grass, neither is their shell so clear; for they are commonly overgrown with barnacles which spoil the shell; and their flesh is commonly yellow, especially the fat.

Hawksbill-turtle are in many places of the West Indies: they have islands and places peculiar to themselves where they lay their eggs, and seldom come among any other turtle. These and all other turtle lay eggs in the sand; their time of laying is in May, June, July. Some begin sooner, some later. They lay 3 times in a season, and at each time 80 or 90 eggs. Their eggs are as big as a hen's egg, and very round, covered only with a white tough skin. There are some bays on the north side of Jamaica where these hawksbills resort to lay. In the Bay of Honduras are islands which they likewise make their breeding-places, and many places along all the coast on the Main of the West Indies from Trinidad de La Vera Cruz in the Bay of Nova Hispania. When a sea-turtle turns out of the sea to lay she is at least an hour before she returns again, for she is to go above high-water mark, and if it be low-water when she comes ashore, she must rest once or twice, being heavy, before she comes to the place where she lays. When she has found a place for her purpose she makes a great hole with her fins in the sand, wherein she lays her eggs, then covers them 2 foot deep with the same sand which she threw out of the hole, and so returns. Sometimes they come up the night before they intend to lay, and take a view of the place, and so having made a tour, or semicircular march, they return to the sea again, and they never fail to come ashore the next night to lay near that place. All sorts of turtle use the same methods in laying. I knew a man in Jamaica that made 8 pound Sterling of the shell of these hawksbill turtle which he got in one season and in one small bay, not half a mile long. The manner of taking them is to watch the bay by walking from one part to the other all night, making no noise, nor keeping any sort of light. When the turtle comes ashore the man that watches for them turns them on their backs, then hauls them above high-water mark, and leaves them till the morning. A large green turtle, with her weight and struggling, will puzzle 2 men to turn her. The hawksbill-turtle are not only found in the West Indies but on the coast of Guinea, and in the East Indies. I never saw any in the South Seas.

The green turtle are so called because their shell is greener than any other. It is very thin and clear and better clouded than the hawksbill; but it is used only for inlays, being extraordinary thin. These turtles are generally larger than the hawksbill; one will weigh 2 or 3 hundred pound. Their backs are flatter than the hawksbill, their heads round and small. Green turtle are the sweetest of all the kinds: but there are degrees of them both in respect to their flesh and their bigness. I have observed that at Blanco in the West Indies the green turtle (which is the only kind there) are larger than any other in the North Seas. There they will commonly weigh 280 or 300 pound: their fat is yellow, and the lean white, and their flesh extraordinary sweet. At Boca Toro, west of Portobello, they are not so large, their flesh not so white, nor the fat so yellow. Those in the Bay of Honduras and Campeachy are somewhat smaller still; their fat is green, and the lean of a darker colour than those at Boca Toro. I heard of a monstrous green turtle once taken at Port Royal in the Bay of Campeachy that was four foot deep from the back to the belly, and the belly six foot broad; Captain Roch's son, of about nine or ten years of age, went in it as in a boat on board his father's ship, about a quarter of a mile from the shore. The leaves of fat afforded eight gallons of oil. The turtles that live among the keys or small islands on the south side of Cuba are a mixed sort, some bigger, some less; and so their flesh is of a mixed colour, some green, some dark, some yellowish. With these Port Royal in Jamaica is constantly supplied by sloops that come hither with nets to take them. They carry them alive to Jamaica where the turtles have wires made with stakes in the sea to preserve them alive; and the market is every day plentifully stored with turtle, it being the common food there, chiefly for the ordinary sort of people.

Green turtle live on grass which grows in the sea in 3, 4, 5, or 6 fathom water, at most of the places before mentioned. This grass is different from manatee-grass, for that is a small blade; but this a quarter of an inch broad and six inches long. The turtle of these islands Galapagos are a sort of a bastard green turtle; for their shell is thicker than other green turtle in the West or East Indies, and their flesh is not so sweet. They are larger than any other green turtle; for it is common for these to be two or three foot deep, and their callapees or bellies five foot wide: but there are other green turtle in the South Seas that are not so big as the smallest hawksbill. These are seen at the island Plata, and other places thereabouts: they feed on moss and are very rank but fat.

Both these sorts are different from any others, for both he's and she's come ashore in the daytime and lie in the sun; but in other places none but the she's go ashore, and that in the night only to lay their eggs. The best feeding for turtle in the South Seas is among these Galapagos Islands, for here is plenty of grass.

There is another sort of green turtle in the South Seas which are but small, yet pretty sweet: these lie westward on the coast of Mexico. One thing is very strange and remarkable in these creatures; that at the breeding time they leave for two or three months their common haunts, where they feed most of the year, and resort to other places only to lay their eggs: and it is not thought that they eat anything during this season: so that both he's and she's grow very lean; but the he's to that degree that none will eat them. The most remarkable places that I did ever hear of for their breeding is at an island in the West Indies called Caymans, and the isle Ascension in the Western Ocean: and when the breeding time is past there are none remaining. Doubtless they swim some hundreds of leagues to come to those two places: for it has been often observed that at Cayman, at the breeding time, there are found all those sort of turtle before described. The South Keys of Cuba are above 40 leagues from thence, which is the nearest place that these creatures can come from; and it is most certain that there could not live so many there as come here in one season.

Those that go to lay at Ascension must needs travel much farther; for there is no land nearer it than 300 leagues: and it is certain that these creatures live always near the shore. In the South Sea likewise the Galapagos is the place where they live the biggest part of the year; yet they go from thence at their season over to the Main to lay their eggs; which is 100 leagues the nearest place. Although multitudes of these turtles go from their common places of feeding and abode to those laying-places, yet they do not all go: and at the time when the turtle resort to these places to lay their eggs they are accompanied with abundance of fish, especially sharks; the places which the turtle then leave being at that time destitute of fish, which follow the turtle.

When the she's go thus to their places to lay the male accompany them, and never leave them till they return: both male and female are fat the beginning of the season; but before they return the male, as I said, are so lean that they are not fit to eat, but the female are good to the very last; yet not so fat as at the beginning of the season. It is reported of these creatures that they are nine days engendering, and in the water, the male on the female's back. It is observable that the male, while engendering, do not easily forsake their female: for I have gone and taken hold of the male when engendering: and a very bad striker may strike them then, for the male is not shy at all: but the female, seeing a boat when they rise to blow, would make her escape, but that the male grasps her with his two fore fins, and holds her fast. When they are thus coupled it is best to strike the female first, then you are sure of the male also. These creatures are thought to live to a great age; and it is observed by the Jamaica turtlers that they are many years before they come to their full growth.

The air of these islands is temperate enough considering the clime. Here is constantly a fresh sea breeze all day, and cooling refreshing winds in the night: therefore the heat is not so violent here as in most places near the Equator. The time of the year for the rains is in November, December, and January. Then there is oftentimes excessive hard tempestuous weather, mixed with much thunder and lightning. Sometimes before and after these months there are moderate refreshing showers; but in May, June, July, and August the weather is always very fair.

We stayed at one of these islands which lies under the Equator but one night because our prizes could not get in to anchor. We refreshed ourselves very well both with land and sea turtles; and the next day we sailed from thence.

The next island of the Galapagos that we came to is but two leagues from this: it is rocky and barren like this; it is about five or six leagues long and four broad. We anchored in the afternoon at the north side of the island, a quarter of a mile from the shore in 16-fathom water. It is steep all round this island and no anchoring only at this place. Here it is but ordinary riding; for the ground is so steep that if an anchor starts it never holds again; and the wind is commonly off from the land except in the night when the land-wind comes more from the west, for there it blows right along the shore, though but faintly. Here is no water but in ponds and holes of the rocks.

That which we first anchored at has water on the north end falling down in a stream from high steep rocks upon the sandy bay, where it may be taken up. As soon as we came to an anchor, we made a tent ashore for Captain Cook who was sick. Here we found the sea turtle lying ashore on the sand; this is not customary in the West Indies. We turned them on their backs that they might not get away. The next day more came up, when we found it to be their custom to lie in the sun: so we never took care to turn them afterwards; but sent ashore the cook every morning, who killed as many as served for the day. This custom we observed all the time we lay here, feeding sometimes on land-turtle, sometimes on sea turtle, there being plenty of either sort. Captain Davis came hither again a second time; and then he went to other islands on the west side of these. There he found such plenty of land-turtle that he and his men ate nothing else for three months that he stayed there. They were so fat that he saved sixty jars of oil out of those that he spent: this oil served instead of butter to eat with doughboys or dumplings, in his return out of these seas. He found very convenient places to careen, and good channels between the islands; and very good anchoring in many places. There he found also plenty of brooks of good fresh water, and firewood enough, there being plenty of trees fit for many uses. Captain Harris came thither likewise, and found some islands that had plenty of mammee-trees, and pretty large rivers. The sea about these islands is plentifully stored with fish such as are at John Fernando. They are both large and fat and as plentiful here as at John Fernando. Here are particularly an abundance of sharks. The north part of this second isle we anchored at lies 28 minutes north of the Equator. I took the height of the sun with an astrolabe. These isles of the Galapagos have plenty of salt. We stayed here but 12 days in which time we put ashore 5000 packs of flour for a reserve if we should have occasion of any before we left these seas. Here one of our Indian prisoners informed us that he was born at Ria Lexa, and that he would engage to carry us thither. He being examined of the strength and riches of it satisfied the company so well that they were resolved to go thither.

Dampier, W. 1697. A New Voyage Around the World; *2007 reprint from Dover Publications, New York, pp. 42–54.*

The Sea Otter and the Sea Cow (1741–1742)

Georg W. Steller

THE SEA OTTER

With warm-blooded sea animals the Bering Island region is more copiously provided. When we arrived there the sea beavers (or sea otters, *Lutris*) were present in large numbers. In November and December we killed them 3 to 4 versts from our quarters at the so-called Bobrovoe Poles (Beaver (i.e. Sea Otter) Field) and Kozlova Ryechka; in January, 6 to 8 versts at Kitova Ryechka (Whale Creek); in February, 20 versts at the Utes and Bolshaya Laida (Large Cliff). In March and April, when the sea otters were driven entirely away from the north side about our quarters, we went overland to the south side and brought the otters by carrying them 12, 20, 30, to 40 versts [*~0.6 miles or 1 km*]. The chase of these animals by us took place in the following manner: These animals at all seasons of the year, more, however, during the winter than in summer, leave the sea for the shore in order to sleep, rest, and play all sorts of games with each other. At low tide they lie on the rocks and the uncovered beaches, at high water on land in the grass or on the snow a half or even a whole verst from shore, though mostly near the shore. In Kamchatka and on the Kurile Islands they never, or at least very rarely, go ashore; so that from this also it is evident that on this island they had never been disturbed by man in their quiet and play. Usually in the evening or at night, in groups of two, three, or four persons provided with long and strong poles of birch wood, we went quietly along the beaches as much as possible against the wind, looking diligently about everywhere. When a sea otter was seen lying asleep one of us went quietly towards him, even creeping when near by. The others in the meantime cut off his passage to the sea. As soon as he had been approached so closely that it was thought he could be reached in a few jumps, the man sprang up suddenly and tried to beat him to death with repeated strokes on the head. However, if he ran away before he could be reached, the other men together chased him from the sea farther inland and gradually closed in on him by running, whereupon, no matter how nimbly and adroitly the animal might be able to run, he would finally tire and be killed. If, as often happened, we came upon a whole herd together, each of us selected the animal that seemed nearest to him, in which case the affair went off still better. In the beginning we needed but scant effort, stratagem, and dexterity, as the whole shore was full of them and they were lying in the greatest security. Later on, however, they learned to know our earpicks so well that, when we spied upon them, we saw them go ashore with the greatest care. They first looked well about them everywhere, turned their noses in all directions in order to catch a scent; and, when after long looking about they had settled down to rest, one would sometimes see them jump up again in fright, look about anew, or go back to sea. Watchers were posted by them wherever a herd was lying. The malicious foxes, by waking them out of their sleep violently or keeping them watchful, also thwarted us. We were therefore compelled to search constantly for new places and to go hunting farther and farther away, also to prefer dark nights to light ones and blustering to quiet weather in order to get them, as our maintenance depended upon it. In spite of all these obstacles, from November 6, 1741, to August 17, 1742, over seven hundred otters were nevertheless killed by us, eaten, and their skins taken along to Kamchatka as tokens. However, as they were often clubbed needlessly and only for the sake of the skin or even frequently left lying about with pelt and flesh when they were not black enough, matters came to such a pass through our wicked persecution of these animals that we nearly lost hope of being able to build a vessel. For in spring, when the stock of provisions had been consumed and the work was to begin, these animals had been driven away for versts on either side of our dwellings. We would gladly have been satisfied with seals, but these were too crafty to venture farther inland, so that it was great luck when we could steal upon a seal.

The sea otter, which, because of the nature of its fur, has been erroneously regarded as a beaver and therefore called Kamchatka *babr*, is a real otter and differs from the river otter only in this, that the former lives in the sea, is almost one half larger, and in the beauty of the fur is more like a beaver than an otter. It is indisputably an American sea animal, occurring in Asia only as a guest and newcomer which lives in the so-called Beaver Sea (Bobrovoe More) from latitude 56° to 50°, where America is nearest and both continents are possibly separated by a channel only 50 miles wide, which, moreover, is filled with many islands, making the transit of these animals to Kamchatka possible in this region, as otherwise they are not able to cross a wide sea. According to information derived from the Chukchi nation I am certain that these animals are to be met with opposite in America from 58° to 66°, and pelts from there have also been received in trade by way of Anadyrsk. That on the Kamchatkan coast, however, no sea otter is to be found above 56° is possibly due to the fact that Kamchatka from there on may extend more northerly but America more easterly, whereby the sea lying between them assumes a greater width and depth than these animals, which only find food on the bottom of the sea and, because they cannot long endure without inhaling air, must not let themselves down to a great depth, are able to cross; particularly as perhaps no islands occur there, which is all the more probable because all islands must be regarded as remnants of the mainland. From 56° to 50° we found sea otters on the islands in sight of the American mainland and on 60° near the mainland, at Cape St. Elias itself, 500 miles east of Kamchatka. Probably this sea otter is the same animal which the Brazilians on the western side of America according to the testimony of Marggraf called *jiya* and *cariguebeju;* and consequently this animal occurs, if not in all, at least in most places on the western as well as the eastern side of America. Accordingly, my former hypothesis would now also seem to be confirmed as a truth, that the sea otters which in winter and spring time arrive in great numbers with the drift ice at the Kamchatkan coasts have been brought hither not only from the mainland of America itself but also mostly from the islands in the Channel which the ice must pass. For I have seen with my own eyes how much these animals like to lie on the ice; and, although on account of the mild winter the ice floes were thin and few, they were nevertheless carried on them, asleep or awake, towards the sea by the ebbing tide.

The sea otter is usually 5 feet long and 3 feet in circumference at the breastbone, where the body is thickest. The largest weighed, with the entrails, 70 to 80 Russian pounds. In shape it resembles an otter, the hind feet only excepted, which are smooth and agree in structure with the hind flippers of seals. The entrails are likewise conditioned as in the otter. The skin, which lies as loose on the flesh as in dogs and shakes all over when the animal is running, so far surpasses in length, beauty, blackness, and gloss of the hair that of all river beavers that the latter cannot be compared with it. The best pelts bring in Kamchatka 20 rubles, in Yakutsk 30, in Irkutsk 40 to 50, and at the Chinese frontier, in exchange for their wares, 80 to 100 rubles. The meat is fairly good to eat and palatable; the females, however, are much tenderer, and, against the course of nature, are most fat and delicious shortly before, during, and after parturition. The suckling otters, which, because of their poor skin, are called *medvyedki*, or young bears, can, because of their daintiness, both roasted and boiled, at any time compete with suckling lambs. The male has a bony penis like sea-dogs (seals) and all other warm-blooded marine animals. The female has two mammae beside the genitals. They copulate in the human manner. Altogether, in life it is a beautiful and pleasing animal, cunning and amusing in its habits, and at the same time ingratiating and amorous. Seen when they are running, the gloss of their hair surpasses the blackest velvet. They prefer to lie together in families, the male with its mate, the half-grown young, or *koshloki*, and the very young sucklings, *medvyedki*. The male caresses the female by stroking her, using the fore feet as hands, and places himself over her; she, however, often pushes him away from her for fun and in simulated coyness, as it were, and plays with her offspring like the fondest mother. Their love for their young is so intense that for them they expose themselves to the most manifest danger of death. When their young are taken away from them, they cry bitterly like a small child and grieve so much that, as we observed from rather authentic cases, after ten to fourteen days they grow as lean as a skeleton, become sick and feeble, and will not leave the shore. In flight they take the suckling young in the mouth, but the grown-up ones they drive before them. If they have the luck to escape they begin, as soon as they are in the water, to mock their pursuers in such a manner that one cannot look on without particular pleasure. Now they stand upright in the water like a man and jump up and down with the waves and sometimes hold the fore foot over the eyes, as if they wanted to scrutinize you closely in the sun; now they throw themselves on their back and with the front feet rub the belly and the pudenda as do monkeys; then they throw the young ones into the water and catch them again, etc. If a sea otter is overtaken and nowhere sees any escape it blows and hisses like an angry cat. When struck it prepares itself for death by turning on the side, draws up the hind feet, and covers the eyes with the fore feet. When dead it lies like a dead person, with the front feet crossed over the breast.

The food of the sea otter consists of marine crustaceans, mollusks, small fishes, a little seaweed, also meat. I have no doubt that if one were not to grudge the expense, a few of these animals could be brought to Russia and made tame; indeed, they would multiply perhaps in a pond or river, for they care but little for the sea water, and I have seen that they stay for several days in lakes and rivers for the fun of it. Moreover, this animal deserves from us all the greatest reverence, as for more than six months it served us almost exclusively as food and at the same time as medicine for the scurvy-stricken.

THE SEA COW

Along the whole shore of the island, especially where streams flow into the sea and all kinds of seaweed are most abundant, the sea cow (*Morskaya korova*), so called by our Russians, occurs at all seasons of the year in great numbers and in herds. After the supplying of ourselves with provisions began to become difficult because of the frightening away of the sea otters from the northern side, we considered ways and means to secure these animals and, because they were near to us, to derive our nourishment more easily from them. On May 21, therefore, the first attempt was made to throw a large manufactured iron hook, to which was fastened a strong and long rope, into this powerful and large sea animal and haul it ashore; but in vain, because the skin was too tough and firm and the hook was much too dull. It was changed in different ways, and several other attempts were made; but these turned out still more poorly, so that the animals escaped from us out to sea with the hook and the rope attached to it. Finally necessity forced us to make preparations for harpooning.

For this purpose towards the end of June the yawl, which had been badly damaged on the rocks in the autumn, was repaired, a harpooner with a steersman and four oarsmen put into it, and a harpoon given to the first together with a very long line, coiled in proper order as in whaling, its other end being held on shore by the other forty men. We now rowed very quietly towards the animals, which were browsing in herds along the shore in the greatest security. As soon as the harpooner had struck one of them the men on shore gradually pulled it toward the beach; the men in the yawl rushed upon it and by their commotion tired it out further; when it seemed enfeebled they jabbed large knives and bayonets into its body until it had lost almost all its blood, which spouted from the wounds as from a fountain, and could thus be hauled on the beach at high tide and made fast. As soon as the water went out again and the animal lay on the dry beach the meat and fat were cut off everywhere in pieces and carried with rejoicing to our dwellings, where the meat was kept in barrels and the fat hung up on high frames. We now soon found ourselves so abundantly supplied with food that we could continue the building of our new vessel without hindrance.

This sea animal, which became so valuable to us, was first seen by the Spaniards in America and described with many intermingled untruths by the physician Hernandez. The Spaniards called it manati, the English and Dutch have named it sea cow. It is found both on the eastern and on the western side of America and has been observed by Dampier with sea bears and sea lions in the southern hemisphere and by me and others in the northern. The largest of these animals are 4 to 5 fathoms (28 to 35 English feet) long and 3½ fathoms thick about the region of the navel, where they are thickest. To the navel this animal resembles the seal species; from there on to the tail, a fish. The head of the skeleton is in general shape not different from the head of a horse, but when covered with skin and flesh it resembles in some measure a buffalo head, particularly as concerns the lips. In the mouth it has on each side in place of teeth two wide, longish, flat, loose bones, of which one is fastened above to the palate, the other to the inside of the lower jaw. Both are provided with many obliquely converging furrows and raised welts with which the animal grinds up the seaweeds, its usual food. The lips are provided with many strong bristles, of which those on the lower jaw are so thick that they resemble quills of fowls and clearly demonstrate by their internal hollowness the structure of the hairs. The eyes of this animal in spite of its size are not larger than sheeps' eyes and are without eyelids. The ears are so small and hidden that they cannot at all be found and recognized among the many grooves and wrinkles of the skin until the skin has been taken off, when its polished blackness reveals the ear opening, which, however, is hardly large enough for the insertion of a pea. Of the external ear there is not the slightest trace. The head is connected with the rest of the body by a short neck not set off from it. On the underside the unusual forefeet and the breasts are worthy of observation. The feet consist of two joints, the extreme end of which has a rather close resemblance to a horse's hoof; they are furnished underneath with many short and closely set bristles like a scratch brush. With these front feet, on which neither fingers nor nails can be distinguished, the animal swims ahead, knocks the seaweeds from the rocks on the bottom, and, when lying on its back getting ready for mating, one embraces the other as with arms. Under these forefeet are found the breasts, with black, wrinkled, two-inch long teats, at the extreme end of which innumerable milk ducts open. When pulled hard these ducts give off a great amount of milk, which surpasses the milk of land animals in sweetness and richness but is otherwise not different. The back of this animal is formed almost like that of an ox. The median crest of the backbone is raised up high. Next to this projection on both sides there is a flat hollow along the back. The flanks are oblongly rounded. The belly is roundish and very distended and at all times stuffed so full that at the slightest wound the entrails at once protrude with much whistling. Its relative size is like the belly of a frog. From the genitals on the body suddenly decreases greatly in circumference. The tail itself, however, becomes gradually thinner towards the flipper, which serves as hind feet; yet immediately in front of the flipper it is still two feet wide. Moreover, this animal has no other fin than the tail flipper and none on the back, in which it differs from the whales. The tail flipper is horizontal as in the whale and the porpoise. The organ of the male is like that of an ox in relative length, almost a fathom long and with the sheath fastened under the navel; in shape and nature it is like that of a horse. The female organ is situated immediately over or before the anus, nearly elongate quadrangular and at the anterior part provided with a strong, sinewy clitoris an inch and a half long.

These animals, like cattle, live in herds at sea, males and females going together and driving the young before them about the shore. They are occupied with nothing else but their food. The back and half the body are always seen out of the water. They eat in the same manner as the land animals, with a slow forward movement. They tear the seaweed from the rocks with the feet and chew it without cessation. However, the structure of the stomach taught me that they do not ruminate, as I had at first supposed. During the eating they move the head and neck like an ox, and after the lapse of a few minutes they lift the head out of the water and draw fresh air with a rasping and snorting sound after the manner of horses. When the tide falls they go away from the land to sea but with the rising tide go back again to the shore, often so near that we could strike and reach them with poles from shore. They are not afraid of man in the least, nor do they seem to hear very poorly, as Hernandez asserts contrary to experience. Signs of a wonderful intelligence, whatever Hernandez may say, I could not observe, but indeed an uncommon love for one another, which even extended so far that, when one of them was hooked, all the others were intent upon saving him. Some tried to prevent the wounded comrade from being drawn on the beach by forming a closed circle around him; some attempted to upset the yawl; others laid themselves over the rope or tried to pull the harpoon out of his body, in which indeed they succeeded several times. We also noticed, not without astonishment, that a male came two days in succession to its female which was lying dead on the beach, as if he would inform himself about her condition. Nevertheless, no matter how many of them were wounded or killed, they always remained in one place.

Their mating takes place in June, after protracted preludes. The female flees slowly before the male with continual turns about, but the male pursues her without cessation. When, however, the female is finally weary of this mock coyness she turns on her back and the male completes the mating in the human manner. When these animals want to take a rest on the water they turn on their backs in a quiet place in a bay and allow themselves to drift on the water like logs.

These animals are found at all seasons of the year everywhere around the island in the greatest numbers, so that the whole population of the eastern coast of Kamchatka would always be able to keep itself more than abundantly supplied from them with fat and meat.

The hide of the sea cow has a dual nature. The outer skin or coating is black or blackish brown, an inch thick and of a consistency almost like cork, full of grooves, wrinkles, and holes about the head. It consists entirely of perpendicular fibers, which lie close upon one another, as in fibrous gypsum. The bulbs of the individual fibers stand out round on the inner side of this coating and fit into delicate cavities in the skin underneath, which thereby almost looks like the surface of a thimble. This outer coating, which can easily be detached from the skin, is, in my opinion, a crust that has coalesced from juxtaposed transformed hairs, which type I have also found in whales. The inner skin is somewhat thicker than an oxhide, very strong and white in color. Under both of these the whole body of the animal is surrounded by a layer of fat or blubber four fingerbreadths thick, after which comes the meat. The weight of this animal with skin, fat, meat, bones, and entrails I estimate at 1200 poods [*43,200 lbs or 19,600 kg*], or 480 long hundredweights. The fat of this animal is not oily or flaccid, but somewhat hard and granular, snow-white and, when it has been lying a few days in the sun, as agreeably yellow as the best Holland butter. The fat itself when boiled surpasses in sweetness and taste the best beef fat; when tried out, it is like fresh olive oil in color and liquidity and like sweet almond oil in taste and is of such exceptionally good flavor and nourishment that we drank it by the cupful without experiencing the slightest nausea. In addition it has the virtue that when taken somewhat often it acts as a very mild laxative and diuretic, for which reason I consider it a good remedy against protracted constipation as well as gallstone and retention of the urine. The tail consists wholly of fat which is much more agreeable even than that found on the other parts of the body. The fat of the calves is entirely like the meat of young pigs; the meat itself, however, like veal. It is boiled through in half an hour and swells up to such an extent that it takes up twice as much space as before. The meat of the old animals is not to be distinguished from beef; but it has this remarkable property that, even in the hottest summer months and in the open air, it will keep for two full weeks and even longer without becoming offensive, in spite of its being so defiled by the blowflies as to be covered with worms all over. This property of the meat would seem to be attributable in part to the diet of the animal. It also has a much deeper red color than the meat of all other animals and almost looks as if it had been reddened by saltpeter. All of us who had partaken of it soon found out what a salutary food it was, as we soon felt a marked improvement in strength and health; this was the experience especially of those among the sailors who until then had constant relapses of scurvy and who until that time had not been able to recover. With this sea cow meat we also provisioned our vessel for the voyage—a problem that we surely should not otherwise have known how to solve.

With regard to the internal structure of this wonderful creature I refer the interested reader to my elaborate description of the sea cow. Here I will only note briefly that the heart of this animal is, contrary to the usual order, divided or double and that the pericardium does not surround it directly but forms a distinct cavity; furthermore, that the lungs are enclosed in a strong tendinous membrane and are situated at the back, as in birds, for which reason it (the sea cow) can remain longer under water without drawing breath. In the third place it has no gall bladder, but only a wide gall duct after the fashion of horses; also its stomach and entrails have some similarity to the intestines of a horse; and, finally, the kidneys, like those of sea calves and sea bears, are composed of very many small kidneys, each of which has its own ureter, pelvis, traps, and papillae, and they weigh 30 pounds and are 2½ feet long. From the head of their manati the Spaniards are said to take out a stone-hard bone, which among druggists goes under the erroneous name of *lapis manati*. This I have vainly searched for in so many animals that I have come to think that our sea cow may be a different kind of these animals. Moreover, it has caused me no little wonder that, notwithstanding that I made careful inquiry about all animals while in Kamchatka before my voyage and never heard anything about the sea cow, nevertheless after my return I obtained the information that this animal is known from Cape Kronotski to Avacha Bay, and that it is occasionally thrown ashore dead. For lack of a special name the Kamchadals have given it the name "cabbage eater."

Steller, G. W. 1925. Steller's journal of the sea voyage from Kamchatka to America and return on the second expedition, 1741–1742; *"Description of Bering Island," pp. 209–237. L. Stejneger, translator. Volume 2 in* Bering's voyages: an account of the efforts of the Russians to determine the relation of Asia and America. *F. A. Golder, editor. American Geographical Society Research Series Number 2, New York, New York, USA.*

Chapters from the Life-Histories of Texas Reptiles and Amphibians (1926)

John Kern Strecker

*A MARAUDER OF THE WOODS (*ELAPHE OBSOLETA CONFINA*)*

The gray pilot snake or "chicken snake," as it is known to the natives of western Louisiana and the eastern half of Texas, is one of the most destructive ophidians in the South. It is a tree-climbing species and an inveterate raider of chicken yards and birds' nests. Angus Gaines, a couple of decades ago a writer of popular natural history articles, was inclined to scoff at the idea that the so-called "chicken snakes," both the black form (typical obsoleta) and its brightly marked southern variety (confina), could swallow several hens' eggs at one meal; but this is something that every Texas farmer knows to be a fact and it is surprising that this man, who really knew a great deal about reptilian creatures and who wrote some excellent articles on their habits, should have doubted it.

I have had many interesting experiences with specimens of the gray phase of obsoleta and will here set down some of them for the edification of herpetologists who are engaged in studying the habits of ophidians, rather than for those who consume most of their waking hours in scheduling the scale formulas of preserved specimens. The gray pilot snake has one besetting sin, and that is gluttony. When it dines either on wild birds, young chickens, or eggs, it is inclined to overeat, becomes sluggish, sleeps off its torpor near the scene of its meal, and often falls into the hands of the indignant farmer, who summarily dispatches it with a stick.

Some years ago, while I was visiting Rev. A. H. Barber and his family on their farm near Refugio, Refugio County, I obtained two unusually large pilot snakes, which were discovered asleep in hens nests, which they had robbed the night before. One of the snakes contained three, the other, four eggs. At Mann's farm, seven miles west of Waco, Texas, one containing three eggs was caught in a hen's nest. The term "chicken" snake would not be a misnomer, even if it were applied to this ophidian merely on account of its woeful lack of reasoning, for it will often swallow china nest eggs, thus demonstrating that its grade of intelligence is quite on par with that of the traditionally stupid hen. A specimen of *Elaphe confina*, found in the Bosque valley by Dr. W. T. Gooch and me, had swallowed a china egg and in attempting to crawl through a fence, it had got caught between two strands of barbed wire, and with part of its length hanging down on each side of the fence, had perished of starvation. The skeleton of this snake, with the egg still held in place by a strip of skin, was for some time exhibited in the Baylor University Museum.

Wild birds suffer greatly from the depredations of the ophidian land pirate. At Mann's farm, a small colony of cliff swallows had located their domiciles under the eaves of a barn. All of the nests had been completed, but the birds had not yet deposited their eggs and were using their little retort-shaped homes as sleeping quarters. One night a pilot snake raided the colony, ate three of the swallows, and the next morning was found coiled around a rafter, sleeping off the effects of its midnight meal. The cries of a scissor-tailed flycatcher once attracted my attention to an *Elaphe* coiled on a limb of a mesquite tree. On being dissected, the snake was found to have eaten a female flycatcher and her eggs.

Birds, however, do not always come off second-best in their encounters with snakes. Sometimes they actually manage to get the best of their foes unaided, although they more frequently attract the attention of human beings to them and thus bring about their deaths. A friend recently told me of an incident which occurred while he was visiting some friends who lived on a farm. A snake, pursued by two angry mockingbirds, glided into the midst of a family circle. In attempting to rob the nest of these birds, it had, aroused the fighting ire of the owners, and in order to save its eyes, was compelled to beat a hasty retreat, which ended in its destruction at the hands of a member of the startled human group gathered in front of the house.

As a rule, *Elaphe confina* is pugnacious, and captive specimens seldom become tame. I had one specimen which I was preparing for shipment. While endeavoring to induce it peaceably to enter the shipping box, it managed to squirm from my grasp and succeeded in crawling into the folds of an old skirt which was hanging from the wall. While recapturing it, it viciously struck my hands five times, in each instance drawing blood. This specimen measured more than five feet in length, ended its career at the British Museum. Another which I discovered in a tall tree, and which was compelled to climb after, struck me twice before I could grasp it by the neck. I probably could not have captured this specimen at all had it not been in a sluggish condition as a result of having dined on some half-fledged mockingbirds.

In central Texas, the pilot snake hibernates about three months of the year. I have found a number of fairly lively specimens as late as the last week in November, and as early as the second week in February. December 18th of an unusually mild winter, I captured, in a hollow rotten stump, one that was still able to show fight. After winter sets in in earnest, they find hibernating places in holes under large fallen trees and usually go down to a depth of several feet. They often share their winter quarters with copperheads, brown snakes, leopard frogs, and toads.

The gray pilot snake is the basilisk of the woods, a terror to all small living creatures. The only warm-blooded animals on which it does not feed are those which are too large for it to master or those which are too clever or too swift for it too capture, but even those swift of foot or flight are often obtained by stealth. Mr. Charles B. Pearre and I experimented with a large pilot snake and some tree lizards. We placed the snake within the hollow of a great stump and dropped the lizards into the hollow one at a time. As soon as a lizard discovered the presence of the snake, it simulated death by turning over on its back. After the snake had been surrounded by half-a-dozen of these "possuming" saurians, we removed it and in a few seconds the lizards all scampered out of the hollow. We tried this experiment several times, using at least fifteen lizards, always with the same result.

Some call the gray pilot snake a "witch snake," on account of the peculiar pattern of the markings of the head. "Woman's-head snake" is another name. If one will draw on his imagination to a considerable extent, he may be able to see in the pattern of light and dark markings, the outlines of a human skull, or a woman's head with long, flowing hair, or an old man's head with long beard. I have seen living and preserved specimens on exhibition, all of which were being shown on account of this peculiarity of marking. When the snake attains a length of as much as twenty-four inches, the bright markings of the head fade and the pattern becomes obsolete. The head markings, of an allied species, *Elaphe laeta*, are very similar to those of the present snake and occasionally a specimen will furnish foundation for the "woman's-head snake" story.

*THE BELL OF THE MARSHES (*HYLA CINEREA*)*

If you are one of those persons who go into ecstasies over beautiful coloring in nature of birds, and flowers, and butterflies, have you ever handled a living specimen of the bell frog? If you have not, no longer talk of symmetry of form and warmth of tone, for you know nothing of such things. On May 24, one of my student friends brought in a magnificent specimen which he had captured on a small tree in the creek bottom. Above it was of a beautiful grass green, with a few rich orange-colored spots on the lower part of the back. The light stripe along its side was yellow, the ear bronze, and the eye golden. The underparts were creamy, with the exception of the inner surfaces of the limbs where the skin was so transparent that the color of the flesh showed through. The throat was yellow, with vertical lines of green, the latter color at times predominating.

There was true harmony of coloring in the marked contrast between the upper and lower surfaces. The frog was placed in a glass jar that the assembled students might study the wonderful ability of the creature to use its digital disks in adaptation to environment. After comfortably clinging to the side of the jar for a few moments, it detected the open mouth of the receptacle and almost immediately leaped for freedom. Its first spring carried it a distance of six feet to the door casing, where, about four feet above the floor, it clung for another moment; then another leap of at least eight feet carried it to the grass in front of the building, and it required the efforts of several persons to recapture it.

Hyla cinerea is not only the handsomest of North American tree frogs, but is also the most alert, graceful, and intelligent in appearance. To some, its abnormally long limbs might make it appear ungainly, but one need only watch its movements a few seconds in order to satisfy himself that the grace of its actions more than compensate for the apparent discrepancy in its proportions. The marked inconsistency between the shades of its coloration and the suggestion of its specific technical name is explained by the fact that the type specimen was a preserved example which had lost the beauty of its life colors through their fading out in spirits.

The name bell frog was applied to this exquisite species on account of its peculiar voice, which has been compared to "the tone of a cowbell heard at a distance." At Laguna Lake, Falls County, where I spent many pleasant days collecting in past years and where these interesting creatures are abundant, its voice was one of the pleasant features of the early night concert produced by the denizens of the marshes and marsh thickets. Proceeding from the heart of a lily-spangled lagoon on a cloudy day, its sound was especially attractive.

Along the Bosque River, in northern McLennan County, a number of bell frogs were discovered clinging to the ends of willow branches, which overhung the water. On Waco Creek, within the city of Waco, several were found on small trees, especially such as were well hooded with vines. At Laguna Lake, they were quite abundant in button willow thickets. Here during the first few cool days of the fall, they were found hiding under logs and fallen branches, in damp places inhabited by small-mouthed salamanders and red efts. They hibernate in such places, though usually, however, they go down to a considerable depth in the ground, or into the hollow hearts of great logs. Hibernation, like death, is a great leveler, and in a hole occupied by wintering cold-blooded vertebrates, can be found the poisonous and the harmless, the cruel and the gentle—copperheads, cottonmouths, pilot snakes, garter snakes, lizards, frogs, and toads, all huddled together in a stiffened mass.

I am of the opinion, after comparing specimens of different sizes labeled with the dates on which they were captured, that it takes the bell frog at least three years to attain to its maximum size. Specimens collected in the fall (October) are less than one-third grown, while young examples collected in May and June, when about one year old, are quite one-third the size of a well-grown adult. The eggs of this species, like those of *Hyla evittata* described by Miss Dickerson, are in "small loose masses, attached to the stems of pond-lily leaves." I am quoting Miss Dickerson's exact language, which could not be improved upon in describing the egg masses of the bell frog. The eggs are laid in April and May (Laguna Lake and Waco). Some of the egg masses of this species contain only about thirty eggs, and I doubt whether any one mass will contain more than sixty. The tadpoles are a trifle darker yellow than those of *Hyla versicolor chrysoscelis* and are comparatively more slender, though larger in size. Those found by me were in shallow bay-shaped depressions in the margins of lagoons.

*SPOTTED AND OTHER TOADS (*BUFO PUNCTATUS; BUFO DEBILIS; BUFO COMPACTILIS; GASTROPHRYNE TEXENSE*)*

The life-histories of only four or five of the sixteen species and subspecies of true toads (*Bufo*), which inhabit North America, have been recorded in detail, while those of several species are absolutely unknown. Five of these are inhabitants of those portions of the state of Texas in which most of my herpetological work has been done. One, *Bufo punctatus*, commonly know as the spotted toad, has always been of particular interest to me on account of its peculiarly alert appearance and odd habits and owing to the fact that it is by far the rarest toad in my home locality. Tracy I. Storer in his excellent work "A Synopsis of the Amphibia of California" (Univ. of Calif. Pub. in Zoology, Vol. 27, pp. 1–342), has published the best account of the life-history of this species to be found anywhere in American herpetological literature. Most of his notes apply to its habits in the states of Arizona and California, and I believe that some of my field observations in central Texas will prove of interest to students of batrachology.

Central Texas specimens are colored much like those from more western localities, but the dorsal surfaces are rather lighter in shade. Stejneger's color description, "malachite green above," applies very well to examples collected in western Texas, but I should describe the color of central Texas specimens as a very light, dull green. Of two examples collected in a gully in the northern part of the city of Waco, the dorsal surfaces were light buff. In some cases the red-tipped tubercles are surrounded by black rings, but usually, in very young specimens, the black is not in evidence. My attention was attracted to the specimens found in the above-mentioned gully by the notes of the males. These were clear, bird-like whistles, repeated at intervals. When handled, the male emits a chirp which sounds much like that of a young bird.

The numerous specimens collected on the above occasion were mostly males and many of these were found lurking around the edges of large stones, where they were signaling for the females. When disturbed, they attempted to escape by climbing the rocky banks of the gully. They were adept in concealing themselves in crevices, and for this reason we lost several. The spotted toad is a permanent resident of this place and here in daylight I captured a number by digging out partially loosened pieces of shale and by dragging along the edges of ledges of stone with a scraper fashioned for that purpose.

Bufo punctatus moves around only after darkness sets in. During the breeding season we found them in the water about nine o'clock p.m. The most of those already mated were in the shallower pools in the bed of the gully. The next morning small strings of eggs were found in these pools. Some of the strings were attached to plant stalks, while others were freely floating among dead leaves or in the shallows. Eight days later, the pools were filled with small blackish tadpoles about one inch (25 mm.) in total length; the head and body measuring about two-thirds of this. Camp has described the tadpoles of *Bufo punctatus* in detail, although at the time he was not absolutely sure that they belonged to this species.

The present year (1926), my untiring volunteer collector, Mr. Walter Williams, brought me an interesting series of young punctatus from San Marcos, Hays County. Similar specimens in alcohol were sent me many years ago from Boquillas, Brewster County, by the late J. D. Mitchell. Williams' specimens were all living and were in fine condition. Several which had not yet absorbed their tails were fifteen millimeters in total length and were light gray in color. The head and body measured 9 mm. Specimens 11 mm. in body length had already lost their tails by absorption and were a uniform dark gray. By the time the head and body measure as much as 15 mm., the adult coloration is assumed, in the case of these particular specimens, a rather dull gray green, with the tubercles of the back tipped with bright red but lacking the black rings which make these so conspicuous when the toad is full-grown. The San Marcos specimens were collected May 29.

Bufo debilis is smaller than the spotted toad and is even more retiring in habits. Like *Scaphiopus couchii* and *Bufo compactilis*, it is possible to obtain this species in quantity only a few days in each year, when they are in their breeding pools or have been driven from their burrows by heavy rains. *Bufo debilis* is a toad of the mesquite timber prairie and with the aid of a flashlight an occasional specimen may be found in such environments even on a dry summer night, but the collector may be literally compelled to walk miles in an effort to find such scattering examples. Like punctatus they move around only after darkness has set in. On our toading expeditions for the past twenty years, only two debilis were found in thickly settled portions of Waco, and these were caught in a section where much mesquite timber had been allowed to stand.

Breeding specimens of this little toad are unusually handsome in coloration. The dorsal surfaces are of a brighter shade of green than those of the spotted species, and the tubercles are tipped with bright yellow and encircled by black rings. It has the same upright bearing as has punctatus, but its flat head and unusually wide and divergent paratoids give it quite a different appearance. Its notes are much like those of the other species referred to but are even shriller.

Debilis breeds in April and May in what are usually termed prairie sinks, i.e., small temporary ponds, and in roadside ditches. I once found a number of them occupying a ditch in which were numerous pairs of *Gastrophryne texensis* in copula, and at another time I caught several in a pool in which were many striped chorus frogs (*Pseudacris trisenata*). The eggs are in small strings and are attached to grass and weedstems. I have unfortunately been unable accurately to trace the development of this species on account of its breeding grounds being so far removed from home. The tadpoles are slightly smaller than those of punctatus, and their metamorphosis is accomplished within a very short space of time. This is very necessary on account of the extremely temporary character of their breeding places. I returned twenty days later to one pond in which I had found debilis breeding and discovered that it was almost perfectly dry, only a few mud holes remaining to indicate moisture. In one of these mud holes were a few belated tadpoles, and in the grass along the banks I found two small toads with tails.

Bufo compactilis is a burrowing toad which in many details of its habits resembles *Scaphiopus couchii*, the common spadefoot of central Texas. Several of the boys who have assisted me in collecting toads call this species the "quacker," on account of the peculiar cry of the male which reminds one of the quacking of a duck, although the intervals between the notes are longer. It has the loudest voice of any of our toads with the exception of *Scaphiopus couchii*, and when an old male specimen is heard signaling for a female, the sound is anything but musical.

A few years ago these toads were almost as abundant within the city limits of Waco as the common toad. During 1924 and 1925, two dry years, only two specimens were found around electric lights, although several toading expeditions were engaged in. The present year, when rainfall was abundant, biological students obtained twenty-two specimens. On a farm located at the point where White Rock creek flows into the Brazos River, many year-old toads of this species were found under boards in damp places near wells at which cows were watered. As these were found soon after heavy showers of rain had fallen, I believe that they were driven out of their shallow burrows by water. They are found abundantly in truck patches and the older specimens occupy burrows very similar to those of Couch's spadefoot with which species they are usually found associated.

Bufo compactilis breeds in May. The following is a description of several batches of eggs discovered by me: eggs (just emitted), spherical, one-sixteenth of an inch in diameter. Color brown, vegetable pole gray. The jelly mass is long and irregular, but cannot well be called a string for it is usually nearly an inch in width in the middle and tapers to a point at each end. These ribbons or masses or whatever one sees fit to call them are attached to grass blades and plant stems in from four to eight inches of water. The eggs are scattered through them without apparent order. Five of these masses were found in one small pool at which was found only one pair of toads in copula. Number of eggs in a mass (estimated), 200 to 450. In the five cases, the total number must have been nearly 2000. The jelly is translucent. At the end of twenty-four hours the embryos are as far advanced as those of toads of the americanus group. The tadpoles are hatched in from 48 to 60 hours.

The tadpoles are shorter and more robust than those of *Bufo fowleri* and *B. americanus*. Color blackish green. They undergo their complete metamorphosis in about eighteen days after leaving the egg, and like all species of tailless amphibian which rapidly transform, leave the water with the tail of the tadpole stage still in evidence. These tails are absorbed within the next two or three days. Completely metamorphosed young toads may be found around the banks of their breeding places for only a short time after they leave the water. In this respect they resemble the young of *Scaphiopus couchii*, the young of both species adopting temporary burrows at a very tender age.

The egg mass of the little Texas toothless frog or toad (*Gastrophryne texensis*) is oblong and irregular in outline. The eggs are light brown with white vegetative pole. Size, less than that of a number eight shot (about one millimeter). The number ranges from 100 to 200 in a mass. The mass is attached to weed stems in from four to eight inches of water. I do not think that there are more than two or three emissions from one female. The note of this species is a shrill, long-drawn "quaw," "quaw," repeated at intervals of several seconds duration. This note is quite different from the loud, explosive "quock" of *Bufo compactilis*.

Strecker, J. K. 1926. Chapters from the Life-Histories of Texas Reptiles and Amphibians. *Contributions from Baylor University Museum Number Eight; pp. 2–12. Reprinted with permission of Baylor University, Waco, TX.*

Account of the Electrical Eels, and the Method of catching them in South America by means of Wild Horses (1820)

Alexander von Humboldt

Real Gymnoti or electrical eels inhabit the Rio Colorado, the Guarapiche, and several little streams, that cross the missions of the Chuyma Indians. They abound also in the large rivers of America, the Oroonoko, the Amazon, and the Meta; but the strength of the current, and the depth of the water, prevent their being caught by the Indians. They see these fish less frequently than they feel electrical shocks from them when swimming or bathing in the river. In the *Llanos*, particularly in the environs of Calabozo, between the farms of Morichal, and the missions *de arriba* and *de abaxo*, the basins of stagnant water, and the confluents of the Oroonoko, the Rio Guarico and the Canos of Rastro, Berito and Paloma, are filled with electric eels. We at first wished to make our experiments in the house we inhabited at Calabozo; but the dread of the electrical shocks of the gymnoti is so exaggerated among the vulgar, that during three days we could not obtain one, though they are easily caught, and though we had promised the Indians two piastres for every strong and vigorous fish.

Impatient of waiting, and having obtained very uncertain results from an electrical eel that had been brought to us alive, but much enfeebled, we repaired to the Cano de Bera to make our experiments in the open air on the borders of the water itself. We set off on the 19th of March for the village *Rastro de abaxo*; thence we were conducted to a stream, which, in the time of drought, forms a basin of muddy water, surrounded by fine trees. To catch the gymnoti with nets is very difficult, on account of the extreme agility of the fish, which bury themselves in the mud like serpents. We would not employ the *barbasco*, that is to say, the roots of the *Piscidea erithryna* and *Jacquinia armillaris*, which, when thrown into the pool, intoxicate or benumb these animals. These means would have enfeebled the gymnoti; the Indians therefore told us, that they would "fish with horses." We found it difficult to form an idea of this extraordinary manner of fishing; but we soon saw our guides return from the savannah, which they had been scouring for wild horses and mules. They brought about thirty with them, which they forced to enter the pool.

The extraordinary noise caused by the horses' hoofs makes the fish issue from the mud, and excites them to combat. These yellowish and livid eels, resembling large aquatic serpents, swim on the surface of the water, and crowd under the bellies of the horses and mules. A contest between animals of so different an organization furnishes a very striking spectacle. The Indians, provided with harpoons and long slender reeds, surround the pool closely; and some climb upon the trees, the branches of which extend horizontally over the surface of the water. By their wild cries, and the length of their reeds, they prevent the horses from running away, and reaching the bank of the pool. The eels, stunned by the noise, defend themselves by the repented discharge of their electric batteries. During a long time they seem to prove victorious. Several horses sink beneath the violence of the invisible strokes, which they receive from all sides in organs the most essential to life; and stunned by the force and frequency of the shocks, disappear under the water. Others, panting, with mane erect, and haggard eyes, expressing anguish, raise themselves, and endeavour to flee from the storm by which they are overtaken. They are driven back by the Indians into the middle of the water; but a small number succeed in eluding the active vigilance of the fishermen. These regain the shore, stumbling at every step, and stretch themselves on the sand, exhausted with fatigue, and their limbs benumbed by the electric shocks of the gymnoti.

In less than five minutes two horses were drowned. The eel, being five feet long, and pressing itself against the belly of the horses, makes a discharge along the whole extent of its electric organ. It attacks at once the heart, the intestines, and the *plexus cæliacus* of the abdominal nerves. It is natural that the effect felt by the horses should be more powerful than that produced upon man by the touch of the same fish at only one of his extremities. The horses are probably not killed, but only stunned. They are drowned from the impossibility of rising amid the prolonged struggle between the other horses and the eels.

We had little doubt that the fishing would terminate by killing successively all the animals engaged; but by degrees the impetuosity of this unequal combat diminished, and the wearied gymnoti dispersed. They require a long rest, and abundant nourishment, to repair what they have lost of galvanic force. The mules and horses appear less frightened; their manes are no longer bristled, and their eyes express less dread. The gymnoti approach timidly the edge of the marsh, where they are taken by means of small harpoons fastened to long cords. When the cords are very dry; the Indians feel no shock in raising the fish into the air. In a few minutes we obtained five large eels, the greater part of which were but slightly wounded. Some were taken by the same means toward the evening.

The temperature of the waters in which the gymnoti habitually live, is about 86° of Fahrenheit. Their electric force, it is said, diminishes in colder waters. The gymnotus is the largest of electrical fishes. I measured some that were from five feet to five feet three inches long; and the Indians assert, that they have seen them still longer. We found, that a fish of three feet ten inches long weighed twelve pounds. The transverse diameter of the body was three inches five lines. The gymnoti of *Cano de Bera* are of a fine olive-green colour. The under part of the head is yellow, mingled with red. Two rows of small yellow spots are placed symmetrically along the back, from the head to the end of the tail. Every spot contains an excretory aperture. In consequence, the skin of the animal is constantly covered with a mucous matter, which, as Volta has proved, conducts electricity twenty or thirty times better than pure water. It is in general somewhat remarkable, that no electrical fish, yet discovered in the different parts of the world, is covered with scales.

It would be temerity to expose ourselves to the first shock of a very large and strongly irritated gymnotus. If by chance you receive a stroke before the fish is wounded, or wearied by a long pursuit, the pain and numbness are so violent, that it is impossible to describe the nature of the feeling they excite. I do not remember having ever received from the discharge of a large Leyden jar, a more dreadful shock than that which I experienced by imprudently placing both my feet on a gymnotus just taken out of the water. I was affected the rest of the day with a violent pain in the knees, and in almost every joint. To be aware of the difference, which is sufficiently striking, that exists between the sensation produced by the pile of Volta and an electrical fish, the latter should be touched when they are in a state of extreme weakness. The gymnoti and the torpedoes then cause a twitching, which is propagated from the part that rests on the electric organs as far as the elbow. We seem to feel at every stroke an internal vibration that lasts two or three seconds, and is followed by a painful numbness.

Gymnoti are neither charged conductors, nor batteries, nor electromotive apparatuses, the shock of which is received every time they are touched with one hand, or when both hands are applied to form a conducting circle between two heterogeneous poles. The electric action of the fish depends entirely on its will; whether because it does not keep its electric organs always charged, or by the secretion of some fluid, or, by any other means alike mysterious to us, it be capable of directing the action of its organs to an external object. We often both tried, both insulated and uninsulated, to touch the fish, without feeling the least shock. When

Mr. Bonpland held it by the head or by the middle of the body, while I held it by the tail, and standing on the moist ground, did not take each other's hand, one of us received shocks, which the other did not feel. It depends upon the gymnotus to act toward the point where it finds itself the most strongly irritated. The discharge is then made at one point only, and not at the neighbouring points. If two persons touch the belly of the fish with their fingers, at an inch distance, and press it simultaneously, sometimes one, sometimes the other, will receive the shock. In the same manner, when one insulated person holds the tail, and another pinches the gills, or pectoral fin, it is often the first only by whom the shock is received. It did not appear to us, that these differences could be attributed to the dryness or dampness of our hands, or to their unequal conducting power. The gymnotus seemed to direct its strokes sometimes from the whole surface of its body, sometimes from one point only.

Nothing proves more strongly the faculty which the gymnotus possesses, of darting and directing its stroke according to its will, than the observations made at Philadelphia, and recently at Stockholm, on gymnoti rendered extremely tame. When they had been made to fast a long time, they killed from afar small fishes put into the tub. They acted at a distance; that is to say, their electrical stroke passed through a very thick stratum of water. We need not be surprised, that what was observed in Sweden, on a single gymnotus only, we could not see on a great number of individuals in their native country. The electric action of animals being *a vital action*, and subject to their will, it does not depend solely on their state of health and vigour. A gymnotus, that has made the voyage from Surinam to Philadelphia and Stockholm, accustoms itself to the imprisonment to which it is reduced; it resumes by degrees the same habits in the tub which it had in the rivers and pools. An electrical eel was brought to me at Calabozo, taken in a net, and consequently having no wound. It ate meat, and terribly frightened the little tortoises and frogs, which, not knowing the danger, placed themselves with confidence on its back. The frogs did not receive the stroke till the moment when they touched the body of the gymnotus. When they recovered, they leaped out of the tub; and when replaced near the fish, they were frightened at its sight only. We then observed nothing that indicated *an action at a distance*; but our gymnotus, recently taken, was not yet sufficiently tamed to attack and devour frogs. On approaching the finger, or metallic points, within the distance of half a line from the electric organs, no shock was felt. Perhaps the animal did not perceive the neighbourhood of this foreign body; or, if it did, we must suppose that the timidity it felt in the commencement of its captivity, prevented it from darting forth its energetic strokes, except when strongly irritated by an immediate contact. The gymnotus being immersed in water, I approached my hand, both armed and unarmed with a metal, within the distance of a few lines from the electric organs; yet the strata of water transmitted no shock, while Mr. Bonpland irritated the animal strongly by an immediate contact, and received some very violent shocks. If I had plunged the most delicate electroscopes we know, prepared frogs, into contiguous strata of water, they would no doubt have felt contractions at the moment when the gymnotus seemed to direct its stroke elsewhere.

The electrical organ of the gymnotus acts only under the immediate influence of the brain and the heart. On cutting a very vigorous fish through the middle of the body, the fat part alone gave me shocks. The shocks are equally strong, in whatever part of the body the fish is touched; it is most disposed, however, to dart them forth when the pectoral fin, the electrical organ, the lips, the eyes, or the gills are pinched. Sometimes the animal struggles violently with a person holding it by the tail, without communicating the least shock. Nor did I feel any when I made a slight incision near the pectoral fin of the fish, and galvanized the wound by the simple contact of two pieces of zinc and silver. The gymnotus bent itself convulsively, and raised its head out of the water, as if terrified by a sensation altogether new; but I felt no vibration in the hands which held the two metals. The most violent muscular movements are not always accompanied by electric discharges.

The action of the fish on the organs of man is transmitted and intercepted by the same bodies that transmit and intercept the electrical current of a conductor charged by a Leyden vial, or Volta's pile.

In wounded gymnoti, which give feeble but very equal shocks, these shocks appeared to us constantly stronger on touching the body of the fish with a hand armed with metal, than with the naked hand. They are stronger also, when, instead of touching the fish with one hand, naked, or armed with metal, we press it at once with both hands, either naked or armed. These differences, I repeat, become sensible only when you have gymnoti enough at your disposal, to be able to choose the weakest; and the extreme equality of the electric discharges admits of distinguishing between the sensations felt alternately by the hand naked or armed with, a metal, by one or both hands naked, and by one or both hands armed with metal. It is also in the case only of small shocks, weak and uniform, that the shocks are more sensible on touching the gymnotus with one hand (without forming a chain) with zinc, than with copper or iron.

Resinous substances, glass, very dry wood, horn, and even bones, which are generally believed to be good conductors, prevent the action of the gymnoti from being transmitted to man. I was surprised at not feeling the least shock on pressing wet sticks of sealing wax against the organs of the fish; while the same animal gave me the most violent strokes, when excited by means of a metallic rod. Mr. Bonpland received shocks when carrying a gymnotus on two cords of the fibres of the palm-tree, which appeared to us extremely dry. A strong discharge makes its way through very imperfect conductors. Perhaps also the obstacle, which the conducting arc presents, renders the discharge more painful. I touched the gymnotus with a wet pot of brown clay without effect; yet I received violent shocks when I carried the gymnotus in the same pot, because the contact was greater.

When two persons, insulated or not insulated, hold each other's hands, and one of these persons only touches the fish with the hand, either naked or armed with metal, the shock is most commonly felt by both at once. It happens, however, also, that, in the most painful shocks, the person who comes into immediate contact with the fish alone feels the shock. When the gymnotus is exhausted, or in a very weak state of excitability, and will no longer emit strokes on being irritated with one hand; the shocks are felt, in a very vivid manner, on forming the chain, and employing both hands. Even then, however, the electric shock takes place only at the will of the animal. Two persons, one of whom holds the tail, and the other the head, cannot, by joining hands and forming a chain, force the gymnotus to dart his stroke.

In employing very delicate electrometers in a thousand ways, insulating them on a plate of glass, and receiving very strong shocks, which passed through the electrometer, I could never discover any phenomenon of attraction or repulsion. The same observation was made by Mr. Fahlberg at Stockholm. This philosopher, however, has seen an electric spark, as Walsh and Ingenhousz had done before him at London, by placing the gymnotus in the air, and interrupting the conducting chain by two gold leaves pasted upon glass, and a line distant from each other. No person, on the contrary, has ever perceived a spark issue from the body of the fish itself. We have irritated it for a long time during the night, at Calabozo, in perfect darkness, without observing any luminous appearance.

Von Humboldt, A. 1820. "Account of the Electrical Eels, and the Method of catching them in South America by means of Wild Horses"; Edinburgh Philosophical Journal *II: pp. 242–249.*

Comments on the Cephalopods Found in the Stomach of a Sperm Whale (1913/1914)

Prince Albert I of Monaco

18th of July 1895

In the time it took to get to the site, the whalers of the second group had struck a course due west in pursuit of the pod which was escaping in that direction. After an initial burst of speed in a vain attempt to escape, the wounded animal was already slowing, dragged down by the harpooner's boat behind it. As I arrived, the first spears thrown by the whalers in tow were finding their mark. Soon thereafter, a breath exploded from the creature's blowhole and the steam issuing forth turned a pale shade of pink, and then went crimson, while the sea itself took on the color of the animal's gushing blood.

From that moment on, we stood face to face with the ordeal of the giant. Its enormous dark body, seemingly asleep, in turn submerged in the bloody sea, teetered heavily. Its large tail beat down violently on the crimson tide floating on the shifting sea, slicing through it and leaving churls of white foam in its wake.

The fifty individuals on my ship, having gathered to the front, perched on boxes and aloft in the rigging, were frozen in a dumb stupor. I myself gripped down to the marrow by the grandeur of the alien spectacle, followed the action as ardently as if witnessing a vision too fleeting to turn away from for an instant, lest it disappear forever. I was moved by the pain of a giant, so grandly manifested, which in the fullness of detail seemed more intense than that of lesser creatures. I felt for this powerful sea traveler, which for centuries perhaps had moved its large mass to all horizons and all depths without fear of enemies, surviving the worst storms and finally meeting its end at the hands of spears wielded by pygmies!

All that spilled blood, that mass of dead flesh seemed to me to be the equivalent of a major disaster such as a giant fallen tree or a shipwreck.

Suddenly the whale ceased thrashing and, as if our arrival had reanimated its brain, headed straight for us at top speed.

In a flash of worry and fear I wondered what effect the shock of that body violently hurled, either on purpose or as a result of a random convulsion, would have against the side of the ship, when it disappeared twenty meters away from us. Would it break the keel of the ship, the rudder, or the propeller with its back or a swipe of its tail? These were the thoughts that went through my mind as I was waiting the ten seconds for the massive beast to reappear on the other side of the halted ship. When it did, it was no longer moving. The whalers approached and continued spearing the whale, death penetrated it from all sides, and the spectators watched, transfixed, gasping with emotion.

The ship and all the actors in the spectacle floated on a one-hectare blanket of blood. Streams of a deeper flaky red crossed the waves, still flowing from the animal, and soon dissolved into the surrounding blood like clouds descending from the mountains and slowly blending into the mists of the valley.

Its massive head appeared by our stern. The lower jaw, left hanging open by the slackened muscle, was bobbing in the waves when I saw the cavernous mouth vomit again and again several Cephalopods, octopus or squid, of fantastic size. Clearly it was the result of the cetacean's final excursion to the depths prior to being harpooned at the surface: a recent mouthful, which had not yet passed through the esophagus.

I understood the scientific value of these objects originating from regions of intermediate ocean depths. They had been protected by their swimming abilities against all our methods of capture, their existence revealing itself sometimes in adventures we think of as fables or myths.

A rowboat was quickly dispatched to collect them, but the density of the precious vomitus kept them hovering just below the surface, suggesting that they might disappear before they could be reached, then I had a flash of inspiration; the cephalopods were still visible at a dozen meters from the ship, not far from the propeller. I gave the order to reverse engines, just enough to send the desired objects into the whirlpool thus created, this maneuver was effective at bringing them up to the surface so they could be scooped with a net from the boat.

The five octopi and squid my laboratory added to its collection in this unforeseen circumstance were studied, as soon as I returned, by Mr. Joubin, professor at the College of Rennes. They are new discoveries, either of species or genus and their appearance in life must have been extraordinary. One in particular, which unfortunately lost its head in the struggle, is of considerable scientific value. Its body, which is at least two meters in length, is of conical shape with a round fin and is partially covered with scales. Another, equally valuable, but whose body is missing, is known to us only by its crown of tentacles, that is to say by its head with its eight arms, each the thickness of a man's arm, covered with suction cups and sharp claws, and as powerful as those of the large terrestrial carnivores.

Already my naturalists had examined the stomach and extracted, from over 100 kilos of half digested remains, a few pieces of giant octopus sufficiently well preserved that they too could be attributed to totally unknown species. We can only guess at what delights this occupation offered its practitioners, as they had to wade through a fermented purple sludge strewn with eyeballs and beaks resistant to gastric juices and emitting the most noxious of odors. Such that, toward the end, their own weakened stomachs exteriorized the convulsions of a storm that wreaked havoc on the contents and echoed, in their small way, the events that the previous day had offered up so many treasures for science.

On my return to the bay, the physical aspect of the place was different. No more gulls flying the skies, no more fish jumping in the water. All had fled from the overwhelming putrefaction. Only men persisted at this struggle in such appalling conditions, some in the name of science, others for their own self-interest.

Mr. Richard then pointed out some marks on the cetacean's lips which had occurred accidentally. They were round and several centimeters wide. When compared to the suction cups that lined the enormous arms of the octopus found in the stomach of the whale, they clearly seemed to be marks left by the powerful suction of these devices, and I instantly had a vision of the colossal fights occurring in the depths of the oceans when this giant mammal prowls the abyss in search of prey.

Was it able, with a few powerful moves, to grab a giant octopus? The eight arms of the latter wrap around its head and latch on with all their suckers, while the rest of the body, stretched by the swallowing ends up breaking at the neck. The body falls into the stomach. But the head with all its tentacles, remains, holding fast to the whale's head, until the suction cups one by one lose grip as death slowly sets in. And if the sperm whale were to attack again and again before its prey's death set in, we could imagine a scene in which the monster's head would be completely engulfed in the gripping tentacles of several cephalopods.

The thought of these strange dramas brought back memories of an incident which occurred on my 1887 voyage aboard *l'Hirondelle*. I was in the middle of the Atlantic, on my way to the Azores, when one fine day, some majestic splashes rose on the horizon of a tranquil sea. We clearly saw that the cause was the frolickings of a colossal creature whose head and body sometimes reared up straight as a tower while the whip of its tail pushed the water in formidable columns of spray.

Soon the sea returned to calm, but the area where the action had taken place retained a white milky surface recognizable from over eight kilometers away, and which could have been either some kind of liquid or simply froth on the water. Despite my best efforts, a headwind kept *l'Hirondelle*, a modest sailing ship, from reaching the spot before the whiteness had dissipated. It stayed visible a long time however, and when finally we were able to reach the site, a freshly detached giant octopus head was waiting for us in the very spot. Described later by Professor Joubin, along with the other cephalopods from my trips, it happened to belong to the same group as a number of the octopi obtained from the sperm whale in this account, denizens of the middle depths of the sea, which are almost completely unknown.

Would it be too bold then, after correlating the facts of that incident with the new information that I just related here, to conclude that in the first experience I was witness to a particularly tragic scene in which a sperm whale wrapped in layers of giant tentacles had surfaced in an attempt to free itself from their grip?

Here is the list of cephalopods. It is important to note that all these cephalopods were pelagic, of large size and mostly of unknown genus and species.

1. *Cucioteuthis unguiculata* (Molina) Steenstrup: Enormous crown of tentacles.
2. *Ancistrocheirus lesueuri* (D'Orbigny et Ferussac) Gray: The body of a large individual and one separate plume probably from the same species.
3. *Lepidoteuthis grimaldii* Joubin: The bodies of two very large individuals.
4. *Histioteuthis ruppelli* Verany: Three large individuals about one meter long, complete and perfectly preserved.
5. *Histioteuthis* sp.: Brachial crowns of two large individuals.
6. *Dubioteuthis physeteris* Joubin: The body of a large individual
7. Series of 54 beaks, all of large size.

Prince Albert I of Monaco. 1913/1914. "Comments on the Cephalopods Found in the Stomach of a Sperm Whale": Résultats des campagnes scientifiques accomplies sur son yacht par Albert Ier, Prince Souverain de Monaco, *no. 17; pp. 471–475. Translated from French by Lajos Szoboszlai (August 2006).*

On collecting at Cape Royds (1910)

James Murray

In making the biological collections, almost every member of our small party lent more or less assistance. Though busy with other occupations, all were ready to bring home to the biologist anything strange which they noticed. Armytage found a sea urchin during a walk on the sea-ice before we had made a landing, and while landing he picked up the first scrap of seaweed. Mr. Shackleton brought in some moss and lichen soon after the *Nimrod* departed [Nimrod *was the ship used by Ernest Shackleton during British Imperial Antarctic Expedition 1907–1909*]. Wild got specimens of the lake vegetation. Adams found a starfish on the beach, and the others in like manner helped when opportunity offered.

All these indications that there was life in the district, coming as they did before the biological work was properly started, were encouraging for the future.

When the dredging operations began there were many willing helpers. Mr. Shackleton, with David, Mawson, and Priestley, were always interested in the dredging and ready to help, not only at hauling the dredge, but at the more disagreeable labour of conveying the collections to the hut. It is no disparagement to others to acknowledge the share which Priestley took in the biological work. Without him the greater part of the collections would not have been made. When the biologist was debarred by some trifling but mastering indisposition from active participation in the dredging during the midwinter weeks, Priestley kept the dredging-holes open, no small labour with the temperature sometimes as low as from minus 30° to minus 40° Fahr. With the assistance of Mr. Shackleton, David, and others, he kept up the dredging and brought home the proceeds to be examined.

Throughout the entire season practically the whole of the arduous labour of digging holes in the sea-ice and of sinking shafts in the lakes fell upon Priestley, and he did the active hauling of the dredge as well. It does not lessen our indebtedness to him to tell that he enjoyed his self-imposed task, and his voice might be heard issuing in light-hearted song from some deep shaft in lake or sea.

The Field of Operations—This was extremely limited. The great majority of the collections were made within a radius of a few miles from our base camp at Cape Royds. Those who went on the long sledging journeys brought back specimens from more distant points. Mr. Shackleton brought lake vegetation and Joyce brought moss from Hut Point (the *Discovery's* winter quarters, twenty miles south of Cape Royds); Priestley brought rotifers, mosses, lichens, and some marine organisms from the neighbourhood of the Ferrar Glacier and the Stranded Moraines, when he visited the west with Armytage's party; and Brocklehurst on the same journey obtained some lichens at an elevation of about 4000 feet at New Harbour Heights; David brought moss and lichen from near Cape Irizar, the most distant point from our camp at which any biological specimens were collected. So far as known, the species were the same in all the localities.

The promontory of Cape Royds, round which most of the collecting was done, is a hilly tract of triangular form, separated from the main mass of Mount Erebus by a valley in which there is a series of small lakes, and terminating at its southernmost point in a bluff rising vertically from the sea to a considerable height. This culminating point of the Cape was familiarly known as Flagstaff Point, from a pole which we erected there for the purpose of signaling to the ship. In the hollow between it and the hut was Pony Lake, and between the lake and the shore was the rookery of Adelie Penguins.

The triangular area is just about a mile in length and half a mile in greatest breadth. It includes many little sharp rocky peaks, composed of kenyte, with ridges of the same rock diverging from the peaks. The valleys are filled with a gravelly debris resulting from the decomposition of the kenyte, and contain many little lakes or ponds. Considerable stretches of morainic material occur. The highest point of land is no more than 300 feet above the sea. The greater part of the shoreline consists of low cliffs with a few small patches of sandy beach. The most extensive of these beaches, known as Black Sand Beach, is about a mile to the north of the hut.

On the shore there is no vestige of marine life, animal or vegetable, such as is found in the littoral zone of other coasts. The beaches are formed of a coarse, hard, black sand, with boulders of kenyte and other rocks. The presence of an ice foot throughout the greater part of the year, and the grinding of ice along the coast when there is open sea, must destroy any living things which attempt to establish themselves. The zone thus kept devoid of life is of no great depth. Standing on the edge of the ice foot at Black Sandy Beach, when the Sound was open, various living things could be seen at a depth of from one to two fathoms. Starfish were commonest in this situation, but a living *Pecten colbecki* was got in equally shallow water at Back-door Bay.

The larger lakes were given distinguishing names. Pony Lake, close beside the hut, formed the exercising ground for the ponies during the long night. A short distance to the north was Green Lake, named from the colour of its ice. A mile north of the hut and close to the shore was Coast Lake, remarkable for its level smooth ice, which would have served for skating and curling. Close by was Clear Lake, named from the clear transparent ice, through which on our first arrival we could see the vegetation growing on the bottom at a depth of several feet. The largest of all was Blue Lake, which most nearly deserved the name of "lake." It was nearly half a mile in length, and filled about half of the valley separating the promontory of Cape Royds from Mount Erebus. The lake was divided into two portions by a very narrow strait, known as the "Narrows," in which the depth was only about three feet. The northern half was deeper, and was found to have a few feet of water under twenty-one feet of ice. The southern half was frozen to the bottom.

Beyond the valley occupied by Blue Lake rise the lower slopes of Mount Erebus. This is a region of rocky ridges and moraines like those of Cape Royds promontory, extending for several miles up the mountain, on the sides of which the moraines have been traced to a height of 1100 feet. Many small lakes occur up to a height of at least 500 feet. Northward and southward, at the distance of a few miles, the rock and moraine give place to crevassed glacier. Beyond the limits of the small area thus briefly described, snowfield and glacier stretch for many miles, offering no support for any living thing, unless it be some of those lowliest organisms which can exist on the surface of the snow itself.

At Cape Barne, two miles south of Cape Royds, there is a district of hill and valley similar to those of Cape Royds, but the hills are higher and the valleys deeper. In this region there are many interesting lakes. In one place there are two concentric curved gullies, both ends of which open to the shore. These are occupied by lakes, one of which, in the gully nearest the shore, is considerably below sea level.

The bay immediately east of the hut was our customary dredging ground. All the dredging was done within a mile of the Cape, in shallow water, nowhere more than 100 fathoms in depth. Farther out the Sound deepened to 300 fathoms or more, but that region remained forever inaccessible.

Collecting in the Lakes—Soon after landing at Cape Royds, on walking across the lake afterwards called Green Lake, some thin films of vegetation of a dull green colour were seen projecting above the surface of the ice. Shortly afterwards Wild found pieces of a similar plant, but of pink or brown colour, exposed on the surface of Clear Lake. At the margin of

Clear Lake, where the ice was transparent, the same plant was seen at a depth of about a foot, of much brighter colour than that exposed at the surface. Pieces of this were cut out with an ice pick, and taken home and melted, when several microscopic animals were found.

This was the introduction to the most prolific source of freshwater life in the district. The plant was found embedded in the ice of nearly all the lakes, and when the smaller ones melted in summer it was seen that it formed continuous sheets over the whole bed of some of them. Everywhere microscopic life swarmed on this weed. The method of collecting during the winter was very simple. A few pieces of ice containing plants were chipped out and taken to the house to thaw. When quite melted the weed was put into a coarse silk net, which was again put inside a very fine silk net, and the whole immersed in a bucket of water. When the nets were violently shaken in the water the microscopic organisms, animal and vegetable, were washed off the weed, and strained through the coarse net into the fine one, from which they could be easily transferred to a bottle. Thus were obtained multitudes of living things for study.

In summer the collecting was still simpler. The weed could be washed in the lake water without the need for preliminary thawing.

In this way we collected continually from the shallow lakes. There were some deeper lakes, which, as it proved, did not melt in summer. In Clear Lake a hole was dug through the ice. We came on water at a depth of about four feet. Here we dredged on the lake bottom, at a depth of seventeen feet. There was a quantity of vegetation brought up, but it was discoloured and dead, and there was no living thing upon it.

Late in the winter, Priestley sunk a shaft in the southern portion of Blue Lake, for the purpose chiefly of observing the temperatures of the ice. At a depth of nine feet some scraps of weed were got, and when thawed a number of living animals were found on it. At a depth of fifteen feet we came on the bottom of the lake, which was here composed of angular fragments. These were covered by a continuous film of yellow weed, and on this also there were numerous living things.

In autumn an attempt was made to use the tow-net in some of the lakes. The lakes being at this time covered by a thick sheet of ice, the net could not be drawn through the water. A hole was dug and some gallons of water taken up with a bucket and poured through the net. At this time the temperature of the air was about zero Fahrenheit, and the net was soon filled with ice. When thawed out there was no living thing found in it. In summer, when the lakes were melted, and the air temperature was about freezing point, the tow-net could be easily used, and good collections were got.

The vegetation of these lakes is so important a feature in the biology of the district that it merits some attention. Portions of it have been submitted to expert botanists, but no report has yet been received, so it cannot be stated to what group of the vegetable kingdom it belongs. Its appearance and method of growth will be described.

In Green Lake and many other lakes it is in the form of sheets, from a few inches to many yards in extent, sometimes continuously covering the bed of a pond from side to side. It varies in thickness, from one-eighth to half an inch or more, and is of a consistency like sodden paper, so that it was not possible to lift up sheets of any considerable size without breaking them. The upper surface is of a bright orange colour, and is coarsely and irregularly wrinkled. It is composed of few or many layers, like superposed sheets of paper. The lower surface of the sheet is of a dirty green colour, and is composed of a tangled mass of many different algae, green and blue-green. The whole mass was slimy to the touch.

The mode of growth differs in different lakes. The broad sheets above described are the commonest form. In Clear Lake it does not form flat sheets, but is coarsely lobed and undulate, and can be seen through the clear ice growing up from the bottom. In Coast Lake the lobed character is carried further, and little dendroid masses of fine lobes can be seen embedded in the ice near the surface. When one of these is cut out and thawed the plant loses its dendroid character and falls down to form flat sheets. When the ablation of the ice of Coast Lake goes so far as to expose part of the lake bed, it is seen to be covered with a deposit of small flakes of the plant, in colour and appearance not unlike used tea-leaves.

In a lake near Cape Barne the ablation of the ice exposed small masses of the weed in which the successive superposed layers made up a thickness of six inches. The layers were very thin and the colour a fine pink.

Large fragments dredged from the bottom of Clear Lake and dried on blotting paper had a glossy surface and ash-grey colour like some of the lichens of the genus *Peltigera.* In most other samples the surface was dull when dried.

Under the microscope the brown weed is seen to be composed of a felt of very fine fibres, crossing one another irregularly in all directions. Usually no definite structure can be detected in the fibres, but Mr. Scourfield noticed some in which all obscure division into cells could be seen.

In some ponds we found another weed of very similar colour and appearance, but in very small quantity. This was definitely composed of moniliform rows of cells of some blue-green alga, very probably of the large olive-green lamina which we got in some streams and ponds in summer. The similarity of the two suggests that the commoner brown weed has in like manner originated in the blue-green filamentous algae generally associated with it. Plausible though the suggestion is, it requires expert investigation before we can decide upon it. The filaments seem inadequate to the production of such masses, being to a large extent in the form of longer or shorter rods.

The shallow lakes were very easily trenched, a few hours work sufficing to reach the bottom. The trenching of the deeper lakes, Clear Lake and Blue Lake, was a more laborious undertaking. The trench is marked out by a draft cut with the icepick, enclosing an area of about six feet long by three feet wide. The whole surface is then picked over to a depth of a few inches, and the chips are shoveled out. A very little chipping seems to make a great depth of chips, and frequent clearing out is necessary, or it becomes impossible to get at the solid ice on account of the loose stuff. Thus alternately picking and shoveling, a few inches at a time, the trench gradually deepens to five or six feet. Up to this time everything is easy. The debris is easily shoveled out with force enough to carry it well clear of the hole. With every increase in depth this becomes more difficult, till at last the chips come showering back on the worker below. It then becomes necessary to have another man to pull up the chips in a bucket, or if one man attempts to do everything, he must interrupt his work below every little while, climb out of the hole and pull up a load of chips. Beyond a depth of five feet it is necessary to construct a stairway. It was usual to have a few broad steps near the top, and lower down to cut niches for the feet alternately at one side and the other.

Blizzards are apt to interfere with the work, filling the trenches with snow. After some experience we learnt wisdom, and covered the trenches with sacking, which was secured with ice picks whenever we had to go away for a time. The most laborious part of the picking was the squaring of the corners. Some of the latest shafts were made round, and of just the diameter at which a human arm can conveniently wield an ice pick. In these the minimum of material had to be removed.

At depths of fifteen feet and more progress becomes very slow. It is necessary to have a ladder to get down. If there are two men the ladder can be drawn up out of the way after one man has gone down. If there is only one the ladder is very much in the way. The man below is in some danger when the bucket is being hauled up, as the breaking of the line would let it fall upon him.

Collecting in the Sea—Mr. Hodgson has given an account of his collecting at Hut Point, and he mentions some of the difficulties which attend the collector in polar regions. Though our location at Cape Royds was only twenty miles north of the *Discovery* winter quarters the local conditions differ very considerably. The temperature appears to be usually ten degrees or more (Fahrenheit) higher than at Hut Point. Being close to the spot where McMurdo Sound opens into the Ross Sea we had open water close by throughout the year. In fact, even in winter the edge of the permanent ice was never more than a mile from our camp. Beyond the fast ice the Sound frequently filled with pack ice stretching as far as eye could see. Sometimes, in a period of calm, the pack was cemented into a solid field by new ice, but this was broken up by every storm, and it was therefore always unsafe to go out on it. The marginal zone of even the permanent winter ice was liable to be broken off in a severe storm.

From this cause marine dredging was confined within a very small area. We could work steadily during winter only in a little bay between Cape Royds and Cape Barne, where the ice formed early and stayed late. Here, as early as the beginning of May, the ice was strong enough to allow us to cut holes and put down traps. The traps were baited, and brought up Amphipods and Molluscs. Some pieces of a dendroid sponge were entangled in the net and from these we got a number of minute molluscs and other animals. A storm early in May broke up the ice and our dredging apparatus went out with it.

On May 11 ice again formed in the Bay, and proved to be permanent, remaining fast till February of the following year. As soon as the ice was strong enough dredging was begun. The first dredging-line was put down while the Bay was open, from the edge of a small area of fast ice which remained near the head of the Bay. Afterwards we had to take advantage of tide-cracks in order to get lines put down. It rarely happened that we found the cracks open and could get the line down without labour. Usually they were filled with new ice to a depth of 6 inches or a foot, and it was by hard labour with ice pick and crowbar that we got a sufficient length open to serve for dredging. Foot by foot as the crack was cleared the rope was forced through, for with the low temperatures new ice quickly forms in the part we have opened. When the rope was through for a sufficient length it was secured at the two ends to bamboo poles, enough slack being paid out to allow the ends to hang nearly vertically, thus avoiding the danger of the rope being frozen in. It was then necessary to dig holes in the ice at the two ends of the rope, through which the dredge could be lowered and drawn up.

The holes were from fifty to one hundred yards apart, but the effective dredging distance was less than that on the ice, as the dredge would always leave the bottom some considerable time before arriving directly under

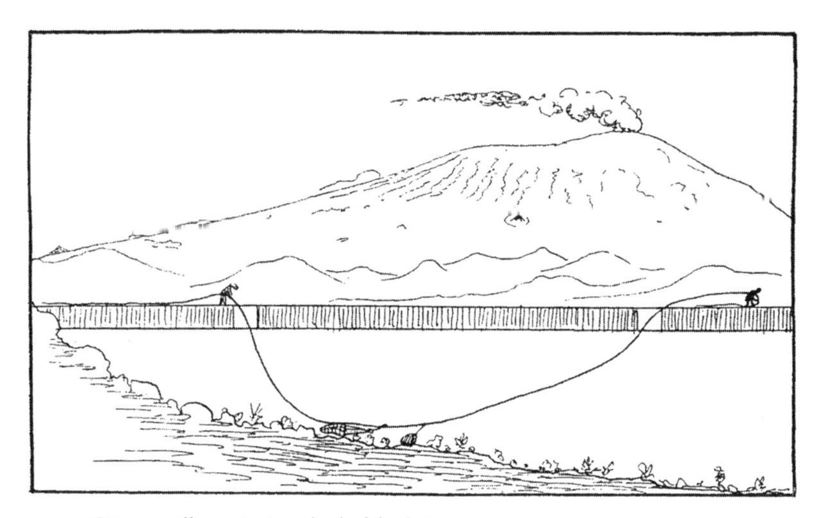

FIG. 1. Diagram illustrating method of dredging.

the opening in the ice. Each time that we wished to dredge the holes had to be reopened with pick and crowbar. They would be frozen over with ice from a few inches to a foot or more in thickness, according to the temperature and the length of time they had been left undisturbed. In cold weather it was not well to leave them for more than a day, and Priestley sometimes opened them, although there was no intention of dredging, in order to lessen the labour next time. The Weddell seals were of assistance in keeping the holes open. They found them useful as breathing-holes and visited them frequently, sometimes arriving in an apparently exhausted condition, to judge from their labored breathing.

In order to avoid dredging too frequently over the same ground it was necessary to cut trenches in the ice alongside the ends of the rope and at right angles to the line joining the two ends. In these trenches the rope could be shifted a yard or so at each time of dredging and so the dredge covered entirely fresh ground. Sometimes the rope was left too long and got frozen in too solidly to be cut out in the usual way. A new hole was then cut close by the old one, and the line was fished up by means of a hook on the end of a long bamboo pole.

The dredge was fixed to the middle of the line so that it could be used in either direction. It was found that it often caught nothing when traveling downhill, so it was usual to haul it downhill and then back again uphill before bringing it to the surface.

When the dredge reached the surface both men went to the opening, the one who had been hauling keeping the line taut to prevent it sinking again. The contents of the dredge, consisting of a thick black mud in which only the larger objects such as sponges, shells, blocks of kenyte, and sea anemones could be distinguished, were emptied into a bucket. This was made from a 4-gallon kerosene tin provided with two wire handles. The bucket was filled nearly to the top with seawater in the hope of enabling the animals to remain alive for some time. At first it was conveyed home, a distance of about a quarter of a mile, by slinging it on a bamboo pole carried on the shoulders of two men. This caused too much splashing, and the bucket was thenceforward carried by the handles.

Generally all the water in the bucket was frozen into a kind of soft sludge before we reached home. It was placed behind the stove to thaw, but so cold was it on the floor of the hut that it often took a day, or even two days, before it was ready for examination.

Any large objects visible on the top of the mud were first taken out. To get out the smaller organisms the thick coherent mud was taken by a handful at a time and put in a small silk net having a mesh of about one-sixteenth of an inch. This was shaken in clean seawater till the fine mud was all washed away. What remained in the net was emptied into plates and picked over. The larger pieces of kenyte and shells, sea urchins, etc., were first separated. Then came the task (very trying in the dim gas-light, which alone penetrated to the biological lab, or the equally dim light of a hurricane lamp) of picking out the smaller things, minute crustacea, shells, etc., requiring the use of a lens to detect them. All these various objects were sorted according to their size and kinds, and stored in bottles.

When we had enough jars filled to occupy one of the compartment boxes provided for the purpose, they had to be removed, as there wasn't room in the hut for them. The first was put under the floor of the hut, as likely to be warmer than the outside air, and to escape filling with snow during blizzards. The air lock under the house was so very difficult of access that the boxes were afterwards put outside, to take their chance of cold and snow. A few jars were broken by the cold, but there was no help for it. The formalin, which was used for certain kinds of animals, suffered a change from the low temperature, becoming milky, and did not again regain its clearness.

When not in use the dredge was left at the bottom of the sea. This kept the rope soft, and the dredge was ready for use whenever the holes were opened. The one-inch lines used lasted throughout the season. This may be partly attributable to leaving them in the sea and never allowing them to freeze solid. The part never immersed was air-dried and flexible.

The Bay, which was the only place where dredging was possible during almost the entire season, was very shallow, the depth varying from seven to eighteen fathoms. The sea bottom was everywhere covered with a deep layer of very fine black mud, in which there were many pebbles of kenyte. While this mud was favourable to certain forms of life it was unfavourable to others, and thus though life was abundant it was restricted to a comparatively small number of species.

Shells of the large mud-loving Mollusc (*Anatine*) were very plentiful (though we rarely found the living animal), as well as of *Pecten colbecki*. The large predaceous Gastropod (*Neobuccinum*) crowded to any bait put down. Dendroid sponges and a large kind of yellow sea anemone adhered to the shells of *Anatina* or to the pebbles of kenyte, and rarely large turnip-like Tunicates came up. Ugly little fishes with enormous heads (*Notothenia*) were grubbing closely among the other organisms. In the mud were numerous worms of many kinds, echini, and multitudes of minute crustacea, etc.

This shallow muddy dredging-ground was used constantly throughout the winter, and at intervals afterwards till February 1909. It was not till the beginning of July that there was an opportunity to dredge over fresh ground. At this time a crack opened which stretched from the cliff of Cape Royds away towards Cape Barne. This crack appeared to be caused by the contraction of the main icefield in McMurdo Sound. At any rate it only opened in cold spells and closed in warmer weather. When the cracks were open the dredge could be put down and dragged a long way without any need to cut through the ice.

Near the Cape this new ground was quite as shallow as the Bay but it was quite free from mud, and the collections differed a good deal in their composition. There were many loose stones near the shore, but as we extended our operations farther and farther out we reached greater depths and there were no more stones. The sea bottom here appears to be covered by a continuous carpet of living things.

The sponges were much more numerous and the siliceous kinds were first obtained. There were Sea-spiders (Pycnogonids), Lace-corals (Polyzoa), Holothurians, File-shells (*Lima*), Alcyonarian Corals, Starfishes and Brittle stars, pretty milky white Nudibranchs (*Tritoniella*), and many other things.

The greatest depth at which we dredged in this crack was about eighty fathoms. The dredge was not left down on the bottom when not in use as we did in the Bay, because the crack was apt to close at any time, and when it did so the one side of the floe was often caused to override the other, which would have snapped the line and lost the dredge.

On July 6, a crack opened from the Penguin Rookery westward out into the Sound, in which we were able to dredge once at 100 fathoms. Nothing strikingly different was obtained. In one haul pretty near the shore, the dredge was filled with the common red starfish, and there was almost nothing else.

Murray, J. 1910. "On collecting at Cape Royds." British Antarctic Expedition 1907–1909, *1:1–15.*

A Submarine Gully in Wembury Bay, South Devon (1934)

John A. Kitching, T. T. Macan, and H. Cary Gilson

INTRODUCTION

Both the intertidal region of the shore, and the sea-bottom at depths of anything over three or four fathoms, have been comparatively well explored biologically; but that part of the sea coast which lies just below the level of low water spring tides is—around the British Isles—almost unknown. It is out of range of ordinary shore collecting, and yet much of it is also unsuitable for detailed investigation from a boat. This is especially true of steep, rocky shores, where a large part of the sublittoral region consists of ledges and vertical or even overhanging rock faces, which would be quite inaccessible even if work from a small boat were attempted. In Sweden Gislén (1930), who employed a professional diver, has made an extensive general survey of the plant and animal associations characteristic of a rocky bottom in the Gullmar fiord. But for the British Isles previous knowledge relates only to the distribution of the larger and more conspicuous algae (Cotton, 1912); although the flora of the German warships, which were sunk in Scapa Flow, has been investigated more closely by Miss Lyle (1929), also with the help of a professional diver.

In the summer of 1931, at the suggestion of Mr. G. A. Steven, of the Plymouth Laboratory of the Marine Biological Association, the writers decided to test the possibility of carrying out a detailed investigation of this hitherto unexplored region by means of a diving helmet. Such a helmet has been used by many workers in the tropics, but has not previously been tried in temperate waters, in view of certain supposed difficulties. Chief of these were: (i) water too cold for an unprotected diver; (ii) poor visibility; (iii) excessive interference by rough weather.

These difficulties were found to be either not effective, or else easily surmountable, and are discussed in detail below.

(i) *Cold.* In July and August near Plymouth we found that we could each stay down for about a quarter of an hour at a stretch fairly comfortably, and at most for twenty minutes. We compensated ourselves for this shortness of time below water by working as a team generally of four people, of whom each went down in his turn; so that an hour's work could be done below water on any one day, and in favourable circumstances more. We found that the use of grease was of little help in combating the cold. Frequent diving led to a greatly increased appetite for sugar and treacle.

(ii) *Visibility.* There has always been plenty of light at the depths at which we have worked (down to six fathoms on the west coast of Scotland). After rough weather the water was often turbid with silt in suspension, but even so with a sandy bottom it was always possible to see clearly objects not more than two feet away.

(iii) *Rough weather.* On shores open to the English Channel the helmet could only be used with safety when the sea was moderately calm. A heavy "ground" swell makes diving impossible, even though the water is otherwise smooth, as it breaks with violence on the rocks. For this reason on an Atlantic coast diving would be difficult, but might be done from a boat. We were able to work under water on about half of the number of days during which operations were in progress, but in settled summer weather it should be possible to dive every day.

In this paper are presented the results of an ecological survey of a small area on the coast near Plymouth, extending from Low Water of Ordinary Spring Tides down to a depth of about 10 feet below this level. The survey was carried out between July 14th and August 14th, 1931, and between July 1st and August 3rd, 1932.

APPARATUS AND METHODS

As the methods employed were modified to suit conditions in our seas, a moderately full account of them is given below.

THE DIVING HELMET AND PUMPS

The helmet which we used is similar to that described and figured by Beebe (1926), with the difference that, on the advice of Professor C. M. Yonge, in order to avoid "seeing double," we had only one window. Such a helmet fits loosely over the diver's shoulders, and the pressure of air inside it, which is maintained from above by means of pumps, keeps the water out. A helmet of this type is comparatively light and easy to handle, and below water it restricts the movements of the diver very little. It is therefore suitable for use when working in crevices and on steep rock faces. Further, its use requires no previous experience, and the diver, who is in no way attached to it, can easily leave it and swim to the surface if anything goes wrong.

A steady supply of air was provided for the diver from two foot motortyre pumps, which were connected through Schræder valves to a T-piece; this last was connected through about 150 feet of garden hose with the helmet. One pump alone could in emergency supply sufficient air for the diver.

THE TELEPHONE

A telephone system was installed, which enabled conversation to go on between the diver and those on shore. The system for speaking from the shore to the diver consisted of an ordinary microphone and earphone circuit. But since carbon granule microphones are liable to give trouble if they get damp, a different system was used for speaking in the reverse direction. The diver spoke into a high-resistance earphone, which was connected at the land end through a transformer to a one-valve amplifier. The output from this was fed into a pair of wireless headphones. In order to keep out the seawater, the earpiece and acting microphone were each nearly filled with paraffin wax and covered with a thin rubber toy balloon.

GUIDE ROPES AND SAFETY LINES

(i) In order to recover the helmet should the diver find it necessary to leave it and come to the surface, and in order to take the weight of the helmet off the diver's shoulders while he was entering or leaving the water, a rope was attached to the top of the helmet itself.

(ii) A rope, securely belayed to the rock at its shore end, and attached to a fairly heavy stone at its free end, was thrown out in the direction in which the diver was intending to go. On this the diver lowered himself and hauled himself up. It also assisted him in finding his way to the required spot, and enabled him to work against a rock wall, suspended at any depth.

(iii) A third rope was used for lowering collecting bags to the diver.

METHODS OF COLLECTING

It seemed essential that data should be quantitative. Accordingly, (*a*) complete collections were made of all animals and plants from small selected areas of rock surface; and (*b*) observations were made of the general distribution of those types of biological community of which the areas were considered typical. In 1931 (July 14th–August 14th) the selected areas were scraped clean with a strong knife into jars. This was a very slow process, and great care was needed to avoid loss of material. Therefore in 1932 (July) we used large bags with folding metal frames in the mouths. These frames were 1 foot × 1 foot or 2 feet × 1 foot, and when closed could be clipped together so as to keep the bags shut. Areas were scraped into these with a paint-scraper. With each of these methods there were some fast-moving animals (e.g. prawns, small fish), which must have escaped; but as this work was directed mainly towards sedentary or slow-moving forms, which make up by far the greater part of the community, this does not matter much. Encrusting Coralline Algæ could seldom be scraped off the rock, and boring animals would not have been collected by this method, so that the collections must be considered incomplete in respect of these types. Although a standard size and shape of area would have been advantageous, we had in practice to take whatever areas were allowed by the configuration of the rock surface.

DESCRIPTION OF ENVIRONMENT

GEOGRAPHICAL ACCOUNT

The locality chosen for a detailed survey by means of diving was in Wembury Bay, South Devon; and the work was done in the gully on the S.E. side of the reef next to and S.E. of "Tomb Rock" (Ordnance Survey, 1913). This gully runs approximately N.E. and S.W., and is formed in the Dartmouth Slates (Ussher, 1912). It has on its N.W. side a nearly smooth rock wall, which is at about 15–20° from the vertical, facing upwards. On the S.E. side the rook surface is irregular and in most places overhanging. This gully is one of a series of similar parallel gullies which have been eroded in this particular way owing to the high angle of dip of the strata. The middle part of the gully has a sandy bottom at a depth of 10 feet below the level of low water of ordinary spring tides. At the two ends of the gully the bottom is strewn with large boulders. At the S.W. end, beyond the boulders, the gully opens out to join the even sandy bottom, which is characteristic of that part of Wembury Bay.

ECOLOGICAL FACTORS

The chief environmental influences which are likely to affect the flora and fauna of this gully are as follows:

(i) *Slope of Surface.* This is believed to be of great biological importance, and has obvious connexions with illumination and with the settling of silt.

(ii) *Light.* Measurements of the light intensity at various positions in the gully were made by Dr. Atkins by means of two cuprous oxide photoelectric cells. A full account of these cells, including the methods of use and the corrections to be applied, is given by Atkins and Poole (1933). While one cell remained exposed to the full light in air, the other was placed or held in any required position below water by the diver. The diver took precautions against shading his cell himself, and when it was in position he notified the shore party by telephone. Thus the illumination at any position was compared with that in air. While such figures only represent the light intensity at one time of day and state of weather and tide, they may nevertheless be used as a basis for comparison. The chief feature is the very low percentage of light which reaches the bottom of the gully. This is due partly to shading by the gully walls, and partly to the opacity of the water, especially near the bottom. The lowest figure obtained—less than 1% of the intensity in air—was from a position in the *Laminaria* forest at the base of the plants, below the "canopy" of fronds. In view of the shading by the gully walls, the absorption coefficient of the water remains uncertain, but in any case it must vary greatly, depending much on the roughness of the sea, and increasing downwards. The upper layers at high tide are probably comparatively clear. The biological significance of illumination in this gully is discussed later.

(iii) *Silt.* Both the presence of silt in the water, and the settling of silt, may be of importance. The amount of silt in the water was very variable, but was greater near the bottom. Sometimes the bottom foot of water was thick with seaweed remains. Silt carried along in the water by waves probably has a scouring action, especially near the bottom. The amount of silt which settles will depend partly on the slope of the rock surface, and partly on changes in the turbulence of the water. We hope in a future investigation to be able to obtain some comparative measurements of the turbulence in different parts of the gully.

(iv) *Wave action.* This part of Wembury Bay is exposed to weather from S.W. by W. to S., and heavy storms occur in winter. It is possible that wave action is more intense on the N.W. wall than elsewhere, owing to the fact that waves rush up the intertidal slope above it, and then fall back with great force. It was observed during diving that the effects of a short choppy sea could scarcely be felt more than a few feet below the surface. But this part of the coast is much more subject to a comparatively long sea or swell, the influence of which extends to the bottom. Some indication is given of the degree of exposure to wave action by the fact that in the intertidal region *Balanus balanoides*, *Chthamalus stellatus* and *Fucus vesiculosus evesiculosus* are present—the last in small patches only, and poorly grown; while Pelvetia, *Fucus platycarpus* and Ascophyllum are absent. This combination is indicative of moderate exposure. It is hoped that later it may be possible, at any rate on some parts of the coast, to use certain common intertidal plants and animals as indicators of the degree of exposure to waves of a locality.

(v) *Exposure to the air.* This factor is not operative except in the intertidal region, which was studied in its lowest part only, merely for comparison with the contiguous infra-tidal region. Exposure to air may subject organisms both to desiccation and to rise in temperature.

(vi) *Variations in oxygen-content of the water, and other chemical factors.* These were not measured, as it is unlikely that areas in a region of so much water movement differ greatly in this respect. It is, however, possible that the fronds of the larger algæ cause sufficient stagnation around their bases to produce an appreciable reduction in oxygen tension. It remains to be seen to what extent the growth of various animals and plants upon the fronds and stipes of some of the larger algæ is to be ascribed to the presence of an environment more favourable than the rock surface itself (e.g. in oxygen tension, temperature, light, silt, turbulence), and to what extent it represents merely an overflow from the normal rock-surface community, induced by the struggle to gain a foothold.

(vii) *Temperature.* This rises near the surface on a calm warm day, but owing to unfavourable weather no adequate data were obtained. A discussion of this matter must therefore be withheld, although it is unlikely that temperature differences have any great significance.

DISTRIBUTION AND ANALYSIS OF TYPES OF COMMUNITY

There were two main associations in the gully, the "Laminaria forest association" (including the "Laminaria association" and the "Himanthalia association" of Cotton (1912)) and the "Distomus-Halichondria association." In this description only the most important species are mentioned. The nomenclature followed is for algæ that was used by Newton (1931), and for animals that given in the Plymouth Marina Fauna (Marine Biological Association, 1931), unless it is stated otherwise.

THE LAMINARIA FOREST ASSOCIATION

This covered all the N.W. wall, and the boulders at the bottom of the gully. It occurred on all upward-facing rock surfaces in this and neighbouring gullies. When fully developed it consisted of:

(a) A "canopy" layer formed by the fronds of one or more species of large brown algae (generally a Laminaria)
(b) An undergrowth of Corallina and other algae
(c) Species associated with the canopy
(d) Species inhabiting the undergrowth and the holdfasts of larger brown algae

THE CANOPY-FORMING ALGÆ

Himanthalia lorea was dominant in a narrow belt about a foot wide around the level of low water of ordinary spring tides. It occurred along the N.W. wall, on the E. side of "Gully" and "Far" rocks, and on the top of "Flat" rock (here above its more usual level). Corallina was present as a thick undergrowth. Himanthalia "buttons" were seen at lower levels, but never the reproductive parts. *Laminaria cloustoni* was also present in this zone. *Saccorhiza bulbosa* was dominant in 1931 in a belt just around the Himanthalia zone on the N.W. wall, and also on boulders in the gully. In 1932 it was almost completely absent from the gully except at the mouth, its place as dominant being taken by *Laminaria cloustoni*. *Laminaria digitata* was confined to the top of Far Rock (at L.W.E.S.T. level) and to ledges at L.W.E.S.T. level on the E. side of Far Rock and the seaward end of the N.W. wall. *Laminaria saccharina* was dominant in all parts of the Laminaria forest association except for the above-mentioned places, and even in these (except for places occupied by *Laminaria digitata*) it was present. On the N.W. wall it was graded in size, being larger (up to 340 grams) near the top. Only small specimens (up to 60 grams) were found near the bottom. On the boulders however it reached up to 280 grams or more; and it was very large (up to 1800 grams) and covered with other growths (*Distomus variolosus* on the holdfast, *Rhodymenia palmata*, other red algæ and *Sertularia operculata* on the stipe) on the top of the narrow ridge N.W. of Far Rock. *Desmarestia aculeata* and *Desmarestia ligulata* occurred fairly plentifully in 1931, but were scarce in 1932. The former was most common on the N.W. wall at about 5 feet from the bottom, while the latter was commoner above this level.

THE UNDERGROWTH

Corallina spp. formed a dense growth all over the N.W. wall except at the bottom. (Upper limit, 1 foot above low water of ordinary spring tides and higher in cracks.) In places it was to some extent embedded in *Halichondria panicea*. From an examination of the samples collected, it appears that *C. squamata* predominated in the Himanthalia zone, but that lower down it was replaced by *C. officinalis*. *Chondrus crispus*, *Chylocladia ovata*, and *Laurencia pinnatifida* were common in the zone of reproductive Himanthalia, and extended down to 1 or 2 feet below the bottom of this zone. (Upper limit, 1 foot above L.W.O.S.T.) *Gigartina stellata* was locally plentiful in this zone. *Rhodymenia palmata* was widely distributed (growing on the rock as well as on stipes of *Laminaria cloustoni*). *Cladostephus verticillatus* was locally abundant on the boulders.

THE POPULATION OF THE CANOPY LAYER

Rhodymenia palmata and *Sertularia operculata* grew on the stipes of the larger plants of *Laminaria cloustoni*. *Obelia lucifera* and *Gibbula cineraria* occurred on the fronds of some plants of *Laminaria cloustoni*. *Patina pellucida* (small individuals only) occurred feeding on fronds and stipes of Laminaria spp., Saccorhiza, and Himanthalia. Large individuals were occasionally found inside cavities, which they had no doubt themselves eaten out, in the holdfasts of large plants of *Laminaria cloustoni*. *Membranipora membranacea* was nearly always present on fronds of *Laminaria digitata*, but only occasionally on those of *L. cloustoni*.

THE POPULATION OF THE UNDERGROWTH

Sessile forms (dealt with approximately in order from the rock outwards) included:

(*a*) An encrusting corralline alga forming a continuous coating over the tops of many of the boulders, and over the bottom of the N.W.

wall. It may have extended upwards, but if so it was concealed by other growths.

(*b*) *Pomatoceros triqueter*, widely distributed, and reaching as far as the upper limit of Corallina; and *Umbonula verrucosa*, occasional. Both of these were fixed to the rock.

(*c*) *Balanus crenatus*, at the bottom of both walls of the gully—fixed to the rock.

(*d*) *Hiatella arctica*, at base of undergrowth and probably burrowing in the rock, but this was not observed.

(*e*) *Sabellaria spinulosa*, occasional on the N.W. wall, but often filling the holdfasts of *Laminaria cloustoni* growing on the boulders. On low horizontal ledges at the foot of the S.E. wall—ledges on which there were a few scattered and small plants of *Laminaria cloustoni*—Sabellaria tubes formed a continuous coating to the rock.

(*f*) *Mytilus edulis* (very small) and *Heteranomia squamula* on and among the undergrowth, but chiefly fixed to Laminaria holdfasts and Corallina.

(*g*) *Schizoporella hyalina*, chiefly in small patches on holdfasts of *Laminaria cloustoni.*

(*h*) *Membranipora pilosa*, on holdfasts of *Laminaria cloustoni* and on the fronds of red algæ.

(*i*) The vegetative "buttons" of Himanthalia—which may be treated as part of the undergrowth—were always at least partly covered with Polyzoa (*Schizoporella hyalina* and *Membranipora pilosa*) on their undersides. On the other hand polyzoa and algæ were never found growing on the upper side of the "button," or on the fertile frond, of a healthy Himanthalia plant; although on one occasion a dead "button" was found to be covered all over with *Membranipora pilosa.*

(*j*) *Bunodactis verrucosa* and *Spirorbis borealis* on Corallina.

(*k*) *Grantia compressa* and *Sycon coronatum* attached to the undergrowth at random.

And motile forms:

(*a*) *Nereis pelagica* and *Platynereis dumerili*, the largest specimens being mostly inside holdfasts; and many other Polychæta errantia.

(*b*) *Jassa* spp.

(*c*) *Elasmopus rapax.*

(*d*) *Eurystheus maculatus* (this occurred in considerable numbers inside the holdfasts of *Saccorhiza bulbosa*); and many other Amphipoda.

(*e*) *Amphipholis squamata.*

THE DISTOMUS-HALICHONDRIA ASSOCIATION

This covered those parts of the S.E. wall which were overhanging, and all overhanging walls in a neighbouring gully. It ranged upwards above the level of L.W.O.S.T. in places. It consisted of:

(*a*) A carpet-like growth of *Distomus variolosus* or of various sponges.
(*b*) Certain species (chiefly red algæ) projecting through (*a*) but with their bases embedded in it.
(*c*) Species living on or among the "carpet."
(*d*) Species associated with (*b*).

CARPET-FORMING SPECIES

Distomus variolosus was dominant on vertical or slightly overhanging rock faces. It formed in many places a continuous sheet. (Upper limit, about 2½ feet above L.W.O.S.T.). *Stolonica somalis* was locally dominant at the seaward end of "Far Rock" in the region of transition from the Distomus-Halichondria association to the Laminaria forest association. Sponges became more important the more the rock face overhung. On surfaces at an inclination of about 60° from the vertical, and facing downwards, sponges (especially *Halichondria panicea*, but also other species) were dominant, and Distomus was almost absent except where the inclined part adjoined a vertical Distomus-covered part.

PROJECTING SPECIES

Red algæ such as *Rhodymenia palmata*, *Cryptopleura ramosum*, and *Myriogramme bonnemaisoni* were widespread below L.W.E.S.T. level on vertical or slightly overhanging surfaces, but they did not grow on the more steeply overhanging rock faces. *Gigartina stellata*, *Laurencia pinnatifida*, and *Lomentaria articulata* were plentiful at and above the L.W.E.S.T. level, and in places grew thickly. (Upper limit 2½–3 feet above L.W.O.S.T.). Hydroids, especially *Aglaophenia pluma* and *Sertularia operculata*, were common.

THE POPULATION OF THE SPONGE-DISTOMUS "CARPET."

Sessile species included:

(a) various Cirripedia, *Umbonula verrucosa*, *Pomatoceros triqueter*— all growing on the rock surface, and generally much overgrown.

Pomatoceros triqueter was very common at about 1 to 2 feet above L.W.O.S.T., under overhanging rocks, and near the upper limit of the carpet-forming species. The upper limit was 3 feet above L.W.O.S.T.

(*b*) *Hiatella arctica*, *Sabella pavonina*, *Dasychone bombyx*, and other Polychæta sedentaria embedded in the sponges or Distomus.

(*c*) *Corynactis viridis*, commonest under shelves in the sponge-covered areas.

(*d*) *Mytilus edulis.*

(*e*) *Heteranomia squamula.*

And motile species:

(*a*) Many Polychæta errantia, including *Nereis pelagica* and *Platynereis dumerili.*

(*b*) Many Amphipoda, including Jassa spp., Elasmopus rapax, Apherusa jurinei, and various Caprellidæ.

(*c*) Amphipholis squamata.

THE POPULATION OF THE PROJECTING SPECIES INCLUDED:

Membranipora pilosa, growing in great profusion on red algæ, especially on *Gigartina stellata. Schizoporella kyalina*, also very plentiful on red algæ, especially on *Rhodymenia palmata. Parajassa pelagica*, nesting in the branches of *Sertularia operculata.* Caprellidæ.

UNDERGROWTH LAYERS

More detailed examination of the undergrowth of the two sides of the gully leads to the subdivision of that of the Laminaria forest association into: (i) Outer undergrowth layer, consisting of larger undergrowth algæ such as *Rhodymenia palmata*, *Chondrus crispus*, etc., and generally rather loosely packed or sparse. (ii) A middle layer of Corallina and holdfasts of Laminaria, which is dense. (iii) A basal layer of species, which are completely adherent to the rock surface, such as *Balanus crenatus*, *Pomatoceros triqueter*, and *Umbonula verrucosa.*

The "carpet" layer of the Distomus-Halichondria association corresponds with the middle layer of the Laminaria forest undergrowth, and it is this layer which harbours most of the animal species. The projecting species of the Distomus-Halichondria association correspond with the outer layer of the Laminaria forest undergrowth; this layer is generally populated chiefly by Polyzoa, and is too loosely packed to give shelter to many motile animals.

DISCUSSION

The differences between the various associations and sub-associations are so sharply marked, and show such close correlation with the differences in the external conditions of the habitat, as to suggest that the nature of these associations is rigorously controlled by environmental factors. The principal difference between the regions colonized by the two main associations is the slope of the rock surface, which will react on the fauna and flora in several ways, viz.:

ABILITY OF SPORES OR LARVÆ TO SETTLE OR DEVELOP

It is possible that some spores or larvæ may be unable to settle on overhanging rock surfaces, or that such positions may be unsuitable for their further development into adult organisms. The possibility that some such unknown factor excludes the larger brown algæ from overhanging surfaces must not be ignored. The fact that the areas scraped clear in 1931 were almost all unrecognizable in 1932 shows that no such difficulty operates against the dominant members of the Distomus-Halichondria association, and that any given area continues to support a similar community even after such drastic interference. This was perhaps to be expected, since depopulation by storms must be a fairly frequent event.

SETTLING OF SILT

It is noticeable that the most important of the species confined to the Distomus-Halichondria association are such as might be damaged by excessive deposition of silt. Unless they are excluded from the Laminaria forest association by competition or by some adverse effect of light, silt may be the responsible factor. It is possibly significant that whereas the stalked sponges were found plentifully on both walls of the gully, the encrusting species were confined to, or only flourished on the vertical or overhanging parts of the S.E. wall. The fact that hydroids were not found on the N.W. wall except on Laminaria-fronds suggests that they may require to be held further off the rock face into a position where the water is more turbulent and silt-deposition less.

Laminaria cloustoni occurred, though poorly and sparsely, on the Sabellaria ledge; but on the upper part of the S.E. wall—a position which is better illuminated—it did not grow at all. While its poorness of growth on the former may possibly have been due to a deficiency of light, its absence from the latter cannot be attributed to this cause. Unless *Laminaria cloustoni* is unable to gain a foothold among or compete with the members of the Distomus-Halichondria association, its absence from the upper part of the S.E. wall must in some way be connected with the slope of the surface. The grading in size of Laminaria plants on the N.W. wall may be due to the greater illumination near the surface; they probably often get torn off by storms, and the Laminaria forest would be regenerated more quickly where the illumination was greater. But there may also be some adverse factor such as silt in suspension affecting those near the bottom. The absence of *Saccorhiza bulbosa* in 1932 may probably be attributed to its being torn off by storms. The Sabellaria ledge at the bottom of the S.E. wall may be regarded as potential Laminaria forest which has failed to develop properly, possibly owing to inadequate illumination.

Of the large brown algæ *Saccorhiza bulbosa* with its capacious holdfast supported by far the largest fauna, as regards both number of species and number of individuals; these included not only many small forms such as amphipods and polychætes, but also small fish and young echinoderms. *Laminaria cloustoni* had a moderate fauna; but Desmarestia spp. and Himanthalia, which have small holdfasts, were in this respect unimportant.

The same types of community, which are found below the low water of equinoctial spring tides, reach a little way above this level, especially under ledges where there is protection from the adverse effects of being left high and dry. Colman (1933) has pointed out that at any time the belt of sea-shore which is within reach of the splashing of waves may be considered as being, for biological purposes, submerged. Thus infra-tidal conditions extend upwards above the level of L.W.E.S.T. by the width of the "splash zone." Many species found at this level may be regarded as really belonging to the infra-tidal region, and as having ranged upwards as far as conditions have allowed. Certain algæ, on the other hand (notably Himanthalia and Laurencia), belong properly to the lowest part of the inter-tidal region, and do not flourish below it. It is possible that illumination is the limiting factor. Strong evidence has been brought forward by Gail (1918) that insufficient illumination limits the downward distribution of Fucus in Puget Sound.

The part played by the basal undergrowth layer (encrusting Coralline algæ, barnacles, *Pomatoceros triqueter*, and *Umbonula verrucosa*) is especially interesting. The occurrence of barnacles both in the intertidal region and at the bottom of the N.W. and S.E. walls suggests that they are only able to live where they are free from competition with various other species. They are able to endure conditions which other species cannot endure, but in open competition become overgrown and smothered. The same is probably true to a less extent of other species of the basal undergrowth layer. Pomatoceros occurred most plentifully at the bottom of the N.W. wall and on vertical or overhanging faces in the lowest part of the inter-tidal region (up to 3 feet above L.W.O.S.T.), near the upper limit of the distribution of the main carpet-forming species. Unlike the barnacles, it is prevented, no doubt by the effects of exposure to air, from colonizing the corresponding inter-tidal levels on the upward facing N.W. wall. From the occurrence of barnacle remains under *Distomus variolosus* and other carpet-forming species, and from an examination of the fauna of Area J(2),—an area which was scraped on the same site as the Area J(l) of the previous year—it seems that *Balanus crenatus* is (at any rate in some cases) an early colonizer of any available space, but that it later succumbs owing to competition. Our inability in 1932 to find any trace of the previous year's scrapings shows that decolonization is very rapid.

REFERENCES

Atkins, W. R. G., and Poole, H. H. 1933. The Use. of Cuprous Oxide and other Rectifier Photo Cells in Submarine Photometry. Journ. Biol. Assoc., N.S., Vol. XIX, pp. 67–72.

Barnard, K. H. 1925. A Revision of the Family Anthuridre. J. Linn. Soc. Zool., Vol. XXXVI, p. 140.

Beebe, W. 1926. The Arcturus Adventure. New York.

Børgeson, F. 1908. The Algæ-vegetation of the Faeroëse coasts. Botany of the Faeroës, Vol. III. London.

Bowerbank, J. S. 1874. A Monograph of the British Spongiadæ. Vol. III, pp. 1–367. Ray Society, London.

Chevreux, E., et Fage L. 1925. Amphipodes. Faune de France, Vol. 9, pp. 1–488.

Collinge, W. E. 1917. A Revision of the British Idoteidæ. Trans. Roy. Soc. Edin., Vol. LI. p. 742.

Colman, J. 1933. The Nature of the Intertidal Zonation of Plants and Animals. Journ. Mar. Biol. Assoc., N.S., Vol. XVIII, pp. 435–476.

Cotton, A. D. 1912. Clare Island Survey. Algæ. Proc. Roy. Irish Acad., Vol. XXXI.

Fauvel, P. 1923, 1927. Polychetes errantes; Polychetes sedentaires. Faune de France, Vol. 5, 1923, pp. 1–488; and Vol. 16, 1927, pp. 1–494.

Gail, F. W. 1918. Some Experiments with Fucus to determine the factors controlling its vertical distribution. Publ. Puget Sound Biol. Station, Vol. II, pp. 139–151.

Gislen, T. 1930. Epibioses of the Gullmar Fjord. Kristinenbergs Zoologiska station, 1877–1927, Upsala.
Lebour, M. V. 1933. The British species of Trivia. Journ. Mar. Biol. Assoc., N.S., Vol XVIII, pp. 477–484.
Lyle, L. 1929. Marine Algæ on some German Warships in Scapa Flow and of Neighbouring Shores. J. Linn. Soc. Bot., Vol. XLVIII.
Marine Biological Association. 1931. Plymouth Marine Fauna. Plymouth.
Newton, L. 1931. A Handbook of the British Seaweeds. London.
Ordnance Survey. 1913. Scale Devonshire, Sheet CXXX, 6. Southampton.
Sars, G. O. 1899. An Account of the Crustacea of Norway. Vol. II, Isopoda, pp. 1–270.
Schmid, O. 1864. Supplement der Spongien des Adriatischen Meeres, pp. 1–48. Leipzig.
Ussher, W. A. E. 1912. The Geology of the Country around Ivybridge and Madbury. Memoirs of the Geological Survey, England and Wales. London.

Kitching, J. A., T. T. Macan, and H. C. Gilson. 1934. "Studies in sublittoral ecology. I. A Submarine Gully in Wembury Bay, South Devon." Journal of the Marine Biological Association *19:677–705. Reprinted with permission of the Cambridge University Press. Note: Reader should visit original publication for the Analysis of Lists.*

A Briefe and True Report of the Newfoundland of Virginia (1588)

Thomas Hariot

To the adventurers, favorers, and well-wishers of the enterprise for the inhabiting and planting in Virginia.

Since the first undertaking by Sir Walter Raleigh to deal in the action of discovering of that Country which is now called and known by the name of Virginia; many voyages having been there made at sundry times to his great charge; as first in the year 1584, and afterwards in the years 1585, 1586, and now of late this last year of 1587. There have been diverse and variable reports with some slanderous and shameful speeches bruited abroad by many that returned from there. Especially of that discovery which was made by the Colony transported by Sir Richard Grenville in the year 1585, being of all the others the most principal and as yet of most effect, the time of their abode in the country being a whole year, when as in the other voyage before they stayed but six weeks; and the others after were only for supply and transportation, nothing more being discovered then had been before. Which reports have not done a little wrong to many that otherwise would have also favoured and adventured in the action, to the honour and benefit of our nation, besides the particular profit and credit which would redound to themselves the dealers therein; as I hope by the sequel of events to the shame of those that have avouched the contrary shall be manifest: if you the adventurers, favourers, and well-wishers do but either increase in number, or in opinion continue, or having been doubtful renew your good liking and furtherance to deal therein according to the worthiness thereof already found, and as you shall understand hereafter to be requisite. Touching which worthiness through cause of the diversity of relations and reports, many of your opinions could not be firm, nor the minds of some that are well disposed, besetled in any certainty.

I have therefore thought it good being one that have been in the discovery and in dealing with the natural inhabitants specially employed; and having therefore seen and known more than the ordinary: to impart so much unto you of the fruits of our labours, as that you may know how injuriously the enterprise is slandered. And that in public manner at this present chiefly for two respects.

First, that some of you which are yet ignorant or doubtful of the state thereof, may see that there is sufficient cause why the chief enterpriser with the favour of her Majesty, notwithstanding such reports; have not only since continued the action by sending into the country again, and replanting this last year a new Colony; but is also ready, according as the times and means will afford, to follow and prosecute the same.

Secondly, that you seeing and knowing the continuance of the action by the view hereof you may generally know and learn what the country is; and thereupon consider how your dealing therein if it proceed, may return you profit and gain; be it either by inhabiting and planting or otherwise in furthering thereof.

And least that the substance of my relation should be doubtful unto you, as of others by reason of their diversity: I will first open the cause in a few words wherefore they are so different; referring myself to your favourable constructions, and to be adjudged of as by good consideration you shall find cause.

Of our company that returned some for their misdemeanor and ill dealing in the country, have been there worthily punished; who by reason of their bad nature, have maliciously not only spoken ill of their Governors, but for their sakes slandered the country itself. The like also have those done, which were of their comfort.

Some being ignorant of the state thereof, notwithstanding since their return amongst their friends and acquaintance and also others, especially if they were in company where they might not be gain; would seem to know so much as no men more; and make no men so great travailers as themselves. They stood so much as it may seem upon their credit and reputation that having been twelve months in the country, it would have been a great disgrace unto them as they thought, if they could not have said much whether it were true or false. Of which some have spoken of more then ever they saw or otherwise knew to be there; othersome have not been ashamed to make absolute denial of that which although not by they, yet by others is most certainly and there plentifully known. And othersome make difficulties of those things they have no skill of.

The cause of their ignorance was, in that they were of that many that were never out of the land where we were seated, or not far, or at the leastwise in few places else, during the time of our abroad in the country; or of that many that after gold and silver was not so soon found, as it was by them looked for, had little or no care of any other thing but to pamper their bellies; or of that many which had little understanding, less discretion, and more tongue then was needful or requisite.

Some also were of a nice bringing up, only in cities or towns, or such as never (as I may say) had seen the world before. Because there were not to be found any English cities, norsuch fair houses, nor at their own wish any of their old accustomed dainty food, nor any soft beds of down or feathers: the country was to them miserable, and their reports thereof according.

Because my purpose was but in brief to open the cause of the variety of such speeches; the particularities of them, and of many envious, malicious, and slanderous reports and devises else, by our own country men besides; as trifles that are not worthy of wise men to be thought upon, I mean not to trouble you withal; but will pass to the commodities, the substance of that which I have to make relation of unto you.

The treatise whereof for your more ready view and easier understanding I will divide into three special parts. In the first I will make declaration of such commodities there already found or to be raised, which will not only serve the ordinary turns of you which are and shall be the planters and inhabitants, but such an overplus sufficiently to be yielded, or by men of skill to be provided, as by way of traffic and exchange with our own nation of England, will enrich your selves the providers; those that shall deal with you; the enterprisers in general; and greatly profit our own country men, to supply them with most things which heretofore they have not provided, either of strangers or of our enemies: which commodities for distinction sake, I call Merchantable.

In the second, I will set down all the commodities which we know the country by our experience does yield of its self for victual, and sustenance of man's life; such as is usually fed upon by the inhabitants of the country, as also by us during the time we were there.

In the last part, I will make mention generally of such other commodities besides, as I am able to remember, and as I shall think behooveful for those that shall inhabit, and plant there to know of; which specially concern building, as also some other necessary uses: with a brief description of the nature and manners of the people of the country.

PAGATOWR [CORN]

A kind of grain so called by the inhabitants; the same in the West Indies is called Maize: English men call it Guinney wheat or Turkey wheat, according to the names of the countries from where the like have been brought. The grain is about the bigness of our ordinary English peas and not much different in form and shape: but of diverse colours: some white, some red, some yellow, and some blue. All of them yield a very white and sweet flower: being used according to his kind it maketh a very good bread. We made of the same in the country some malt, whereof was brewed as good ale as was to be desired. So likewise by the help of hops thereof may be made as good Beer. It is a grain of marvelous great increase; of a thousand, fifteen hundred and some two thousand fold. There are three sorts, of which two are ripe in an eleven and twelve weeks at the most: sometimes in ten, after the time they are set, and are then of height in stalk about six or seven feet. The other sort is ripe in fourteen, and is about ten feet high, of the stalks some bear four heads, some three, some one, and two: every head containing five, six, or seven hundred grains within a few more or less. Of these grains besides bread, the inhabitants make victual either by parching them; or seething them whole until they be broken; or boiling the flour with water into a paste.

The ground they never fatten with muck, dung or any other thing; neither plow nor dig it as we in England, but only prepare it in sort as followeth. A few days before they sow or set, the men with wooden instruments, made almost in form of mattocks or hoes with long handles; the women with short peckers or parers, because they use them sitting, of a foot long and about five inches in breadth: do only break the upper part of the ground to raise up the weeds, grass, and old stubs of corn stalks with their roots. Then which after a day or two drying in the Sun, being scraped up into many small heaps, to save them labor for carrying them away; they burn into ashes. And whereas some may think that they use the ashes to better the ground; I say that then they would either disperse the ashes abroad; which we observed they do not, except the heaps be too great: or else would take special care to set their corn where the ashes lie, which also we find they are careless of, and this is all the husbanding of their ground that they use.

Then their setting or sowing is after this manner. First for their corn, beginning in one corner of the plot, with a pecker they make a hole, wherein they put four grains with that care they touch not one another, (about an inch asunder) and cover them with the mould again: and so through out the whole plot, making such holes and using them after such manner: but with this regard that they be made in ranks, every rank differing from other half a fathom or a yard, and the holes also in every rank, as much. By this means there is a yard spare ground between every hole: where according to discretion here and there, they set as many beans and peas.

The ground being thus set according to the rate by us experimented, an English Acre containing fourty pearches in length, and four in breadth, does there yield in crop of corn, beans, and peas, at the least two hundred London bushels: When as in England fourty bushels of our wheat yielded out of such an acre is thought to be much.

I thought also good to note this unto you, if you which shall inhabit and plant there, may know how special that country corn is there to be preferred before ours. Besides the manifold ways in applying it to victual, the increase is so much that small labor and pains is needful in respect that must be used for ours. For this I can assure you that according to the rate we have made proof of, one man may prepare and husband so much ground (having once borne corn before) with less then four and twenty hours labor, as shall yield him victual in a large proportion for a twelve months, if he have nothing else, but that which the same ground will yield, and of that kind only which I have before spoken of: the said ground being also but of five and twenty yards square. And if need require, but that there is ground enough, there might be raised out of one and the self same ground two harvests or outcomes; for they sow or set and may at any time when they think good from the middle of March until the end of June: so that they also set when they have eaten of their first crop. In some places of the country notwithstanding they have two harvests, as we have heard, out of one and the same ground.

TOBBACO

There is an herb which is sowed a part by itself and is called by the inhabitants uppówoc. In the West Indies it has diverse names, according to the several places and countries where it grows and is used: The Spaniards generally call it Tobacco. The leaves thereof being dried and brought into powder: they use to take the fume or smoke thereof by sucking it through pipes made of clay into their stomach and head; from whence it purgeth superfluous phlegm and other gross humors, opens all the pores and passages of the body: by which means the use thereof, not only preserves the body from obstructions; but also if any be, so that they have not been of too long continuance, in short time breaks them: whereby their bodies are notably preserved in health, and know not many grievous diseases wherewithal we in England are oftentimes afflicted.

The uppówoc is of so precious estimation amongst them, that they think their gods are marvelously delighted therewith. Whereupon sometime they make hallowed fires and cast some of the powder therein for a sacrifice: being in a storm upon the waters, to pacify their gods, they cast some up into the air and into the water: so a weir for fish being newly set up, they cast some therein and into the air: also after an escape of danger, they cast some into the air likewise: but all done with strange gestures, stamping, sometimes dancing, clapping of hands, holding up of hands, and staring up into the heavens, uttering therewithal and chattering strange words and noises.

We ourselves, during the time we were there used to suck it after their manner, as also since our return, and have found many rare and wonderful experiments of the virtues thereof; of which the relation would require a volume by itself: the use of it by so many of late, men and women of great calling as else, and some learned physicians also, is sufficient witness.

OF BEASTS

Deer, in some places there are great store: near unto the sea coast they are of the ordinary bigness as ours in England, and some less: but further up into the country where there is better feed they are greater: they differ from ours only in this, their tails are longer and the snags of their horns look backward.

Conies, those that we have seen and all that we can hear of are of a grey colour like unto hares: in some places there are such plenty that all the people of some towns make them mantles of the fur or flue of the skins of those they usually take.

Saquenúckot and Maquówoc; two kinds of small beasts greater then conies which are very good meat. We never took any of them ourselves, but sometime eat of such as the inhabitants had taken and brought unto us.

Squirrels, which are of a grey colour, we have taken and eaten.

Bears which are all of black colour. The bears of this country are good meat; the inhabitants in time of winter do use to take and eat may; so also sometime did we. They are taken commonly in this sort. In some lands or places where they are, being hunted for, as soon as they have spied of a man they presently run away, and then being chased they climb and get up the next tree they can, from whence with arrows they are shot down stark dead, or with those wounds that they may after easily be killed; we sometime shot them down with our caleevers.

I have the names of eight and twenty several sorts of beasts which I have heard of to be here and there dispersed in the country, especially in the main: of which there are only twelve kinds that we have yet discovered, and of those that be good meat we know only them before mentioned. The inhabitants sometime kill the lion and eat him: and we sometime as they came to our hands of their wolves or wolfish dogs, which I have not set down for good meat, least that some would understand my judgment therein to be more simple than needs, although I could allege the difference in taste of those kinds from ours, which by some of our company have been experimented in both.

OF FISH

For four months of the year, February, March, April and May, there are plenty of Sturgeons: And also in the same months of Herrings, some of the ordinary bigness as ours in England, but the most part far greater, of eighteen, twenty inches, and some two feet in length and better; both these kinds of fish in those months are most plentiful, and in best season, which we found to be most delicate and pleasant meat.

There are also Trouts, Porpoises, Rays, Oldwifes, Mullets, Plaice, and very many other sorts of excellent good fish, which we have taken and eaten, whose names I know not but in the country language; we have of twelve sorts more the pictures as they were drawn in the country with their names.

The inhabitants use to take them two manner of ways, the one is by a kind of weir made of reeds which in that country are very strong. The other way, which is more strange, is with poles made sharp at one end, by shooting them into the fish after the manner as Irishmen cast darts; either as they are rowing in their boats or else as they are wading in the shallows for the purpose.

There are also in many places plenty of these kinds which follow.

Sea crabs, such as we have in England.

Oysters, some very great, and some small; some round and some of a long shape: They are found both in salt water and brackish, and those that we had out of salt water are far better than the other as in our own country.

Also Mussels, Scallops, Periwinkles, and Crevises.

Seekanauk, a kind of crusty shellfish which is good meat, about a foot in breadth, having a crustier tail, many legs like a crab; and her eyes in her back. They are found in shallows of salt waters; and sometime on the shore.

There are many Tortoises both of land and sea kind, their backs and bellies are shelled very thick; their head, feet, and tail, which are in appearance, seem ugly as though they were members of a serpent or venomous: but notwithstanding they are very good meat, as also their eggs. Some have been found of a yard in breadth and better.

OF THE NATURE AND MANNERS OF THE PEOPLE

It rests I speak a word or two of the natural inhabitants, their natures and manners, leaving large discourse thereof until time more convenient hereafter: now only so far forth, as that you may know, how that they in respect of troubling our inhabiting and planting, are not to be feared; but that they shall have cause both to fear and love us, that shall inhabit with them.

They are a people clothed with loose mantles made of Deer skins, and aprons of the same round about their middles; all else naked; of such as difference of statures only as we in England; having no edge tools or weapons of iron or steel to offend us withal, neither know they how to make any: those weapons that they have, are only bows made of Witch hazel, and arrows of reeds; flat edged truncheons also of wood about a yard long, neither have they any thing to defend themselves but targets made of bark; and some armors made of sticks whickered together with thread.

Their towns are but small, and near the sea coast but few, some containing but 10 or 12 houses: some 20, the greatest that we have seen have been but of 30 houses: if they be walled it is only done with barks of trees made fast to stakes, or else with poles only fixed upright and close one by another.

Their houses are made of small poles made fast at the tops in round form after the manner as is used in many arbories in our gardens of England, in most towns covered with bark, and in some with artificial mats made of long rushes; from the tops of the houses down to the ground. The length of them is commonly double to the breadth, in some places they are but 12 and 16 yards long, and in other some we have seen of four and twenty.

In some places of the country one only town belongeth to the government of a Wiróans or chief Lord; in other some two or three, in some six, eight, and more; the greatest Wiróans that yet we had dealing with had but eighteen towns in his government, and able to make not above seven or eight hundred fighting men at the most: The language of every government is different from any other, and the farther they are distant the greater is the difference.

Their manner of wars amongst themselves is either by sudden surprising one an other most commonly about the dawning of the day, or moon light; or else by ambush, or some subtle devices: Set battles are very rare, except if fall out where there are many trees, where either part may have some hope of defense, after the delivery of every arrow, in leaping behind some or other.

If there fall out any wars between us and them; what their fight is likely to be, we having advantages against them so many manner of ways, as by our discipline, our strange weapons and devices else; especially by ordinance great and small, it may be easily imagined; by the experience we have had in some places, the turning up of their heels against us in running away was their best defense.

In respect of us they are a people poor, and for want of skill and judgment in the knowledge and use of our things, do esteem our trifles before things of greater value: Notwithstanding in their proper manner considering the want of such means as we have, they seem very ingenious; For although they have no such tools, nor any such crafts, sciences and arts as we; yet in those things they do, they show excellence of wit. And by how much they upon due consideration shall find our manner of knowledges and crafts to exceed theirs in perfection, and speed for doing or execution, by so much the more is it probable that they should desire our friendships and love, and have the greater respect for pleasing and obeying us. Whereby may be hoped if means of good government be used, that they may in short time be brought to civility, and the embracing of true religion.

THE CONCLUSION

Now I have as I hope made relation not of so few and small things but that the country of men that are indifferent and well disposed may be sufficiently liked. If there were no more known than I have mentioned, which doubtless and in great reason is nothing to that which remains to be discovered, neither the soil, nor commodities. As we have reason so to gather by the difference we found in our travels: for although all which I have before spoken of, have been discovered and experimented not far from the sea coast where was our abode and most of our traveling: yet sometimes as we made our journeys farther into the main and country; we found the soil to be fatter; the trees greater and to grow thinner; the ground more firm and deeper mould; more and larger champions; finer grass and as good as ever we saw any in England; in some places rocky and far more high and hilly ground; more plenty of their fruits; more abundance of beasts; the more inhabited with people, and of greater policy and larger dominions, with greater towns and houses.

Why may we not then look for in good hope from the inner parts of more and greater plenty, as well of other things, as of those which we have already discovered? Unto the Spaniards happened the like in discovering the main of the West Indies. The main also of this country of Virginia, extending some ways so many hundreds of leagues, as otherwise then by the relation of the inhabitants we have most certain knowledge of, where yet no Christian Prince has any possession or dealing, cannot but yield many kinds of excellent commodities, which we in our discovery have not yet seen.

What hope there is else to be gathered of the nature of the climate, being answerable to the land of Japan, the land of China, Persia, Jury, the lands of Cyprus and Candy, the South parts Greece, Italy, and Spain, and of many other notable and famous countries, because I mean not to be tedious, I leave to your own consideration.

Whereby also the excellent temperature of the air there at all seasons, much warmer then in England, and never so violently hot, as sometimes is under and between the Tropics, or near them; cannot be unknown unto you without farther relation.

For the wholesomeness thereof I need to say but thus much: that for all the want of provision, as first of English victual; excepting for twenty days, we lived only by drinking water and by the victual of the country, of which some sorts were very strange unto us, and might have been thought to have altered our temperatures in such sort as to have brought us into some grievous and dangerous diseases: secondly the want of English means, for the taking of beasts, fish, and foul, which by the help only of the inhabitants and their means, could not be so suddenly and easily provided for us, nor in so great numbers and quantities, nor of that choice as otherwise might have been to our better satisfaction and contentment. Some want also we had of clothes. Furthermore, in all our travels which were most special and often in the time of winter, our lodging was in the open air upon the ground. And yet I say for all this, there were but four of our whole company (being one hundred and eight) that died all the year and that but at the latter end thereof and upon none of the aforesaid causes. For all four especially three were feeble, weak, and sickly persons before ever they came thither, and those that knew them much marveled that they lived so long being in that case, or had adventured to travel.

Seeing therefore the air there is so temperate and wholesome, the soil so fertile and yielding such commodities as I have before mentioned, the voyage also thither to and fro being sufficiently experimented, to be performed thrice a year with ease and at any season thereof; and the dealing of Sir Walter Raleigh so liberal in large giving and granting land there, as is already known, with many help and furtherances else: (The least that he has granted has been five hundred acres to a man only for the adventure of his person): I hope there remains no cause whereby the action should be misliked.

If that those which shall thither travel to inhabit and plant be but reasonably provided for the first year as those are which were transported the last, and being there do use but that diligence and care as is requisite, and as they may with ease: There is no doubt but for the time following they may have victuals that is excellent good and plenty enough; some more English sorts of cattail also hereafter, as some have been before, and are there yet remaining, may and shall be God willing thither transported: So likewise our kind of fruits, roots, and herbs may be there planted and sowed, as some have been already, and prove well: And in short time also they may raise of those sorts of commodities which I have spoken of as shall both enrich themselves, as also others that shall deal with them.

And this is all the fruits of our labours, that I have thought necessary to advertise you of at this present: what else concerns the nature and manners of the inhabitants of Virginia: The number with the particularities of the voyages there made; and of the actions of such that have been by Sir Walter Raleigh therein and there employed, many worthy to be remembered;

as of the first discoverers of the Country: of our general for the time Sir Richard Grenville; and after his departure, of our Governor there Master Rafe Lane; with diverse other directed and imployed under their government: Of the Captains and Masters of the voyages made since for transportation; of the Governor and assistants of those already transported, as of many persons, accidents, and things else, I have ready in a discourse by itself in manner of a Chronicle according to the course of times, and when time shall be thought convenient shall be also published.

This referring my relation to your favourable constructions, expecting good success of the action, from him which is to be acknowledged the author and governor not only of this but of all things else, I take my leave of you, this month of February 1588.

Hariot, T. 1588. A Briefe and True Report of the Newfoundland of Virginia. *1972 reprint from Dover Publications, New York; pp. 7–33.*

Initiation

Nancy Knowlton

What does it mean to be a great naturalist? Let's start with a passage about Hutchinson (Riley 1971, pg. 177, with my apologies for the now off-putting use of "men" for "people").

> Once upon a time, in the land of the ecologists, there were wise men who catalogued the creatures of the earth and drew up calendars of the marvelous panoply of life that paraded before them through the seasons. And people believed them, for they were wise men of careful vision, though they understood not whereof they spoke, some merely saying what they observed was of God's will and others speculating that pH might be of significance.
>
> Then one came, wiser than most, who said that every fact should be useful, but no fact is useful unless it is used. Putting this premise to the test, he marshalled every fact he knew, and it was a mighty assemblage, and he began to reason as to what might be the wellsprings of the intricate relations he observed among the creatures of the earth. He arose from his bed early, wandering over the earth and rowing over the lakes, observing with loving understanding, and he pondered through the day and into the night. He began to arrive at truths which he hammered so finely that he was able at length to say, about many things, "This is almost certainly so."

Now, Hutchinson was unrivaled not only in his command of facts but also in his ability to make connections among them. When Jeremy Jackson (full disclosure—my husband) went to talk to him about his thesis on the clams living in tropical seagrass beds, he perked up when Jeremy told him that members of the family Lucinidae dominated these communities. "Lucinids have hemoglobin—that will be important," he said. "But clams don't have hemoglobin," Jeremy replied. "Aah, but lucinids do," intoned Hutchinson, drawing on his seemingly limitless store of information about the natural world.

I make no claims to such a comprehensive mastery, but in my own way I have made a point of grounding hypotheses and their tests in natural history. I have spent countless hours underwater, many of them spent looking and thinking, undistracted by collecting data or taking photographs. Some of my best ideas have come during these aquatic sojourns, and not only because of the elevated partial pressure of oxygen! I have often told students that to really understand how a reef works, you must live underwater for a year (or at least hours a day for months at a time). In doing so, you develop a sixth sense of the rules of the game and their exceptions, the understanding of which underpins most great conceptual advances in ecology, animal behavior, systematics, and evolution. And other fields of biology for that matter—consider the title of the biography by Keller (1983) of the great geneticist Barbara McClintock: *A Feeling for the Organism*. Intuition based in natural history inspires one to persevere when the experimental results don't quite seem right—e.g., for me, to keep plugging away at genetic analyses of three cryptic species of corals, because natural history made it clear that they were different, even if the mitochondrial COI gene by itself initially didn't.

The following articles are early contributions in natural history from twentieth-century leaders in ecology and evolutionary biology, written before they were widely recognized as such—even giants start out small. The subjects and approaches are diverse, and although the latest was published over forty years ago, they still have some lessons for us today.

STUDIES OF INDIVIDUAL SPECIES

Many of the articles in this section represent studies of individual species: Connell, fourteen months of trapping brush rabbits in California; Paine, undated observations during his postdoctoral studies on feeding in a marine gastropod; Wilson, about a year of observations and experiments on South American fire ants; and Odum, a few days of observations on roosting starlings in Illinois. The papers provide a variety of possible reasons for spending time documenting and thinking about the natural history of a species, ranging from lack of information on a common species, to interest in particular phenomena that a species illustrates well, to the role that natural history plays in generating patterns, to the sheer joy of following the daily lives of creatures.

Joseph Connell (1923–) introduces his article (Connell 1954) by stating that very little had been published on the brush rabbit, one of the commonest game mammals of the chaparral of California. In the course of his study he set 2251 traps, tattooed and took a variety of data on forty rabbits (size, sex, weight, reproductive condition, external and blood parasites), and twice scoured the area for rabbit pellets. This was quite a lot of work, and it was carefully done; throughout the paper he concerns himself with potential biases in the methods, and he provides as well a summary of the vegetation of the area. Nowhere, however, does he comment on the possible broader significance of the data he presents. Yet this was a period when Connell was thinking hard (as all good graduate students do) about what his research direction was going to be. In an essay about his Citation Classic 1961 paper on barnacles (Connell 1992, pg. 8) he states: "At the time, I was doing an MA degree, studying a population of reclusive rabbits, a system not conducive to applying any of the methods described by Deevey." Connell goes on to say how the Deevey paper eventually led him to Hatton's pioneering studies on barnacles, and the rest, as they say, is history.

In contrast, Robert Paine's (1933–) paper begins with a clear statement of goals (Paine 1963, pg. 1): "An adequate description of the ecological properties of a species should include its trophic relationships in terms of recognition of both predators and potential prey, and the quantitative and qualitative results of such behavior." Though he switched to different predators, his interest in predation in the marine environment continued for decades. The approach in this paper, however, was very different from what has typified much of his renowned later experimental work on Tatoosh Island—here, he presents laboratory and field studies of feeding, laboratory studies of track following (ah for the days when one could publish an entire page figure that consists of a large loop with the notations of "animals started here" at one end and "*Bulla* eaten here" at the other!). But one of the reasons that the Paine school of marine ecology is so successful is that the experiments are done in the context of a comprehensive understanding of the natural history of the organisms involved—a deep feeling for the rocky intertidal of Tatoosh Island underpins it all.

Edward Wilson (1929–) had both a species and a problem in mind—the invasion of the South American fire ant, which occurred in his home state (Wilson 1951). With a close eye for detail, he first noticed the appearance of a new variety of imported fire ant, quantified the geographic patterns of variation in color pattern and size, and inferred its genetic underpinnings and possible origins from these observations and laboratory experiments, work that would later lead to the recognition that not one but two species of fire ants had invaded the southern US. He notes in his autobiography (Wilson 1994) his exhilaration in learning that natural history could be of both general interest and practical use, aided by his now legendary ability to make natural history vivid. For example, he once described attempts to eradicate the fire ants as "the Vietnam of entomology" (Wilson 1994, pg. 117). Sadly, comparable analogies are all too available today.

The paper by Eugene Odum (1913–2002) records the effects of a severe winter storm on a flock of starlings (Odum and Pitelka 1939). The storm caused mortality of about 4% of the flock (726 dead birds were tabulated), which they attributed (thanks to the weather data, which they provide) to high winds (48 mph) and rain, followed by a sharp drop in temperature (from 60° F to below freezing in 5 hours). Odum had a long-standing interest in birds (his thesis, part of which was published in *Science*, concerned the heart rates of wild birds), but this paper was clearly not part of a systematic study. Rather, it represents a good example of opportunistic science in the form of a natural experiment. We all on some level know that occasional extreme events, too infrequent to be an organizing structure for a research grant, can nevertheless be among the most important of all shapers of the natural world. But there is nothing like a real-life demonstration to bring the point home. This happened to me in August 1980, when Hurricane Allen swept along the north coast of Jamaica. In one night, the storm taught all of us there why the reefs were not entirely dominated by the rapidly growing but fragile branching acroporid corals—these were decimated by the storm, but standing tall and relatively unscathed were the massive head corals, that in retrospect clearly depended on the fifty- to hundred-year mega-storm events to persist. Odum, over the course of his career, became more focused on the natural history of ecosystems than of individual species, and he grew to be one of the most famous and influential ecologists of his era, thanks to his classic text *Fundamentals of Ecology* and his legacy in the founding of the Savannah River Ecology Laboratory, the University of Georgia Marine Institute on Sapelo Island, and the Institute of Ecology. He is best known as the modern father of ecosystem ecology, and in my field, his study of the coral reefs of Eniwetok is still well known. His approach was holistic, and led him eventually and naturally to concerns about sustainability of human activities, a problem that of course remains with us today.

EXPEDITIONARY NATURAL HISTORY

Expeditions taken early in a career have long been the archetypal inspiration of later insights, and two such expeditions are described in this section: that of Elton to Spitsbergen and that of Mayr to Papua New Guinea. The importance of expeditions is hardly unique to these authors—other sections of this book draw on trips by Darwin, Bates, Wallace, von Humboldt, Beebe, and others. Those in this section are utterly different in style and purpose, but together they portray the scientific and nonscientific benefits to any naturalist of spending time away from home.

Charles Elton (1900–1991) went on several expeditions to the north early on that strongly influenced him (Hardy 1968, MacFadyen 1992, Southwood and Clarke 1999), and he clearly had an eye for natural history details. Between 1922 and 1975, his publications concern a staggering array of animals, including (here listed roughly chronologically) rabbits, natterjack toads, water mites, beetles, butterflies, gape-worms, freshwater copepods, aphids, hoverflies, mammal parasites, birds, voles, dung flies, marine snails, biting flies of humans, and grizzly bears (as well as one paper on the honey of Chinese sumac and another on the origin of arctic soil polygons) (see biography by Southwood and Clarke 1999). Of course, Elton is best known for his classic *Animal Ecology*, which he wrote in eighty-five days in a flurry of youthful inspiration and which lays out many of the topics still studied by ecologists today. But in just their second sentence Southwood and Clarke attribute the brilliance of Elton to his deep interest in natural history, and the same conclusion appears in other biographies (Hardy 1968, MacFadyen 1992). The paper reproduced here on the insects of Spitsbergen (Elton 1925), based on expeditions in 1921, 1923 and 1924, reads in large part like a field notebook, with detailed observations of what was seen when and where. Yet like many field notebooks it reveals the underlying question that intrigued Elton. As he notes, glaciation did for Spitsbergen what lava did for Krakatoa, and he wants to know, "How has the new fauna got there?" Elton only very briefly alludes to the "extremely strenuous conditions" and "very bad surfaces," but his adventures included falling up to his neck through the ice and being saved only by his rucksack (Southwood and Clarke 1999).

Ernst Mayr (1904–2005) was also an eminent thinker rooted in natural history (Bock 1994), with many scholarly books and articles to his credit. Mayr's article in this volume (Mayr 1932), which describes his first and only expedition in 1928–1929, is mostly travelogue. There is a plenty of description of bird and plant life of New Guinea, with the occasional nugget of generalization thrown in (e.g., the greater abundance of endemics at high altitudes). We learn at least as much about the people as about the wildlife of the region (some parts make a reader of today cringe, but overall the descriptions reflect Mayr's admiration of their endurance and natural history knowledge). Mayr himself is a leading character in the travelogue. Thus, unlike Elton's, Mayr's article gives a real sense of the personal difficulties and dangers, which included vertical ascents, hair-raising logistical challenges, serious illness (infected legs, malaria, pneumonia, arsenic poisoning), floods, and pillaging by half-starved native dogs. We also learn a lot about Mayr the person; in describing his preparations he states (pp. 83–84), "when I arrived in New Guinea, I knew not only the name of every bird I might collect, but also whether it was rare, or desirable for my collection, and whether it showed any peculiarities of particular interest to science." At the end (pg. 97) he concludes that "the daily fight with unknown difficulties, the need for initiative, the contact with the strange psychology of primitive people, and all the other odds and ends of such an expedition, accomplish a development of character than cannot be had in the routine of civilized life."

Expeditions these days are almost always tamer affairs, what with antibiotics and satellite phones to help stave off catastrophe (but read about and see the photographs of Mike Fay's central African megatransect (Fay and Nichols 2005) to remind yourself that true exploration, with all its original challenges, is still possible on the planet). But regardless of the danger, the importance of exposing yourself to places of bewildering strangeness or diversity, with a prepared but open mind, remains. In 2005 I had an expeditionary experience on the coral reefs of the Northern Line Islands, where at the least disturbed sites, 85% of the total fish biomass consisted of apex predators (Knowlton and Jackson 2008). For all of us, it was like traveling in a time machine, allowing us to see firsthand what reefs were like prior to the human impacts that are nearly universal now—one does not have to be young to benefit from an expedition. Nor is it necessary to travel far from home; Wilson (1994) notes the importance of Rock Creek Park in Washington, D.C., for his development, and Elton described the hedgerows that run between fields as one of the last really big remaining nature reserves in Britain (Hardy 1968). The ecological importance of human dominated landscapes has also recently reemerged in conservation biology (e.g., Pereira and Daily 2006, Sekercioglu et al. 2007).

ASSEMBLING NATURAL HISTORY PATTERNS

Biodiversity can be so overwhelming that, inspiring though it is, it can verge on intellectual noise, and so the essence of the fields of ecology and evolution is to elucidate the patterns in this noise. The last three articles in this section take what was known about some aspect of distribution and diversity and try to make sense of it: Hutchinson for the biogeography aquatic insects, Grant for patterns of cooccurrence of closely related birds on islands and the mainland, and Jackson for intertidal mudflat clams.

Like Elton, G. Evelyn Hutchinson (1903–1991) wrote papers on a large array of topics—in addition to many papers on freshwater insects and other limnological topics, he published articles on the branchial gland of cephalopods, the blood of hagfish, the biochemistry of a starfish, rhinoceros behavior, the Burgess Shale fossils, ocean geochemistry, psychobiology and human sexual selection, and science, religion, and art. There is little hint in the paper reproduced here (Hutchinson 1931) of the enormous contributions he would make to key conceptual problems in ecological theory, many of which are still with us today (e.g., the nature of the niche and patterns of species diversity). But the paper resembles Elton's in that he was curious about how organisms get to where we find them today. Hutchinson drew on collections of and correspondence from colleagues, much as Darwin did in the previous century, and inferred that a group of water boatmen, buoyant and typically freshwater insects restricted to the Americas, were salt tolerant and thus capable of reaching outposts in Hawaii and China via ocean currents. The relative importance of dispersal versus continental drift in explaining why similar species can be found so widely scattered across the globe remains a central topic of biogeography to this day.

Peter Grant's (1936–) paper on birds (Grant 1966) was published shortly after the first publication by MacArthur and Wilson (1963) on island biogeography. Both articles invoke a dynamic process of immigration and extinction to shape the patterns seen today, but instead of focusing on numbers of species, Grant looked at the types of species found together on islands, using several levels of analysis. In particular, he tested the idea that although many species can arrive, extinction will tend to eliminate newcomers that are too ecologically similar to those already resident (or less commonly, the newcomer will eliminate the resident). His predictions were largely upheld: the proportion of congeners was lower on islands (particularly small ones) than on the mainland, and differences in bill length between potential competitors tended to be exaggerated on islands compared to the mainland. The fact that bill shapes and size so accurately reflect diet made birds a wonderful subject for this analysis, and Grant (with his wife, daughters, and many assistants and students) went on to exploit this for the most iconic of organisms, the Galapagos finches (or Darwin's finches, although Darwin, naturalist par excellence that he was, did not initially appreciate their significance). What the Grants have documented is arguably the best example we have of evolution in the wild in action, based on observations of many thousands of birds, each part of a genealogy extending over tens of generations embedded in the records of changing environmental conditions over several decades. As Jonathan Weiner notes in his wonderful book about their work, *The Beak of the Finch* (Weiner 1994), the results are astounding for the precision with which predictions match reality. For example, using models based on heritabilities of beak dimensions, and seed abundances and size distributions, they predicted that the width of the bill of one species should have fallen from 8.86 mm to 8.74 mm between 1984 and 1987, and so it did. Natural history can be a very exact science, no "close enough for jazz" or "quick and dirty" here.

Finally, the paper by my husband, Jeremy Jackson (1942–), concerns the much finer scale distributions of two intertidal clams of Long Island Sound, and inferred movement in life (and in death) from the patterns observed (Jackson 1968). This was a term paper for a class taught by Hutchinson when Jackson was a first-year graduate student in geology at Yale. As an aside, term papers can serve as surprisingly important entrees into the world of scientific publishing—my own first publication was also a term paper originally (I will always be grateful to John Maynard Smith for encouraging me to publish it), and Yale must have been a great place for writing term papers in those days, as Jackson's first two were both published in *Science*. Jackson documented the simple fact that otherwise similar species with different life histories had very different spatial and size frequency distributions, but it got the attention it did (not only the original publication but two letters challenging the results) because it undermined the assumption that size frequency distributions could be used to determine the degree to which fossil assemblages had been transported (and thus altered in composition in the process). That is, fossils (and their living counterparts) are not beans in a bean bag—their natural history matters—a fact not undermined by recent interest in using null models to determine what proportion of any pattern can be due simply to chance acting on ecologically equivalent species (e.g., Hubbell 2001). Much of Jackson's work now concerns marine conservation, but he has never abandoned his interest in the past (inspired by his father, who was a maritime historian)—ranging from Neotropical evolution over the last 10 million years to the impacts of people on ocean life over the much more recent past. And Hutchinson's hemoglobin? Though its general importance in reducing environments was clear to Jackson, it would take someone with access to better microscopes than those available to him in Jamaica to discover that like many deep-sea vent organisms, his lucinids host sulfur-oxidizing bacteria in their gills.

LESSONS FOR BUDDING (AND BUDDED) NATURALISTS

It is always tempting to look for hints of future directions in the early work of pioneers. There is plenty of evidence of that in the articles reproduced here—Mayr and Grant on birds, Wilson on ants, Paine on marine predators. But people don't necessarily stick with their earliest efforts, often for quite good scientific reasons, as noted above for Connell, but for a variety of other reasons as well. Jackson, for example, swore off bivalve research after switching to infaunal communities and digging his way through 18 metric tons of mud for his dissertation.

Wilson, in his autobiography, describes a series of passions and accidents and intellectual instincts that led to his devotion to ants. His first vivid natural history memory is that of a jellyfish, but a fishing accident left him with no sight in one eye (but excellent close-up vision in the other), and in adolescence he lost his upper register hearing, ruling out birds and frogs. Having settled on insects due to his sensory limitations, he then ruled out butterflies (too well known) and flies (because Czechoslovakia, the source of needed long black insect pins, was in a war zone in 1945). Thus he settled on ants, which could be preserved in locally available alcohol. At the end of his book (Wilson 1994, pg. 364), however, he writes: "The smaller the organism, the broader the frontier and the deeper the unmapped terrain . . . If I could do it all over again . . . in the twenty-first century, I would be a microbial ecologist." In sum, a set of interests dictated in turn by practicality, chance, and biological intrigue.

The articles also clearly show that while taxonomy is ultimately important to being a good naturalist, this does not necessarily mean learning the name of everything out there. Elton thanks the taxonomists that identified his flies and aphids, Paine thanks the taxonomists that identified his opisthobranchs, and Connell thanks the taxonomists that identified his parasites. The importance of having some idea of what to expect is clearly what being a good naturalist entails, because it helps one to recognize the unusual (as the earlier quote from Mayr about his preparation for his trip indicates). But as Wilson (1994, pp. 11–12) notes: "Hands-on experience at the critical time, not systematic knowledge, is what counts in the making of a naturalist. Better to be an untutored savage for a while, not to know the names or anatomical details. Better to spend long stretches of time just searching and dreaming." Though he points to childhood as the best time for this, anyone who retains his or her childlike curiosity can join the naturalist club.

Finally, how "useful" is natural history? Riley (1971, pg. 177) in his homage to Hutchinson continues:

> And then there began to be parlous times in the land of the ecologists, for man made the environment to be sick, and some said there was no redress from this scourge. There arose a cry in the land for relevant research to cure the sickness. Yet those who understood the message would also know

> that beyond relevant research is the further and greater goal of research to find out what is relevant, and the only hope for our world is to continue, as we have been taught, to seek unfettered truth.

Riley is alluding to the long-standing tension between what are often called basic and applied research. I actually feel that the distinction is far grayer than the dichotomy implies, and that research suffers when artificial boundaries are created. In founding the Center for Marine Biodiversity and Conservation at the Scripps Institution of Oceanography, I was inspired by the Bell Labs—a place where some scientists were working on the next telephone and others on the next Nobel Prize (there have been six in all). Closer to my expertise, the initial discoveries about the diversity of symbiotic dinoflagellates living with corals—now so relevant in a warming world—was basic natural history (who lives where and with whom), albeit abetted by the techniques of the human genome toolkit. In an era of ever increasing human impacts on the planet, all natural history is both basic and applied, serving as a baseline for the future changes that will inevitably occur (Knowlton and Jackson 2008).

What about natural history information being collected today? Although it is more difficult now to publish basic descriptive information in journals without framing it as the test of a hypothesis, the need to make the information available in some forum is just as true now as it was then—otherwise, how can the Hutchinsons of the world find and use the information? The last ten years have seen an explosion of information in nontraditional formats, made possible by the web. This diversity of sources is actually not a new phenomenon—Darwin depended utterly on it, and Elton wrote that "the professional ecologist has to rely and always will have to rely, for a great many of his data, upon the observations of men like fishermen, gamekeepers, local naturalists and, in fact, all manner of people who are not professional scientists at all" (quoted in Hardy 1998, pg. 6). Today the growing field of historical ecology relies critically on such informants—even the writings of pirates such as William Dampier. How to organize the information still remains a challenge, however. Even when electronically available, thousands of hits on Google needing to be tracked down do not solve this problem, and much important information remains buried in unpublished form. Initiatives like the Encyclopedia of Life (http://www.eol.org/), the Biodiversity Heritage Library (http://www.biodiversitylibrary.org/), and the digitization of museum records are urgently needed to make natural history information more broadly available. Indeed, the vast collections of natural history museums cover places and times that no single scientist could ever replicate; in some cases, sadly, museums are all that are left of the natural history of a region, and thus provide us with an invaluable window into times past.

I began with a quote about Hutchinson, and end with a quote by him. I think it captures one of the central truths about natural history—that great ecologists and evolutionary biologists are almost always grounded in a deep love for the details of nature. These details are a source of inspiration and puzzlement, and ultimately new ideas and answers, even for those best known as theoreticians, such as Robert MacArthur, from whose eulogy this passage is drawn:

> Modern biological education, however, may let us down as ecologists if it does not insist, and it still shows too few signs of insistence, that a wide and quite deep understanding of organisms, past and present, is as basic a requirement as anything else in ecological education. It may be best self-taught, but how often is this difficult process made harder by a misplaced emphasis on a quite specious modernity. Robert MacArthur really knew his warblers. (Hutchinson 1975, pg. 516).

In 1924, Elton argued in the *School Science Review* for the teaching of field zoology, "an almost unheard of innovation at that time" and a message still important today (MacFadyen 1992, pg. 500). Indeed, in this era of meta-analyses and model organisms, the editors and I hope that this message of why natural history matters comes through loud and clear in the papers that follow.

LITERATURE CITED

Bock, W. J. 1994. Ernst Mayr, naturalist: His contributions to systematics and evolution. Biology and Philosophy 9: 267–327.

Connell, J. H. 1954. Home range and mobility of brush rabbits in California chaparral. Journal of Mammalogy 35: 392–405.

Connell, J. H. 1992. The profound effect of a pioneer French ecologist. Current Contents 24: 8.

Elton, C. S. 1925. The dispersal of insects to Spitsbergen. Transactions of the Entomological Society of London 73: 289–299.

Fay, M. (author), and M. Nichols (photographer). 2005. Last Place on Earth. National Geographic Society, Washington, D.C.

Grant, P. R. 1966. Ecological compatibility of bird species on islands. American Naturalist 100: 451–462.

Hardy, A. 1968. Charles Elton's influence in ecology. Journal of Animal Ecology 37: 3–8.

Hubbell, S. P. 2001. A Unified Neutral Theory of Biodiversity and Biogeography. Princeton University Press, Princeton.

Hutchinson, G. E. 1931. On the occurrence of *Trichocorixa kirkaldy* (Corixidae, Hemiptera-

Heteroptera) in salt water and its zoo-geo-graphical significance. American Naturalist 65: 573–574.
Hutchinson, G. E. 1975. Variations on a theme by Robert MacArthur. In Ecology and Evolution of Communities, M. L. Cody, J. M. Diamond, eds., 492–521. Harvard University Press, Cambridge.
Jackson, J. B. C. 1968. Bivalves: Spatial and size-frequency distributions of two intertidal species. Science 161: 479–480.
Keller, E. F. 1983. A Feeling for the Organism: The Life and Work of Barbara McClintock. Henry Holt and Co., New York.
Knowlton, N., and J. B. C. Jackson. 2008. Shifting baselines, local impacts, and global change on coral reefs. PLoS Biology 6: e54.
MacArthur, R. H., and E. O. Wilson. 1963. An equilibrium theory of insular zoogeography. Evolution 17: 373–387.
MacFadyen, A. 1992. Obituary: Charles Sutherland Elton. Journal of Animal Ecology 61: 499–502.
Mayr, E. 1932. A tenderfoot explorer in New Guinea. Natural History 32: 83–97.
Odum, E. P., and F. A. Pitelka. 1939. Storm mortality in a winter starling roost. Auk 56: 451–455.
Paine, R. T. 1963. Food recognition and predation on opistobranchs by *Navanax inermis* (Gastropoda: Opisthobranchia). Veliger 6: 1–9.
Pereira, H. M., and G. C. Daily. 2006. Modeling biodiversity dynamics in countryside landscapes. Ecology 87: 1877–1885.
Riley, G. A. 1971. Introduction. Limnology and Oceanography 16: 177–179.
Sekercioglu, C. H., S. R. Loarie, F. O. Brenes, P R. Ehrlich, and G. C. Daily. 2007. Persistence of forest birds in the Costa Rican agricultural countryside. Conservation Biology 21: 482–494.
Southwood, R., and J. R. Clarke. 1999. Charles Sutherland Elton. Biographical Memoirs of Fellows of the Royal Society of London 4: 131–146.
Weiner, J. 1994. The Beak of the Finch. Alfred A. Knopf, New York.
Wilson, E. O. 1951. Variation and adaptation in the imported fire ant. Evolution 5: 68–79.
Wilson, E. O. 1994. Naturalist. Island Press, Washington, D.C.

Home range and mobility of brush rabbits in California chaparral (1954)

Joseph H. Connell

Abstract—*The brush rabbit,* Sylvilagus bachmani, *is one of the commonest game mammals in the chaparral areas of California. It is widely hunted for sport and for food, yet surprisingly little has been published on its natural history. This study was undertaken to learn something of the mobility of brush rabbits in a natural chaparral habitat near Berkeley, California. Trapping and marking of animals in a more or less isolated unit of habitat was the principal method of study. The period of trapping extended from March 1948, to May 1949; 40 rabbits were marked. Incidental data on population density, breeding period, and physical condition of trapped animals were obtained and are summarized in this report.*

Area of study—The area used for this study is a canyon located in the Berkeley Hills immediately southeast of the University of California campus; its position lies at the head of Dwight Way in Berkeley. The main drainage is to the west with steep sides rising from 300 to 800 feet elevation on the western spurs, which unite at about 1,200 feet at the east. Water is available only in late winter and spring in the main creek bed. The drainage basin contains about 30 acres, near the center of which the study area of approximately 11.4 acres is located. The latter area, composed of chaparral and adjoining grassland, is the only part of the basin suitable for occupancy by brush rabbits. The other principal vegetation types are grass and trees. Light grazing of the area had been permitted in past years but was eliminated several years before the period of study.

The brush area lies mostly on the north-facing slope, with coyote brush (*Baccharis pilularis*) the dominant shrub. Coffee-berry (*Rhamnus californica*) is common, with poison oak, cow parsley (*Heracleum lanatum*) and elderberry (*Sambucus glauca*) in the more humid draws. Bush monkey flower (*Diplacus aurantiracus*) occupies edges and dry openings in the brush. Along the main creek bed is a riparian association of willow (*Salix* sp.) with other brush of the before-mentioned species and a ground cover of blackberry (*Rubus vitijolius*).

The grass type is dominant on hilltops and on the xerophytic south-facing slope of the canyon. Principal components of the type are native and introduced grasses (*Avena*, *Bromus*, *Festuca*, etc.), and various forbs of which thistles (*Cirsium* sp., *Centaurea* sp.), tarweed (*Madia sativa*), soap plant (*Chlorogalum pomeridianum*), bur clover (*Medicago hispida*) and filarees (*Erodium* sp.) are among the dominants. The grass area has been burned each year in August for the past few years by the local fire department. Mixed in the herbaceous growth is scattered bushes of California sagebrush (*Artemisia californica*), poison oak (*Rhus diversiloba*) and lupine (*Lupinus* sp.).

Trees occur in the lower streambed, mostly California laurel (*Umbellularia californica*) with a few coast live oaks (*Quercus agrifolia*) and buckeyes (*Aesculus californicus*). A few small laurel trees are scattered in the brush on the north-facing slope. In the lower parts of the gulch (where, incidentally, no trapping was done) various exotic trees such as *Sequoia*, *Eucalyptus* and *Acacia* have been planted. Additional brush species found there include hazelnut (*Corylus rostrata*), creambush (*Holodiscus discolor*) and wild rose (*Rosa gymnocarpa*).

In general, the rabbits used the brush area for cover and feeding, the grass area adjoining the brush for feeding, and the wooded areas for transit only, since few pellets were found there. The broad grass area above the brush seem first to provide an effective barrier to the ingress and egress of rabbits from the area, but subsequently runways were found linking the brush area in the upper canyon to brush in the next canyon south. Trapping in the spring these runways failed to capture any migrating rabbits, but the numbers of pellets and runways seemed conclusive evidence of some movement in and out of the area.

Except for the writer, relatively few people visited the area. There is one poorly defined path up the main creek bed. Only those rabbits that died in the traps were removed from the study area. No hunting or snap trapping was done.

Trapping procedures—Homemade Wellhouse traps, 6" × 6" × 20" with one end closed, and Young traps measuring 4½" × 5" × 18", open at both ends, were used. The best bait seemed to be rolled barley since the large-sized particles did not fall through the hardware cloth of the traps as commercial rabbit pellets did. The dry condition made the barley more constant in appeal than fresh bait that quickly dried.

Prebaiting to allow all the population to become familiar with the traps is recommended by Chitty and Kempson (1949) and others. This was tried, first putting bait in likely locations, then with a trap wired open with bait. When the trapping began, no more success was achieved at prebaited locations than at others. Sometimes a trap placed in a new location caught a new rabbit the first night. The data show that two females and five males were caught initially in traps placed in position the first night, without prebaiting. The traps had caught rabbits previously in other locations, so were "seasoned." Four other rabbits were caught the second or third nights after placing the traps. Therefore, of the forty rabbits caught in this study, one-fourth were caught for the first time in newly placed traps. It appears that unfamiliarity with traps does not deter rabbits from entering. Unsuccessful trapping was probably due to incorrect placing of traps rather than to wariness of the animals or lack of prebaiting.

The rabbits were marked by tattooing a number inside both ears with a steel quill pen using Higgins Eternal Ink. This method is especially suitable for species like *Sylvilagus bachmani* and *Sylvilagus audubonii* that have little hair inside the ear. The tattooing, as recommended by Allen (1939), caused no infection or damage and lasted in one case over three years. For handling, the animals were placed a cloth bag with a draw-string opening. Individual animals were weighed, marked (when first trapped), sexed, measured, and inspected for condition and parasites at each handling. No valid criterion for age was found in either the weights or measurements.

Some rabbits injured themselves in the traps, lacerations and broken toes being fairly common. Nine deaths can be attributed to trapping. One rabbit was killed by a spotted skunk (*Spilogale gracilis*) that entered one of the Wellhouse traps. Eight died, presumably from exposure coupled with "shock."

OBSERVATIONS

Validity of trapping in tracing movements—The three main factors affecting the validity of this trapping study were the irregularities of time and trap positions, the acquiring of a "trap habit" by some individuals, and the general question of the trap line as a means of sampling. Trapping was heaviest in the spring. In no month were traps set more than 15 nights, usually less than 10 (Table 1).

The positions of the traps were changed approximately every two and one-half weeks, with about one-third of the traps being moved at each moving period. The most successful traps were usually not shifted. This procedure allowed some individuals to develop a "trap habit," which resulted in more captures. An analysis of trap success as compared to the length of time that the trap was left in one position is shown in Table 2. Traps left more than three weeks in position were more successful than those shifted more frequently. The difference in success between leaving a trap three weeks in one place rather than five or six weeks is so slight, however, that no real inaccuracy would appear to result from the practice of leaving a trap in a favorable position for several weeks.

The real danger from this practice is that an individual might become attracted to bait and so engage in an unnatural pattern of movement. Four rabbits seemed to have developed a trap habit. Female number 9 was caught 14 times in one trap within eight months, and seven times

TABLE 1. Summary of trap nights in relation to vegetation types

Type	MAR	APR	MAY	JUNE	SEPT	OCT	NOV	DEC	JAN	FEB	MAR	APR	MAY	JULY	TOTALS	%
Grass-herb	–	–	–	–	67	66	8	16	3	77	93	54	86	1	471	21
Edge of brush	20	112	71	36	89	220	67	106	56	85	86	52	158	11	1169	52
Brush interior	–	–	–	–	–	7	11	39	54	140	128	53	160	9	611	27
Monthly totals	20	112	71	36	156	293	86	161	113	302	307	169	404	21	2251	
Percent of total trap nights	1	5	3	2	7	13	4	7	5	13	14	8	18	1		

TABLE 2. The effect of trap success of leaving a trap in position; only those positions were made were tabulated

No. trap nights in same position	1–20	21–30	31–40
No. of positions	31	17	11
Total trap nights	334	417	411
No. of captures	57	97	90
Success: Captures/Trap nights	0.17	0.23	0.22

in other traps. Male number 21 was captured six times in two months at one trap, then nine times in two months at another. Female number 11 was caught in the same trap six times (out of eight trap nights) within 18 days; male number 35 likewise was taken five times (six trap nights) in nine days.

Whether a trap line gives a valid indication of animal numbers in an area is a much-argued question. The steepness of terrain and density of brush made the quadrat method of trap deployment too difficult in this area. Various workers have criticized both line and quadrat methods, but the more systematic quadrat system is usually agreed to be better. Calhoun's recent cooperative study uses the trap line as a good, easily analyzed method. Stickel (1948a) points out, however, that a trap line can capture any individuals that are within the distance of their "cruising radius" of the line, hence it follows that such a line samples an area equal to the cruising radius (or one-half the diameter of an animal's home range) on both sides. Since this radius is usually smaller for females (as it is for *Sylvilagus bachmani*), the actual area sampled is different between sexes of the same species. Likewise the radius tends to be larger in habitats where food is scarce, so the area sampled will also differ in contrasting habitats or in the same habitat at different seasons of the year.

Another aspect of the problem is that, postulating equal numbers of males and females in an area, the greater area of "male sampling" will result in more males being caught by a given trap line. Of course, in an area such as the present one, with definite habitat boundaries, all the animals in the area should be caught if the period of trapping is long enough. Later discussion will show that such was probably the case in this study.

Restriction of movement by any arrangement of traps is emphasized by Hayne (1949). Mohr (1947) advises shifting of a trap line to prevent such artificial restriction. Stickel (1948b) showed, conversely, that animals do not tend to be attracted into a new area by bait. In this study, conducted in a somewhat isolated area over a year's period, the problems of trap lines *per se* seem to be compensated for by the intensity and length of trapping.

The pellet census—As a check on the trapping, pellet counts were made during May and June 1949. This method has been advocated by MacLulich (1937) and Eadie (1948) as a relative census method. My concern was not census however, but rather checking the distribution of use in various parts of the area. Deterioration of pellets is assumed equal throughout the area. The distribution of the counting stations was determined by marking ten lines across the area at random distances apart, then arbitrarily distributing 100 stations at random distances along the lines, the number of positions on each line being proportional to its length across the irregular brush area. All distances were taken from a book of random numbers. At each station an area of one-quarter square meter was cleared of all litter and the rabbit pellets counted. The results are discussed in the next section.

Habitat preferences—The rabbit activities of real significance revealed in this study can be classed into three groups; habitat preferences, movements between captures, and individual home ranges. A preceding section sets forth the limitations that should be kept in mind when considering the conclusions drawn about these activities.

Preference for various parts of the habitat was judged by the intensity of use of the grass and continuous brush areas, as reflected in trap success, pellet counts, and location of computed centers of activity. In addition, the line of juncture or edge between grass and brush was treated as a distinct type of habitat, and defined as the 10-foot-wide strip extending 2½ feet into the grass and 7½ feet into the somewhat scattered brush that makes up the chaparral border. No trapping was done under tree cover so this habitat is not considered.

For each individual caught in three or more different locations, a "center of activity" was computed as outlined by Hayne (1949). The difference in trapping methods necessitated a modification of his scheme as follows. A grid was superimposed on the plotted capture points of an individual and each point was assigned coordinates. Since the traps had been placed at the different capture points for varying lengths of time, the computed trap success ratio was the only way to show relative intensity of use of each point. Therefore at each position the number of captures of that individual was divided by the number of nights the traps were set there. Then each coordinate for the position was multiplied by the

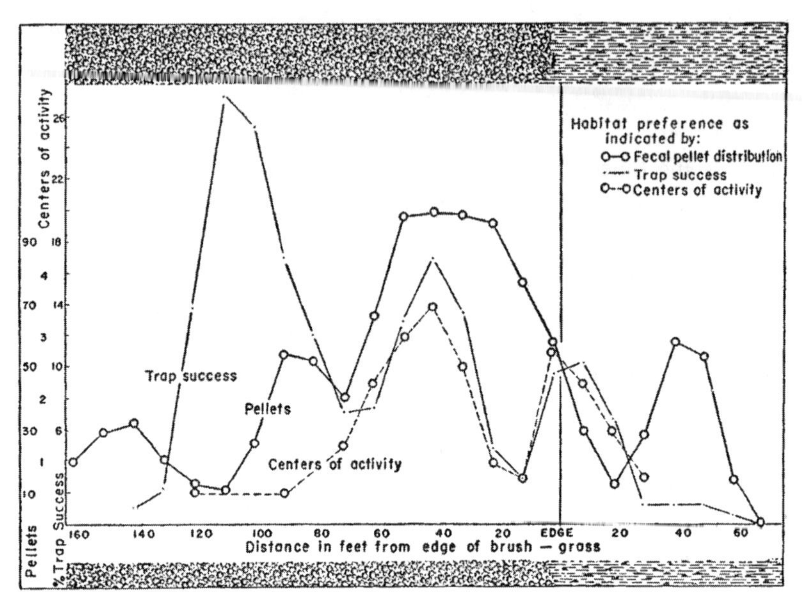

FIG. 1. Use of the chaparral and grass as indicated by fecal pellet occurrence, success of trapping, and distribution of computed centers of activity.

trap success figure, giving two *position-intensity values* for this position. The sum of these values (horizontal and vertical coordinates separately) for all points of capture gave two total position-intensity values, which, when divided by the sum of the trap-success figures, gave a resultant average set of coordinates that could be plotted as a point on the map. This point, embodying both intensity of use and geographical position, is the *center of activity* for the individual. The position of the center of activity in relation to different parts of the habitat is used as an indication of preference.

In Figure 1, the results of the three methods for judging habitat preference are graphically compared. The horizontal coordinate is linear distance measured from the edge of the brush. Both trap success and location of centers of activity show a concentration of use near the brush edge, with a rapid decline out in the grass area. However the pellet counts indicate an area of activity extending to about 40 feet from the edge, with some pellets occurring even 70 feet out. That this obvious activity was not reflected by trapping may be a result of an added wariness by the rabbits in the unprotected areas, with avoidance of unfamiliar objects such as traps.

Use of the interior of the brush seems to be irregular. In all three graphs, a peak of activity is shown 40 feet inside the edge. The low activity 15 to 20 feet inside may indicate a zone used only when moving from the greater protection of the continuous brush to the edge for feeding activity. The peak of activity shown by trap success at 100 feet inside the brush is not indicated by the other two analyses. This may reveal a flaw in trap-revealed activity if animals enter traps more readily in highly protected areas. No other explanation is forthcoming for this apparent contradiction in my data. In any event, some brush rabbits used all parts of the chaparral zone, which was only 340 feet across at the widest point.

Figure 2, A illustrates the trends of trap success by months in the 10-foot-wide edge strip, and the grass and brush areas, as units, on either side. The peak of most successful trapping shifted from the grass area and adjoining edge in winter into the brush by March. This may be correlated with the winter growth and subsequent spring desiccation of the annual herb vegetation, with consequent greater use of the brush in the drier season. To observe the spring shift of rabbits from the grassy edge toward watercourses deep in the brush, three individual ranges were plotted. Most of the captures near grass were made from September to February. Captures deeper in the chaparral were mostly made from March through May. These three were chosen because of the striking regularity with which they showed this tendency. However the others, though less clear-cut than these three, contributed to the general curve illustrated in Figure 2, A.

The three peaks of trap success shown in Figure 1, when analyzed for each sex, show three equal peaks for females, but for males show increasing success from the edge inward with the high peak 100 feet in the brush produced mostly by male captures. The center of activity of male brush rabbits seems to be farther from the edge and closer to stream courses indicating a preference for damper sites deep in the chaparral.

At this point it may be well to consider whether conclusions drawn from a small brush tract with a relatively large edge circuit can be applied to those areas composed of expanses of continuous chaparral. The fecal pellet counts showed that all the brush area was used to some degree. However, the activity peaks at the edge and in the herbaceous vegetation would seem to suggest a decided preference for a discontinuous brush-herb situation. Mixed chaparral and grass is probably better habitat for brush rabbits than continuous chaparral.

Day-to-day movements—Short-term movements between capture points may give some idea of the daily mobility of brush rabbits, as distinguished

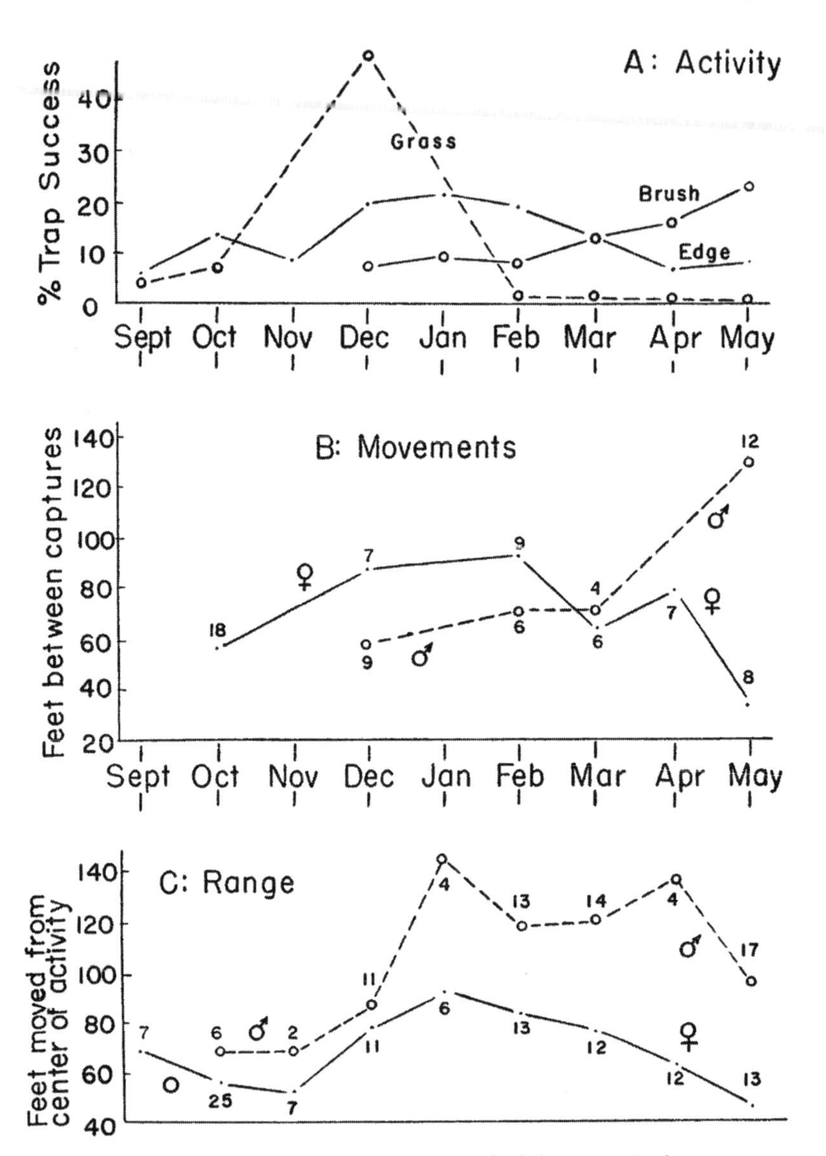

FIG. 2. Seasonal changes in behaviour: A. Preference for habitat type; B. short term movements; C. distance ranged from center of activity. The numeral at each point indicates the number of measurements that were averaged to obtain the value.

from seasonal shifts just discussed. Distances between successive capture points, when less than 10 days had elapsed between the captures, were averaged and plotted as shown in Figure 2, B. Females made their longest moves from December to February, males in February–March and in May, when females moved least. The care of young may curtail female movements in the spring. Summarizing the day-to-day movements between captures, females averaged 66 ± 10 feet and males 81 ± 15 feet, a non-significant difference.

Individual range measurements—An individual home range is defined by Burt (1940) as the area traversed in the normal activities of food gathering, mating, and caring for young. Whether the area delineated by plotting a set of capture points on a map adequately approximates this range is highly debatable. However, the attitude of Stickel (1946) that "actual movements of the animals in the given area in the given period are the item of importance in attempting to derive figures on populations," may be held as an aim in the in the following discussion.

Ranges of the females overlap much less than those of the males (three overlaps as compared to 11 for the 23 rabbits plotted.) If all 40 rabbits are included, there are only five cases of female ranges overlapping as compared to 13 for the males. Five females had completely separate home ranges, but no male enjoyed this isolation. Female brush rabbits seem to be arranged in a semi-territorial pattern, each female occupying a fairly discrete area. However, since no direct observations were made, it is not known whether these areas are defended.

Size of home range was computed by averaging distances of capture from the computed center of activity. For 10 females and 10 males caught before April 1949, in three or more positions, radii were drawn from the computed center of activity to the capture points. When these data are summarized for the whole period of study, the females averaged 69 ± 8 feet, the males 115 ± 10 feet from their respective centers of activity. In other words, the home ranges of males are significantly larger than those of females.

Since ranges may change with season and with other aspects of behavior, the distance ranged from center of activity was summarized by months, as shown in Figure 2, C. The male range is larger at all times, but much more so during the breeding season. The decline in range size among males from April to May is in apparent contradiction to the concurrent lengthening of male short-term moves, shown in Figure 2, B. A possible explanation is that at this time the seasonal shift across the diameter of the individual range is being made, from grass edge deeper into the chaparral (Fig. 2, A). Although the shift may occasion fairly long movements, these are not movements away from the yearly center of activity, but in the vicinity of it. The spring change in female range is more gradual.

Burt (1940) used the greatest diameter of range as one indication of relative size. A summary of this measurement for the 10 females and 10 males used above gave 169 ± 22 feet for the females and 290 ± 31 feet for the males, the difference again being significant.

A possible areal indication of home range might be the area of a circle whose radius is the average distance from the center of activity. The resultant areas are 0.34 acres for the females, 0.95 acres for the males; 0.64 for both sexes together.

Blair (1943) considered that an animal had been trapped over most of its range if further captures did not increase the area indicated by plotting the capture points. In this study, four individuals, all females, were trapped one or more times after their range, as revealed, by trapping, had ceased to increase. Averaging the figures for these four female rabbits showed that for three months and five captures their range did not increase further in size. The distance moved from center of activity was 63 feet, a figure which falls within the significant range for all females, 69 ± 8 feet. In summary, the ratio of female to male values was 0.6:1 for both distance from center of activity and greatest diameter of range; the area ratio was 0.36:1.

Breeding season—The increase in size of the testis of the male was the measure used to indicate the onset of the breeding season. Although the limits of the breeding season are probably controlled by the condition of the female, the only quantitative data secured were male testes sizes. Of the females that were trapped, none was recognized as being pregnant, and examination of nipples at each capture did not reveal nursing by females.

The measurement of testes of live rabbits in the field was possible only after the testes had descended. Testes were not found in the fall or winter until February 1, when, after a lapse in trapping of 18 days, four males were captured with descended testes. According to these data, the onset of the breeding condition in males can be placed in the last half of January in 1949. Figure 3, A shows monthly averages of testis length. The peak of size in April, though based on fewer measurements, is significant. This measurement of length is the entire scrotal contents, which includes the mass of epididymis at the posterior end.

No immature rabbits were captured during the study, for the reason that trapping was suspended in summer. The density of the brush allowed very few sight observations and no small young were seen.

In the first week of May 1949, after a long period with no new captures, six unmarked rabbits were taken. Comparison of weight and also hind foot and ear measurements to others with a year-capture record showed

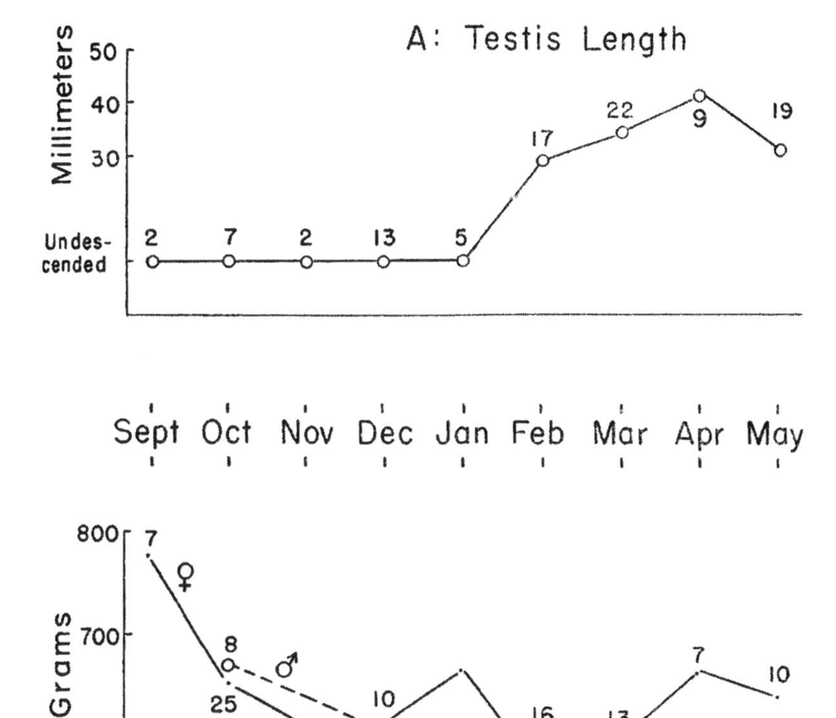

FIG. 3. Seasonal variation in testis size and individual weight: A. Male breeding condition; B. weight difference between the sexes. The numeral at each point indicates the number of measurements that were averaged to obtain the value.

no significant difference, although only measurements taken in May 1949 were compared. This seems to show either that the young grow very quickly, or that these were not young, but immigrants. If the latter assumption is accepted, there is reason to suspect a considerable "shuffle" of adult rabbits during the breeding season.

Population size—During the winter trapping period, unmarked individuals appeared regularly, indicating that the population had not all been marked. By the latter half of March and throughout April 1949, only previously marked rabbits were caught, although intensive trapping continued. From this evidence it may be postulated that the rabbit population of the area, gradually marked through the autumn and winter, had been completely marked by late March. In May, new individuals appeared, possibly by immigration connected with breeding season activity.

Analyzing the captures made during the spring months, ten individuals were caught either during the month of April, or both before and after. Since it is assumed that all individuals were marked, these ten can be taken as the adult population on the area in April.

To estimate numbers for other times of the year, a curve representing the rate of disappearance of rabbits from the trapping record was constructed. A table of months after first capture was set up and all individual records filled in, then added together for each month. For example, of 34 rabbits marked, 25 were recaptured at least a month after marking, 17 two months later, and so on. Figure 4, A illustrates this curve. The causes of disappearance may have been emigration as well as death, so that this curve must not be taken as a true record of death rate. However, it was felt that the rate of depopulation, regardless of its natural causes, was the important measurement. Besides the uncontrollable factor of inadequate trapping, the seven trap deaths may be reckoned as an unnatural influence. Five of these occurred within two months, the other two at 10 and 13 months, after marking. Therefore the trap deaths exaggerate the steepness of the curve mostly in the first two months.

The curve serves as an approximation of the yearly population cycle, with a peak in late spring when the adult population is augmented by grown young, a rapid die-off during the dry summer and early autumn, (mostly, I presume, of growing young), and a slower death rate during the winter when the rains bring a better food supply of green annual vegetation. We may take the April population as the minimum adult population of the year, just before the first crop of young have reached full size. The minimum point on the curve "A" of Figure 4 is reached about month number eleven, which would be roughly equivalent to April. The population for September, just before the winter rains begin, can be approximated by the following ratio, using the curve values of month eleven and month four, for April and September respectively:

$$\frac{\text{Sept. pop.}}{\text{April pop.}} = \frac{\text{Month 4 value}}{\text{Month 11 value}};\ \frac{\text{Sept. pop.}}{\text{April pop.}} = \frac{13}{5};\ \text{and Sept. pop.} = \frac{13\text{x}10}{5} = 26$$

The population density may be computed, taking the brush area plus a strip of grass 7 feet wide around the edge, since Figure 3 indicates that

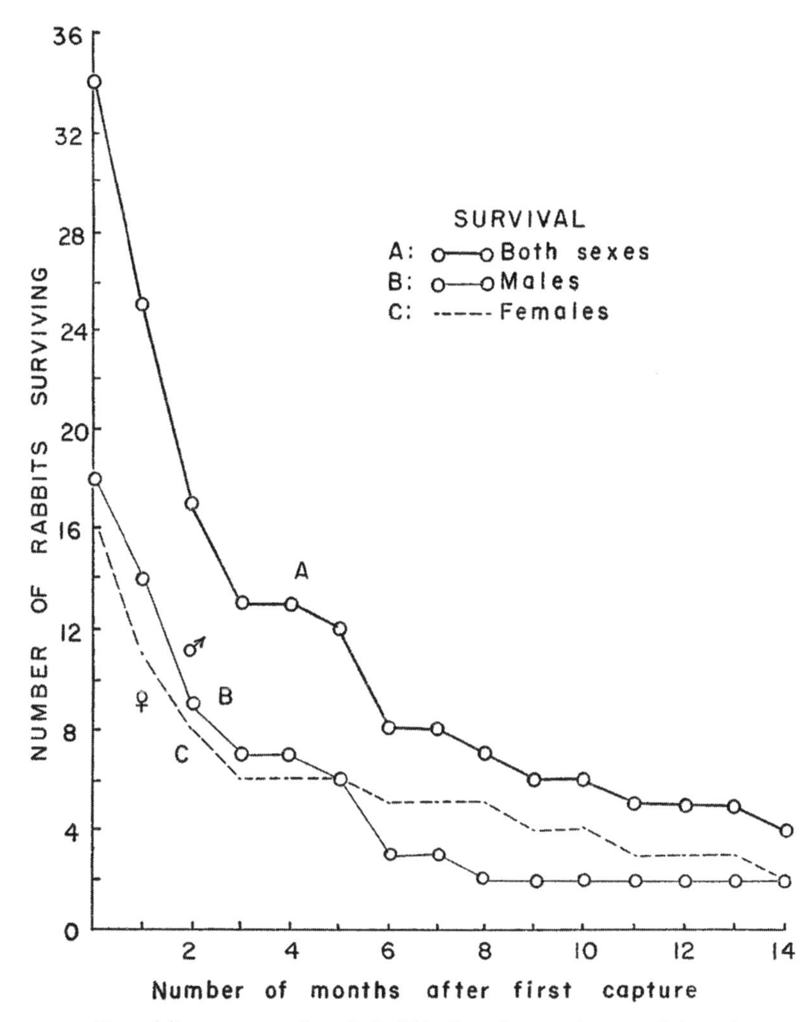

FIG. 4. Rate of disappearance of marked rabbits from the trapping record throughout the trapping period: A. Both sexes; B. males; C. females.

the grass was effectively used to this distance. This area is 11.5 acres, so 26/11.5 equals 2.3 rabbits per acre of chaparral-grass mixture, or a range of .44 acres per rabbit in September.

The mortality rates for males and females differ, as shown by the curves "B" and "C" in Figure 4. The rate of disappearance of males was greater during the early months, possibly indicating that males may escape trapping by their habits of wide ranging and possible emigration. A greater susceptibility to "shock disease" may be indicated by the fact that of eight rabbits dying in traps, five were males. No blood tests were made to diagnose possible hypoglycemia, but the shock convulsions resembled those described by Green and Larson (1938)—retraction of the head, kicking movements of the legs while the animal lay on its side, extension of the hind legs and almost immediate rigor mortis in this position.

There is some evidence that most of the summer mortality is in the young of the year, since of the seven adult rabbits trapped in April and May of 1948, five were recaptured in September.

Certain individuals lived to a relatively advanced age. During the summer following this study, A. S. Mossman trapped the area and removed four rabbits marked in this study. These plus one from the original period, yielded the following longevity data for rabbits in the natural environment: three females of ages 2¼, 1¾ and 1 year each; two males of 1½ and 1¼ years each. The two longest-lived females were kept in captivity an additional year, so that their total known life spans were 3¼ and 2¾ years each. In addition, the rabbits were full grown when marked, so that an additional increment of from two to eight months is probable, increasing the longest span to at least 3½ years.

Weight changes—The weight of each rabbit was taken with a spring balance at every capture. The animals generally remained quiet in a drawstring cloth bag. They were weighed to the nearest ten grams.

Figure 3, B shows the monthly average weights for each sex. After the February onset of the breeding season the weights of males dropped significantly, while the females increased, due perhaps to pregnancy. In both sexes, the heaviest animals were trapped in early fall. The overall average weights were 583 ± 8 grams for males, 637 ± 7 grams for females. This includes 76 male weighings, 101 female.

Some extreme examples of individual variation were noted. Within two days, changes of 65 to 95 grams, in both sexes, were recorded. One female weighed 800 grams in September, 540 in November, 730 in April, and 650 in May.

Parasites—Ectoparasites were collected at each capture and a blood smear taken from the ear. The blood slides were examined by Dr. Betty Davis and were found to contain no parasitic organisms. The ectoparasites consisted of fleas in the pelage, ticks (usually attached to the skin in the head and neck region), and mites in the ear canal and in the genital region. At present writing the fleas have not been identified; the ticks proved to be *Ixodes* sp. and *Haemophysalis leporis-palustris*; the mites were *Trombicula californica.*

Autopsies of rabbits that died in the traps yielded some endoparasites. In all rabbits, intestinal tapeworms were found as well as nematode stomach worms. No evidence of degeneration or spotting of the liver was noted. In two live individuals, bot fly larvae (*Cuterebra*) were found under the skin in the neck area. In one of these, a female, the larva had matured and dropped out in the trap. This same female had bot larvae twice, in October and again in December; its behaviour was sluggish at these times, but after these infestations it appeared normal for the remaining five months of its capture. A male, caught by A. S. Mossman, a year after this study, had two large bot larvae in the neck region. The individual died four days after capture; since there was much neck bleeding the bots were assumed to be the main contributing cause of death.

Acknowledgements—In the identification of the ectoparasites, Mr. Douglas Gould and Mr. G. H. Rohrbacker of the Agricultural Experiment Station at the University of California were very helpful; Dr. Betty S. Davis and Dr. Marietta Voge of the Department of Zoology kindly examined the blood smears and parasitic worms, respectively. I am especially indebted to Dr. A. Starker Leopold of the Museum of Vertebrate Zoology for his advice and encouragement throughout the study.

SUMMARY

From March 1948, to May 1949, 40 brush rabbits were trapped, marked, and released in a more or less isolated patch of chaparral near the University of California campus, Berkeley. Subsequent movements of the animals were traced by retrapping.

Home ranges of most of the animals centered about 40 feet inside the edge of the brush, but rabbits were trapped throughout the chaparral and the presence of fecal pellets indicated that, at least in winter, the rabbits foraged into the adjoining grassland up to 70 feet from the brush. During the dry season little use was made of grassland; at that time the population seemed to shift deeper into the chaparral. Males moved about more freely than females with the result that the computed average home range for males was 115 feet in radius, and for females, 69 feet. Home ranges of males overlapped, whereas those of females seemed to be discrete with little or no overlap.

By early spring no unmarked rabbits were caught, and it was assumed that the whole population had been marked. Rabbits known to be still living at that time (April) were 10 in number, which constituted a density of 0.9 animals per acre on the 11.5-acre study area. Many animals marked in fall and early winter had disappeared from the population by April. Construction of a curve, based on the rate of population shrinkage, permitted an estimation of the September population at about 2.3 rabbits per acre.

Breeding season, as revealed by testis development in the male, began in February and peaked in April. Yet no immature rabbits were taken in the traps up to early May when trapping was suspended each year. Six unmarked rabbits appeared in the traps in May, but these seemed to be immigrating adults rather than young.

LITERATURE CITED

Allen, D. L. 1939. Michigan cottontails in winter. Jour. Wildlife Mgt., 3: 307–322.

Blair, W. F. 1943. Populations of the deer mouse and associated small mammals in the mesquite association of southern New Mexico. Contrib. Lab. Vert. Biol., Univ. Michigan, no. 21. 40 p.

Burt, W. H. 1940. Territorial behaviour and populations of some small mammals in southern Michigan. Misc. Publ. Mus. Zool., Univ. Michigan, no. 45: 58 p.

Calhoun, J. B. 1948. North American census of small mammals. Release No.1, Rodent Ecology Project, Johns Hopkins Univ. 9 p.

Chitty, D., and D. A. Kempson. 1949. Prebaiting small mammals and a new design of the live trap. Ecology, 30: 536–542.

Eadie, W. R. 1948. Shrewmouse predation during low mouse abundance. Jour. Mamm., 29: 35–37.

Green, R. G., and C. L. Larson. 1938. A description of shock disease in the snowshoe hare. Amer. Jour. Hygiene, 28: 190–212.

Hayne, D. W. 1949. Calculation of size of home range. Jour. Mamm., 30: 1–18.

MacLulich, D. A. 1937. Fluctuations in the numbers of the varying hare (*Lepus americanus*). Univ. Toronto Studies, Biol. Ser. no. 43, 1–136.

Mohr, C. O. 1947. Table of equivalent populations of North American small mammals. Amer. Midl. Nat., 37: 223–249.

Stickel, Lucille F. 1946. Experimental analysis of methods for measuring small mammal populations. Jour. Wildlife Mgt., 10: 150–159.

——— 1948a. The trap line as a measure of small mammal populations. Jour. Wildlife Mgt., 12: 153–161.

——— 1948b. Effect of bait in live trapping *Peromyscus*. Jour. Wildlife Mgt., 12: 211–212.

Connell, J. H. 1954. Home range and mobility of brush rabbits in California chaparral. Journal of Mammalogy *35: 392–405. Reprinted with permission of American Society of Mammalogists, Allen Press, Inc.*

Food recognition and predation on opistobranchs by *Navanax inermis* (1963)

Robert T. Paine

An adequate description of the ecological properties of a species should include its trophic relationships in terms of recognition of both predators and potential prey, and the quantitative and qualitative results of such behavior. Among higher organisms, gastropods offer many advantages in understanding these phenomena; they possess a variety of sensory mechanisms capable of detecting specific chemical stimuli (Kohn, 1961), and many species are characterized by a dietary eclecticism (Graham, 1955). The present paper describes the fundamental features of predation by the cephalaspidian *Navanax inermis*, 1862, on other opisthobranchs, especially the method of prey detection, and potential defensive adaptations of these prey species.

This work was done during the tenure of a Sverdrup Postdoctoral Fellowship at Scripps Institution of Oceanography, La Jolla, California, and probably would not have been completed without the taxonomic counsel of J. R. Lance, who identified most of the opisthobranchs. Thanks are also due to R. Stohler, R. Rosenblalt, and T. S. Park for identifications of the prosobranchs, fishes, and amphipods, respectively; J. Lythgoe, P. B. Taylor, R. Ford, and B. Harrison for collection of many of the nudibranchs used in the feeding experiments; and A. J. Kohn for critically reading the manuscript.

DISTRIBUTION OF NAVANAX

Navanax inermis, hereinafter referred to as *Navanax*, ranges from Monterey Bay at least to Ensenada, Baja California (MacGinitie and MacGinitie, 1949). In my experience maximum size (*ca.* 20 cm or 300 gm wet weight) and abundance are reached in shallow protected bays where *Navanax* occurs from the tidal zone to depths of 15 to 25 feet. Although such bays must be considered the "typical" habitat, small individuals (5 to 6 cm length, up to 8 gm wet weight) were found regularly on rocky intertidal areas where positioning of boulders results in areas relatively protected from severe surf action and surge. These sites, however, do not appear able to support permanent populations because seasonal shoreward movements of sand alter the bottom topography by filling in protective depressions and crevices. The oceanic subtidal distribution is enigmatic. Scuba divers at the Scripps Institution of Oceanography have rarely, if ever, taken *Navanax* in depths of 30 to 120 feet in the vicinity of La Jolla, although diving frequently. On the other hand, R. Ford, on an initial dive off the Coronado Islands, found one individual at 80 feet, and W. D. Clarke (pers. comm.) has observed *Navanax* in numbers in 30 to 50 feet off Santa Barbara. *Navanax*, then, should be considered a potential predator of acceptable organisms inhabiting bays and some oceanic environments from the intertidal zone to a depth of about 100 feet, and also of any more oceanic organisms washed into these areas.

FEEDING BEHAVIOR

When *Navanax* makes contact with a suitable prey species it is swallowed whole, being sucked into *Navanax*'s mouth along with much water and debris. The latter are subsequently expelled through grooved channels in the buccal apparatus while fleshy finger-like protuberances help retain the prey. The prey is simply digested without the aid of trituration, and its jaws, radulae, and shell, if any, are defecated unaltered. (Incidentally, this provides a good preparation of delicate nudibranch jaws and radulae.) The field diet was determined from such unaltered remains.

The chemosensory acuity of many gastropods permits them to recognize and then locate prey by distance chemoreception (Stehouwer 1952, Braams & Geelen 1953, Kohn 1959, Blake 1960, Cook 1962). Kohn (1961) has drawn a distinction between this type of behavior and contact chemoreception in which prey recognition is based on taste of gustation. The observations reported below suggest that, on the basis of present knowledge, *Navanax* is unique among gastropods in relying entirely on contact chemoreception to locate acceptable prey.

It has been known for some time that *Navanax* feeds on other opisthobranchs, especially *Bulla* and *Haminoea* (Johnson and Snook, 1927; MacGinitie and MacGinitie, 1949; Ricketts and Calvin, 1952). Therefore, to emulate natural conditions 3 to 4 cm of clean sand were placed in aquaria and water added to a depth of five to ten cm. A *Bulla*, or occasionally *Haminoea* or *Hermissenda* was then tediously guided around the central portions of the container, being forced to avoid the walls. Because the cephalic shield of a moving *Bulla* is almost invariably just under the sand surface, sand grains or other small debris are bound in mucus, passed dorsally and posteriorly and eventually form a hollow tunnel marking the animal's past movements. In quiet bays these mucus tunnels formed by *Bulla*, *Haminoea* and *Aglaja* are conspicuous features of undisturbed bottoms. Upon completion of the experimental mucus tunnel, a *Navanax* would be released and observed, and the results mapped as in Figure 1. When encountering the mucus trail *Navanax* deviates from its former course and immediately begins following the trail. Two organs, called by Marcus (1961, p. 66) the "inner fold of head shield" are kept in contact with the mucus trail in such a fashion that one, and usually both, are over it. If one of these presumed chemoreceptors loses contact with the trail, the predator's movements are appropriately corrected. In this manner the spoor is followed accurately.

The *Bulla* by appropriate guidance can be forced to leave a trail, such as in Figure 1, in which after some meandering, it approaches to within 2–3 cm of an older portion of the trail. *Navanax*, released at the trail's beginning, invariably bypassed its intended prey while following its trail, although the prey was within a short distance. In repetitions of this simple experiment, none of 9 *Navanax* deviated from the mucus trail. In addition, *Navanax* was not seen to orient to water currents flowing over *Bulla* or *Haminoea*, although apparatus similar to that used by Stehouwer (1952) and Kohn (1959) was tried. The conclusion seems warranted that prey is located only after direct contact, or contact with its mucus trail; distance chemoreception does not appear to operate in obtaining food.

Navanax does not orient to low ridges or depressions in its environment, but moves over these in haphazard manner, either when it is or is not tracking prey. Also, only opisthobranch mucus appears capable of eliciting the tracking since trails formed by the prosobranch gastropods *Olivella biplicata*, *Nassarius tegula* and *Conus californicus* were not followed. Ability to distinguish opisthobranch from prosobranch mucus, or failure to respond to the latter, has an obvious selective advantage in that trails of both varieties characterize environments in which *Navanax* seeks prey. If, as demonstrated below, mainly opisthobranchs are devoured, tracking prosobranchs will increase energy expenditures and decrease food-getting deficiency.

Upon contact with a trail, *Navanax* is just as likely to turn away from the prey as toward it. If the trail is followed away from the prey, a characteristic "searching" behavior is observed at its end. Once contact is lost, *Navanax* swings its head back and forth in small arcs, and eventually may even turn itself around. Since an individual responds to its own mucus, as well as the mucus of other *Navanax* individuals, such behavior obscures

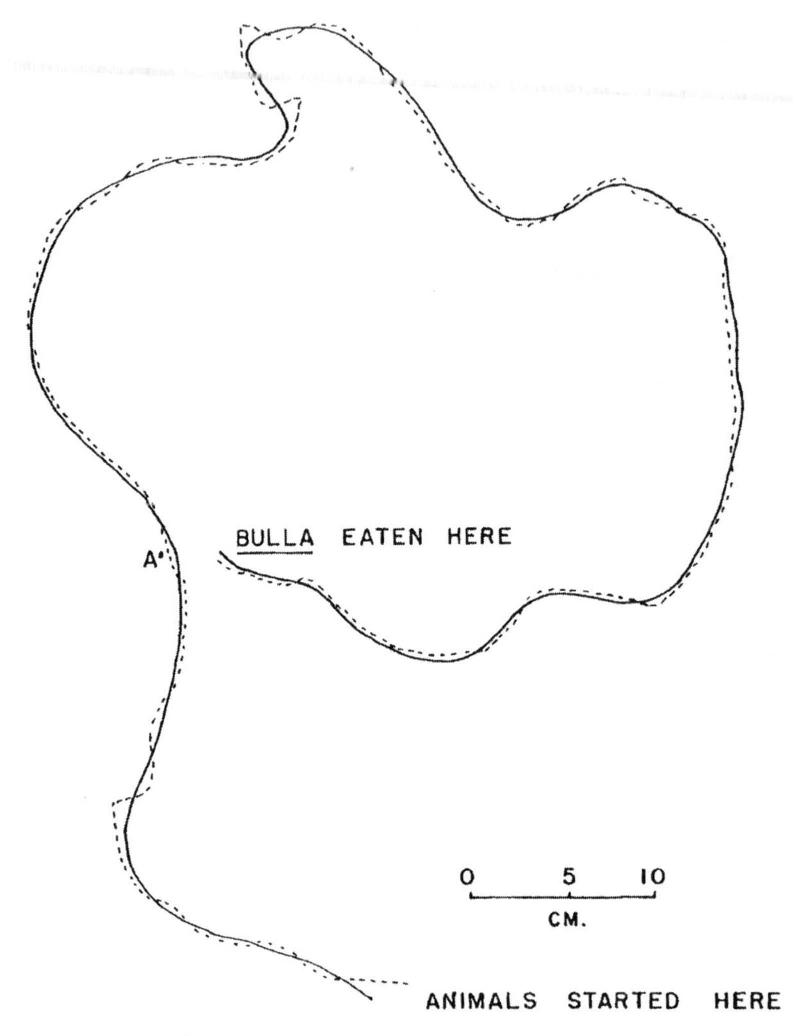

FIG. 1. A typical example of the ability of *Navanax* (------) to follow the mucus-impregnated sand trail of *Bulla* (——). At point "A" the *Navanax*, although within two to three cm of its intended prey, will continue to follow the trail away from its prey.

the original trail so that it may be lost. The success of corrective behavior in this situation is low.

Finally, some comments should be made on the relationship of size and behavior. Small individuals (2–12 cm length) almost invariably would track suitable prey, and their behavior could be demonstrated predictably, however, larger *Navanax* at times unaccountably failed to respond to fresh *Bulla* mucus, suggesting the presence of an unrecognized age or size specific behavioral pattern (such as mating behavior) superimposed on the food getting requirements.

FOOD IN NATURE

In nature, *Navanax* feeds on opisthobranchs, and to a lesser extent on a variety of other invertebrates and fishes as represented by identifiable hard parts in *Navanax* fecal pellets. The species list (Table 1) has not been quantified because diet depends on, among other things, prey availability, and such a quantification would only indicate the author's choice of sampling areas. It should be emphasized, though, that on a dry weight basis, opisthobranchs constituted more than 95% of all food eaten.

In Table 1, ten of the 29 species of natural prey are not opistobranchs but of these, only prosobranch gastropods, *Cystiscus pyriformis* and *Barleeia* sp., and one fish, were eaten frequently. Neither of these prosobranchs as discussed below, were swallowed voluntarily by *Navanax* in the laboratory, suggesting that they probably were eaten accidentally in the field. The opportunity for this to happen exists naturally since *Barleeia* and *Cystiscus* along with *Bulla* abound on mats of vegetation over which *Navanax* hunts. These snails are probably engulfed in the following manner. When *Navanax* first encounters a trail of acceptable mucus, they will often "taste" it, swallowing in the process some of the local substrate. These snails, then, may be taken during these "tastings," or during the actual ingestion of a selected prey species.

In collections of *Navanax* from the Flood Control Channel, Mission Bay, San Diego, *Barleeia*, an operculate snail, was often (as many as 17) found living in the fecal pellets, having survived the passage through *Navanax's* gut; rarely was a dead individual encountered. If *Barleeia* were offered to *Navanax* in the laboratory, they were not eaten, and, in fact, *Navanax* did not appear to recognize them as food. Therefore, *Barleeia* were counted and then rolled into strands of *Bulla* foot, which slightly starved *Navanax* ingested readily. By varying the number of *Bulla* eaten prior to the swallowing of this package, the rate of passage through the gut could be altered. Table 2 gives the percentage survival of *Barleeia* as a function of the time of passage, indicating this small snail readily survives the normal term of such treatment.

Navanax have also defecated up to 25 partially digested *Cystiscus* (a non-operculate prosobranch), which on the basis of laboratory observations do not appear to be eaten voluntarily either. Ten *Cystiscus* were confined for two days with each of 5 different *Navanax*, the equivalent

TABLE 1. Tabulated observations on the field and laboratory diet of *Navanax*. In Column 1 the (*) indicates species eaten occasionally while (**) denotes prey species swallowed frequently. Columns 2, 3, and 4 report laboratory observations in which a variety of opisthobranchs were offered to *Navanax*. Columns 5 and 6 report data substantiating the division of some prey species into the categories "soft" and "hard" bodied.

		Laboratory Diet				
	FIELD DIET OCCURRENCE	NUMBER EATEN	NUMBER OFFERED	% CONSUMED	% WATER	SAMPLE SIZE
Opisthobranchia						
Doridacea						
Cadlina flavomacalata		2	2	100	70.3	1
Cadlina marginata		0	3	0	81.3	2
Glossodoris californiensis		4	4	100	88.6	1
Glossodoris macfarlandi		1	4	25	89.6	1
Rostanga pulchra		0	11	0	70.2	1
Archidoris montereyensis		0	1	0	85.5	3
Anisodoris nobilis		0	2	0		
Diaulula sandiegensis		0	19	0	77.7	4
Aegires albopunctatus		0	20	0	71.0	3
Polycera atra	**	16	16	100	89.8	8
Laila cockerelli		2	19	10.5	69.7	3
Triopha carpenteri	*	4	4	100	91.1	1
Triopha maculata	**	26	26	100	89.0	16
Crimora coneja		2	2	100		
Acanthodoris rhodoceras	**	2	2	100	90.5	3
Onchidoris hystricina		0	4	0	74.6	1
Hopkinsia rosacea		1	17	5.9	76.7	7
Trapania velox		1	1	100		
Corambe pacifica		4	4	100		
Dendrodoris albopunctata	*	0	13	0	78.3	9
? *Thordisa*		0	10	0	78.9	1
unidentified mottled dorid		0	10	0	78.7	4
unidentified yellow dorid		0	2	0		
unidentified white dorid		0	1	0		
Dendronotacae						
Tritonia festiva		1	1	100		
Dendronotus frondosus	*	6	6	100		
Melibe leonina		2	2	100		
Ariminacea						
Antiopella barbarensis		4	4	100	93.2	2
Dirona albolineata		2	3	66.7		
Dirona picta	**	8	8	100	92.1	3
Eolidacea						
Flabellina iodinea		1	22	4.5	90.3	8
Capellinia rustya		17	17	100		
near *Catriona*	*					
Hermissenda crassicornis	**	73	73	100	90.3	15
Phidiana pugnax		3	3	100		

	FIELD DIET OCCURRENCE	Laboratory Diet				
		NUMBER EATEN	NUMBER OFFERED	% CONSUMED	% WATER	SAMPLE SIZE
Tergipes sp.		1	1	100		
Facelina sp.		1	1	100		
Aeolidia papillosa		2	2	100		
Spurilla chromosoma	*					
unidentified eolid, Point Loma		1	1	100		
Other Opisthobranchia						
Bulla gouldiana	**	95	95	100		
Haminoea virescens	**	118	118	100		
Aglaja sp.	**	16	16	100		
Navanax inermis	**	13	13	100	89	27
Chelidonura phocae		1	1	100		
red and white cephala-spidean		1	1	100		
Aplysia californica	*	2	2	100		
Pleurobranchus cali-fornica	*					
Pleurobranchaea sp.		0	6	0	93.7	2
Aeteon punctocaelatus	*	1	13	7.7		
Turbonilla sp.	*					
Retusa sp.	*	1	3	33.3		
Hermaeina smithi	*					
Prosobranchia						
Barleeia sp.	**					
Cysticus pyriformis	**					
Erato vitellina	*					
Mitrella carinata	*					
Annelida						
Nereidae	*					
Arthropoda						
Caprellidae	*					
Ischyroceridae	*					
Pleustidae	*					
Spirontocaris sp.	*					
Vertebrata						
Gobiidae	*					
Porichthys myriaster	**					

of 20 *Cystiscus* days association with each predator. Only one was eaten, although the predators often went through the preliminary attack motions. On contact *Navanax* usually withdrew sharply, and on several occasions was seen to rapidly eject a previously swallowed *Cystiscus*. At the termination of this experiment, all *Navanax* fed in the usual fashion on the nudibranchs *Antiopelle* and *Hermissenda*, suggesting that their feeding abilities were not impaired. Like the *Barleeia*, *Cystiscus*, and quite possibly

TABLE 2. Percentage survival of *Barleeia* sp. After varying intervals of time in the digestive tract of *Navanax*

Hours in gut	NUMBER USED	% SURVIVAL
18	113	100
22	56	96
40	64	77
70	26	46
94	2	50

all prosobranchs appearing in *Navanax* fecal pellets probably have been swallowed accidentally; nereid worms, amphipods, and shrimp probably enter in *Navanax's* diet similarly.

The gobiid fish, and especially *Porichthys myriasta*, were found in fecal pellets occasionally from December 1961 to February 1962 but in 3 collections from the Flood Control Channel, 20 July to 4 August 1962, *Porichthys* formed the bulk of the *Navanax* diet, with up to 13 fish, 20 to 30 mm in length, in each *Navanax*. These fish did not have the appearance of carrion; their eyes and photophores seemed fresh despite their recent exposure to *Navanax's* digestive fluids. It is improbable that any item consumed in these numbers, simultaneously forming the bulk of the food ingested, is eaten accidentally. Hubbs (1920) mentions that *P. notatus*, a sluggish fish of retiring habits, migrates into shallow bays in spring where it buries itself just under the sand; in so doing, it or similar fishes become accessible to *Navanax*. The radical departure from the normal diet cannot be explained at present, but it is the author's opinion that the piscivorous habit of *Navanax* is neither fortuitous behavior, nor represents the consumption of carrion. *Navanax* thus joins *Conus* (Kohn, 1956) as a piscivorous gastropod.

LABORATORY DIET

Navanax can be retained readily in the laboratory. Therefore, a wide variety of local opisthobranchs from shallow and deep water were offered to *Navanax* in plastic containers of about 500 cc capacity. Prey specimens were placed with small to medium sized (2 to 10 cm length) *Navanax*, and the ensuing events recorded. This is perhaps the most direct way of observing food preference, the judgment as "acceptable" or "not acceptable" being governed by whether a particular prey species was swallowed and then retained by the predator. These observations were made over a 6 to 7 month period, and involved 75 to 100 different *Navanax*; as such they

TABLE 3. The breakdown by species within various higher categories of opisthobranchs, of those organisms either eaten or not eaten in the laboratory by *Navanax*.

	NUMBER EATEN	NUMBER NOT EATEN
Eolidacea	8	1
Doridacea	9	15
Arminacea	3	0
Dendronotacea	3	0
Other opisthobranchs	7	3

probably represent a modal behavior for the species. The results tabulated by prey species in Table 1 and by higher categories in Table 3 confirm the field data that *Navanax* actively preys on other opisthobranchs. This is then the first predator known to the author in which postlarval nudibranchs form an appreciable portion of the diet.

DEFENSIVE ADAPTATIONS OF THE OPISTHOBRANCH PREY

Table 3 indicates that not all opisthobranch groupings are eaten equally. Opisthobranchs have several defense mechanisms including nematocysts, acid secretions, other secretions, spicules, behavior, including swimming, and coloration (Thompson, 1960a). These will now be examined with reference to those species consumed by *Navanax*. Where these can be shown to be an inefficient predation deterrent against *Navanax* one must assume that they prove effective against other kinds of predators.

Nematocysts: Although the possession of nematocysts by some eolids is well known (Grosvenor, 1903; Thompson, 1960a), they appear to be an ineffective deterrent. All eolids offered were typically consumed except *Flabellina* in which only 1 out of 22 specimens was eaten. In addition *Flabellina* proved to be the only eolid that could be maintained with *Navanax* in large, running water aquaria. Although *Flabellina* swims with ease, such evasive action would not prevent *Navanax* from consuming it if it were an "acceptable" food item. Its relative immunity must be laid to other causes, as yet unknown.

Acid secretions: Only two "tectibranch" species (*Pleurobranchea* sp., pH 1 and *Acteon punctocaelatus*, pH 3) and one nudibranch (unidentified mottled dorid, pH 1) were demonstrated (by touching their bodies with pH paper) to secrete acid, presumably similar to that described by Thompson & Slinn (1959). All the remaining species large enough to be tested proved to be neutral. None of the acid-secreters were eaten by

Navanax in the laboratory, with on exception, and observations on the predator at the moment of prey contact, in which *Navanax* either withdraws abruptly or rolls its anterior end into its body, suggest that acid secretion is extremely effective. *Navanax* has been observed to swallow, and then immediately regurgitate, *Acteon*, a shelled opisthobranch known to have toxin and acid secreting glands (Fretter & Graham, 1954). This delayed response supports Thompson's (1960b) belief that acid secretion is a discontinuous process, active release of acid being invoked only upon disturbance.

The similarity in behavior of *Navanax* on contact with both *Cystiscus* and *Acteon*, a known acid-secreter, suggests that the former also produces a noxious, probably acidic, secretion. However, both of these species are eaten by *Navanax* in nature (Table 1). It is likely that degree of prior stimulation of the prey, strength of the acid produced, or time since last feeding of the predator all influence the interaction. The presence of an external shell in *Acteon* and *Cystiscus* may also prove important, since this will increase the interval between contact by *Navanax* and release of acid, hence increase the probability of being swallowed.

A single *Navanax* also defecated two shells of *Pleurobranchus californicus*. Little is known about this latter species, but by analogy with another member of the genus (Thompson & Slinn, 1959), it is apt to secrete strong acid. This observation suggests that the "trophic rogue" discussed below may not be limited in occurrence, to the single specimen observed in the laboratory.

Spicules: Thompson (1960a) has suggested that the spiculate texture of the dorid mantle may decrease the animal's attractiveness as fish food. Another explanation of the defensive value of these spicules is now proposed.

Navanax was offered 24 dorid species, which divide into 2 distinct groups on the basis of their dry weight to wet weight relationships (Table 1). These data were obtained by weighing a damp-dried animal ("wet" weight) and then oven drying that individual at 100°C ("dry" weight). Because dorid spicules are a hydrated amorphous calcareous gel (Odum, 1951) a weight value, comparable to that of other animals cannot be obtained at 100°C, and the present values are reported as "dry" weight. With two exceptions, all dorids in which about 90% of their wet weight is water were eaten, and conversely, those in which only 70–80% is water were not.

When *Navanax* feeds, the prey are sucked into the mouth. Larger prey individuals are ingested only if they can be molded to conform to a size

Navanax can swallow. In this respect "soft" bodied dorids are similar to eolids, dendronotaceans, arminaceans, and some other opisthobranchs. On the other hand, spicules provide a certain body rigidity which prevents the "hard bodied" grouping from being treated in this manner. In addition, body rigidity, coupled with the broad foot of such dorids, makes them difficult to dislodge from flat surfaces, especially by a predator which can only ingest whole prey and possesses no teeth, jaws or similar structures.

However, the relationship between *Navanax* and its dorid prey is not this obvious. In the laboratory feeding experiments it was noted that the predators would often turn away, ignore, or make a minimal effort at swallowing "hard" bodied dorids even if the individual was small enough and had been placed on its dorsal surface. Such behavior, characterizing contact with *Aegires*, *Rostanga*, *Hopkinsia*, etc., suggests that other information is involved. The logical explanation is that some glandular secretion, distasteful or repellent to *Navanax*, has been emitted, and has been recognized by the predator. In this regard, *Navanax* made no attempt to swallow most "hard" dorids and the "soft" bodied nudibranchs *Flabellina* (90.3% water) and *Glossodoris macfarlandi* (89.6% water) enjoyed a certain immunity. The consumption of *Cadlina flavomaculata* (70.3% water) warrants further investigations.

Other Secretions: The heavily glandular surface epithelium and copious mucus secretations of many opisthobranchs are potential defense mechanisms (Thompson 1960a), although their efficacy generally remains undemonstrated. An exception is the report of a seemingly volatile, neutral, poisonous secretion emitted along with much mucus by *Phyllidia varicosa* Lamarck, 1801 (Johannes, 1963). The failure of *Navanax* to attack or eat *Flabellina* and *Glossodoris macfarlandi* suggests the presence of an effective non-acidic secretion although it must be noted that all non-spiculate, heavily glandular Arminacea and Dendronotacea species were eaten without hesitation.

Coloration and Behavior: Little information exists with respect to the efficiency of these categories of defense mechanism. *Navanax* has eyes, yet gives every appearance of being guided entirely by its chemosensory abilities and in fact seems oblivious to the striking coloration of some nudibranchs, consuming with equal relish the dull and the gaudy. Prey color is thus unlikely to effect this particular predator-prey interaction. Although most of the eolids wriggled violently in the presence of *Navanax*, such behavior seemed of little avail. In the field it may serve to dislodge them into currents capable of moving them from the immediate vicinity of this predator.

OTHER OBSERVATIONS

The general dominance of shelled opisthobranchs, especially *Bulla* and *Haminoea*, in *Navanax*'s diet indicates that these forms enjoy no special immunity. A shell may impart some protection when a particular prey individual is attacked by a *Navanax* too small to successfully engulf it. The rigidity due to the shell, then, functions in the same fashion and with similar results as the internal spicules.

The above records and observations probably indicated the usual course of, and limits of, predation by *Navanax* on other opisthobranchs. A single individual was discovered, though, which did not respond in the usual fashion, and in its 15 days of captivity ate everything which, by the observer's prior experience, should have been avoided. This trophic rogue consumed the only individuals (Table 1) of *Flabellina*, *Acteon*, *Glossodoris macfarlandi* and *Laila* observed to have been eaten. It appeared to be morphologically intact, and showed no signs of previous injury. Conceivably it represents an adaptational safety factor, insuring a third or fourth level consumer species, already characterized by a high degree of dietary selectivity, against a sudden disappearance of its preferred food species.

Finally, it should be emphasized that description of a species' normal diet, based solely on gut analyses from preserved material, without supplementary observations of living specimens, is not without risks. Many species, including *Navanax*, capture prey by sucking or tearing at them, unavoidably engulfing some adventitious materials. The role of these in the predator's nutrition is unknown (though in the present case is known to be minimal). In addition, though gut analyses will be representative of what has been swallowed, they will not reflect accurately the species' preferred diet.

SUMMARY

The cephalaspidean *Navanax inermis* subsists primarily on other opisthobranchs which, after being selected are swallowed whole. A smattering of small prosobranchs, shrimp, worms, and amphipods in its diet are probably engulfed accidentally. Fish, notably *Porichthys myriaster*, are also eaten, at times in such quantity that the act appears voluntary.

Navanax locates its prey by contact (not distance) chemoreception by following mucus-impregnated opisthobranch sand trails. It does not respond to prosobranch mucus. *Navanax* appears to be the first species known to prey heavily on postlarval nudibranchs, and only the second reported piscivorous gastropod.

Among potential nudibranch defense mechanisms, nematocysts, coloration, and behavior provide little protection from *Navanax*, whereas acid secretion especially, and other glandular secretions as well, may be extremely effective. Shelled forms enjoy no special immunity. The spicules of dorids, in making the nudibranch body more rigid, are an adequate defense mechanism. *Navanax* eats "soft" bodied (90% water) dorids such as *Polycera*, *Triopha* spp., *Acanthodoris*, *Trapania*, and *Glossodoris californicus*, but not most "hard" bodied (70 to 80% water) ones. All but one eolid species, and all arminacean and dendronotocean species offered to *Navanax* in the laboratory were also consumed.

LITERATURE CITED

Blake, J. W. 1960. Oxygen consumption of bivalve prey and their attractiveness to the gastropod *Urosalpinx cinerea*. Limn. Ocean. 5:273–280.

Braams, W. G. and H. F. M. Geelen. 1953. The preference of some nudibranchs for certain coelenterates. Arch. Neerl. Zool. 10:241–262.

Cook, E. F. 1962. A study of food choice of two opisthobranchs, *Rostanga pulchra* McFarland and *Archidoris montereyensis* (Cooper). Veliger 4:194–196.

Fretter, V. and A. Graham. 1954. Observations on the opisthobranch mollusc *Acteon tornatilis* (L.). Journ. Mar. Biol. Assoc. U.K. 33:565–585.

Graham, A. 1955. Molluscan diets. Proc. Malac. Soc. London 31:144–159

Grosvenor, C. H. 1903. On the nematocysts of eolids. Proc. Roy. Soc. London 72: 462–486.

Hubbs, C. L. 1920. The bionomics of *Porichthys notatus*. Am. Nat. 54:380–384.

Johannes, R. E. 1963. A poison-secreting nudibranch (Mollusca: Opisthobrancha). Veliger 5:104–105.

Johnson, M. E. and H. J. Snook. 1927. Seashore animals of the Pacific Coast. The MacMillan Co., New York.

Kohn, A. J. 1956. Piscivorous gastropods of the genus *Conus*. Proc. Nat. Acad. Sci. 42: 168–171

——— 1959. The ecology of Conus in Hawaii. Ecol. Monogr. 29:47–90.

——— 1961. Chemoreception in gastropod molluscs. Am. Zool. 1:291–308

MacGinitie, G. E. and N. MacGinitie. 1949. Natural history of marine animals. McGraw-Hill Book Co., Inc. New York.

Marcus, E. 1961. Opisthobranch mollusks from California. Veliger 3 (Supplement): 1–85.

Odum, H. T. 1951. Nudibranch spicules of amorphous calcium carbonate. Science 114:395.

Ricketts, E. F. and J. Calvin. 1952. Between Pacific tides. Stanford University Press, v–xiii; 3–502; Stanford, California.

Stehouwer, H. 1952. The preference of the slug *Aeolidia papillosa* (L.) for the sea anemone *Metridium senile* (L.) Arch. Neerl. Zool. 10:161–170.

Thompson, T. E. 1960a. Defensive adaptations in opisthobranchs. Journ. Mar. Biol. Assoc. U. K. 39:123–134.

——— 1960b. Defensive acid-secretion in marine slugs and snails. New Scientist 8: 414–416.

Thompson, T. E. and D. J. Slinn. 1959. On the biology of the opisthobranch *Pleurobranchus membranaceus*. Journ. Mar. Biol. Assoc. U. K. 38:507–524.

Paine, R. T. 1963. Food recognition and predation on opistobranchs by Navanax inermis. Veliger *6: 1–9. Reprinted with permission of the California Malacozoological Society, Inc.*

Variation and adaptation in the imported fire ant (1951)

Edward O. Wilson

INTRODUCTION

The South American fire ant *Solenopsis saevissima richteri* is one of the insect pests which has become recently established in the Southeastern United States. In 1949 it was reported from Florida, Alabama, and Mississippi (Wilson and Eads, 1949). Its known distribution in these three states at that time is shown in Figure 1. Many other populations have been found during 1950 in Louisiana, Mississippi, Alabama, and Georgia by members of the Federal Bureau of Entomology and Plant Quarantine stationed at Mobile, Alabama (*in litt.*); these are mostly very young and limited to nurseries which receive shipments of plants from the Mobile area.

The first published record of this ant in the United States was made by W. S. Creighton (1930), who found it initially in urban Mobile in 1925. Creighton was told by H. P. Loding, a local and reliable amateur naturalist, that the ant had first appeared in the bayfront area of Mobile around 1918, had been pushed north of the city by the subsequently invading Argentine ant, *Iridomyrmex humilis*, and had later re-entered its original range. As late as 1928 Creighton found *richteri* still confined to a comparatively small area extending from the northwestern part of the town to Spring Hill (*in litt.*). In 1932 L. C. Murphree, scouting Argentine ants in Alabama, recorded *richteri* from several localities in Mobile and Baldwin Counties, including Whistler, St. Elmo, and Fairhope (Murphree, 1947).

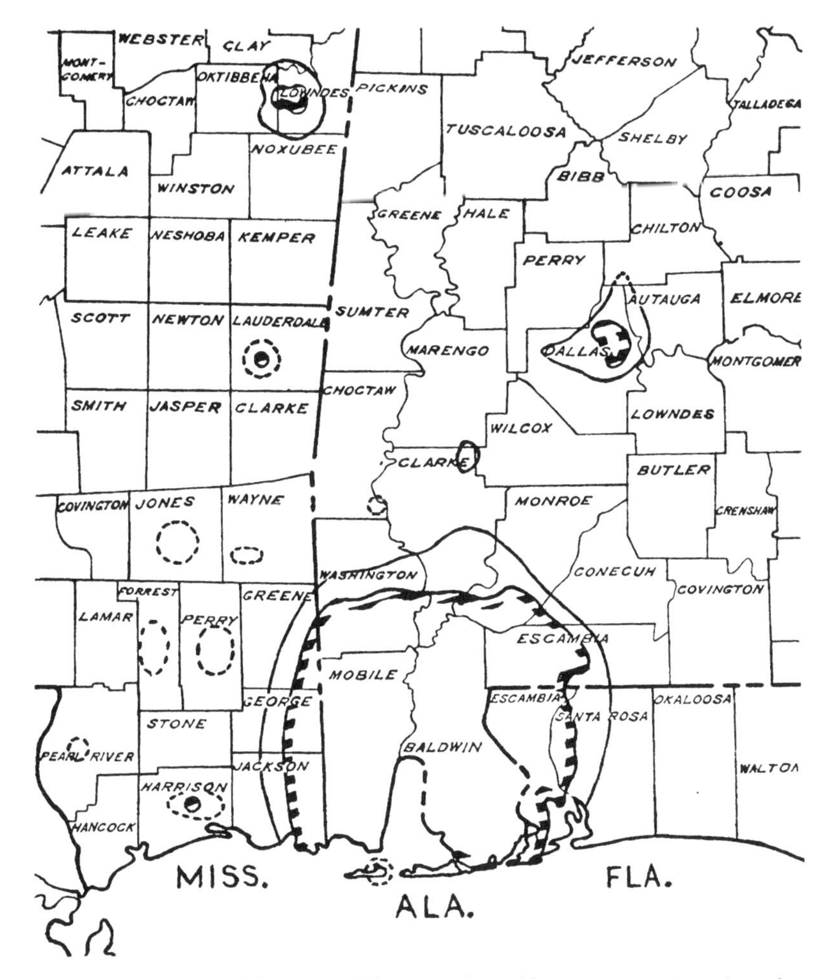

FIG. 1. Known range of the imported fire ant in the Gulf States in 1949. Inner barred lines enclose areas of heavy infestation; outer unbarred lines enclose the extreme range. Dotted circles indicate small populations the ranges of which were not accurately determined. The Choctaw County population and the populations in southern Mississippi were reported by the Federal survey crew (*in litt.*, 1949).

The first comprehensive study of this ant was undertaken during the period of March to July 1949, by Wilson and Eads (1949) under the auspices of the Alabama State Department of Conservation. The ant was shown to be a versatile but erratic pest, doing extensive damage to seeds and young seedlings of a variety of crops. Part of the data used in the following study was presented in that report. The U. S. Department of Agriculture initiated its own study in July 1949, and has continued it up to the time of this writing (August, 1950).

During the spring survey of 1949 there was observed an unusual amount of variability in color from nest to nest. This variation included one extreme blackish phase referable to the typical subspecies *richteri*, one extreme reddish phase referable to no described form, and intermediates between the two. Furthermore, the phase referable to *richteri* appeared to be mostly limited to part of the periphery of the range (see fig. 4). Since Creighton in 1930 described the forms he found in Mobile as typical *richteri* and has later stated (*in litt.*) that this form was the predominant one during that early period, the possibility of the present predominance of the extreme reddish phase representing an important change in the population must be considered. The following report is an attempt to analyze and explain the variation in this light from the point of view of the entire population.

Most of the field data presented in this report was obtained during the 1949 survey. Experiments with living colonies and morphological studies were conducted during the following year at the University of Alabama.

Grateful acknowledgment is made to Mr. J. H. Eads for his invaluable collaboration in the first survey of the ant, to Dr. R. L. Chermock for continuous assistance and advice, and to Dr. W. S. Creighton for the hitherto unpublished information concerning the early infestation.

STUDY OF THE VARIATION

Solenopsis saevissima richteri, as defined by W. S. Creighton in 1930, is apparently the southernmost race of a highly variable South American ant. Creighton described its range as extending from Uruguay south to the state of Rio Negro in Argentina, west to the Andes, and northwest almost to Bolivia. Another form, *S. saevissima* var. *quinquecuspis*, is found over part of the range of *richteri* in eastern Argentina, but it has been distinguished from *richteri* only on the basis of a highly variable color character and its validity is very questionable. *Solenopsis s. saevissima*, the typical subspecies, ranges with no great variability from the Guianas to the southern part of the state of Minas Geraes, Brazil. A vast zone of intergradation, containing a complexity of described forms, extends from Minas Geraes west through Bolivia to the Andes, south to Uruguay in the east, and south to the oases area of Argentina in the west. The taxonomic picture here is a very confused one; further study will probably reveal some of the forms to be intergrades of *saevissima* and *richteri* while some may be shown to represent distinct species. A thorough discussion of the diagnostic characters and the known ranges of these forms has been presented by Creighton (1930). The extreme reddish phase, which plays so important a role in the Gulf States population, has not been formally described from South America, but it is possible that it exists there as a submerged element (see Discussion). The whole Gulf States population has been referred to in this report as *Solenopsis saevissima richteri* strictly as a matter of convenience. This has been its popular designation in reports and correspondence up to the present time. For reasons discussed later it is believed that the new form should not be given immediate formal taxonomic recognition.

The population found in the Gulf States exhibits a great deal of variation. In order to facilitate a more exact study, the extreme variants and their intermediates were divided into six arbitrary phases according to the color of the workers. These can be described as follows:

1. Extreme dark. Ground color of alitrunk and head piceous brown; a light brownish orange to light brownish fulvous stripe covering approximately the anterior three-fifths to four-fifths of the dorsum of the first gastric segment in all but the minimas and smallest medias; a similar stripe on the venter of the first gastric segment but the rest of the gaster piceous brown, the condition seen in all the phases. This is the form best referable to Forel's description of *richteri.*

2. Dark intermediate. Ground color of alitrunk and head dark brown but not piceous, and with no reddish tinge evident; stripe on dorsum of gaster similar to above.

3. Intermediate. Ground color dark reddish brown, lighter than dark intermediate phase and approximately intermediate between the two extreme phases; stripe on dorsum of gaster similar to two darker phases.

4. Light intermediate. Ground color medium reddish brown; stripe on gaster covering approximately one-half to two-thirds of the anterior surface, darker than in above phases, posterior border frequently indistinct.

5. Dark red. Ground color light reddish brown; stripe on dorsum of gaster covering approximately the same area as in light intermediate phase but darker, being very close to the color of the head and alitrunk and distinguishable only in contrast to the piceous brown of the rest of the

TABLE I

<table>
<tr><th></th><th>WORKERS</th><th>QUEENS</th><th>MALES</th></tr>
<tr><td>Extreme Dark</td><td>1.264</td><td>2.682</td><td rowspan="2">2.712</td></tr>
<tr><td>Dark Intermediate</td><td>1.236</td><td>2.635</td></tr>
<tr><td>Intermediate</td><td rowspan="2">1.130</td><td rowspan="2">2.584</td><td rowspan="4">2.604</td></tr>
<tr><td>Light Intermediate</td></tr>
<tr><td>Dark Red</td><td>1.095</td><td rowspan="2">2.499</td></tr>
<tr><td>Light Red</td><td>1.038</td></tr>
</table>

gaster, its posterior border indistinct; stripe present only in larger medias and soldiers.

6. Light red. Ground color light reddish brown; stripe on gaster absent or present only in largest soldiers, very similar when present to that of dark red phase.

During the survey three colonies were found which could not correctly be assigned to any of the above phases. One of these, found near Gulf Shores, Baldwin Co., Ala., had workers with the ground color of the dark intermediate phase but with little or no gastric stripe. These are referable to Creighton's definition of *Solenopsis saevissima* var. *quinquecuspis*. The other two, found near Gulf Shores and Fairhope, Baldwin Co., Ala., had workers all of which possessed very nearly the same color as callows of the light red phase. The queens of all the phases are very similar in coloration to the largest workers and exhibit the same variation. The males are all uniformly black.

Morphological studies of the six phases revealed significant differences only in size. It was found that in workers, queens, and males, the darker phases are larger on the average than the lighter phases. As demonstrated in Figures 2 and 3, the successive intermediates tend to show successive differences in size, this being particularly pronounced in the worker caste. [Note: Figure 2 is omitted here because it was peripheral and poorly used in the original text. Eds.] All the measurements shown were made of the alitrunk as seen in profile, from the dorsal base of the pronotal collar to the dorsum of the junction of the propodeum and petiole. The alitrunk was used because of its rigidity and the ease with which it is measured. To correlate the variability of the head with that of the alitrunk, the heads in addition to the alitrunks of fifty light red workers and fifty extreme dark and dark intermediate workers were measured in profile from the base of the clypeal spines to the extreme occipital border. No significant divergence was noted in the two sets of measurements nor could the two groups of ants be separated on the basis of head-alitrunk proportion.

In nearly all the areas where many of the phases were observed together it was noted that the darker forms tend to build larger and proportionally taller mounds. Ten mounds each of light red and extreme dark-dark intermediate phases were measured in an open field several miles south of Theodore, Mobile Co., Ala., in June 1949, and the following differences recorded:

Light red: The smallest mound had a base diameter of 13" × 13" and a height of 5"; the largest had a base diameter of 24" × 22" and a height of

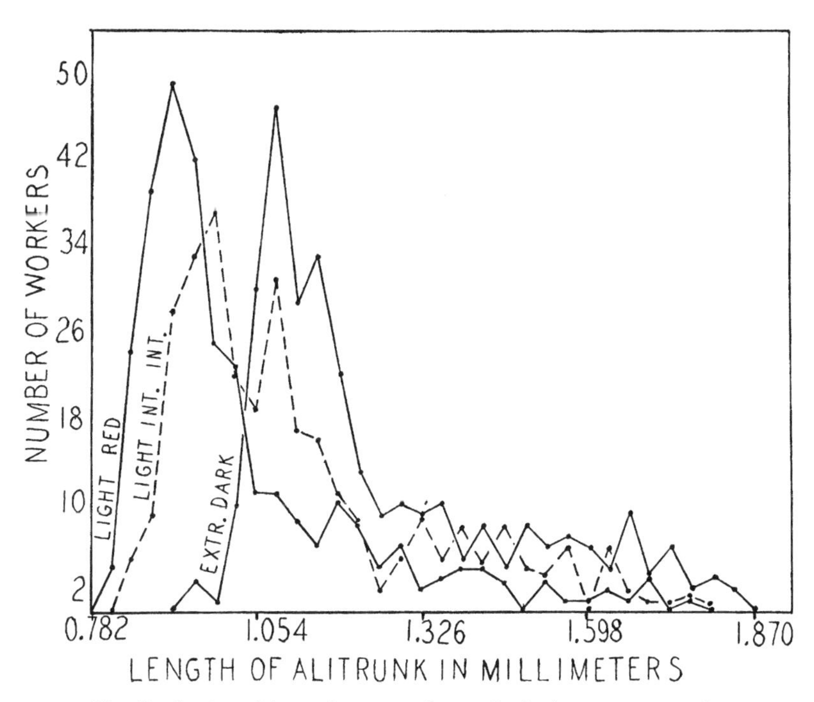

FIG. 3. The distribution of the workers according to size in the two extreme phases and two of the intermediate phases. Three hundred workers from four localities were measured in each group.

11". Height varied from 3" to 13". The overall average of the base diameters was approximately 24" × 22" and of the heights, 7".

Extreme dark and dark intermediate: The smallest mound had a base diameter of 22" × 20" and a height of 10"; the largest had a base diameter of 39" × 24" and a height of 14". Height varied from 10" to 18". The overall average of the base diameters was approximately 27" × 24" and of the heights, 14".

All of the phases tend to accumulate bits of vegetable detritus, small pieces of charcoal, and small pebbles on the surfaces of their mounds, but this is especially noticeable in the darkest three or four phases. Occasionally mounds of the darkest phases are found nearly covered by a thin layer of this debris. In areas where they occur together it is often possible to tell with reasonable accuracy from a distance which are the darker and which are the lighter phase mounds, judging from the size, shape, and outer surface.

In order to plot the distribution of the color variants 10 colonies each

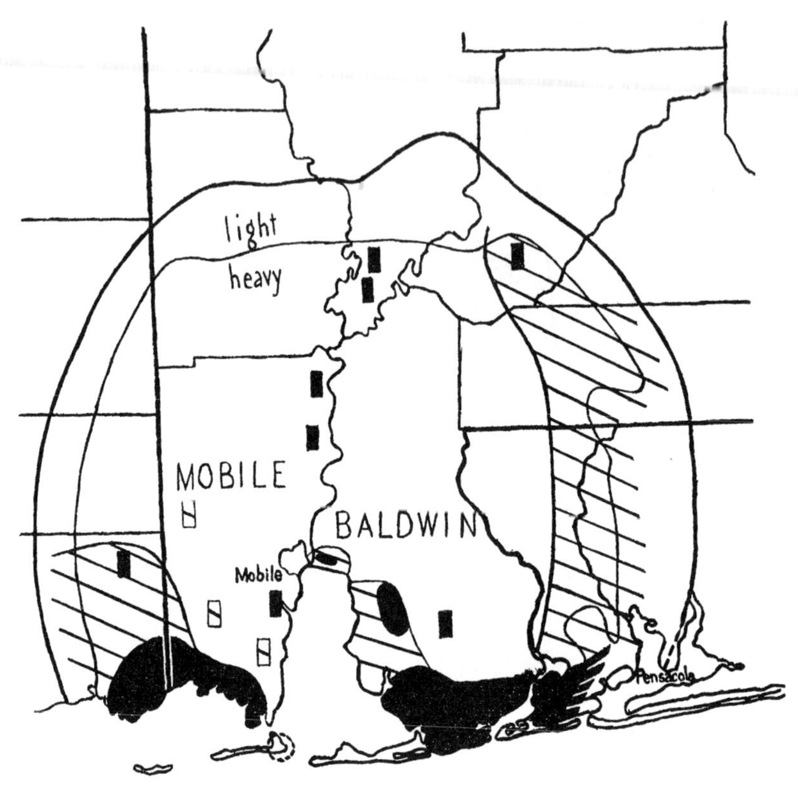

FIG. 4. Distribution of the color phases in the main area of infestation. Intense shading represents areas with incidence of four darkest phases greater than 20 per cent; hatching represents areas with incidence of these phases 5–20 per cent. Small rectangles represent small isolated dark phase populations.

in 84 relatively random localities over the main infested area were classified according to the six color phases. To supplement this a total of 193 colonies were classified in 18 other localities where ten nests could not be found together. Figure 4 shows the approximate distribution of the phases as determined by this survey. It will be noted at once that the darker phases are limited to the southern and eastern portions of the periphery of the range, to a small part of the western portion of the periphery, and to a number of small, isolated areas throughout the range. It is interesting to add here that the isolated areas are in nearly every case centered around marshy fields or grassland, but not all such situations contain darker phases. The Artesia and Meridian population in Mississippi are apparently homogeneously composed of the darkest one or two phases. In the Selma population the darker phases are scarce and irregularly distributed, and none at all have been found in the Thomasville population.

The populations in Louisiana, Georgia, and southern Mississippi have not been carefully classified according to the color phases by the writer. However, Mr. G. H. Culpepper, of the Federal survey crew, has very kindly studied the color variation in these populations and has reported that with the exception of the one at Meridian all contain a predominance of light phase colonies (*in litt.*). He has expressed the opinion that all but those at Meridian and New Iberia are apparently small and recent in origin.

It is believed that the variation studied in the Gulf States has a genetic basis. This conclusion is based on the following observations:

1. In many areas where both extreme phases and their intermediates occur, colonies of nearly all the phases may be found in the same immediate area, sometimes within a few feet of one another, apparently under nearly identical conditions.

2. Variation within individual colonies is very slight, and in none of the nests examined were there found workers which covered more than two adjacent phases.

3. It would be very difficult to rationalize the distribution of the variants according to environment or colony age. In the main area of infestation the darker forms are mostly limited to the southern and eastern portions of the periphery, while the population at Artesia, in northern Mississippi, is apparently composed entirely of darker forms. The Artesia and Selma populations are both located in the clearly defined "Black Belt," but the Selma population contains very predominantly the two lightest phases.

4. Colonies of light red and dark intermediate phases have been maintained in the laboratory under a variety of conditions without appreciable change in the color of the original workers or those reared in artificial nests. Workers have been reared at temperatures above 30° C and below 20° C; others have been heated excessively and chilled. Still others have been reared variously at substarvation and near-optimum conditions. Minima workers of the dark intermediate phase produced under substarvation conditions tend to be lighter in color but are still distinguishable from those of the lightest phases.

5. The brood of two light phase queens adopted by dark intermediate phase workers in the laboratory and one adopted by *Solenopsis geminata* workers developed into workers with the color of their mothers. In each case young, recently fecundated queens were introduced into groups of twenty to thirty workers. These were maintained in Fielde nests modified by the addition of plaster-of-paris chambers and were fed honey, dogfood, and miscellaneous insects. It has been found that occasionally groups of workers, especially those from depauperate colonies, accepted

alien queens readily, but in the majority of cases could be induced to do so only after being chilled to immobility from several hours to several days. Many remained hostile regardless of treatment.

DISCUSSION

It is apparent that the Gulf States population of *Solenopsis saevissima richteri* has undergone a marked change during the period of 1929 through 1949. The original population was at least mostly composed of the darkest phases, as stated by Creighton. As late as 1929 this remained true (Creighton, 1930 and *in litt.*). In 1932 L. C. Murphree collected the ant from five localities in Mobile and Baldwin Counties, Ala., and judging from his description (1947), the material he collected must be assigned to one or both of the two darkest phases. In 1941 the writer observed a large number of colonies along the bayfront of urban Mobile and in several areas in the western part of the city, all of which belonged to the lightest two or three phases. By 1949 the darkest phases could be found in some abundance only along part of the periphery in the main area of infestation and even there were outnumbered by the lightest phases. The Selma and Thomasville populations were predominantly light phase, the Meridian and Artesia populations at least predominantly dark.

Two approaches may be used to explain this peculiar recent distribution, one considering the history of the population, the other the possible climatic preferences of the extreme forms. Figure 5 shows the approximate rate and direction of spread of the ant since its introduction, as based on the observations of W. S. Creighton, the records of L. C. Murphree, and the estimates of 65 residents of the main infested area (Wilson and Eads, 1949). It appears that by 1934, while the dark phases may still have been dominant, the ant spread south over the southern portions of Mobile and Baldwin Counties, a range now co-occupied by another introduced ant, *Brachymyrmex heeri obscurior*. Estimates place the origin of the Artesia population around 1935 and the Meridian population around 1940. The Selma population was not in any case estimated to have originated before 1944 nor the Thomasville population before 1948. The light phases were apparently responsible for the ant's explosive spread to the north in the main area of infestation. The partly peripheral distribution of the darkest, presumably oldest phases seems to indicate that they were pushed outward by the expanding light forms, in a pattern somewhat similar to that first demonstrated by Matthew (1939) for the primitive members of some groups of mammals, that is, with the most primitive forms at the periphery and the most recently evolved toward the center. The relatively

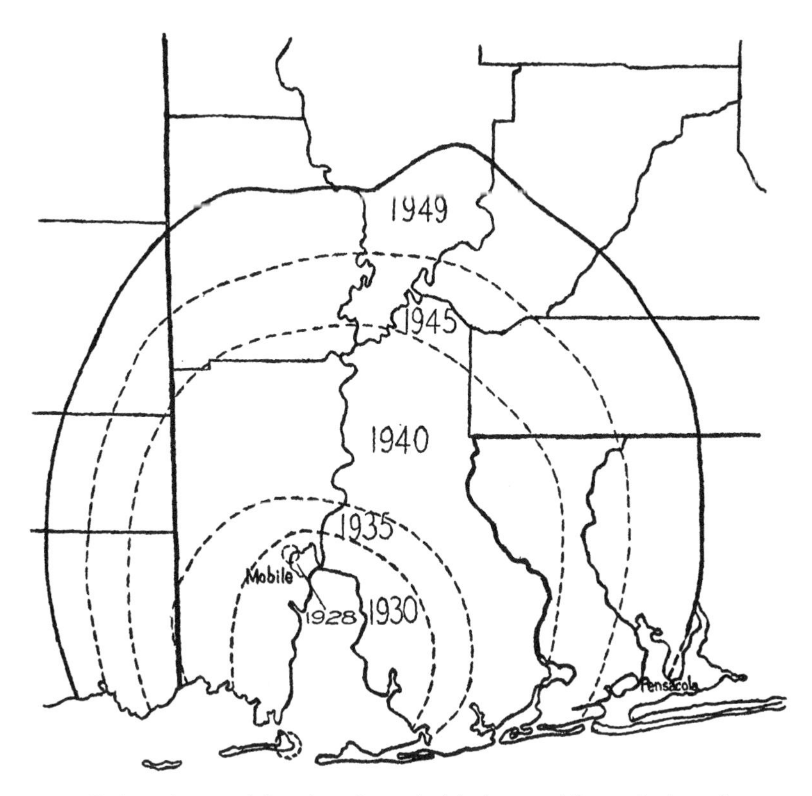

FIG. 5. Estimated rate and direction of spread of the imported fire ant in the main area of infestation.

homogeneous ecological conditions in and around the area of infestation, coupled with the absence of any significant geographical barriers, has allowed an even and steady spread of the ant since its introduction (see fig. 5). It has also resulted in the relatively clear preservation of the concentric pattern produced by the spread of the light forms. The darkest phases were best able to survive along the coast, along the eastern periphery, and in isolated spots through the infested area. Why marshy areas should suit these forms is not known. Large numbers of light phase nests often occur in the same places, with as many as seventy nests to an acre, and competition must be intense. Over much of the southern, continuous part of the range of the darkest phases, as around Gulf Shores, these forms occur in as wide a variety of situations as the lighter phases. The absence of light phases at Artesia and Meridian might be explained on the basis of early origin, that is, derivation from the early, nearly homogeneous stock. The lack of conspicuous success on the part of these populations, as compared with that of the younger Selma and Thomasville populations, and the southward expansion of the early population in the main area of infestation seem to indicate that the darker phases are poorly adapted to the climate of the Gulf States.

The present predominance of the light phases suggests an adaptive change within the population. The manner by which these phases have largely replaced the dark phases has been, in the opinion of the writer, partly rapid expansion in range and population size and subsequent dilution of the dark phase genes, resulting in a diversity of intermediate forms; it has also been partly through considerable populational pressure and elimination in direct competition. An indication of this latter process can be seen in the scarcity of dark and intermediate phases over much of the range formally occupied by the dark phases. Another indication can be seen in the scarcity of the native fire ants, *Solenopsis xyloni* and *S. geminata* and of the Florida harvesting ant, *Pogonomyrmex badius* in the infested areas. Colonies of the imported fire ant are extremely antagonistic to these ants as well as to alien colonies of their own species. During the survey no colonies of *Solenopsis xyloni* and very few of S. *geminata* were found in the heavily infested areas. *Pogonomyrmex badius* was more common but quite sporadic in distribution. These three species reach relative abundance around the periphery of the imported fire ant's range. Another important consideration in this light is the incidence of the phases. Although exactly random samples in quadrats were not taken during the survey, approximately twice as many nests of the two darker phases were found as those of the two intermediate phases in the counts taken. Assuming that the phases represent nearly equal arbitrary divisions of the successive genotypes, it is possible that this difference in incidence is a result of replacement by the light forms through direct competition as opposed to genetic dilution. The relative importance of these two modes of replacement can only be estimated on the basis of available data, but it seems safe to say that both play an important role.

When considering the significance of the light phases as an adaptive replacement, the possible role of the Argentine ant, *Iridomyrmex humilis*, must be studied. This ant reached its peak in the latter part of the 1920's and during that time succeeded in eliminating nearly all of the native ants in the Mobile area (Creighton, 1950). By 1932 it had been partly controlled and the native ants had begun to infiltrate the infested area. It was this time also that *Solenopsis saevissima richteri* began its initial spread to the south. It would seem quite possible that the decline of the Argentine ant, and not the replacement of the dark phases by the light phases, was responsible for the fire ant's rapid spread to pest proportions. However, several observations have been made which seem to indicate that the Argentine ant had a minor influence on the phenomenon under consideration. One is that the Argentine ant appears to offer the imported fire ant little serious competition at the present time, even in areas where the former are abundant enough to affect the native ants (Wilson and Eads, 1949). The two species thrive together in some areas of Mobile. Another is that the Argentine ant is primarily an urban dweller; it never was able to blanket the wide variety of rural situations in which the imported fire ant thrives (*Ibid.*). There is no reason why the fire ant could not have spread by way of these situations, even assuming that the Argentine ant was able to hamper it during that ant's peak. Finally, the predominantly dark phase Artesia population and the predominantly light phase Selma population can be critically compared, since both are under very similar conditions. The latter is less than half as old as the former and yet much larger and denser. Both are situated mostly in rural areas, and neither could have been greatly affected by the Argentine ant since their inception.

In considering the possible genetic basis of the variation, there are several peculiarities in the expression of the characters which deserve mention. One of these is the small amount of variability observed in intergrade colonies. This, coupled with the great variability existing from nest to nest, cannot he explained on the basis of a single mutation. Admitting that the phases are totally arbitrary, that they overlap, and that occasional individuals can be assigned to an adjacent phase, at least four or five distinct genetic groups clearly exist; the workers of the six phases can be divided on the basis of size into five statistically significant units. Part of the difficulties are removed by explaining the variation on the basis of multiple factors or multiple alleles, which could control successional variation of the type observed. Of these two possibilities, multiple alleles seems to be the more tenable. From only three alleles a total of six combinations, or genotypes, is possible. Also, the greatest number of genotypes which could be produced by a single cross is three; the small amount of variability in single colonies studied in the field may well indicate such a limited number of genotypes. On the other hand, a dihybrid cross involving multiple factors showing no dominance would make possible nine genotypes and phenotypes, all of which, including the extremes, could be produced by a single heterozygous cross.

The close correlation between color and size variation and the present selective advantage the light phases hold seems to indicate a pleiotropic expression of whatever alleles are involved, producing a relatively clear-cut combination of characters for both extreme color forms. It would seem unlikely that these combinations of characters could maintain their identity in the extensive intergradation that has occurred unless they were very closely linked.

Even assuming multiple factors or multiple alleles, the small amount of variability within individual colonies is difficult to explain by the familiar laws of heredity. Provided that the queen is heterozygous and diploid, it would seem that the phenotypes of workers produced by an intermediate queen would be mixed according to random assortment. It appears very likely that the reason why this fails to occur in the imported fire ant (and in many other species of ants which exhibit the same peculiarity in their color varieties) entails an aberrant hereditary mechanism. One such possible mechanism is maternal determinism, a condition demonstrated in the snail *Limnaea peregra* by Diver, Boycott, and Garstang (1925). In this organism the direction of spiraling in the shell is determined by the genotype of the mother, regardless of the genotype of the offspring. On the other hand, in *Solenopsis* a mechanism involving haploidy in the queen and parthenogenesis is apparently rendered untenable, because unfertilized eggs are capable of developing at the most only to the late larval or pupal stage. Mature larvae were noted in fourteen out of fourteen nests containing only young alate queens; one of these nests produced several worker pupae, but since there was no absolute proof that all the queens were not fertilized, general conclusions concerning thelytoky should be avoided. One nest containing alate queens raised from larvae in the absence of males produced only larvae over a long period of time. This was noted in fourteen artificial nests out of fourteen which contained only virgin queens; mother queens in other nests kept up a prolific production of workers. Unfortunately, the interesting and significant problem of the genetic basis of the variation will prove to be a very difficult one to solve. For all practical purposes the colony is the individual in the life cycle of the ant, and with present knowledge of caste determination each generation of virgin queens will have to be preceded by hundreds or thousands of workers.

The origin of the light phases is another problem which can only be conjectured on the basis of available data. Three possibilities are treated below.

1. The light phases could have been a submerged recessive element in the genotype of the original invaders. However, this view is considerably weakened by the apparent homogeneity of the Artesia and Meridian populations, which were probably established while the dark phases were still dominant. Also, the original invaders were probably few in number;

it seems unlikely that they could have contained all the variability present today, unless this were to appear conspicuously in the early population.

2. The light phases could have been derived through mutations in the early, dark phase population. This possibility must always be considered, because the population mutation rate has been potentially tremendously high. Even in the early population, each of the many thousands of colonies were producing hundreds to thousands of sexual forms the year round. Today the number of colonies in the main infested area easily numbers into the millions (Wilson and Eads, 1949).

3. The light phases could have been derived from one or more later introductions. Despite the fact that nothing exactly comparable to the light red phase has ever been described, examination of representative material of most of the South American forms of *Solenopsis saevissima* has convinced the writer that the light red phase could have originated anywhere in the great zone of intergradation from southern Brazil to Argentina. Variation in this area is too great and as yet too poorly defined to dismiss the idea that the light phases were derived anywhere but in the Gulf States.

In final analysis the phenomenon can be described as an adaptive replacement, a shift in population dominance from one genetically distinct form (darkest phases, or *richteri* s. str.) to another (lightest phases). At present the population is quite unstable; it cannot be said to have reached an adaptive peak. Although it appears that a complete replacement is under way and a homogeneous population is in the making, new forms such as the two anomalous phases (see discussion of variation) may yet rise to dominance. The origin of the new form is not known. Because of this instability and uncertain origin, it does not seem wise to accord the light phases immediate formal taxonomic recognition, although by strict definition it might appear to some authorities to constitute a new subspecies. Otherwise, the question of the origin is an academic one only. Whether indigenous or introduced the light phases have functioned as a favorable mutation appearing in the population; they seem to have progressed very much as such a mutation would be expected to progress. As an example of this type of phenomenon the adaptive replacement in the Gulf States population is especially noteworthy. This is due to three significant conditions:

1. The new forms are easily identified and their history and present distribution can be carefully studied because of a distinct color character.

2. The adaptive change within the population has occurred so recently as to aid greatly the study of its history. The new forms constitute an almost vertical evolutionary change which has taken place in less than twenty years. Their history illustrates the extremely rapid rate with which

such a change, which can be interpreted as an initial step in subspeciation, can occur in a population of insects. Similar changes have been observed in populations of the scale insects *Aonidiella aurantii*, *Coccus pseudomagnoliarum* and *Sassieta oleae* (Quayle, 1938), and of the codling moth, *Carpocapsa pomonella* (Hough, 1934, and Boyce, 1935).

3. The very uniform terrain of the Gulf Coast area has allowed an identifiable preservation of what is apparently a concentric distribution of the old and new forms. That the dark forms have been pushed out along the periphery is still evident in their distribution today.

This populational change is indicative of the way that evolution can proceed in any population of ants. However, the rapidity with which this has occurred is probably rarely equaled by that in endemic species. Too little is known about North American ants at the present time to determine precisely the rates of evolution, but it appears that most holarctic genera have been evolving at a considerably lower rate than such better known groups of insects as the Rhophalocera. For instance, postglacial relics are rare or absent, and subspeciation initiated by glacial isolation seems to have progressed to a relatively slight degree. The accelerated evolution in the imported fire ant probably has the same origin as the explosive spreads of imported insects as a whole: the removal of the biotic pressures (parasites, competition with other insects, etc.) which controlled it in its native environment. If this is true, then the populational change in the imported fire ant can be regarded only as a swiftly enacted replica of the normal evolutionary processes and not as typical subspeciation.

SUMMARY

1. The imported fire ant, *Solenopsis saevissima richteri* Forel, is the southernmost race of a widespread and highly variable South American ant. It was introduced into the port of Mobile, Alabama, sometime around 1918 and by 1949 had spread to parts of Florida, Mississippi, and Louisiana.

2. A great deal of color variation from nest to nest has been noted in the Gulf States population. This includes an extreme blackish phase referable to the original description of *richteri*, an extreme reddish phase referable to no described form, and intermediates between the two. This color variation is correlated with differences in size of the ants and in appearance and proportion of their nests.

3. The variation has a genetic basis. It is suggested in this study that the variation can be explained most readily on the basis of multiple pleiotropic alleles.

4. The history of the variation has been determined as follows:

The darkest forms, or *richteri* s. str. were the ones originally dominant from the time of the ant's introduction until at least 1929 and probably sometime after 1932. The origin of the new form is not known, although it is believed that it originated either through mutation within the population or through a second introduction. In 1949 it was by far the dominant form. It had apparently replaced the typical *richteri* partly by rapid expansion and subsequent genetic dilution and partly through natural selection by direct competition. Its predominance in the main population and in at least two smaller isolated populations has evidently been responsible for a much greater success of the species. In the main population in 1949 the typical *richteri* was mostly limited in distribution to portions of the periphery of the range, forming with the new form roughly the concentric pattern of Matthew's modified Age-and-Area hypothesis.

5. The new form has been interpreted as functioning, regardless of its origin, as a favorable mutation introduced into the population. Its rise to dominance has constituted an extremely rapid, nearly vertical evolutionary change.

LITERATURE CITED

Boyce, A. M. 1935. The codling moth in Persian walnuts. J. Econ. Ent., 28: 864–873.

Creighton, W. S. 1930. The new world species of the genus *Solenopsis.* Proc. Amer. Acad. Arts and Sci., 66: 39–151.

——— 1950. The ants of North America. Bulletin of the Museum of Comparative Zoology at Harvard College, 104.

Diver, C., A. E. Boycott, and S. Garstang. 1925. The inheritance of inverse symmetry in *Limnaea peregra.* J. Genet., 15: 113–200.

Hough, W. 1934. Colorado and Virginia strains of codling moth in relation to their ability to enter sprayed and unsprayed apples. J. Agric. Res., 48: 533–553.

Matthew, W. D. 1939. Climate and evolution. Special Publications of the New York Academy of Science, 1: 1–223.

Murphee, L. C. 1947. Alabama ants, description, distribution, and biology, with notes on the control of the most important household species. Unpublished Master's thesis, Mississippi State College.

Quayle, H. J. 1938. The development of resistance in certain scale insects to hydrocyanic acid. Hilgardia, 11: 183–225.

Wilson, E. O., and J. H. Eads. 1949. A report on the imported fire ant *Solenopsis saevissima* var. *richteri* Forel in Alabama. Special mimeographed report to the Director of the Alabama State Department of Conservation.

Wilson, E. O. 1951. Variation and adaptation in the imported fire ant. Evolution *5: 68–79. Reprinted with permission of Wiley-Blackwell.*

Storm mortality in a Winter Starling roost (1939)

Eugene P. Odum and Frank A. Pitelka

Reports of large-scale mortality among birds resulting from severe weather conditions are frequent in ornithological literature. Kendeigh (Ecolog. Monogr., 4: 342–352, 1934) summarizes a number of these, pointing out the environmental factors concerned. However, accounts of avian catastrophes do not often include attempts to determine the amount of mortality or to measure the causal factors. The ultimate value of records of such incidents can be much augmented by accompanying them with accurate climatic and habitat data. The present report is concerned with mortality among roosting 'blackbirds' at Urbana, Champaign County, Illinois, during a regional rain and windstorm on the night of February 9, 1939. Additional data were obtained after subsequent shooting into the roost on February 25–26, and March 3 or 4, 1939.

The Roost—During recent years, swarms of 'blackbirds' have roosted in the northern half of a thirteen-acre grove, locally known as the 'forestry,' planted seventy years ago on the south campus of the University of Illinois. The tract is surrounded by more or less open country on south and west sides. Usually the birds massed in a belt of white pines on the northwestern side, but sometimes a part of the flock settled to the east in deciduous trees, which are sheltered from prevailing winds by the pines. The density of white pines is about 250 per acre, that of deciduous trees (chiefly green ash) somewhat less; undergrowth is entirely absent.

During the winter of 1938–39, to the best of our knowledge, four species were represented in the roost: Starling (*Sturnus vulgaris vulgaris*), Cowbird (*Molothrus ater ater*), Bronzed Grackle (*Quiscalus quiscula aeneus*), and Red-wing (*Agelaius phoeniceus* spp.). The latter three species usually occur in central Illinois in small numbers throughout the winter. The number of birds occupying the roost at the time of the storm was estimated at 25,000. Of these probably not more than 3% were cowbirds and grackles. Redwings, of which only a few individuals were present, were not represented in the mortality.

Storm Mortality—On the morning of February 10, following the storm, dead birds were found on the ground over most of the roosting area and also to the east, northeast, and north, even to a distance of several blocks in these directions. In exposed places and on the windswept 'forestry' grounds, the carcasses were frozen hard to the ground. They were most numerous in the peripheral areas of the roost in the directions mentioned. Actual counts made by the writers are as follows:

TABLE I

Starling	570	78.5%
Cowbird	93	12.8%
Bronzed Grackle	63	8.7%
Total	726	100.0%

The grackles and cowbirds were found chiefly under a small group of tall Norway spruces somewhat apart from the main roosting site in a seemingly more protected location. A number of crippled Starlings, at least twenty-five, were observed during the counting. Eight white pines and six or seven deciduous trees were down as a result of the storm.

The total mortality as a result of the storm exceeded the figure given since dead birds were scattered widely beyond the roost chiefly to the east and northeast, but it probably did not pass 1,000. While this is only an approximate 4% of the roost, it is appreciable enough to warrant investigation especially since accurate weather data are available.

Post-Storm Roost Composition—The proximity of the roost to habitations having proved objectionable, shooting into the roost was carried out on February 25 and 26, causing the birds to shift to evergreens in a cemetery south of the University campus, where further shooting was done on March 3 and/or 4. The counts following the shooting can be regarded as random samples, and the percentages given are believed to be representative of roost composition after storm mortality.

The cowbirds and grackles in the second shooting are believed to be remnants of the wintering population since no northward migration of early spring birds was detected up to these dates, a point which seems to be supported by the relative constancy of the sex ratio (approximately 2 ♂ : 1 ♀) among the Starlings over the week period (February 25–March 4) during which initial shifting and migratory movements might have been expected. The variation in sex ratio is comparable to that found by Hicks

TABLE 2

(Dates of shooting)	FEBRUARY 25–26		MARCH 3–4	
Starling	631	97.1%	361	98.6%
Cowbird	9	1.4%	3	0.8%
Bronzed Grackle	8	1.2%	2	0.6%
Red-wing	2	0.3%	–	–
Totals	650	100.0%	366	100.0%

(Bird-Banding, 5: 103–118, 1934) during the same period in an extended study of Ohio Starling roosts.

Weather Data—The accompanying graphs illustrate the actual weather conditions on February 9 and 10, 1939, before, during, and after the storm. All data except relative humidity, which was obtained at the University Woods, a few miles east of Urbana, were provided by the University of Illinois meteorological station on the campus. For comparison and emphasis of extremes, the temperature graph is accompanied by a daily fluctuation line based on the hourly means for the month of February 1939. The graph of maximum wind velocity shows the approximate trend and does not include minor variations from minute to minute. Unfortunately, hourly readings of rainfall are not available; however, amounts together with period of duration are shown. The greater part of the 0.3 in. of rainfall between 11.00 p.m. on February 9 and the following morning fell during the high wind between 11.00 p.m. and midnight.

Analysis—A study of these graphic records reveals two critical periods which were probably responsible for the mortality: (1) high wind and warm rain storm from 11 o'clock to midnight; (2) sharp drop in temperature to below freezing accompanied by high wind during the remainder of the night. Which of these was the more important would be difficult to state, but probably both contributed.

Undoubtedly many birds were killed outright as a result of the high wind (48 miles per hour) as evidenced by the number of cripples, the distance of some birds from the roost, and the obviously mangled condition of others lying about the wind-fallen trees. At the height of the storm about 11.30, one Starling crashed through the window of a nearby house (several hundred feet northeast of the roost).

On the other hand, the drop in temperature from a remarkable high of 60° F. to below freezing in five hours likewise seems to be important.

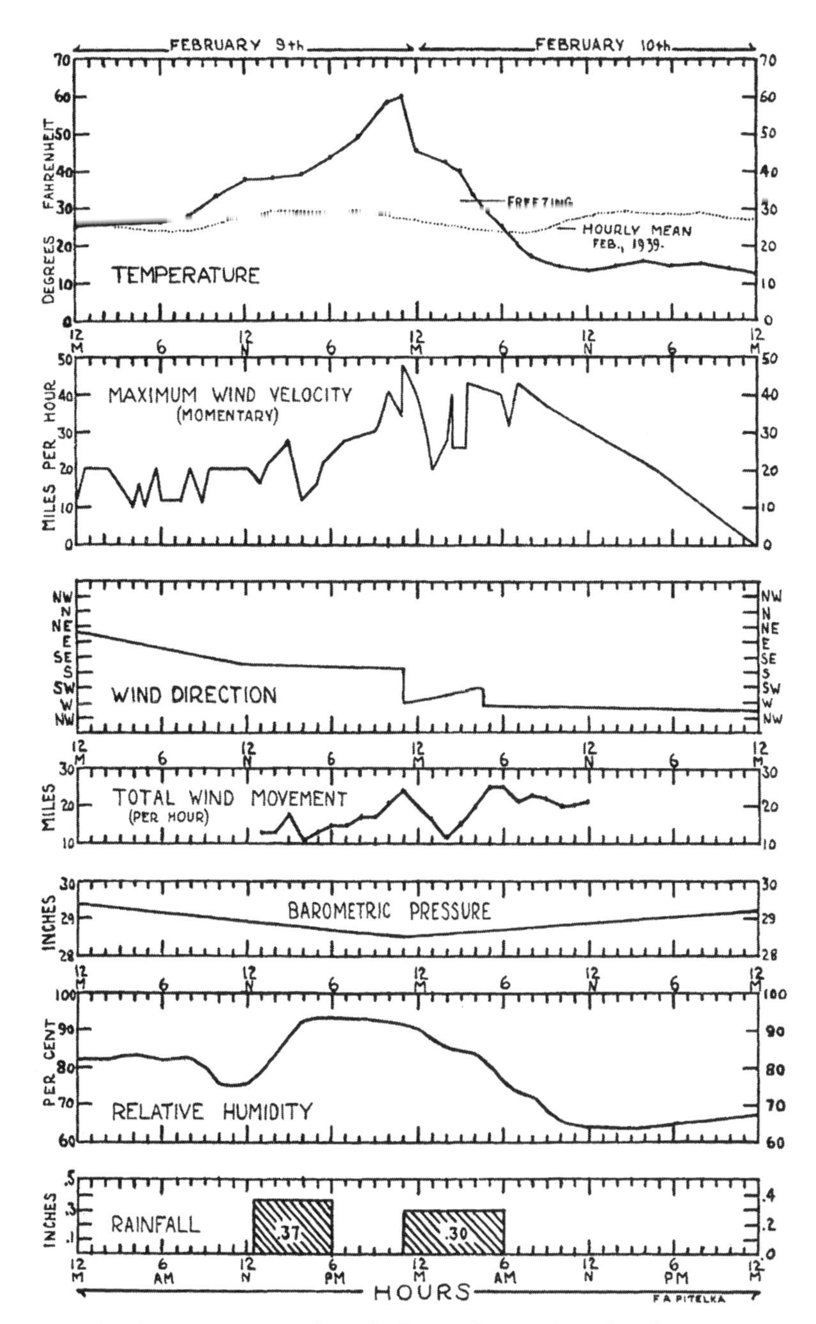

FIG. 1. Graphic representation of specific climatic data coordinated on the same time scale, correlating possible factors involved in mortality of birds during storm period.

Increased vulnerability of birds to low temperatures would seem to be caused by thorough wetting especially during the night when they are without food. It was evident that many birds perhaps dislodged by wind, were later killed by cold. Positions of some of the birds indicated attempts to find protection under logs and in pockets at the bases of trees, where they died and froze. From this angle of the probable mortality causes, it is interesting to note the report of a Starling which attempted to roost on the warmer inner sill of an opened window about 1 a.m. and again at about 3 a.m. approximately half a block east of the roost; the bird vigorously ruffled its feathers and attracted the attention of the occupant of the room each time. Obviously dislodged by the wind and soaked by the rain, it was having difficulty drying its feathers and finding a protected spot in which to roost.

The proportion of Bronzed Grackles and Cowbirds to Starlings in the total storm mortality was certainly much greater than that in the total roosting flock as indicated by observation and especially by results of later shooting (an average 2% cowbirds and grackles from random sample shooting as compared with 21.5% killed in the storm). This together with the fact that the grackles and cowbirds were roosting in a more protected spot would indicate a differential mortality among the three species, these two being considerably less resistant and dying in proportionally greatest numbers. Several grackles and cowbirds were frozen in the Norway spruces (see above) without having fallen to the ground, and the greater number of them were lying directly below the evergreens, indicating that they had fallen out rather than been blown out.

SUMMARY

1. A 'blackbird' roost was subjected to a driving wind and rainstorm from relatively unprotected southwest and west sides.

2. The wind reached a maximum velocity of 48 miles per hour and was followed by a sharp drop in temperature and continued high wind.

3. The destructive combination of weather conditions occurred when a warm southeast wind suddenly shifted to the southwest and west.

4. A mortality of approximately 4% resulted in the 'blackbird' roost, the size of which was estimated at 25,000 birds.

5. A differential mortality among the species of the roost was apparent; the proportion of Grackles and Cowbirds to Starlings in the total mortality was considerably greater than that in the total roosting flock.

Conclusions—In an analysis of storm mortality, the time of occurrence, intensity, duration, and relation of specific weather factors must all be considered. Such examination of weather records provides a basis for more accurate judgment of the causes of mortality than can be made from simple observation alone. Furthermore, comparisons with records of subsequent severe storms can be made; these may reveal more clearly the factors most critical to winter roosting flocks.

The writers wish to express thanks to E. J. Koestner and C. T. Black for assistance in making counts of birds; to Mrs. D. H. Mills, J. M. Speirs, F. C. Bellrose, and M. J. Feldman for confirmatory counts; to Mr. H. P. Etler of the University of Illinois Meteorological Station for kindnesses in making weather data available; and to Mr. R. W. Lorenz of the Department of Forestry, University of Illinois, for data on the tract occupied by the roost.

Odum, E. P., and F. A. Pitelka. 1939. "Storm mortality in a Winter Starling roost." Auk *56: 451–455. Reprinted with permission of University of California Press.*

The dispersal of insects to Spitsbergen (1925)

Charles S. Elton

INTRODUCTION

This paper contains various observations made by myself and other members of the Oxford University Expeditions to Spitsbergen, during 1921, 1923, and 1924. I wish to thank the following gentlemen, who have kindly given me much information and other help: Messrs. F. G. Binney, W. B. Carslake, H. M. Clutterbuck, F. A. Montague, K. S. Sandford, members of the Oxford sledging parties in 1924, who gave me the data about flies and Aphids on North-East Land; Mr. J. E. Collin, for identifying the flies; Mr. F. Laing, for identifying the Aphids; Mr. E. Koefoed, for notes on flies observed on Hope Island in 1924; Mr. Krogness, for allowing me to consult the weather charts for August 1924, while I was in Tromso; Dr. Otto Stapf, for information about the distribution of the spruce in Europe. Finally, I wish to express my thanks to Professor

E. B. Poulton, for his kind help and encouragement during the preparation of this paper.

THE SCANTY INSECT FAUNA OF SPITSBERGEN

The insect fauna of Spitsbergen is remarkably poor. There are only two orders of insects which are present in sufficiently large numbers to provide food for other animals. These are the Diptera (flies, gnats, etc.) and Collembola. The former are mostly Chironomids, with aquatic larvae. In addition to these (Summerhayes et al. 1923) there are three species of beetles, a moth, a sawfly, several species of parasitic Hymenoptera, a caddis fly, an Aphid, and a certain number of bird and mammal parasites (lice, fleas, and bird-lice). Now the poverty of this fauna is not only due to its high arctic character, but also to the geographical isolation of Spitsbergen, as a comparison with Greenland in regions of similar climate will show. One notes, for instance, the complete absence of bees, butterflies, and Tipulid flies in Spitsbergen. The recent geological history of the Spitsbergen archipelago appears to be as follows (see Nansen 1921): the islands were connected to each other and to the continent of Europe by land before the Pleistocene Ice Age, when all these regions sank much below sea-level, with the result that Spitsbergen became isolated from Europe by the Barents Sea. This isolation has remained to the present day. At the time of maximum glaciation the preglacial flora and fauna was wiped out by a thick covering of ice, such as still exists on North-East Land. When the ice retreated, the islands were entirely barren, and inasmuch as there has been no land connection with Europe since that time, the present flora and fauna must have reached them by other means. That this conclusion is correct is shown by the striking gaps in the mammalian fauna of Spitsbergen. The lemming, arctic hare, ermine, and musk-ox are absent, and the only true land mammals are the arctic fox and the reindeer. The last two are known to make long journeys over the sea-ice in winter (in fact it has been proved that the Spitsbergen reindeer has come over from Asia), while the other animals do not. There is thus a close analogy between Spitsbergen and Krakatoa, the former having had its flora and fauna blotted out by ice, and the latter by volcanic lava. In both cases the question arises: How has the new fauna got there?

METHODS OF DISPERSAL

There are several methods by which insects might reach Spitsbergen: 1. Wind, or flight, or both. 2. Driftwood. 3. Drift-ice. 4. Ships. 5. Birds. These factors will be considered in turn.

(a) Hover-flies and Aphids on North-East Land, August 1924.

North-East Land is the second largest island of the Spitsbergen group; it is about the same size as Wales, and is almost entirely covered by a thick ice cap, which rises to some 2300 feet above sea level. There are patches of barren rocky land round the coast, which possess an exceedingly scanty fauna. The only insects which live on this coast area are a few flies and Collembola.

A very curious case of dispersal on a large scale was met with by the Oxford sledging parties on the ice cap of North-East Land in August 1924. I myself was not sledging and am therefore indebted for the following facts to members of the various parties, viz. Messrs. F. A. Montague and W. B. Carslake of the Northern Party, Messrs. H. M. Clutterbuck and K. S. Sandford of the Central Party, and Mr. F. G. Binney of the Southern Party. It is worthwhile giving the detailed experience of these parties, although exact localities, distances and heights are not yet available. (For the present purpose extreme accuracy in these data is unnecessary.)

Northern Party. August 6th. The party were camped on the top of the North-East Land ice cap, i.e. on the N.W. corner of the island, at about 2300 feet above sea level. There had been light south winds earlier in the day, but at 10 p.m. a strong gale sprang up from the south, and lasted six hours.

August 7th. They were still in the same place, and there was mist and fairly brisk rain with a strong south breeze.

August 8th. The party started to travel towards the south (over old ground traversed a few days before), and after they had gone about two miles, reaching a long south slope, a large black Aphid was seen sitting on the snow. After this, Aphids were met with all day on an eight-mile march south, about one to every thirty or forty yards, either on the level or on a slight south slope. They were resting on the surface of the snow, not buried, and with the wings mostly dry, about 80 per cent being alive. Specimens of these were collected. During the day one large yellow and black hover-fly was also seen and collected; it was walking on the snow, with the wings more or less dry. Later in the day another hover-fly was also seen, but was rather bedraggled. One specimen of some large Tipulid fly was seen by Mr. Carslake on the snow. The weather during this day was cold, damp, and misty.

August 9th. During this day the party marched five miles south, still high up on the icecap, and they saw black Aphids still, but about one-third as thickly, and about 90 per cent dead. There were no hover-flies.

August 10th. No more insects were observed, and during the day a blizzard came on, which lasted until August 13th. This must have destroyed all the insects.

Central Party. August 5th. The party was encamped near the head of Wahlenberg Bay, a deep fjord running into the west coast of North-East Land; they were about fifteen miles from the north side of the Bay and about 1000 feet up on the ice cap. (Camp 9 = "Amen Camp.") At this time there were no insects on the snow.

August 6th. A reconnaissance was made, and they returned to camp by 1 p.m., no insects having been seen during the morning. There was a wind from the S.W., and the barometer was falling.

August 7th. Still at Amen Camp. Wind from the S.W. Barometer falling very low. Heavy rain on the night of the 7th.

August 8th. At 1 a.m. Mr. Clutterbuck went out and saw one yellow and black hoverfly sitting on the snow, and secured it. At 7 a.m. the party started to return in a westerly direction along the side of the ice cap, still at about 1000 feet. Large numbers of hoverflies and black Aphids were seen during the whole of that day, on a march of from five to ten miles. The insects were sitting on the snow, alive but rather sluggish. Another hoverfly and some Aphids were collected. The wind was now rather indefinite.

August 9th. Starting from "Blue Lagoon Camp" (No. 10) the march was continued, but no more hover-flies were seen, although there were still a few black Aphids about. August 10th. The party left "Ice Cone Camp" (No. 11) in the morning, and saw no more insects. At 2 p.m. they were stopped by a blizzard, which lasted for three days, and must have destroyed all the insects.

Southern Party. On the early morning of August 7th (midnight to 8 a.m.) no insects were recorded by Mr. Binney. But while marching at the same time on the 8th (12.30 to 8.30 a.m.) both yellow and black hover-flies and black "flies" were seen. The latter were obviously the Aphids, although no specimens were collected. The place where the insects were seen was on the top of the ice cap some miles from the east coast, between Isis Point on the East Coast and Wahlenberg Bay on the west.

I should like to emphasize the fact that these sledging parties were traveling under exceedingly strenuous conditions and over very bad surfaces. This made the collection of specimens and taking of notes no easy task. Therefore the careful and scientific way in which the data about these insects was recorded reflects great credit on the men who did the work.

Base Camp. I myself was working on Reindeer Peninsula, Liefde Bay, North Spitsbergen (about forty miles from North-East Land), at this time, and observed no hover-flies or Aphids in a long day's excursion on August 8th. Therefore the fly swarms must have passed to the northeast or else at greater heights.

It is clear from the facts detailed above, that the insects must have arrived on the icecap some time during August 7th. All the observations agree with this conclusion. The numbers of these flies and Aphids must have been enormous, running into hundreds of thousands or even millions, for all three parties to have seen so many at the same time. The insects appear to have been blown in a broad belt over the island. Now, where did they come from and how did they get to North-East Land?

The hover-flies (three females) have been identified by Mr. Collin as *Syrphus ribesii*, which is widely distributed in Europe, occurring, amongst other places, in the extreme north of Sweden and in Iceland, but has never been recorded from Spitsbergen or Greenland. The Aphids are winged females, and have been identified by Mr. Laing as *Dilachnus piceae*, the food-plant being the spruce, *Picea* (*P. excelsa*, etc.). Mr. Collin informs me that although there are no actual records of the larva of *Syrphus ribesii* preying upon this Aphid, it is quite likely that it does so. It is only reasonable to suppose that the hover-flies and Aphids came from the same district, in view of their vast numbers and the almost complete absence of other insects in the swarms.

There are no large Tipulids in Spitsbergen. The only *Syrphus* occurring in Spitsbergen is *S. tarsatus*, which lives in a few favoured spots on the west coast of West Spitsbergen (Collin 1923). With it is found a small species of Aphid (not *Dilachnus piceae*). We know from the work of foreign naturalists in Bellsound and Icefjord, and the ecological survey carried out by the Oxford Expeditions in Icefjord and other parts of Spitsbergen, that there are no places where *S. ribesii* or *D. piceae* do or could occur. The fact that there are no trees proves that *D. piceae* must have come from some other country. To Dr. O. Stapf, through Professor Poulton, I am indebted for the following information about the distribution of the spruce in Europe:

> The northern Picea of Scandinavia and Russia is *P. obovata*, which is treated by some as a variety of *P. excelsa*. Its northern limit runs from Saltdalen in Norway (67° 10' N.) over the Sultjelma to the upper Muonio river and thence to Lake Enare and the Kola Peninsula, but there is an outpost in the Varangerfjord (69° 30' N.) and another at Karasjok (68° 30' N.). Neither of these outposts is likely to supply those 'immense numbers' of the Aphid you speak of, nor is it likely that they came from the Norwegian coast at Saltdalen or the Swedish stations. At Saltdalen, at

any rate, there are only a few trees of it. The tree is much more common in the Kola Peninsula (say at 67° N.), and I would suggest this to be the origin of the Aphid. It does not occur in Novaya Zemlya, which would be perhaps a little nearer.

It is probable then that the Aphids, and therefore presumably the *Syrphi*, came over from the Kola Peninsula to North-East Land, a distance of over 800 miles in a straight line. This idea is confirmed by the following facts about the weather conditions prevailing at this time over Spitsbergen, the Barents Sea and North Europe. During August 6th, 7th, and 8th, there was a very large depression which moved over Spitsbergen from the Atlantic. As a result of this there were strong south and southeast winds blowing from Europe over Spitsbergen and more especially North-East Land. This is precisely what would be required to carry the flies from Kola Peninsula to North East Land. Mr. Krogness, of the Tromso Geophysical Institute, kindly allowed me to consult the synoptic charts for these days, and from the wind directions and forces it appeared that the insects would take from twelve to twenty-four hours to do the journey.

This view of the track followed by the insects is confirmed in an interesting way by Mr. Koefoed, who was engaged on fishery research in the Barents Sea in 1924. On August 13th Mr. Koefoed was able to land on Hope Island, a small, rather inaccessible, and extremely barren island belonging to the Spitsbergen group, and lying between North-East Land and Europe. It is about 250 miles from North Cape in Norway, and some 600 miles from the Kola Peninsula. On the island Mr. Koefoed saw four specimens of a large yellow and black *Syrphus*, walking about over the snow-free ground. One of these was collected, and is now in the possession of Mr. L. Natvig of the Oslo Museum, Norway. The position of Hope Island shows that these flies very likely formed part of the swarms that reached North-East Land on August 7th, and if this is so, they must have remained alive there for over a week.

The cause of these swarms of insects remains to be considered. In August 1869 Dunning (1869) recorded "countless swarms of *Syrphi*" at Walton-on-the-Naze, on the Essex coast, and another observer noticed similar swarms at Margate on the same day. The Essex swarms consisted mainly of *S. ribesii*, *S. corollae*, and *S. pyrastri*. It is clear, then, that hover-flies are subject to occasional great increase in numbers. Now, I have shown (Elton 1924) that the periodic abundance of mammals and birds in Northern Europe is mainly controlled by climatic factors acting at regular intervals. The explanation of the hover-fly swarms may be that good years cause unusual abundance of Aphids, upon which the hover-fly larvae feed. This idea is supported by the North-East Land episode, and also by the fact that ladybirds (*Coccinella punctata* and *bipunctata*) were unusually abundant in Essex and Yorkshire in 1869, since ladybirds, like the hover-flies, feed on Aphids.

It may be noted that Amundsen, while on the north coast of Arctic America, noted (Amundsen 1908) the arrival of gnats during a storm from the southeast. This would be a similar case to that of the North-East Land insects.

(b) Local migration of Fucomyia parvula against the wind.

On July 12th, 1924, examples of the seaweed-fly *Fucomyia parvula* were found in the shingle above high-tide mark, near the Expedition's Base Camp (south side of Reindeer Peninsula, Liefde Bay, North Spitsbergen). On this day none were flying, and they were hiding under the stones, and when exposed to light crawled at once very actively down into cracks among the stones, although the wings were expanded. The air temperature was about 10° C. in the shade, and the weather was cold and cloudy.

On July 23rd, i.e. eleven days later, numbers of these flies could be seen near the shore, flying quite fast against a light south wind. (One or two could also be seen being blown very fast in the opposite direction.) The sun was shining, and it felt quite hot, although the air temperature was only about 5° C. in the shade. The migration had no relation to the position of the flies' normal habitat (drifted seaweed).

Clearly the reactions of the flies to light had changed from positive to negative between July 12th and 23rd. A similar reversal of phototropism was noticed on board the expedition ship "Polarbjorn" in the case of the fly *Leria modesta*. The larvae were living among herring refuse in the ship's hold, and began to pupate in large numbers in July. On, July 26th there were large numbers of newly emerged adults; hiding in the darkest corners, some with wings still unexpanded. On July 27th they were flying about outside in the sun. In the case of *Fucomyia* the reversal of phototropism was probably due to rise in insolation temperature, since in the arctic this is far more important than air temperature; in the case of Leria, to the changed physiological state of the flies. These examples illustrate how the reactions of insects may be reversed in nature by change in the animal or in the environment. Similar changes seem to occur on a much larger scale in locusts, causing them to migrate. It does not seem unreasonable to suggest that unusual climatic conditions in certain years may cause locusts to migrate in the same way that seasonal changes did in the *Fucomyia*, by reversing the normal tropisms.

(c) Local migration of blowflies.
Blowflies do not usually occur in Spitsbergen except near the coal settlements, and they have probably been brought there by ships. On July 3rd, 1924, numbers of blowflies (*Phormia groenlandica*) were found emerging from the carcase of a whale at Green Harbour, in Icefjord. On July 5th several examples were seen flying past at a distance of one to two miles from the whale, while dozens were to be seen flying round the carcass itself.

(d) Local migration of Chironomid flies.
On sunny days in the summer one usually sees a good many Chironomids flying along singly, and this habit of "voluntary migration" seems to be common to many species, e.g. *Diamesa poultoni*, flying over snow-covered ground in early July 1921, on Prince Charles Foreland, and also at Green Harbour.

DRIFTWOOD

Quantities of driftwood are carried by the polar current over to Spitsbergen, from the north of Europe and Siberia. I have never succeeded in finding anything living in it, but it may occasionally act as a dispersal agent. Logs become frozen into the ice pack, and animals or their resting stages might thus be transported long distances without being in contact with salt water.

DRIFT-ICE

The same drift that carried the wood, carried the pack ice from Siberia to Spitsbergen. Collembola might be carried on ice floes, since some species, e.g. *Agrenia bidenticulata*, are in the habit of walking out over snowfields (probably in order to eat wind-blown pollen).

SHIPS

The sealing ships which visit Spitsbergen carry with them a small insect fauna. On the "Terningen" (1923) were a good many flies (*Leria modesta*). On the "Polarbjorn" (1924) were several species of flies (*Leria*, *Limosina* and blowflies), a beetle (*Enicmus*) and a spider (*Leptyphantes leprosus*). The flies breed in the remains of herrings which are carried as cargo in the winter months to the Roman Catholic countries of southern Europe.

The spider lives on the *Limosina*. It is thus apparent that the dispersal of insects to Spitsbergen is partly dependent on religious movements. So far, however, none of these species seem to have established themselves in Spitsbergen, except blowflies.

BIRDS

(a) The insects transported to Spitsbergen by birds include a flea (*Ceratophyllus vagabundus*), occurring in the down of goose nests and possibly reintroduced every summer, and numerous mallophaga living on birds themselves. (b) Birds may sometimes assist in the dispersal of free-living insects. On Bear Island during June 1921, kittiwake gulls could be seen carrying lumps of moss for some miles to their nesting cliffs. In their nests a good many flies and Collembola occur, which have obviously been brought there by the gulls in the moss. (c) At Salmon Lake, Wijde Bay, North Spitsbergen, there is a colony of Arctic Terns, which nest on some of the small islands in the lake. On these islands in August 1924 there were enormous numbers of a mite, *Bdella littoralis* Linn. Now this mite normally occurs only at or below high-tide mark on the seashore, where it is very common. Here it was living on islands in a freshwater lake. Further, it occurred only on those islands where the terns were nesting, and not on the others. Since Arctic Terns often settle on, or nest on, shingle near the shore, where *B. littoralis* abounds, running about actively over the stones, it seems certain that the mites must have been accidentally carried to the islands by terns. This example illustrates the manner in which insects, e.g. Collembola, might also be dispersed by birds.

CONCLUSION

The concrete examples given above would make it appear that wind, flight, birds, and ships, are mainly responsible for the dispersal of the present fauna of insects to Spitsbergen. In the case of insects like flies, "voluntary migration" seems to play an important part, just as among butterflies and locusts.

SUMMARY

1. The insect fauna of Spitsbergen is very scanty.
2. This scantiness is due to the high arctic character of the islands and to geographical isolation from Europe by sea.

3. The possible methods of dispersal of insects to Spitsbergen are: (1) Wind, or flight, or both, (2) Driftwood, (3) Drift-ice, (4) Ships, (5) Birds.
4. Examples of the action of some of these factors are given:
   (a) *Syrphus ribesii* and *Dilachnus piceae* and Tipulid fly, blown to North-East Land from Europe.
   (b) Local "voluntary migration" of *Fucomyia frigida*, blowflies, and Chironomid flies.
   (c) Fleas, mallophaga on birds.
   (d) Gulls carrying insects in moss.
   (e) Mites (*Bdella littoralis*) transported by Arctic Terns.

REFERENCES

Amundsen, R. 1908. The Northwest Passage. vol. 2, p. 208: London.

Collin, J. E. 1923. Diptera (Orthorrhapha, Brachycera and Cyclorrhapha) from Spitsbergen and Bear Island. Ann. Mag. Nat. Hist., ser. 9, vol. 11, p. 116.

Dunning, J. W. 1869. Proc. Ent. Soc. London, Nov. 15th, 1869, p. 25.

Elton, C. S. 1924. Periodic Fluctuations in the Numbers of Animals. Brit. J. Exp. Biol., vol. 2, p. 119.

Hoel, A. 1915. D'ou vient le renne du Spitsberg? Bull. Soc. Geogr. Paris, vol. 30, p. 443.

Nansen, F. 1921. Spitsbergen. p. 42: Leipzig.

Summerhayes, V. S., and Elton, C. S. 1923. Contributions to the Ecology of Spitsbergen and Bear Island. J. Ecol., vol. 11, p. 214.

Elton, C. S. 1925. The dispersal of insects to Spitsbergen. Transactions of the Entomological Society of London *73: 289–299. Reprinted with permission of Wiley-Blackwell.*

A tenderfoot explorer in New Guinea (1932)

Ernst Mayr

I faced one of the biggest decisions of my life when, at the International Zoological Congress at Budapest in 1927, Lord Rothschild, of the Zoological Museum in Tring, the largest private collection in the world, asked me if I would undertake an expedition to New Guinea for him and Dr. L. C. Sanford, trustee of the American Museum of Natural History. I was barely twenty-three years old, had never been on an expedition before, and all that could be said in my favor was that I had had many years of experience in the study of European birds, and, what counts more, had the ambition and untiring enthusiasm of youth. My mind was therefore made up quickly. I said, "Yes" to Lord Rothschild's proposition and started immediately with my preparations.

Anyone about to undertake an expedition should possess a thorough knowledge of the animal-life occurring in the region he plans to visit. Thus, before starting for the field, I went to several of the large European museums and worked through their New Guinea collections, with the result that when I arrived in New Guinea, I knew not only the name of every bird I might collect, but also whether it was rare, or desirable for my collection, and whether it showed any peculiarities of particular interest to science. Equipped with this knowledge, I departed from Europe feeling a good deal more confident than when I had agreed to the expedition.

After a pleasant journey through the Mediterranean, the Suez Canal, the Red Sea, and the Indian Ocean, I arrived in Java to make the final arrangements. The Zoological Museum, of the Department of Agriculture in Buitenzorg, Java, assisted me by the loan of two Javanese birdskinners and one plant collector. These "mantris" were of invaluable service to me and proved themselves faithful and hardworking companions during the six months I stayed in Dutch New Guinea.

After a beautiful trip through the East Indian Archipelago, which gave me an opportunity to get acquainted with such interesting places as Bali, Celebes, and the Moluccas, I arrived on the 5th of April 1928, in Manokwari, the capital of Dutch New Guinea. What a thrill I had when I saw the towering Arfak Mountains, rising abruptly to an altitude of 9000 feet on the other side of the Dorei Bay! The summits were hidden in clouds—clouds that envelope the higher mountains of New Guinea for many weeks during the rainy periods of the year.

Manokwari, the largest settlement in northern Dutch New Guinea, has a white population of twelve, a fact that indicates somewhat the backwardness and wildness of the country. There are no railroads, no motorcars, not even horses and mules in this part of New Guinea. All the carrying is done by the natives.

In all tropical countries the mountains possess an animal-life strangely different from that which is to be found in the lowlands and hills. The lowland seems to be much more affected by the going and coming of forms from neighboring places, and we find many recent invaders, while the mountains are the homes or refuges of the primitive types and perhaps the original inhabitants of the whole area. This is true of mankind as well as of animals. In the lowland of New Guinea we find the Melanesians—tribes that are related in culture and language to the Malayans in the West and the Polynesians in the East. In the mountains we find the Papuans, a very primitive type of mankind, in my opinion inferior in their culture to any other human race, including even the aboriginals of Australia.

The same is true of the bird-life. In the lowland we find mostly species and genera which are distributed over wide parts of the Indo-Australian region, while in the mountains we find endemic genera with no close relatives anywhere. It is here in the mountains that we meet the choicest of the birds of paradise. It is in the mountains that we find the most beautiful parrots and some of the most peculiarly developed members of the honey suckers. To make a thorough investigation of these mountain forms in the different ranges of northern New Guinea was the main object of my expedition.

I did not spend much time in the lowland, and after I had bought the necessary provisions, I said "good-bye" to the small white colony in Manokwari and sailed across the bay to Morni on the southeastern foot of the Arfak Mountains. For several months thereafter I did not speak a European word, using Malay in all my conversations until my return from exploring the Arfak Mountains. In Momi I sent out word to the surrounding villages asking for carriers, and with the help of the Malayan district officer I succeeded, after several days, in assembling a caravan of about fifty.

No one who has traveled in Africa can imagine the carrier difficulties in New Guinea. The race is small and, considering the roughness of the country and the bad condition of the bush-trails, the carriers refuse to take loads weighing more than thirty pounds. Only in exceptional cases could they be persuaded to take two man loads. It required a good deal of figuring to cut down the outfit into such small loads.

After much hard and noisy gesticulation, the long string of carriers finally departed. We soon left the vicinity of Momi village with its secondary growth, its native gardens and open clearings, and entered the virgin forest that spreads over the alluvial plain of Momi River. The noise of the forest edge with its numerous parrots, starlings (*Mino*), leatherheads (*Philemon*), and New Guinea magpies (*Cracticus*), was exchanged for a deep silence only occasionally interrupted by the voice of a thickhead (*pachycephala*) or a fly-catcher (*Monarcha*). From the treetops we heard now and then the deep *oo-oo-oo* of forest pigeons, but suddenly all these voices were overpowered by a melodious whistling series of calls.

"Boeroeng goening," whispered the interpreter excitedly, and, leaving the trail, we began cutting our way through the vines and shrubs with bush knives. Soon we reached the foot of a medium-sized tree on which—I shall never forget this exciting moment—I saw for the first time in my life the display of the yellow bird of paradise (*Paradiscea minor*). Two males in full plumage and a few immature males displayed and went through all the eccentric and acrobatic movements of this performance. I had no time on this day to study any details, but full of joy and satisfaction at this impressive sight I continued the march.

In order to get more opportunities to collect data on the vertical distribution of birds in different altitudes, I had chosen a route into the mountain which had been used by but few parties before me, but which led through a well populated district with villages at all altitudes between sea level and 6000 feet. The powerful chief Basi, of the Manikion district, and I were in the lead as we penetrated deeper and deeper inland, following the Momi River.

In Momi I had been told many stories about the treachery of the mountain tribes, and I was somewhat worried about what might happen during the coming weeks. Occupied with these thoughts I was suddenly startled by a noise that began at the end of my caravan and ran along the string of carriers toward me, increasing rapidly until it became a bloodcurdling series of screams and yells. I was frightened, feeling certain that this was a signal to attack, and I expected every moment to feel the knives of the carriers in my back. I looked cautiously back to Basi, but he, apparently guessing my worries, assured me there was no danger. As it turned out, it was really the war cry of the Manikion tribe, but on this occasion it was uttered only to inspire the energy of the carriers. With increasing experience I grew surer of myself, but on this first occasion I showed that I was a thorough greenhorn.

Late in the afternoon, after long and strenuous wading in a rocky riverbed, we arrived at the first camping place at the foot of the real ascent. My tent was set up, my carriers built their leafhouses, and the regular camp life developed. The few birds collected during the march had to be prepared, and soon darkness fell on my first night in the tropical forest. What this means only the man who has witnessed a tropical night himself, can appreciate. No words can describe the concerts produced by the cicadas, locusts, treefrogs, and night birds, a symphony of peculiar and deeply impressive harmony. Listening and dreaming, I lay awake for a long time in spite of the fatigue caused by the march and all the exciting experiences of the day.

When I awoke the next morning, camp life was in full swing again. Breakfast was served quickly, and soon the loads were packed, and we started on a long and strenuous climb toward the summit of the Taikbo Mountain, 4000 feet above us. The trails of these mountain tribes have no similarity to those familiar to us in civilized countries. There are no zigzags, but the trails go straight up the slopes, steep as they may be. Only when the crest is reached does the road become somewhat easier. I had plenty of occasion to admire the stamina of these mountain people who, although inferior to the white man in their physique, are superior in heart and lungs. Despite their loads they could set a pace that I was hardly able to follow. Before we had climbed a thousand feet the appearance of the forest had begun to change. The number of forest giants began to decrease, the undergrowth that was rather open in the lowland forest grew thicker. Thorny rotan palms entangled the shrubs with their long vines, and a few tree ferns and epiphytes were in evidence. Above two thousand feet this change of formation was quite obvious.

The bird-life also showed a change. Many of the species leading the concert in the lowland forest disappeared and new voices could be heard. The weather likewise began to change. The temperature was lower and soon we reached the zone that is enveloped in clouds after ten o'clock in the morning. The wind blew the fog through the forest, moisture collected on leaves and twigs, and big, heavy drops fell to the ground. No wonder that moss and lichens grow luxuriously in this atmosphere, and, although the true moss forest is decidedly higher up, moss was quite abundant on the trees and on the ground even at less than three thousand feet altitude. The moist fog gradually changed into rain and, at the request of my carriers, I decided to camp at the highest water-place, which was still considerably below the summit. Big fires were built to warm my carriers. I could imagine how cold they felt, being entirely naked but for a narrow strip of cloth around their loins. The temperature went down to 19° C. but singing and talking, and with very little sleep, the carriers sat around their fires during the night.

In the beautiful, clear sunlight that greeted us the next morning, the forest looked entirely changed. The birds were much more active and were singing all about, and, knowing that I would reach my first real permanent station on that day, I started collecting. One of the first birds shot was the "superb bird of paradise," one of the most beautiful creatures of the New Guinea mountains. The breast is covered by a shield, composed of little, glossy, green scales. On the neck is a large, black, velvety crest, which can be spread out during the display, as are the tail-coverts of a peacock. Another welcome addition to my collection was *Drepanornis*, another bird of paradise with an extraordinary, long, sickle-shaped bill. After a short climb I reached the top of Taikbo Mountain, but here, as in most cases in New Guinea, reaching the summit was a disappointment, for the heavy forest did not permit any view.

The path that led down to the Siwi Valley was just as steep as the one that led from the coast up to the summit, and after a hurried descent we reached a little creek in the Siwi Valley. Now signs of population became apparent, the forest became lighter, and soon we entered a large clearing, the area of Siwi village. These Arfak Mountain villages do not quite come up to our ideas of a village. Siwi, like other places I visited afterward, consisted of single houses, 200 to 500 feet above the bottom of the valley and located on both sides. It covered an area of several square miles. Most of the houses are within calling distance of one another, and whenever there is any news to tell, the valley is filled with shouts. The arrival of a white man was occasion enough to make this signal service work full blast, and soon the natives began to come down from their houses to help in establishing my camp. In order to have water handy, I decided to camp on the shore of the little creek, on a place high enough to be safe from floods. I was warned against these floods and before long I had an opportunity to witness such a spectacle.

I was very glad that we had finally reached the camping place, for I had a bad infection on my leg and was hardly able to walk. When swimming in Manokwari, I had hurt my foot on a coral reef, and such coral wounds heal very slowly, even if properly attended. I, however, had had no time to attend to the wound, and the marching in riverbeds on the previous days and constantly wet shoes had done their part to make the infection really dangerous. My whole leg was swollen and weeks passed before I could do any strenuous collecting.

While I paid off my carriers (they received 25 cents a day), the natives started to clear away the secondary growth, to erect my tent, and to build houses for my baggage and my Javanese "mantris." The arrival of my party was like a big festival, for the natives made little huts for themselves all around my camp, sang and danced, and the boys and younger men especially stayed in my camp even over night and for several days afterward.

The river bottom, where the camp was situated, was at an altitude of 2400 feet, but all around mountains rose steeply to 4800 feet. The real forest had been destroyed by the natives in the valley and on the lower slopes, and had been replaced by native gardens, savanna-grass ("alang-alang") and open secondary growth. There I found grass finches (*Munia*) and parrot finches (*Erythrura*), wren-warblers (*Malurus*), and even (in winter quarters) a Siberian sedge-warbler (*Locustella fasciolata*), a bird that puzzled me considerably and which I regarded as a new species for a long time, knowing that it was none of the known New Guinea birds. The Arfakers are hunters rather than farmers. Most of the fieldwork is done by the women, while the task of the men is to cut down the forest. But they prefer to take bow and arrow and wander about and hunt birds or, if luck is favorable, even "big game,"—cassowaries and pigs. Except for some phalangers and other marsupials, a few small rodents and bats are the only indigenous mammals on this island. Pigs have been introduced, but are in a more or less wild state all over the island. To get the desired meat, the New Guinea hunter has to look for smaller game, so he devotes his attention to the abundant bird-life. From his early youth he has learned to know the voices and habits of the birds, and I was amazed at the exact knowledge of the life-histories of the birds these natives possessed. Almost every species had its own name, and they even distinguished some species which have been confused with others by some systematists on account of their similarity.

The natives are wonderful shots with bow and arrow, and soon learned to handle the several small shotguns which I had brought with me. After I had acquired the vocabulary of their bird names I had only to send out my hunters with orders to secure certain species and I was sure to get them. I thus succeeded in obtaining a collection of unusual quality, consisting of the rare and desirable species and lacking the great number of common birds so often found in the collections of inexperienced travelers.

The joy over the success of my collecting activity was a great help to me in overcoming the many difficulties that sometimes almost crushed my energy and willpower. The rainy season was not yet quite over and on some days the fog and rainstorms prevented collecting completely. The drying of the skins was also quite a task, as the air was saturated with moisture and the sun was not seen very often. Half-starved native dogs broke into my tent during the night and managed to get away unharmed with a few skins, thoroughly poisoned with arsenic. I never saw anything so thin and shabby as these dogs, which are related to the Australian dingo and do not bark. All these things, however, were only minor difficulties. What was much worse was that most of my boys fell sick, and all at the same time. One developed arsenic poisoning on his hand, and his whole arm swelled so that he could not work. The other of my mantris had malaria and alarmed me by his fantastic speeches in his delirium, while another helper fell sick with pneumonia. His life was saved only by the most careful nursing day and night. My sore foot had not healed and my plant collector was also laid up with a big tropical sore, so that my camp really resembled a hospital more than a collecting station.

Few people can realize what a strain it was for me to have to overcome alone all these difficulties, with no companion to talk to. Every situation was new to me and required careful consideration, especially the handling of the natives, who are very touchy and have many taboos that must be respected. On the other hand, they showed amazingly little imagination. I remember a little incident that happened during an eclipse of the moon. The moon became more and more covered by shadow, it grew darker and darker, but the natives showed no signs of interest or excitement. I asked them if they had no myth about it. I told them the myths of our own country and the myths believed by the Chinese and Javanese, and asked them what they considered as the cause of the sudden darkness. Not getting any response to my questions, I really became quite excited in my efforts to get some information about their belief.

Suddenly one of the men slapped my shoulder in a fatherly fashion and said soothingly:

"Don't worry, master, it will become light again very soon."

That cured me and I never again tried to acquire any information that was not given willingly. Their realism toward the mysteries of nature was sometimes quite appalling.

After my foot was better I started out again on excursions, but did not do much collecting. My main interest was to get some data on the habits of the New Guinea birds. Aside from the marvelous displays of the birds of paradise, which have been described in NATURAL HISTORY by R. H. Beck and Lee S. Crandall, there are many things in the New Guinea forest to arouse the interest of the naturalist. There are the cassowaries, large ostrich-like birds that run along the mountain slopes with an amazing agility. The male is less brightly colored than his mate, and apparently hatches the eggs as well as rears the young, rather unusual for birds, although this reversal is common to several other species. There are the megapodes or brush turkeys which bury their eggs in large heaps of leaves where the fermentation of the organic material incubates them, and the pigmy parrots (*Micropsitta*), not longer than one's thumb, which climb on trees as woodpeckers do and feed on termites. There are a score of beautiful doves and splendidly colored parrots. Once in a while I found a flowering tree where dozens of birds were feeding on the nectars. Here I found some of the rarest birds—birds that I had never seen before nor saw again. Exquisite colors are found not only on the birds, but all the orders of animals seem to compete for the prize of beauty. Butterflies and beetles show unsurpassed splendor of coloration, and the plants also deserve mention. There is no place on the earth where so many species of orchids have been found in a given area as in New Guinea. However, all this beauty is hidden away in the luxury of plant growth, and the real loveliness of these creations of this strange island continent becomes apparent only when in the hand of the collector.

After many weeks of hard work that continued until a late hour every night, I succeeded in getting a fair representation of the fauna and flora of this hill region, and the desire grew to penetrate farther into the interior, higher up in the mountains. To accomplish this was no easy task, and I was not yet sure what dangers and difficulties would await me. In this region, more than in any other place in New Guinea where I collected afterward, there were many rumors in circulation about the dangerous mountain people who are quiet and peaceful during the daytime but go out to kill during the night. Even in recent years several police boys had been killed during government expeditions, and I had been most earnestly warned not to go too far inland. I personally did not take much notice of these warnings, but it was a difficult task to persuade my companions to follow me.

I arranged with Basi to call carriers from the Misemi district, but just after the messengers had gone, I fell ill with influenza and was in bed for several days with high fever. The carriers finally arrived and we broke up our camp, but after I had gone a few steps, I fainted, weakened by the fever. I was in a rather desperate situation, as my carriers wanted to go ahead, and I did not know if I ever could get them again if I let them go to their villages now. We finally agreed to a two-days rest and I departed on the 25th of May. On the first day we had to climb the 4200-foot high divide between the Siwi and Ninei valley, and every step was a struggle for me, my heart being very weak. I arrived in Ninei (2800 feet), more dead than alive.

The next day we followed the Duga River up to the foot of Mt. Moendi, and then I camped, not being able to climb that mountain the same day. On both banks of the river were signs of former floods and I therefore decided to establish my camp not less than fifteen feet above the river. Some of my carriers laughed about my caution and made their camp closer to the water, only about ten feet above its normal level. In the late afternoon it began to rain, and after darkness the rain increased to a downpour of such violence as I had never witnessed before. The river had a very strong fall, and at low water fell in cascades over the boulders and rocks. But this rain changed the peaceful creek into a boiling torrent.

With the thundering noise of cannon tremendous rocks were torn away from their foundations and carried down stream. The night was pitch dark. Suddenly I heard a terrific yelling and screaming, and then a score of trembling and soaking wet natives rushed shouting into my tent.

"Master," they cried. "Our camp is flooded, and all our belongings have been carried away by the water."

I was worried lest the flood should damage my expedition outfit, and rushed outside to inquire the state of affairs. When I reached the camp of my carriers, I found to my surprise that the water level was now about five feet below the camp. The only explanation I can think of is that the camp had been swept by a “tidal” wave caused by the breaking of some dam built by fallen trees farther up the river. Fortunately, nobody had been drowned.

The “gang” was the biggest I ever had. All counted there were about 120 natives and many more carriers than loads. This never happened to me again. A large percentage were women, who are much stronger than the men. They have to do all the carrying of the firewood and fieldfruit and are therefore trained for this job. It was a strange experience for me until I got used to it, to see the women carry my loads while their husbands accompanied them with the babies in nets on their backs.

The next morning the difficult climb up Mt. Moendi began. The lower slopes are covered by alang-alang, growing on wide stretches from which the natives had burned the original mountain forest in order to make their fields. Higher up at an altitude of about 5000 feet the moist mountain-forest began. All trees, shrubs, and vines were covered by a stratum of moss several inches thick, and in places on the ground the moss was more than a foot thick. Most of the trees in this moss forest are rather small, but grow close together. Several birds not encountered on my former collecting stations were found here.

Approaching the summit, we found the trees gradually being replaced by shrubs and open grassland spots, a formation which I will describe in connection with the visit to Mt. Hoidjosera, where I found it much better developed. The summit of Mt. Moendi, at 6300 feet, is the watershed. The valley into which I was descending belonged to the system of rivers that flow into the McCluer Gulf on the south coast of New Guinea.

My path led gently down along a crest into the Ditchi Valley. This was not my original intention. I wanted to descend toward the northwest, but the natives claimed that there were no villages in that region, at least none occupied at that season of the year. So we turned southwest, and finally reached the bottom of the Ditchi Valley at about 3500 feet altitude. At 4000 feet altitude on the other slope of the valley was the village of Ditchi consisting of a few scattered houses, where I established camp on May 22.

This village had never before been visited by a white man. Behind the village two mountains (Mt. Wamma and Mt. Lehoema) rose to an altitude of approximately 6000 feet. These two mountains were my chief collecting grounds during the next weeks. As in Siwi, I wrote down a list of all the native names of the birds, and as soon as I received specimens I was able to identify them by their scientific names. By this method I was sure not to leave out any. At the same time I secured fine collections of plants.

In spite of all the intensive collecting I could not procure all the species in the Ditchi region that were known from the Arfak Mountains. I therefore desired to penetrate still farther inland and establish a collecting station in a higher place. The area that I fancied was the Anggi Lakes, which are situated at an altitude of more than 6000 feet and had been visited by several naturalists previously. But their bird-life had never been studied and I expected some interesting discoveries. The difficulty, however, was how to get there! All the previous parties were accompanied by a troop of soldiers, as the Anggi natives were reputed to be great warriors and anything but friendly toward whites. I heard many stories of murders that had been committed in late years, and I was trying to find a safe way to reach their villages, when chance finally came to my assistance. It turned out that one of the Siwi-men had a sister married to one of the Anggi chiefs. So I sent him up to get an invitation. I reckoned that their curiosity to see a white man would be greater than their suspicion and fright against me, the usual root of all fights. And I knew that I would be perfectly safe if I came to their village as a guest.

On the evening of the 8th of June, Wakil, my messenger, came back from the Anggi lakes and brought with him the chief and ten carriers. After an exchange of the usual formalities, we agreed to start for his village the next morning. As I had only ten carriers (except for Wakil and my fourteen-year-old interpreter Kapal, who spoke five languages, nobody from Siwi or Ditchi wanted to go with me), I had to cut down my luggage to the most necessary items. I left two of my Javanese behind me, but one of the bird skinners accompanied me.

The road was bad as always in New Guinea. First we had to climb down to the Ditchi River, then up again 1500 feet on the opposite slope, and as soon as we had reached the ridge we went down another valley. In all, we crossed about six or seven such valleys, tributaries of the Issim River, and when we finally reached the "village" (two houses) of Dohunsehik in the late afternoon, I was thoroughly tired from climbing, although the net gain of altitude was only about 600 feet.

We were now directly at the foot of Mt. Hoidjosera, which we had to climb before coming down to the lakes. I decided, therefore, the next morning, to give out cartridges to my hunters to secure some specimens in the alpine zone. But when I opened the cartridge-load I noticed to my horror that I had packed the wrong case and left all the small cartridges for my bird-guns in Ditchi. What to do? The only solution was to send back one native to fetch the cartridges and join me later on at Anggi. I would spend the first two days, until this boy arrived, in shooting large birds and collecting plants.

At 7:30 we started, for Anggi while one boy left for Ditchi. We gained altitude rapidly and the forest soon took on a very mossy character. Above 5000 feet we reached the ridge, and here there became evident a plant formation like that I had encountered already in a lesser degree on Mt. Moendi.

The forest opened up and was replaced by a brushy heather mixed with low conifers. Many of the shrubs, especially the rhododendron trees, were in flower and made this day's walk a very pleasant experience. On the other hand, the birdlife was disappointing. I did not meet a single species of bird that I had not met already in lower altitudes. There is a decided change of faunal zones at 4500 feet altitude, and as I had collected in this higher zone on Mt. Lehoema and Mt. Wanna, it was perhaps only natural that I did not make any new discoveries on this mountain. About noon we reached the summit of Mt. Hoidjosera, which means in the Manikion language: "The place where the pig fell." Despite much questioning, I was unable to learn the story on which this name is based.

From the summit I had a magnificent view, as it was unusually clear. In the west stood Mt. Lina (about 8600 feet), the highest summit of the Arfak region. To the south was the Issim Valley, which sends its waters to the south coast of New Guinea, and to the north were the two Anggi lakes, the male and the female, as the natives call them. The two lakes are separated by a ridge approximately 1400 feet in height, and are two entirely different basins; one is the origin of a river that flows to the north coast, and the other of a river that flows eastward to the Geelvink Bay.

After a quick descent I reached the shore of Anggi gidji (the male lake) about two o'clock, and established camp in the village of Koffo. Shortly after five o'clock the boy I had sent back for the cartridges arrived. I could hardly believe my eyes as he handed the cartridges to me. In one day he had made the three-days' march, including at least 8000 feet of actual climbing. He said he had been running hard for most of the day. I cite this case as an example of the marvelous stamina and climbing ability of these mountain natives. In the lowland, however, I had no difficulty in keeping pace with them.

The five days I spent on the lake easily surpasses all my New Guinea memories. The beauty of the landscape, the splendid scientific success (I discovered in the reeds and grasslands on the edge of the lake several species of birds either new to science or at least new for New Guinea) and the hospitality of these primitive and supposedly savage natives made me very loath to leave. When my party departed, all the women and girls of the village were lined up along the road, shedding copious tears, according to a custom widely distributed over New Guinea. However, as it was the first time that I experienced this proof of hospitality, I was deeply touched, and felt almost like joining in.

We returned the way we had come and after a short stay in Dohunsehik, where I wanted to get a specimen of the rare longtailed bird of paradise (*Astrapia*), I arrived in Ditchi. Here I found my Malayans in good health and spirits, much to my relief.

In order to meet the next mail steamer, I returned to the coast immediately, where I left my Malayans, while I took a canoe to Manokwari.

After thirty-five hours of continuous paddling I was back in civilization. Tired, unshaved, dirty, and sunburnt, I was invited immediately on my arrival to board the Dutch marine survey ship, the latest word in European luxury, to tell about my adventures. What a contrast!

Looking back on my first expedition, I value more than the discovery of many specimens and facts new to science, the education that it was for me. The daily fight with unknown difficulties, the need for initiative, the contact with the strange psychology of primitive people, and all the other odds and ends of such an expedition, accomplish a development of character that cannot be had in the routine of civilized life. And this combined with a treasury of memories, is ample pay for all the hardships, worries, and troubles that so often lead us to the verge of desperation in the scientific work that takes us into "the field."

Mayr, E. 1932. "A tenderfoot explorer in New Guinea." Natural History *32: 83–97.*

On the occurrence of *Trichocorixa kirkaldy* (Corixidae, Hemiptera-Heteroptera) in salt water and its zoo-geo-graphical significance (1931)

G. Evelyn Hutchinson

In spite of the considerable literature on the mechanisms by which animals and plants might be dispersed, too little attention has been given to the physiological feasibility of the methods of distribution invoked to explain wide and discontinuous ranges. The following records appear to indicate a case where dispersal by ocean currents is within the limits of physiological possibility and may be legitimately offered as an explanation of a very wide and remarkable distribution.

Trichocorixa is a genus of water boatmen of wide distribution in South, Central, and southern North America. Several forms are recorded from the West Indies and the genus would appear to have a distribution typical of many groups of Central or South American origin were it not for a single species which ranges right across the Pacific from California to China. The following notes deal with this form, *T. wallengreni*, and its close Eastern ally, *T. verticalis*.

During fisheries investigations in Delaware Bay in 1929 Mr. Albert E. Parr, Curator of the Bingham Oceanographic Foundation, Yale University, obtained two living male specimens of *Trichocorixa verticalis* associated with typical marine planktonic organisms in tow-nettings taken at stations 48, north of Brandywine Shoal, salinity 24.90 per mille (June 18, 1929) and 63, salinity 29.34 per mille (June 18, 1929). Although drowned flies and other insects were frequently met with in the surface plankton of this region, no specimens of living and apparently healthy insects other than these two corixids were obtained. *T. verticalis* occurs commonly in ponds near the sea in Cape May County, N. J., and is recorded from Connecticut, Pennsylvania, Georgia and the West Indian Islands of Cuba and St. Thomas (Lundblad, 1929a).

Mr. Richard M. Bond, Bishop Museum fellow of Yale University, has forwarded for determination a number of specimens of the closely allied *Trichocorixa wallengreni* taken in "strong brine from salt works at Elkhorn Slough," Monterey County, Cal. (10th November, 1930). Both sexes as well as immature individuals occurred in this locality, which otherwise is inhabited only by the typical halobionts *Dunaniella*, *Artemia* and *Ephedra*. *Trichocorixa wallengreni* was originally described from California, but recently Lundblad (1929b) has shown that *Corixa blackburni* from Hawaii is synonymous and has also recorded the species from Shanghai. This transpacific distribution is probably unique among waterbugs; the Hawaiian records strongly suggest that it is to be explained by dispersal across the Pacific Ocean, rather than by an Alaskan-Siberian land bridge. *T. wallengreni* or its eggs might possibly be transported in damp salt, but the species has clearly been established for some time in Hawaii (*C. blackburni* was described by Buchanan-White in 1877) and a natural method of dispersal seems more probable. Since it is clear that the species can stand salinities considerably above those of the sea it is not inconceivable that specimens might travel by the Northern Equatorial Current from California to Hawaii, and from Hawaii to China. Insects of this family being less dense than water when surrounded by their air bubble, this method of distribution would involve a minimum of effort. Since the above was written, Dr. H. B. Hungerford kindly informs me that he has specimens of *Trichocorixa* from the Galapagos Islands and that he has frequently received specimens of the genus "from saline waters. The exact salinity of the water, however, has never been available."

LITERATURE CITED

Buchanan-White, F. 1877. Descriptions of Heteropterous Hemiptera Collected in the Hawaiian Islands by the Rev. T. Blackburn. No. I. *Ann. Mag. Nat. Hist.* (Ser. 4), xx, p. 110.

Lundblad, O. 1929a. Über einige Corixiden des Berliner Zoologischen Museums. *Archiv. f. Hydrobiol.*, xx, p. 296.

Lundblad, O. 1929b. Beitrag zur Kenntnis del' Corixiden II. *Entom. Tidskrift*, 1929 (1), p. 17.

Hutchinson, G. E. 1931. On the occurrence of Trichocorixa kirkaldy *(Corixidae, Hemiptera-Heteroptera) in salt water and its zoo-geo-graphical significance.* American Naturalist *65: 573–574.*

Ecological compatibility of bird species on islands (1966)

Peter R. Grant

INTRODUCTION

The size of an island largely determines how many species of animals it supports (Preston, 1962; Wilson, 1961). How does the number of species become adjusted to island area? To judge from the success in predicting the numbers of species from a knowledge of the circumscribed area of islands (e.g., Grant, 1965a; Hamilton & Armstrong, 1965; Preston, 1962), the supply of species to islands is not short. Therefore, the adjustment of species number to island area must operate principally through extinction (see also MacArthur and Wilson, 1963; Mayr, 1965). Elton (1958) has commented upon the relative instability of communities composed of a small number of species (such as island communities), and has attributed the large fluctuations in the number of individuals of a species to the lack of checks and balances in the form of predators, parasites, etc. When such fluctuations bring a population to a low level, random extinction is a hazard; and this is one possible means by which species are eliminated. In addition, adverse climate (Asahi, 1962), food shortage (Christian, Flyger, and Davis, 1960; Lack, 1954), disease (Mayr, 1963, p. 75), or intraspecific behavior effects (Chitty, 1960; Christian and Davis, 1964) may contribute to the decline of populations, leading either directly or indirectly to extinction. Evidence is presented in this paper to support an additional view, that the ecological incompatibility of similar species of birds (and other animals) has been an important factor in the process of adjustment of numbers of species to island area. Some aspects of this theme have been treated by Lack (1947), Mayr (1963), and others. Here it is developed from a consideration of two consequences of a theory which accounts for the prevalence of long bills and tarsi among island birds (Grant, 1965b).

Two terms require an explanation: To contrast species of the same genus with those of different genera the terms homogeneric (= congeneric) and heterogeneric are employed. The currently employed term congener is used as a short form of homogeneric species.

THE PAUCITY OF CONGENERS ON ISLANDS

It has been suggested that on islands natural selection has favored birds with long bills and tarsi for the exploitation of a wide range of environmental resources, due to the absence of potentially competitive species (Grant, 1965b, 1966a; Watson, 1962). If this is correct, it is to be expected that pairs of homogeneric species would be rare. This follows from the contention that homogeneric species are usually more similar ecologically than heterogeneric species (Elton, 1946; Lack, 1944; Moreau, 1948), due to the former possessing more similar adaptive systems (cf., Lack, 1965). The contention has been questioned by Williams (1953, 1964), but his argument has been refuted by Hairston (1964). It may therefore be accepted as a useful generalization. Homogeneric species do occur infrequently on islands (e.g., Mayr, 1931; Rand and Rabor, 1960) as a consequence of the rarity of all species, since islands contain fewer species than an equivalent area of mainland (Preston, 1962). It is more important to ask if there is a paucity of homogeneric species even in relation to the paucity of total species.

This is not easy to answer because of the difficulty of assessing which region of the mainland is to be compared with the island. Furthermore, whereas the number of species on islands is reasonably well known, the number in the appropriate mainland area is not because the exact limits of distribution of many mainland species are not known. Nevertheless an answer can be given with data from the Tres Marias Islands and adjacent mainland Mexico. These four islands are close together and may be considered as a unit (Grant, 1965a, p. 65). They support a large number of species and a more or less uniform, relatively undisturbed, habitat, and their faunal affinities are unequivocally with the adjacent mainland on the same latitude 50 miles away. The known number of species of land birds breeding on the islands and a conservative estimate of the number of the species in an equivalent part of the mainland (same area and range of altitude, similar habitat, etc.) are given in Table 1. The proportion of congeners is significantly less on the islands than on the mainland (χ^2 test, df 2, $P < 0.05$). Thus the expectation of paucity of congeners is met. The small proportion of homogeneric species in the Tres Marias bird fauna may indicate that the coexistence of ecologically similar species is more difficult on these islands than in an equivalent area of mainland. It is noteworthy that one member of each genus represented by two species on the Tres Marias is a migrant, and is on the islands for less than half a year (Grant, 1966b).

Is this finding of general relevance to islands? It probably is, although there are reasons for believing that the difference in proportion of congeners between mainland and island assemblages is not always as large as in this Mexican situation. First, it appears that the proportion of congeners increases with size of island (Fig. 1). This is probably due to the way in which area influences the coexistence of similar species directly (e.g., the effect upon population size) and indirectly (the effect upon variety of food

TABLE 1. The number of species per genus of land birds on the Tres Marias Islands and in an equivalent area on the adjacent mainland of Mexico.

	SINGLES	PAIRS[1]	TRIPLETS[2]	TOTAL SPECIES	% CONGENERS
Islands	30	2	0	34	11.8
Mainland	83	11	5	120	30.8

Island data taken from Grant and Cowan (1964), mainland data from Friedman et al. (1950), Miller et al. (1957) and author's observations.

[1] Island genera are *Myiarchus* and *Vireo*; mainland genera are *Buteo, Columbigallina, Amazona, Glaucidium, Trogon, Centurus, Cissilopha, Thryothorus, Turdus, Icterus, Sporophila*.

[2] Genera are *Falco, Amazilia, Tyrannus, Myiarchus, Vireo*.

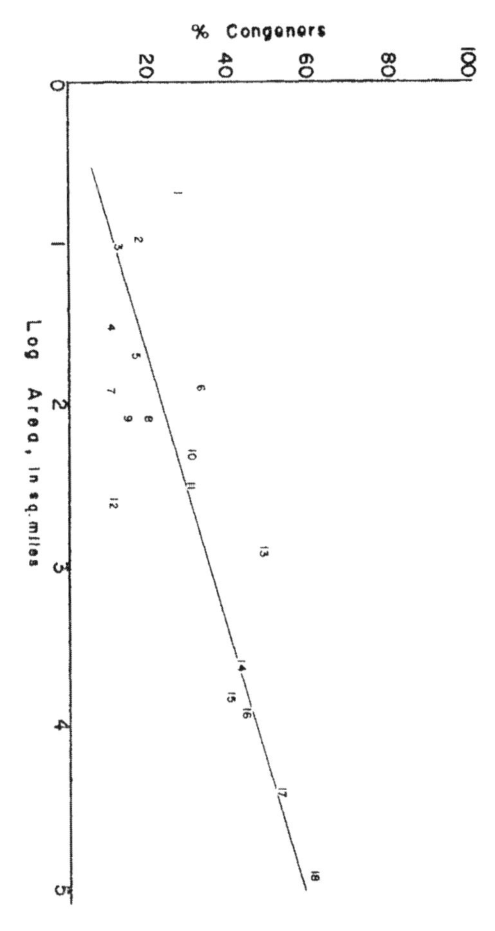

FIG. 1. The relationship between proportion of homogeneric bird species and island area.

Symbols, and sources of information: 1 Mujeres, 2 Holbox (Paynter, 1955); 3 María Cleofas, 4 María Magdalena (Grant, 1965b); 5 Principé (Amadon, 1953); 6 Moheli (Benson, 1960); 7 María Madre (Grant, 1965b); 8 Anjouan, 9 Mayotte (Benson, 1960); 10 Cozumel (Paynter, 1955); 11 Gran Comoro (Benson, 1960); 12 São Tomé, 13 Fernando Po (Amadon, 1953); 14 Corsica (Peterson, Mountfort, and Hollom, 1954); 15 Vancouver Island (Munro and Cowan, 1947); 16 Sardinia (Peterson et al., 1954); 17 Newfoundland (A.O.U. Check-list, 1957); 18 Great Britain, exclusive of Ireland (Hollom, 1955).

The islands, situated alone or in small groups and fairly close to mainland, were chosen from several parts of the world to give a good distribution of sizes.

TABLE 2. The number of species per genus of land birds on the Vancouver Island and in an equivalent area on the adjacent mainland of Canada.

	SINGLES	PAIRS[1]	TRIPLETS[2]	QUARTETS	TOTAL SPECIES	% CONGENERS
Island	56	11	4	1	94	40.4
Mainland	65	11	3	4	112	41.9

Data taken from Munro and Cowan (1947) and the A.O.U Checklist (1957). The genus *Zonotrichia* has been treated *sensu* Paynter (1964).

and habitat, which determines the opportunity for ecological segregation and coexistence). Hence there are only small differences in proportion of congeners between large islands and equivalent mainland, as in the only other example subjected to analysis here (Table 2). Second, the number of species is too small on very small islands and equivalent mainland to contain many congeners in either region with the result that a comparison yields little difference in proportion of congeners. Third, the degree of isolation may influence the proportion of congeners on islands, just as it influences the number of species (MacArthur and Wilson, 1963). For instance, islands close to the mainland, such as Fernando Po, which is isolated by about 20 miles, tend to have a large proportion of congeners; while those well isolated, such as Sao Tome (ca. 175 miles), tend to have a small proportion (Fig. 1). This suggests that the more isolated an island is, the greater must be the difference in ecological preference between a newcomer and those animals already present for the former to become established, perhaps because the more isolated islands have the least varied habitat. Alternatively, since Fernando Po was almost certainly connected with the mainland "in the not too distant past," whereas Sao Tome was never connected (Amadon, 1953, p. 401), the difference in proportion of congeners may simply be a product of the difference in manner of island formation. The likelihood of sustained coexistence would appear to be greater if the species had coexisted on that same land before the island was formed, than if they had colonized the island sequentially. Whatever the reasons for the disparity between near and far islands, and both explanations may be involved, it will obviously affect the comparison with the mainland. Finally, the comparison will be affected by latitude if the proportion of congeners varies with latitude on the mainland, but not on islands. This is possible because the total number of species in an area does vary with latitude on the mainland (Klopfer and MacArthur, 1961), but apparently does not on islands (Preston, 1962), at least at low latitudes.

Therefore, the evidence indicates that paucity of congeners is characteristic of islands, particularly those of small to medium size. A more exhaustive survey is needed to confirm this and to examine more carefully the influence of environmental factors on the number of congeners (cf. Hamilton and Armstrong, 1965). In addition, it would be worth examining the possibility that the magnitude or frequency of structural and behavioral changes in species of birds has also been greater on islands of small to medium size than on those of large size, especially as Mayr (1965) has produced indirect evidence to the contrary.

THE DIVERGENCE OF CONGENERS ON ISLANDS

If species change in morphology and behavior in the absence of ecologically similar species on an island, it is logical to expect no change in the presence of a sympatric congener or else change in one or both in a mutually divergent direction. The length of the bill is a convenient and appropriate structure to consider, since it reflects to some degree the nature of the diet (Kear, 1962; Schoener, 1965). North America and Mexico is a suitable region to consider because of the availability of bill length data. In contrast to the large total number of species occurring on offshore islands in North America and Mexico, there are few pairs of sympatric homogeneric species on those islands. Of these pairs only three have been isolated sufficiently for each member to have evolved into a form which is recognized as taxonomically distinct from its mainland relative. All of these island pairs meet the expectation. In one instance the difference in bill length between a pair is approximately the same in the two regions, and in the other two instances the difference is greater on the island than on the mainland (Table 3). Unfortunately, bill length data are not available for those species whose mainland and island forms are placed in the same subspecies.

TABLE 3. Percentage differences in bill length between homogeneric species of birds occurring on islands (where endemic) and adjacent mainland.

	MAINLAND	ISLAND
Centurus aurifrons and *Centurus pygmaeus*[1]	48	64
Parus atricapillus and *Parus hudsonicus*[2]	28	49
Hylocichla guttata and *Hylocichla ustulata*[2]	9	13

Data taken from measurements of adult male specimens, given in Aldrich and Nutt (1939), Burleigh and Peters (1948), Ridgway (1907, 1914). Percentage differences in this and subsequent tables refers to a difference expressed as a percentage of the smaller measurement.

[1] Cozumel Island and adjacent mainland of Mexico.

[2] Newfoundland and adjacent mainland of Canada.

Data from the Tres Marias Islands are not strictly comparable since, of the two pairs of homogeneric species there, only one species is subspecifically distinct from its mainland relative. Nevertheless they conform to expectation; differences in bill length between the congeners are large on the mainland (cf. Hutchinson, 1959), and are equally large on the islands (Table 4). Bill lengths have not undergone change on the islands. In addition the diet (Table 4), method of feeding, feeding heights, and perch dimensions used do not differ appreciably in the two regions (Grant, 1965b, and unpublished data).

Some heterogeneric species are as similar to each other ecologically as are homogeneric species, and presumably they show the same adaptations to living sympatrically as do homogeneric species. It is therefore not surprising to find, on the Tres Marias Islands, examples of mutual divergence in bill length among those heterogeneric species, which, on the basis of observations of feeding and analyses of gut contents, are known to be ecologically similar. For example, two unrelated species, *Platypsaris aglaiae* (Cotingidae) and *Piranga bidentata* (Fringillidae), were found to be feeding on similar proportions of similar animal and vegetable food in partly different habitats on the mainland, and were feeding on radically different proportions of similar food in the same habitat on the islands

TABLE 4. Differences in diet and bill length between Tres Marias Island and mainland pairs of homogeneric species.

	PERCENTAGE OF BILL LENGTH[1] DIFFERENCES	*Diet*[2]		
		N[3]	ANIMAL	VEGETABLE
Mainland				
Myiarchus tyrannulus	38	19	89.5	10.5
Myiarchus tuberculifer		16	97	3
Island		15	83	17
Myiarchus tyrannulus	34	8	100	0
Myiarchus tuberculifer		32	88	12
Mainland		20	67.5	32.5
Vireo hypochryseus	20	20	80	20
Vireo flavoviridis		40	65.5	34.5
Island				
Vireo hypochryseus	21			
Vireo flavoviridis				

[1] Calculated from data of adult male specimens in Grant (1965b).

[2] % representation calculated from gizzard contents of specimens collected in the months March–July (Grant, 1964).

[3] Sample size.

TABLE 5. Bill and diet characteristics of *Platypsaris aglaiae* and *Piranga bidentata* from the Tres Marias Island and adjacent Mexican mainland.

	Platypsaris aglaiae			*Piranga bidentata*		
	BILL LENGTH[1]	BILL WIDTH[1]	DIET[2]	BILL LENGTH[1]	BILL WIDTH[1]	DIET[2]
Mainland	12.22±0.10	7.61±0.10	55A, 45V	13.23±0.10	7.82±0.05	76A, 24V
Islands	11.55±0.05	6.96±0.04	89A, 11V	13.59±0.06	8.50±0.06	30A, 70V
% difference	−5.5	−9.3	−34V	+2.7	+9.8	+46V

[1] Mean measurements of samples of adult male specimens, in millimeters, with one standard error; taken from Grant (1965b).

[2] A = animal matter, V = vegetable matter; % representation, calculated from gizzard contents of specimens collected in the months March–July (Grant, 1964)—*Platypsaris aglaiae*, 26 mainland and 23 islands specimens; *Piranga bidentata*, 25 mainland and 10 island specimens.

(Table 5). Correspondingly, their bill lengths have diverged; and changes in shape of bill have also accompanied the changes in diet. Two species of hummingbirds (Trochilidae), trophically similar to each other but quite different from all other species on the islands, have diverged to the point of nonoverlap of bill length (Fig. 2). On the mainland these are sympatric with at least three other species of hummingbirds, which have longer or shorter bills. In contrast to these and other examples of divergence on the Tres Marias Islands, and to yet others of no change, there are no obvious examples of convergence in bill length between ecologically similar heterogeneric species for which reliable data are available.

Even when homogeneric species on islands are not necessarily the most closely related forms to those in the same genus occurring on the mainland, such as occurs in archipelagos, bill length differences tend to be the same or greater among the island forms (Schoener, 1965). Similarly, on oceanic islands whose species cannot be compared with mainland relatives, homogeneric species tend to differ markedly in bill length when sympatric (Amadon, 1950; Lack, 1947).

Allopatric congeners, because of their spatial separation, need not take foods of different sizes or types, so it is to be expected that differences in bill length are not as great as those between sympatric congeners. Completely allopatric distributions of homogeneric species occur on large islands, and data for these are scarce. Benson (1960) lists several examples of endemic congeners on the Comoro Islands, which are segregated by habitat to various degrees, and gives bill length data for three pairs. These data indicate that there is an inverse relationship between degree of difference

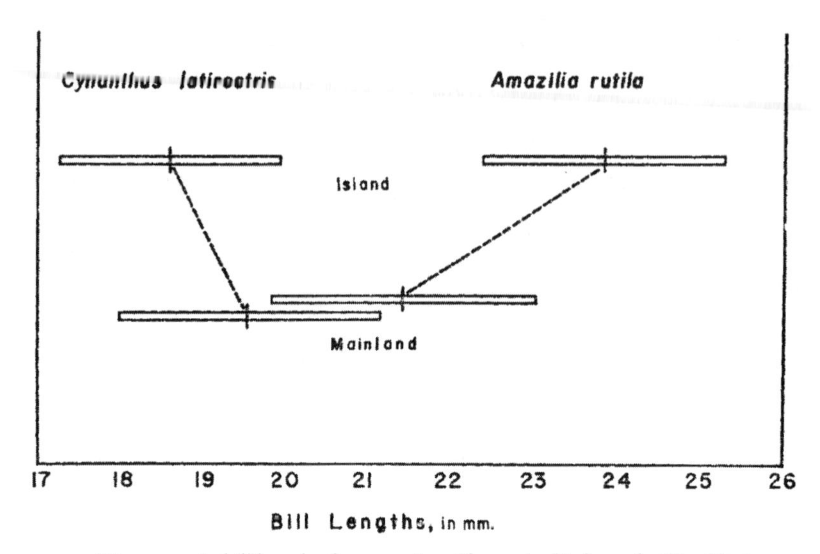

FIG. 2. Divergence in bill length of two species of hummingbirds on the Tres Marías Islands, Mexico.

Samples of adult male specimens only, with two standard deviations shown either side of the mean. Samples: *C. latirostris* 37 mainland and 20 island specimens, *A. rutila* 44 mainland and 27 island specimens.

TABLE 6. Habitat and bill length differences between pairs of homogeneric species of birds on the Comoro Islands.

Species	ISLAND	HABITAT	% DIFFERENCES IN BILL LENGTH
Zosterops senegalensis *Zosterops mouroniensis*	Moheli	Completely exclusive	8
Microscelis madagascariensis *Microscelis crassirostris*	Gran Comoro	Partly exclusive	20
Nectarinia notata *Nectarinia humbloti*	Gran Comoro	Completely inclusive	54

Data taken from Benson (1960).

in bill length and degree of difference in habitat occupancy (Table 6). How widespread this relationship is on islands is not known.

ECOLOGICAL INCOMPATIBILITY AND EXTINCTION

The small proportion of congeners on islands, and the large differences between those present, suggest that coexistence of similar forms is usually more difficult on an island than on the mainland, perhaps due to effects mediated by the restricted nature of the environment (Hutchinson, 1959). Furthermore, it suggests that the arrival on an island of a form similar to one already established will not lead to successful colonization because of the lack of sufficient "ecological space." Consider the arrival on an island of a species which differs relatively little from an established species. It may colonize successfully if there is opportunity for at least partial allopatry, as happens with the two *Hylocichla* (= *Catharus*) species (Dilger, 1956) referred to in Table 3, and with *Tyrannus melancholicus* and *Myiarchus tyrannulus* on each of the Tres Marias Islands (the only conspicuous instance of partial allopatry there: unpublished observations). It may also colonize successfully if the two species exploit the vegetation in the same habitat differently, as may happen with the two *Parus* species (see Dixon, 1961) referred to in table 3, and perhaps with the *Hylocichla* species as well (Dilger, *loco cit.*). Otherwise, in the absence of marked ecological differences, the newcomer will either replace the established species (cf., Lidicker and Anderson, 1962) or, as is more likely, it will fail to establish itself, particularly if its previous absence has allowed the established species to extend its range of activities into the zone occupied on the mainland by the invader (e.g., Lack and Southern, 1949; Svardson, 1949; Watson, 1964).

Evidence of the elimination of one of two species may be sought in the nature of the absence of congeners from islands, as follows. The degree of difference between sympatric congeners on an island may be taken as an approximate measure of the necessary amount of difference permitting coexistence in that environment. This may then be used to examine how many pairs of congeners, from those available on the mainland, and with one member only on the island are excluded from coexisting on the island. Although the absence of one member of a pair whose difference falls below this level may be due alternatively to a failure to reach the island, perhaps owing to poor inherent powers of dispersal, it seems unlikely that resident species of the same genus will differ much in dispersal abilities. Therefore absence is an indication of a failure to coexist with those present. The Tres Marias Islands are the only ones suitable for this analysis: they have probably been in existence as islands for a long time (Grant, 1965a), the opportunities for allopatry on any one island are almost nil, the numbers of congeners available in the adjacent mainland is known reasonably accurately, and measurements of bill length of all species are available. There are 11 species which occur with one or more congeners on the mainland but alone on the islands. The mainland species constitute 14 pairs of congeners. Both members of a pair are usually resident or, in a few cases, both are summer residents only. Of the 14 mainland pairs of homogeneric species (Table 7), 12 differ in bill length by less than the

TABLE 7. Percentage differences in bill length between homogeneric species of west Mexican birds. In each instance the first species only occurs on the Tres Marias Islands.

Species	% DIFFERENCES
Buteo jamaicensis and *Buteo nitidus*	18
Columbigallina passerina and *Columbigallina talpacoti*	4
Amazona ochrocephala and *Amazona finschi*	15
Amazona ochrocephala and *Amazona albifrons*	37
Amazilia rutila and *Amazilia beryllina*	17
Amazilia rutila and *Amazilia violiceps*	1
Trogon elegans and *Trogon citreolus*	6
Tyrannus melancholicus and *Tyrannus vociferans*	19
Tyrannus melancholicus and *Tyrannus crassirostris*	21
Myiarchus tuberculifer and *Myiarchus nuttingli*	2
Thryothorus felix and *Thryothorus sinaloa*	4
Turdus rufo-palliatus and *Turdus assimilis*	1
Vireo hypochryseus and *Vireo gilvus*	12
Icterus pustulatus and *Icterus wagleri*	15

Data of adult male specimens only, taken from Friedman et al. (1950), Ridgway (1902, 1904, 1907, 1911, 1916). The *Myiarachus* and *Vireo* species not present on the Tres Maria Islands are compared with the most similar congener. Only those genera with at least one species on the islands are included in this table.

21% which separates members of the most similar island pair of homogeneric species, the vireos (Table 4). It is inherently unlikely that two species in the same genus have reached the islands only twice (*Myiarchus* and *Vireo* species) and that, coincidentally, these happen to be two of the four pairs of most different homogeneric species. Therefore, the absence of at least some of the missing 12 species may be attributed to a failure to coexist upon arrival with those present.

The absence of one member of each of the two congeneric pairs which differ in bill length by more than 21% or more requires an explanation. In one instance this may be due to the presence of an ecologically similar species in a different genus. The absence of *Tyrannus crassirostris* from the islands may be due to the presence of the similar flycatcher *Myiarchus tyrannulus* (difference of only 7%; calculated from data in Ridgway, 1904), as well as to the presence of the congener *T. melancholicus* (difference of 21%). Similarly, the absence of the other *Tyrannus* species, *T. vociferans*, may be due as much to the presence of *Myiarchus tuberculifer* (difference of only 3%) as to the presence of *T. melancholicus* (difference of 19%); and the absence of the hummingbird species *Amazilia beryllina* may be due as much to the presence of *Cynanthus latirostris* (difference of 3%) as to the presence of its congener *Amazilia rutila* (difference of 17%). This explanation cannot be invoked, however, to account for the absence of *Amazona albifrons*, which is similar to no other species on the island except *A. ochrocephala*, and whose bill is as much as 37% shorter than that of its congener (Table 7). It is possible that larger differences are necessary between species of large birds (Schoener, 1965) whose populations are relatively small, such as these parrots, (Grant, 1966c) than between the species of smaller birds such as the flycatchers and vireos. This is likely to apply particularly to the sexually dimorphic hawks, including the *Buteo* species (Table 7). Even so, the 37% difference between the parrot species is considerable; possibly, therefore, *Amazona albifrons* has never reached the islands. Its absence is the only one for which an explanation in terms of ecological incompatibility is not readily forthcoming. In some instances where only one member of a genus is present on the islands and has undergone change in bill length, it is now more different from a second member of the genus than is its mainland relative (e.g., *Amazilia* spp.). Hence, opportunities for successful colonization by the second member would appear to be better now than at the time of initial colonization. However, if the theory accounting for the changes in bill lengths of island birds is correct, a larger range of food sizes is exploited by the island form as a result of adaptation; and this will still tend to preclude the establishment of a second species.

SUMMARY

The difficulties of coexistence on islands for ecologically similar species of birds are indicated by the paucity of homogeneric species, and by the tendency for their bill lengths, and hence their feeding characteristics, to diverge.

Members of the most similar pair of homogeneric species on the Tres Marias Islands differ in bill length by 21%. This is greater than the difference between members of 12 out of a total of 14 mainland pairs which are each represented by one species only on the islands. It is suggested that in some instances both members of the mainland homogeneric pairs have reached the islands, but that one member has been eliminated due to interactions arising from mutual ecological incompatibility. It is further suggested that this process has been important in the adjustment of number of species to island area.

ACKNOWLEDGMENTS

Parts of this article were read by G. E. Hutchinson and R. H. MacArthur, the whole was read by I. A. McLaren. I gratefully acknowledge their encouragement and criticism.

LITERATURE CITED

Aldrich, J. W., and D. C. Nutt. 1939. Birds of eastern Newfoundland. Sci. Publ. Cleveland Mus. Nat. Hist. 4:13–42.

Amadon, D. 1950. The Hawaiian Honeycreepers (Aves, Drepaniidae). Bull. Amer. Mus. Nat. Hist. 95:155–262.

———. 1953. Avian systematics and evolution in the Gulf of Guinea. Bull. Amer. Mus. Nat. Hist. 100:397–431.

A.O.U. Check-list (5th ed.) 1957. Lord Baltimore Press, Maryland.

Asahi, M., 1962. Ecology and social integration of domestic rabbits introduced into a small islet. J. Biol. Osaka City Univ. 13:119–160.

Benson, C. W. 1960. The birds of the Comoro Islands: Results of the British Ornithologists Union Centenary Expedition 1958. Ibis 103b: 5–106.

Burleigh, T. D., and H. S. Peters. 1948. Geographic variation in Newfoundland birds. Proc. Biol. Soc. Wash. 61:111–126.

Chitty, D. H. 1960. Population processes in the vole and their relevance to general theory. Can. J. Zool. 38: 99 113.

Christian, J. J., and D. E. Davis. 1964. Endocrines, behavior and population. Science 146:1550–1560.

Christian, J. J., V. Flyger, and D. E. Davis. 1960. Factors in the mass mortality of a herd of Sitka deer, *Cervus nippon.* Chesapeake Sci. 1:79–95.

Dilger, W. C. 1956. Adaptive modifications and ecological isolating mechanisms in the thrush genera *Catharus* and *Hylocichla.* Wilson Bull. 68:171–191.

Dixon, K. L. 1961. Habitat distribution and niche relationship in North American species of *Parus,* p. 179–216. In W. F. Blair (ed.), Vertebrate speciation, a University of Texas Symposium.

Elton, C. S. 1946. Competition and the structure of animal communities. J. Anim. Ecol. 15:54–68.

———. 1958. The ecology of invasions by animals and plants. Methuen, London.

Friedmann, H. 1950. The birds of North and Middle America. Bull. U. S. Nat. Mus. 50 (Pt. 11): 1–793.

Friedmann, H., L. Griscom, and R. T. Moore. 1950. Distributional checklist of the birds of Mexico. Pt .1. Pacific Coast Avifauna 29:1–202.

Grant, P. R. 1964. The significance of some insular characteristics in birds. PhD. thesis, University of British Columbia.

———. 1965a. A systematic study of the terrestrial birds of the Tres Marias Islands, Mexico. Postilla, Yale Peabody Mus. Nat. Hist., No. 90, p. 1–106.

———. 1965b. The adaptive significance of some size trends in island birds. Evolution 19:355–367.

———. 1966a. Further information on the relative length of the tarsus of land birds. Postilla, Yale Peabody Mus. Nat. Hist., No. 98, p.1–13.

———. 1966b. The taxonomic status of the Yellow-green Vireo, *Vireo flavoviridis* (Cassin), from the Tres Marias Islands, Mexico, with additional notes. Acta Zool. Hungarica 12: 77–82.

———. 1966c. The density of land birds on the Tres Marias Islands of Mexico, 1. Numbers and biomass. Can. J. Zool. 44: 391–400.

Grant, P. R., and M. T. Cowan. 1964. A review of the avifauna of the Tres Marias Islands, Nayarit, Mexico. Condor 66: 221–228.

Hairston, N. G. 1964. Studies on the organization of animal communities. J. Anim. Ecol. 33 (Suppl.): 227–239.
Hamilton, T. H., and N. Armstrong. 1965. On environmental determination of insular variation in bird species abundance in the Gulf of Guinea. Nature 207: 148–149.
Hollom, P. A. D. 1955. The popular handbook of British birds. Witherby, London.
Hutchinson, G. E. 1959. Homage to Santa Rosalia or Why are there so many kinds of animals? Amer. Natur. 93: 145–159.
Kear, J. 1962. Food selection in finches, with special reference to interspecific differences. Proc. Zool. Soc. London 138:163–204.
Klopfer, P. H., and R. H. MacArthur. 1961. On the causes of tropical species diversity; niche overlap. Amer. Natur. 95: 223–226.
Lack, D. 1944. Ecological aspects of species-formation in passerine birds. Ibis 86: 260–286.
———. 1947. Darwin's finches. University Press, Cambridge.
———. 1954. The natural regulation of animal numbers. Clarendon, Oxford.
———. 1965. Evolutionary ecology. J. Anim. Ecol. 34:223–231.
Lack, D., and H. N. Southern. 1949. Birds on Tenerife. Ibis 91:607–626.
Lidicker, W. Z., Jr., and P. K. Anderson. 1962. Colonization of an island by *Microtus californicus*, analyzed on the basis of runway transects. J. Anim. Ecol. 31:503–517.
MacArthur, R.H., and E. O. Wilson. 1963. An equilibrium theory of insular zoogeography. Evolution 17:373–387.
Mayr, E. 1931. Birds collected during the Whitney South Sea Expedition, XVI. Amer. Mus. Novit., No. 502, p. 1–21.
———. 1963. Animal species and evolution. Bellknap Press, Cambridge, Mass.
———. 1965. Avifauna: turnover on islands. Science 150:1587–1588.
Miller, A. H., H. Friedmann, L. Griscom, and R. T. Moore. 1957. Distributional checklist of the birds of Mexico. Pt. 2. Pacific Coast Avifauna 33:1–436.
Moreau, R. E. 1948. Ecological isolation in a rich tropical avifauna. J. Anim. Ecol. 17: 113–126.
Munro, J. A., and M. T. Cowan. 1947. A review of the bird fauna of British Columbia. Brit. Col. Prov. Mus., Dept. Education special publ. No.2, p. 1–285.
Paynter, R. A., Jr. 1955. The ornithogeography of the Yucatan peninsula. Bull. Yale Peabody Mus. Nat. Hist. 9:1–347.
———. 1964. Generic limits of *Zonotrichia*. Condor 66:277–281.
Peterson, R. T., G. Mountfort, and P. A. D. Hollom. 1954. A field guide to the birds of Britain and Europe. Collins, London.
Preston, F. W. 1962. The canonical distribution of commonness and rarity. Pts. 1 and 2. Ecology 43:185–215; 410–432.
Rand, A. L., and D. S. Rabor. 1960. Birds of the Philippine Islands; Siquijor, Mount Malindang, Bohol and Samar. Fieldiana, Zool. 35: 223–441.
Ridgway, R. 1902. The birds of North and Middle America. Bull. U. S. Nat. Mus. 50 (Pt. 2): 1–834.
———. 1904. The birds of North and Middle America. Bull. U. S. Nat. Mus. 50 (Pt. 3): 1–801.
———. 1907. The birds of North and Middle America. Bull. U. S. Nat. Mus. 50 (Pt. 4): 1–973.
———. 1911. The birds of North and Middle America. Bull. U. S. Nat. Mus. 50 (Pt. 5): 1–859.

———. 1914. The birds of North and Middle America. Bull. U. S. Nat. Mus. 50 (Pt. 6): 1–882.
———. 1916. The birds of North and Middle America. Bull. U. S. Nat. Mus. 50 (Pt. 7): 1–543.
Schoener, T. W. 1965. The evolution of bill size differences among sympatric congeneric species of birds. Evolution 19:189–213.
Svardson, G. 1949. Competition and habitat selection in birds. Oikos 1:156–174.
Watson, G. E. 1962. Three sibling species of *Alectoris* partridge. Ibis 104: 353–367.
———. 1964. Ecology and evolution of passerine birds on the islands of the Aegean Sea. PhD. thesis, Yale University.
Williams, C. B. 1953. The relative abundance of different species in a wild animal population. J. Anim. Ecol. 22:14–31.
———. 1964. Patterns in the balance of nature and related problems in quantitative ecology. Academic Press, New York.
Wilson, E. O. 1961. The nature of the taxon cycle in the Melanesian ant fauna. Amer. Natur. 95:169–193.

Grant, P. R. 1966. Ecological compatibility of bird species on islands. American Naturalist *100: 451–462. Reprinted with permission of University of Chicago Press.*

Bivalves: spatial and size-frequency distributions of two intertidal species (1968)

Jeremy B. C. Jackson

Abstract—*Individuals of* Mulinia lateralis *are randomly distributed over a homogeneous area (0.25 square meter) of a mudflat. Second-year individuals of* Gemma gemma *also are randomly distributed, but its total population is aggregated because of its ovoviviparous habit. As expected for two species having different life histories, their size-frequency distributions are very different, the indication being that the nature of a size-frequency curve may not be a reliable index of the degree of transport or integrity of a fossil assemblage.*

The horizontal spatial distributions of two species of intertidal bivalves, *Mulinia lateralis* Say, 1822, and *Gemma gemma* Totten, 1834, were studied in an inlet off Long Island Sound, west of Guilford, Connecticut; they were collected on a small mudflat at an arbitrarily determined spot 10 m shoreward of low water, an area of 0.25 m² being chosen to ensure homogeneity of environment throughout the sample. A metal grid with cells 5 by 5 by 3.5 cm deep was used to divide the sampling area; 100 such cells covered 0.25 m². The sample from each cell was washed through a 1-mm sieve, and the bivalves were counted, the lengths of their shells being measured in increments of 0.1 mm.

The frequency distributions of the individuals per cell were compared with the expected random frequency of the binomial distribution (1) by the chi-square goodness-of-fit test. Populations were taken as random when the observed distribution did not differ from the expected at the 95-percent level of confidence.

Fisher's variance-ratio statistic (2) was computed for each of the species distributions (live and dead), and the criterion of randomness was again set at the 95-percent level of confidence. This duplication of tests was necessary in order to avoid difficulties in interpolation of binomial values for "half individuals" in the dead-shell distributions.

The results of the analyses (Table 1) show that *M. lateralis* was always randomly distributed as was to be expected for a population in a homogeneous environment—one in which biological parameters are not important in determination of the distribution of individuals of the population. Thus each individual is independent of all others within the area observed, and the arrival of new individuals is independent of those either already present or entering the area at the same time; this condition is provided for *Mulinia* by its planktonic larval stage and consequent random settling within a limited area of uniform substrate. Moreover, *Mulinia* is somewhat unusual among the suspension-feeding bivalves, at least in the adult stage, in its apparent ability to withstand the high content of silt and clay in mudflat sediments; thus it can utilize the very high concentrations of organic matter characteristic of these waters, so that intraspecific competition for food does not occur.

The total *G. gemma* population was aggregated for $N = 100$ but randomly distributed when the cells were lumped for the binomial calculations ($N = 25$). Lumping produced subareas too large for detection of the true aggregated distribution, and therefore randomness was not rejected. Apparently the 5- by 5-cm cell gives an appropriate sample for study of these species. Randomness was rejected in both instances by the more sensitive variance-ratio test.

Aggregation was not expected, as the environment appeared to be homogeneous. However, examination of the distribution of *Gemma* in relation to its reproductive habits and to the size-frequency data explains the anomaly. This species is ovoviviparous, and gravid females commonly

TABLE 1. Variance, means, and spatial distributions for living and dead populations of *Mulinia lateralis* and *Gemma gemma*. Aggr., aggregation.

Year class	NOS. (N, n)	VARIANCE	MEAN	*Spatial distribution*	
				VARIANCE RATIO	BINOMIAL
		Live *Mulinia lateralis*			
	100, 27	0.26	0.27	Random	Random
	25, 27	0.66	1.08	Random	Random
		Dead *Mulinia lateralis*			
	100, 54.5	0.38	0.55	Random	
	25, 54.5	2.46	2.18	Random	
		Live *Gemma gemma*			
All	100, 575	11.83	5.75	Aggr.	Aggr.
All	25, 575	53.67	23.00	Aggr.	Random
1st	100, 434	7.72	4.34	Aggr.	Aggr.
2nd	100, 141	1.66	1.41	Random	Random
		Dead *Gemma gemma*			
All	100, 729.5	282.83	29.18	Aggr.	
All	25, 729.5	51.02	7.29	Aggr.	
1st	100, 465	19.52	4.65	Aggr.	
2nd	100, 264	7.70	2.64	Aggr.	

contain as many as 300 young (3); moreover it has only one reproductive season in northern waters, and the juveniles are released in June and early July (3). Two classes, 1st- and 2nd-year, are easily distinguishable (Fig. 1), and a length of 2.8 mm appears to be a reasonable separation point. Of the 575 live individuals collected, 434 (75 percent) were in the 1st-year class. The 2nd-year class were near the end of their lifespan (average age in November, almost 1.5 years).

The distribution of the 2nd-year class was random; that of the 1st-year class, aggregated. This fact can be explained by the comparatively recent release of the young bivalves by their mothers and by their failure to disperse (or to have been dispersed) since "spawning." Therefore clustering of 1st-year individuals, still close to their mothers, gives an overall aggregated distribution to the population that is greater than the degree of aggregation of the 1st year class alone.

It is not clear why the 2nd-year population should be randomly distributed, unless distribution results from lateral movements associated with burrowing; *Gemma* is a moderately efficient burrower and consequently can live just below the sediment surface in a tidal environment. Moreover, mortality should be spatially random for individuals of anyone age or

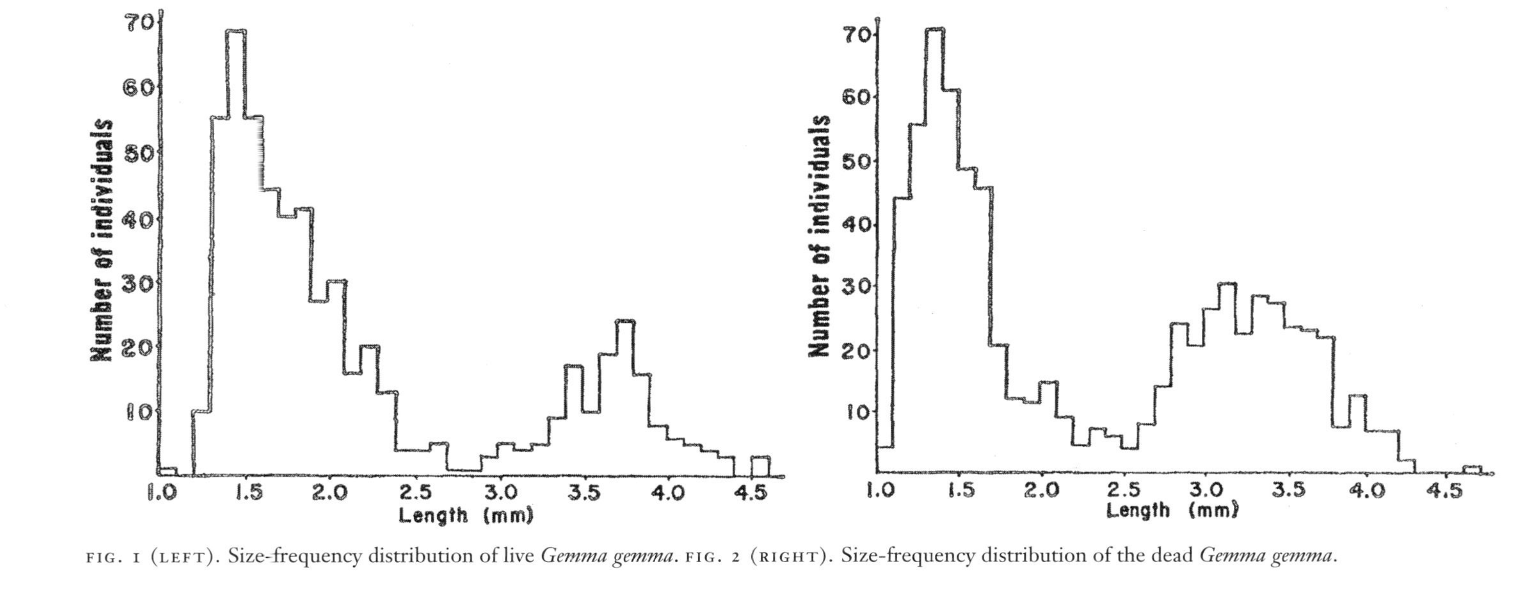

FIG. 1 (LEFT). Size-frequency distribution of live *Gemma gemma*. FIG. 2 (RIGHT). Size-frequency distribution of the dead *Gemma gemma*.

sex, and so should not alter the original aggregate distribution unless the population is reduced to very low densities.

Tidal currents and storms are undoubtedly effective in the transport and reworking of dead shells, as are shown by extensive accumulations of remains of both species on the shores of the inlet. However, the effects on each species are different, for the dead shells of *Mulinia* remain randomly distributed on the flat whereas those of *Gemma* become strongly aggregated relative to the living population. This difference may result from the difference in size and shape (and hence hydrodynamic properties) of the shells of the two species, or from the much greater degree of articulation of dead *Gemma* than of *Mulinia* (or from both factors). However, if tidal currents or storms cause the aggregation of dead *Gemma* shells, they cannot also cause the random distribution of the live 2nd-year population.

The size-frequency distribution of dead *Gemma* shells (Fig. 2) resembles that of live individuals at the time of sampling, but the modes of both classes are shifted slightly to the left. For both year classes, the peaks correspond to the period between early summer and early fall, which is their time of most-rapid growth and highest mortality (3). So, although the size-frequency distribution of the living population changes each year from bimodal to unimodal and back again, that of the dead-shell assemblage is probably constant because of the very different seasonal mortality rates. This fact may relate to greater vulnerability to environmental stress at times of more rapid growth or of reproduction (or of both), or to mass kills during periods of abnormally high temperatures. Evidently winter freezing is no major threat to this small species, probably because of burrowing during cold periods—often to depths of 7 to 8 cm within the sediment (4). The possible correlation of bimodal or polymodal size-frequency distributions with ovoviviparity merits further research; if this is a common phenomenon it could help considerably toward interpretation of the life histories of ancient species.

The population density of *Mulinia* was too low in the sampling area for significant analysis of size, but measurement of 1723 single valves from another part of the same mudflat yielded a strongly positively skewed curve, reflecting very high juvenile mortality for this species. This mortality probably results from poor adaptability by *Mulinia* to the very turbid waters and soft substrate of the mudflat environment before a comparatively large shell is formed (5). On the other hand, *Gemma* is well adapted to this environment from the beginning of its free-living existence, having an inhalent siphonal membrane (3).

It seems that size-frequency distributions of bivalves, even roughly similar forms from the same environment, may differ greatly because of simple and basic variations in their life histories. Therefore, contrary to certain conclusions (6), generalizations on the paleoecological significance of one sort of size-frequency distribution or another seem inappropriate without some idea of the life histories of both species involved. Consequently no one curve or set of curves serves as a reliable indicator of past transport or of the degree of compositional integrity in a fossil assemblage, or of both.

REFERENCES AND NOTES

1. M. A. Buzas, *Contrib. Cushman Found. Foraminiferal Res.* 19 (part 1), 1 (1968); Harvard University Computer Laboratory, *Tables of the Cumulative Binomial Probability Distribution* (Harvard Univ. Press, Cambridge, Mass., 1955).
2. R. A. Fisher, *Statistical Methods for Research Workers* (Oliver and Boyd, Edinburgh and London, 1925).
3. O. P. Sellmer, *Malacologia* 5, 137 (1967).
4. N. H. Bradley and P. Cooke, *U.S. Fish Wildlife Serv. Fishery Bull.* 137 (1959), p. 305.
5. S. J. Levinton, personal communication.
6. G. Y. Craig and G. Oertel, *Quart. J. Geol. Soc. London* 122 (487, part 3), 315 (1966).

7. I thank M. A. Buzas and J. Levinton for valuable suggestions and critical reading of the manuscript, and D. Meyer for help in sampling.

Jackson, J. B. C. 1968. "Bivalves: spatial and size-frequency distributions of two intertidal species." Science *161: 479–480. Reprinted with permission from AAAS.*

Intuition

Shahid Naeem

The charge before the natural scientist is immense. How does one make sense of a three-and-a-half-billion-year-old, trillion-ton network of possibly one hundred million interacting plant, animal, and microbial species, each consisting of hundreds to quadrillions of individuals, all intimately linked in a complex system that is shaped by cosmological, physical, and chemical processes that span scales from the nanosecond to billions of years and from angstroms to thousands of kms? The answer is to select a scale of investigation, immerse oneself as deeply as one can into nature's inevitable abyss of details, observe keenly, grasp a generalization, and surface before drowning in the details.

Once one emerges with a robust generalization or, more simply, a broadly applicable idea about nature that is wholly coherent with the facts surrounding it, its fate in the scientific community depends entirely on timing. If one emerges with a generalization that is consistent with current paradigms, then it will be absorbed by the scientific community. If one emerges during the heyday of a paradigm shift and one's discovered generalization catalyzes the synthesis that breaks the deadlock between thesis and antithesis, it will be hailed by the scientific community as a significant achievement. If, however, one emerges with a generalization ahead of its time, in the absence of any intellectual framework to give it context, then it will suffer one of three fates. It might be ignored because its significance simply cannot be grasped by anyone (e.g., Gregor Mendel and genetics). It might be dismissed because it appears to have no rational explanation (e.g., Alfred Wegener and continental drift). Or it might be understandable, but enigmatic—no one can make head or tail of it (e.g., Barbara McClintock and jumping genes). These are not mutually exclusive fates. An idea ahead of its time may in fact suffer all three fates (e.g., Stanley B. Prusiner and the prion—its significance was unclear; though theoretically possible, it flew in the face of the Central Dogma; and it was just too strange).

This section explores a set of papers by a small number of scientists whose discoveries were ahead of their time. There are many we might have included, such as the aforementioned Mendel, Wegener, McClintock, and Prusiner, but we wanted to draw attention to some lesser-known discoveries relevant to ecology, evolution, and the environment. Though seemingly very different from one another, each has in common with the others an intense immersion into detail followed by the discovery of a powerful generalization that was ahead of its time. We have drawn these examples to cover a broad range of scales, from the microscopic to the biosphere, from the short term to paleoecological, itself suggesting that there is no corner of research that does not have the capacity to radically inform and shape our thinking.

We begin with Frederick II (1194–1250), king of Sicily, Cyprus, and Jerusalem, Holy Roman Emperor, polyglot, patron of the arts, mathematics, and science, a participant in the Sixth Crusade, and key figure in bringing the advances of Arabian science to the Western world; not your typical natural philosopher. Of all his accomplishments, he may never have imagined he might be featured as a giant in the natural sciences, but his treatise on falconry represented a radical departure from Aristotelian thinking (Wood and Fyfe 1943). In a way, his ideas were 400 years ahead of its time if we consider the seventeenth and eighteenth centuries the key period in which Aristotelian thinking gave way to the Scientific Revolution.

Key to the Aristotelian approach to studying nature was seeking logical generalizations consistent with other constructs of nature, and avoiding natural variation in form and habit that obfuscated those underlying constructs. Rather than plumbing the richness of nature to uncover generalities as we do today, one sought generalizations by stripping away detail. Frederick II found this unsatisfactory, for he knew that falconry required understanding the importance of differences in the ecology of raptors and their prey. Zoological illustrations of the period were typically general, providing little detail for species identification (Schramm 2001). In contrast, Frederick II's illustrated volume on falconry presented birds in sufficient detail to be identified to species.

Frederick began with the observation that "birds, like all other animals, may be divided into various species in accordance with their sexual relations, their parturition, their methods of procuring food, their variations in residence at different seasons of the year . . . and their change in diet." Thus, although Frederick II retains the Aristotelian grouping of waterfowl, land, and neutral birds, he explicitly rejects other constructs and notes that closer inspection identifies groupings within groups. This predates Jan Baptist van Helmont's (1577–1644) demonstration that air was composed of subgroups of different gasses, then a radical departure from dominant thinking. Frederick II, for example, rejected the Aristotelian generalizations that birds that fly well walk poorly, or that swimming and flying abilities are inversely related. He realized that useful generalizations about birds derive from understanding what are essentially, in modern parlance, their niches. One can sense what side he would come down on in the current discussions of neutral versus niche theory (e.g., Bell 2000, Hubbell 2001, Nee and Stone 2003), no doubt finding patterns in distribution and abundance explained by small numbers of parameters uninformative about falconry. It is interesting to note that Frederick II was not motivated by rewards for science, then nonexistent and probably not terribly important to a king. He was motivated by his love for falconry, exemplified by his sarcastic reply to the Mongols' request that he submit and become their vassal, that if he did, he could perhaps be their falconer (Schramm 2001).

As discussed earlier, by the 1600s, the Aristotelian influence over Western thinking was waning and the Scientific Revolution was in full swing. The Dutch tradesman, Antony van Leeuwenhoek (1632–1723), had no particular title or any character even remotely in common with Frederick II except for a shared passion for detail, though for van Leeuwenhoek it was microscopy. In the paper we selected (from van Leeuwenhoek 1800), van Leeuwenhoek explores the biology of spiders despite their being commonly held to be lowly creatures. The more he explored, the more he discovered, the deeper he dove, his instrumentation being the only limitation he faced. In this paper, for example, note how in his determination to get a better look at web manufacturing, he succeeds in pinning a spider on its back to get a better look and discovers that the filaments are actually made up of multiple strands. This revelation only whetted his appetite for more information: "Not content with these observations, I sat about devising means of keeping the threads separate, as they issue from the Spider's body, so that I might be able to give some representation of the inconceivable fineness . . . But yet, this fineness cannot by any efforts of pen or pencil, be adequately described." His desire to go deeper and deeper into the detail is plain, and one can only imagine his frustration with the low resolving power of his rudimentary microscope. His microscopes, as advanced as any of the time, were extraordinarily difficult to use, with very low depths of resolution and great sensitivity to vibration, and inducing extraordinary eyestrain.

Van Leeuwenhoek emerged with the generalization that the slim fibers of a spider's web gains its strength by being made up of smaller ones twisted together—now a common finding for biomaterials. Perhaps more important are his back-of-the-envelope calculations for considering the scale of a juvenile spider's web filament in comparison to a human hair. This approach is quintessential van Leeuwenhoek, who devised numerous ways for quantifying what he observed; he is credited with playing an important role in the growth of animal demography (Egerton 1968). While he finds the marvelous detail of spiders a testament to the wisdom of the Almighty Creator, one can sense that he would still rather have the necessary instrumentation and quantitative observations to confirm it.

Like van Leeuwenhoek, Jan Swammerdam (1637–1680) was one of an extraordinary number of scientists in Holland who contributed to the Scientific Revolution. Many of the burning questions of the seventeenth century concerned vital processes such as growth, reproduction, and physiology, especially as they related to human biology. The anatomical amphitheaters played important roles in the development of this science, serving as much as cultural centers as they did laboratories (Rupp 1990). Swammerdam was one of many associated with the anatomical theaters of Holland. He was a gifted anatomist, especially skilled at using the injection of hot colored waxes to highlight vessels in organs, and he made exquisitely detailed drawings that made him a valued collaborator in human anatomical research, though insect biology was his true passion. Swammerdam contributed significantly to advances in understanding the generation of life, still thought by many to take place spontaneously (Cobb 2006). While van Leeuwenhoek and Swammerdam are both examples of early microscopists who examined insects in detail, and may have made each other's acquaintance when they were both in Amsterdam (Parker 1933), they would later take different sides on the theory of human reproductive biology. Swammerdam worked with Johannes van Horne (1621–1670), and through a set of meticulous dissections and drawings they concluded that eggs in the follicles of women's ovaries migrated by peristalsis to the womb, where they developed into babies. Thus, humans, and mammals in general, had eggs just like fish, amphibians, lizards, and birds. The male's contribution, semen, was what deposited the ethereal soul in the egg. Van Leeuwenhoek, however, discovered sperm in semen and so much were sperm like the animalcules he observed elsewhere that he proposed that sperm made their way to the ovaries where they fed and grew to look like eggs, but eventually made their way to the womb were they developed into babies (Cobb 2006).

The paper selected here (from Swammerdam 1758), however, is representative of Swammerdam's greater passion: detailed observations of invertebrates, insects in particular. This is where his contributions are best known and, one might suggest, more enduring, perhaps because his ideas here never strayed too far from his observations. Combining microscopy with dissection, then a novel approach to exploring nature, coupled with superb illustrations, he refuted the idea that the different stages of holometabolous insects were different eggs. He demonstrated that the parts of adult insects could clearly be seen in the penultimate nymphal stage and in the pupa; thus the pupa was not a second egg (Erezyilmaz 2006). In the study reproduced here, Swammerdam's passion for detail at the population or hive level is as extraordinary as his passion for detail when examining single specimens. Most evident is how his understanding of the basic life cycle, egg to larva to nymph ("worms") to adult, provided him with keen insights into the nature of the beehive. His accurate elucidation of the social structure of the hive and its demography are testament to the power of careful and quantitative observation and the importance of understanding life history patterns when studying insects.

A great passion for a single group of organisms that invariably narrows one's view of nature can actually help reveal broad generalizations. Edouard Fischer-Piette (1899–1988), like Swammerdam, was intensely focused on a single group of organisms, in this case the mollusks. Fischer-Piette's life must have been steeped in malacology, for his brother, mother, father, and both his grandfathers were prominent European malacologists and the *Journal de Conchyliologie* was managed by his family between 1856 and 1979 (Marmont and Ponder 2004). Fischer-Piette's specific contributions to malacology were far ranging, including treatments of the systematics and natural history of mollusks from the Indian Ocean, India, Madagascar, Somalia, Ethiopia, Mozambique, Arabia, Morocco, Brazil, Ecuador, and the Galapagos. His most remarkable study, however, and the one considered here (Fischer-Piette 1935), concerned the invertebrate communities of Décollé Point, Saint-Lunaire in his home country of France.

Focusing on the mussel *Mytilus edulis*, he was intrigued by its densities cycling between near mono-dominance to nearly complete disappearance in bands along the entire intertidal zone, only to return again to mono-dominance. From his detailed observations, Fischer-Piette carefully constructed a mechanistic hypothesis: the mussels decline due to predation by a snail (*Nucella lapillus*); barnacles (*Semibalanus balanoides* and *Chthamalus stellatus*) become abundant in the absence of interference competition from the mussels; but then when the snail switches to feeding on the barnacles, mussels can return. One cannot help but be amazed at Fischer-Piette's seven-step hypothesis that includes ideas now familiar to ecologists as equilibrium communities, alternative stable states, succession, climax communities, the interplay between competition and predation in governing distribution and abundance, predator-prey cycles, and predator functional responses. This is an astonishing panoply of general ecological constructs that spans over eight decades of ecological thinking from the 1920s to today. Many of these ideas came into heightened scrutiny in the heyday of experimental ecology in which marine ecology would play a dominant role (see Sarah Baker's paper, below). Quite possibly, Harry Hatton, the author of a once obscure paper (Hatton 1938) that strongly influenced Joseph Connell's (1961) transformatory study, was inspired by Fischer-Piette, with whom he coauthored a paper in 1932 (Hatton and Fischer-Piette 1932).

Immersion in detail, as we see in Frederick II, van Leeuwenhoek, and Swammerdam, is not a guarantee of robust generalization. The egg and sperm theories that van Leeuwenhoek and Swammerdam each contributed to clearly indicate that if one strays from what one can see (they both had incomplete information about human reproduction, but ventured to fill in the blanks), then one's generalizations may not stand the test of time. There is also much in the natural world that cannot be seen, making immersion in detail necessarily incomplete. Evolutionary processes, for example, require time travel to perfectly document past events. Many ideas ahead of their time occur only because some detail pointed the researcher in the right direction, more or less, and they filled in the blanks in the absence of known mechanism. Darwin never observed natural selection (few have, except maybe Rosemary and Peter Grant and their team of researchers who have watched the Galapagos finches in enormous detail for decades [Grant and Grant 2007]) and his subscription to the theory of particulate inheritance may have prevented him from seeing the genetic mechanism. Wegener got continental drift right, even though the mechanism he proposed, *Polflucht* or pole-fleeing, based on centrifugal force due to the Earth's spin, was incorrect (the actual mechanism of seafloor spreading would be developed only in the 1950s and '60s). Where would we be without venturing a hypothesis even in the absence of well-documented mechanism? Another example is that of William Healey Dall (1845–1927), a superb malacologist and one whose detailed studies led him to recognize the importance of rapid change in phyla via heterochrony (changes in the timing of developmental events), the validity of stasis in evolution (e.g., lack of evolutionary change over long periods was real, not flaws in the fossil record), the concept of a species as an equilibrium state in which the majority of traits resisted change, and the importance of biogeography in interpreting evolution.

Rather than a paper rich in detail, we provide here his idea paper (Dall 1877) based on his detailed studies (he published over 1600 papers in his lifetime). The logic is impeccable and it concludes, "It seems as if the preceding reasoning might serve as a key to many puzzling facts in nature, and perhaps deprive catastrophists of their most serviceable weapon." Catastrophists were those who evoked environmental catastrophes when seeking to explain gaps in the fossil or evolutionary record. Dall predates by a little less than a century the punctuated equilibrium debates that dominated evolutionary theory in the 1970s (Lindberg 1998). As Lindberg (1998) notes, however, Dall's contribution to the American neo-Lamarckian movement—Lamarckians believed in evolution by the inheritance of acquired traits—may be the reason Dall's contributions to evolutionary theory did not attract the attention they deserved (Lindberg 1998)—a sort of throwing out of the baby with perceived tainted bathwater. Was Dall ahead of his time, or was his idea on saltatory evolution overshadowed by his subscription to an unrelated, unpopular, and unsupported idea?

Like Dall, Alfred Russel Wallace (1823–1913) also tackled the catastrophism and patterns in distribution and abundance, but from a geographic perspective. Invoking catastrophes to explain gaps in the variety of forms and functions observed in flora and fauna was used much as it was to explain gaps in the fossil record. Gaps in variety (e.g., no intermediates between feathered and unfeathered creatures) were contrary to the theory of the Great Chain of Being. This theory, as described by Arthur Oncken Lovejoy (1873–1962), consisted of three parts (Stevens 1994). It held that species of every possible form exist or existed on Earth (plentitude), that intermediates exist between all seemingly discontinuous forms (continuity), and that species reflect a natural order described crudely as a continuous chain from simple form and function to complex (the *scala natura*). Gaps were also contrary to the idea that the Almighty could create a species that would go extinct or the idea that species were put on Earth at different times. Stevens (1994) suggests that the *scala natura* may have been largely abandoned by the mid-nineteenth century, but the ideas of continuity and plentitude still prevailed. The religious perspectives that species were put on Earth all at one time with designs that would surely not allow them to go extinct, however, may have made geography an uninteresting dimension to the study of biological diversity, even though biogeography was first developed in the seventeenth century by Bernhardus Varenius (1622–1650) and was represented by the works of Alexander von Humboldt (1769–1859) and Phillip Lutley Sclater (1829–1913) (Rotramel 1973). These early biogeographers, however, mostly described where species were found, sometimes exemplifying the glory of God's work, without any particular theorizing as to why species were where they were.

Wallace traveled the world in search of specimens, the sale of which supported him and his family, as well as fed his passion for natural history (Camerini 1996). In this paper (Wallace 1857), we can see how his acute observations of patterns in the distribution and abundance of species throughout his travels, perhaps sharpened by both his financial and intellectual interests in specimens (not unlike Frederick II's acute observations of birds motivated by his passion for falconry), led him to recognize the preeminence of evolution as a predictor of the geography of species, predating the emergence of modern biogeography by nearly a century if we consider the work of Darlington in 1957 to be the turning point (Briggs 1966). The observation that two habitats in proximity to one another and similar in all habitat features would nevertheless have different flora and fauna if they were geographically isolated, defined a nonrandom pattern (i.e., Wallace's line) and was used to support hypothetical biogeographic processes that logically explained them, in this case constrained patterns of emigration and colonization. One can see the debates on vicariance versus centers of origin that would later rage in evolutionary biogeography (e.g., McCoy and Heck 1983, Wiley 1988) foreshadowed in Wallace's generalization that "it is quite unnecessary to suppose that new species have ever been created 'perfectly dissimilar in forms, habits, and organization' from those which have preceded them; nether do 'centers of creation,' which have been advocated by some, appear either necessary or accordant with facts."

One might expect that we would place Lynn Margulis (1938–) after the microscopist van Leeuwenhoek, given the microbial nature of her work, or perhaps closer to Vernadsky, given her current passion for the Gaia hypothesis as it relates to the biosphere, but her radical idea (Margulis 1968) led to a transformation of our perception of the evolution of life on Earth, so her work has more in common with Dall and saltatory evolution and Wallace and evolutionary biogeography. In contrast to some of our previous authors, Margulis immersed herself in biological detail that was impressive not so much for its depth as for its breadth. Combining atmospheric chemistry, metabolism, cell biology, and a knowledge of all the major forms of life on Earth, she comes to the inescapable conclusion that eukaryotes are not advanced prokaryotes, but symbiotic organisms hosting their ancestral prokaryotic mutualist partners in their cytoplasm. This idea shifted the major divisions of life from plant and animal to prokaryote and eukaryote. In collaboration with the great ecologist Robert H. Whittaker (1920–1980), Margulis helped devise the five kingdom scheme (Whittaker and Margulis 1978) that structured biological thought for decades. Not until the hydrogen hypothesis (Martin and Müller 1998) would the central idea of endosymbiosis be reconfigured, following the three domain model whose basic idea originated in the studies of Carl Woese and George Fox, where the distinct prokaryotic divisions of eubacteria and archaebacteria were identified (Woese and Fox 1977). Currently, the plot of life's origins has become much, much thicker with at least four possible hypothetical evolutionary routes to amitochondriate eukaryotes crossed by three possible hypothetical routes from amitochondriate to mitochondriate eukaryotes, all of which preceded the Cambrian explosion, in which the vast diversity of eukaryotic life evolved. Margulis's insights are squarely at the root of these radical reorganizations in the way we think about life on Earth.

We turn now to the experimentalists. In contrast to the observational methods that typify the research above, Abraham Trembley (1710–1784) conducted a series of experimental studies that were ahead of their time *because* they were experimental. Largely forgotten in today's introductory textbooks of biology, he is nevertheless widely recognized as the "father" of experimental biology (Schiller 1974). The paper here is a letter (Trembley 1742), translated from the French, describing Trembley's studies on *Hydra*, a freshwater cnidarian. He did many experiments, but the most intriguing is in which he reached into the mouth of the nearly microscopic *Hydra* and inverted it, bringing its gastrodermis, essentially its stomach, to the outside and moving its ectodermis, or its skin, to the inside. Because he studied the ability of the *Hydra* to regenerate severed parts and the ability for parts of different individuals to fuse, his work is often cited as predating studies in embryology, regeneration, and tissue grafting (Alejandro 2000), while his polyp-inversion experiment is more of a curiosity, though even this odd experiment would later be useful in understanding the fundamental differences in biological organization between diploblastic and triploblastic animals. His first experiment, however, was the polyp-inversion experiment and the motivation for it is fascinating. Borrowing a quote from his paper as provided by Lenhoff and Lenhoff (1989, p. 4) who have done much to revitalize interest in Trembley (Lenhoff and Lenhoff 1984, 1986, 1989),

> Since I was unsuccessful in finding a substance in which I could place the Polyps and which would serve to nourish them, I thought of inverting them so that the external surface of their skins would form the walls of

their stomachs. I had very little confidence that I would see this experiment succeed, but I did not think it proper not to try it.

What is key here is the double negative, "I did not think it proper not to try it."

Lenhoff and Lenhoff (1989) describe Trembley as an organismic biologist, one who studies an organism to learn more about it as opposed to a "problem-oriented" biologist who uses an organism to answer a question. But Trembley did more than study *Hydra* for *Hydra*'s sake. Like Frederick II and van Leeuwenhoek, Trembley immersed himself in the details of the organisms in order to uncover generalizations that tell us something fundamental about nature.

While experimentation became the sin qua non of most fields of science, ecology remained stubbornly descriptive. Among the earliest of experiments was that of Sarah M. Baker (1887–1917), whose 1909–1910 papers in the *New Phytologist*, a journal started by Alfred Tansley in 1902, explored in great detail whether differential tolerance to dessication stress was what determined zonation in marine algae. Her equipment was no more than jars within which she placed algae and varied their exposure to drying. Baker passed away not long after her 1909–1910 papers, but a paper by her published just before her death (Baker 1913) showed how much more sophisticated she would get in experimentation, building elaborate apparati to explore the effects of formaldehyde vapors on seedlings. One is reminded of Frederick II's recognition that multiple factors determine the biology of organisms, though his work was observational. A clear quest to disentangle the biotic (competitive tradeoffs) from the abiotic (in this case, desiccation) mechanisms that determine patterns in distribution foreshadowed ecology's near obsession with these issues during 1970s and '80s; specifically, she clearly identified stress tolerance as the primary regulatory factor in the high intertidal, and competition in the low intertidal, fifty years before such ideas would become popular (Robles and Desharnais 2002). These little-known studies by Baker predate the experimental studies by the likes of Joseph Connell, Jane Lubchenco, and Bruce Menge that would be hugely influential in community ecology (Robles and Desharnais 2002).

Our final group considers larger-scale issues, where the synthesizing of biotic and abiotic process in our ecological and evolutionary thinking is critical. Two concepts, the ecosystem and the biosphere, are important at larger scales and we provide examples of work by the founding figures working at these scales.

Except for a brief sojourn into psychology, Alfred G. Tansley's (1871–1955) career was one of tireless devotion to the promotion of plant ecology as a discipline, including the establishment of the journal *New Phytologist* in 1902, serving as its editor for thirty years, and serving as the British Ecological Society's first president in 1913 (Golley 1993). The eight-part study of the chalk grasslands represents the typical exhaustive surveys of vegetation of its time, the first four parts having been authored by Tansley and R. S. Adamson, with parts five through eight authored by others. In part 2, the importance of biotic (e.g., rabbit grazing) and edaphic factors has become apparent and foreshadows his famous conceptualization of the ecosystem (Tansley 1935). Much of Tansley's work was transformational and his contributions continue to impact the fields of botany and ecology. Here we have excerpted a section of his work where he departs from the often dry, yet useful, description of vegetation zones to outline the effects of rabbit grazing on chalk grassland plant assemblages. His precocious account of the nonlinearity of rabbit grazing effects predated the formation of the intermediate disturbance hypothesis by fifty years (Jackson 1981).

We close with what some might find a curious selection, a paper by Vladimir I. Vernadsky (1863–1945). This paper (Vernadsky 1945) has in common with the others that it is ahead of its time, though the jury is still out on whether it is, in fact, just another idea that should find its place in the gigantic wasteland of ecological ideas that never found support, were flat out wrong, or just utter rubbish. Vernadsky published extensively on biogeochemistry, mostly in Russian, with few of his works translated into English. He was a brilliant biogeochemist whose contributions to this field have earned him much deserved recognition. Though brought to the attention of Anglophone science by G. Evelyn Hutchinson in his review of elemental biogeochemistry (1943), it was not until 1945, the year of Vernadsky's death, that "Some Words about the Noösphere," translated by his son, was published in the *American Scientist.* Thus most English-speaking natural scientists know about his more radical ideas than his specific biogeochemical research.

Like all the authors in this section, Vernadsky had immersed himself in the details of his subject, as described in the passage, "During all those years, wherever I resided, my thoughts were directed toward the geochemical and biogeochemical manifestations in the circumambient nature, the biosphere." He claims, "I left aside as much as I could all philosophical aspirations and tried to rest only on firmly established scientific and empiric facts and generalizations." The question is whether Vernadsky stayed the course, or wandered into untestable terrain. When positing something untestable, like the homeostasis of the biosphere that is critical to the strong Gaia hypothesis (Barlow and Volk 1992, Huggett 1999, Wilkinson 1999), one's generalization gets relegated to things cosmic or things that are articles of faith, belief, or opinion. There is a common tendency to cite Vernadsky in support of Gaia, which conflates the Noösphere with Gaia, but the Noösphere may not be so cosmic and unpalatable an idea in light of the changes occurring in our modern world.

Vernadsky is most known for his promotion of the concept of the biosphere (Smil 2002), an idea he credits to Jean-Baptiste Lamarck (1744–1829), who, ironically, like Vernadsky, is often dismissed, in spite of numerous robust and influential ideas, because of his idea that the inheritance of acquired traits was the mechanism of origination. This is reminiscent of the price William Dall paid for subscribing to American neo-Lamarckianism; this position cost him some recognition as one of the best malacologists and early contributors to saltatory evolutionary theory. Is it possible that Vernadsky's remarkable career of rigorous work on biogeochemistry will be forgotten because of his less palatable theory of the noösphere?

Noösphere is the idea that the biosphere is governed in part by the collective mental activities of humans ("noösphere" comes from the Greek for "mind sphere"). Life on Earth is as much a matter of physical and biological processes as it is the mental processes of humans. He sees the noösphere, unfortunately, as a logical extension of the zoological concept of cephalization, that platyhelminth innovation of centralizing sensory organs in the front of one's body, which utterly transformed the shape of the biota, leaving behind radial symmetry and bringing into the world a near infinite menagerie of creatures that can seek out resources or avoid being sought out as a resource. This anterior concentration of sensory organs forming the head was the beginning of the formation of a brain—hence Vernadsky's peculiar linking of cephalization to noösphere. But if we can forgive van Leeuwenhoek and Swammerdam their ideas about human reproduction, Dall his neo-Lamarckianism, Margulis her Gaia, and Wegener his pole-fleeing land masses, we might similarly look past Vernadsky and cephalization.

Mikhail Gorbachev (1990), in his address at a Forum on Environment and Development held in Moscow, 15–19 January 1990, said of Vernadsky, "He created a theory of the biosphere and raised the problem of turning the entire medium mankind lives in into a sphere ruled by reason." The concept of the noösphere becomes cosmic if one thinks of it as a sort of giant organismic brain. That would be uninteresting because, like strong Gaia, it would lack any mechanism for its evolution, let alone rigorous testing. If, however, we consider activities such as the United Nations Kyoto Protocol, which attempted to regulate atmospheric carbon by regulating human activity, or the Montreal Protocol for regulating chlorofluorocarbons to protect the stratospheric ozone layer by regulating human activity—are these not reasoned attempts to regulate geochemical processes at the level of the biosphere? Consider further the call by Lubchenco et al. (1991) for ecology to reorganize around a framework dubbed the Sustainable Biosphere Initiative. Let us also consider that Vitousek et al. (1997) would declare that we must reorient our science around the idea that we are living in a human-dominated planet. Consider the suggestion of Kareiva et al. (2007) that there is essentially no significant portion of the biosphere that is free of human influence, so scientists should turn their attention to helping humankind wisely domesticate nature. Finally, consider that the *Proceedings of the National Academy of Sciences* has added a section devoted to sustainability science (Clark 2007), a science that examines human development in relation to sustaining Earth's "life support system." Are these and other activities to prevent collapsing fisheries and emerging diseases, control invasive species, or reduce the loss of biodiversity reasoned or scientifically informed activities designed to influence ecosystem and biospheric processes? Are these all telltale signs of a Vernadskyian call to rule the biosphere by reason? In which case, the admittedly somewhat unpalatable construct of the noösphere becomes an idea half a century ahead of its time.

CONCLUDING THOUGHTS

After considering this set of papers, one will be left with a few thoughts about the science of those who were ahead of their time. First, do robust generalizations in ecology, evolution, and environmental biology require a bout of intellectually unfettered immersion into the details of nature (is this what graduate school is about)? Whether a king's obsession with falconry, a tradesman's excessive preoccupation with microscopy, a tutor's fascination with *Hydra*, or a collector in quest of biotic treasures, a near blind passion for one's subject may be the first step to uncovering robust theory in Nature's endless detail.

Second, if immersion in detail is important, then it is interesting to ponder the consequences of our increasing tendency to make papers shorter and to relegate the details that once densely populated the methods, results, and discussion sections to electronic archives read by few. By today's standards, most of these papers would not be acceptable without reducing their text by two-thirds or more—stripping away the detail that actually makes them interesting.

Third, if an idea is ahead of its time, meaning that it emerged in advance of the intellectual framework that could provide it necessary context, is there any mechanism to ensure against premature extinction of the idea? Like species, ideas evolve—they have ancestry and pedagogy, but also aberrant properties reminiscent of mutations, and their fitness is governed by the scientific community, which acts like an environmental filter selecting those ideas that will be assimilated, prosper, or be driven extinct. Perhaps this is where the detrimental nature of intellectual chauvinism in scientific communities has its worst effects.

This is not to say that founders of ideas that were ahead of their time go unrecognized. Van Leeuwenhoek was made a fellow of the Royal Society in 1680. Trembley was elected to the Royal Society in 1743 and was recipient of the Copley Medal, then one of the highest honors for a scientist. Likewise, Wallace was elected to the Royal Society and received the same medal. Tansley too was made a member of the Royal Society and was knighted in 1950 shortly before he passed away, and Dall was elected to the National Academy of Sciences in 1897. Margulis was awarded the National Medal of Science in 1999. Vernadsky has obtained numerous awards in his home country and elsewhere in Europe and, while he was slow to be recognized in the United States, *Time* magazine's (29 January 1945) obituary read, "One of the world's great scientists died in Moscow last fortnight. He was almost unknown in the U.S. Vladimir Ivanovich Vernadsky was a cosmic thinker who founded a cosmic science which he called 'biogeochemistry'—the study of life against the vast background of geological time." In fact, one may have to become well established by a body of work before for one's more exotic ideas can survive to become reclassified by history as ideas ahead of their time. Sarah Baker and Edouard Fischer-Piette have not received wide recognition for their contributions, nor even become well known in marine ecology.

Finally, is there anything that distinguishes generalizati[illegible]ns or ideas ahead of their time from a virtually endless set of other i[illegible] may contribute little, are outright wrong, or just plain wack[illegible]
quinarianism, particulate inheritance, the superorganis[illegible]
ecosystem, the inheritance of acquired traits, phlogist[illegible]
and the many recurring forms of eugenics and gene[illegible]
all born of the same processes. Are there telltale si[illegible]
a better job of recognizing ideas ahead of their ti[illegible]

We are forgiving of the ancients like van Lee[illegible]

dam for their ideas on reproduction, and concentrate on the discoveries they made that remained coherent with accumulating biological knowledge. To the American neo-Lamarckians, the inheritance of acquired traits was an idea ahead of its time, but it fell once again into disfavor and perhaps cost Dall some additional time before his other evolutionary ideas took hold. Conversely, when it comes to Gaia, because Margulis is one of its champions and is a living testament to science's inability to recognize an idea that is ahead of its time, one worries that dismissing Gaia could prove to be a second embarrassment of the scientific community. Where will Vernadsky's concept of the noösphere be placed in the spectrum of ideas—an idea ahead of its time, or at odds with what we know to be true, or just the fanciful meanderings of an overactive mind?

Perhaps every idea, good or bad, has been foreseen by past scientists who reported without prejudice what the weight of the evidence told them, but sometimes found no home for their ideas in the science of their times. Whether one's idea is absorbed, ignored, or forgotten is controlled by the vagaries of the evolution of scientific thought and the society of scientists who seem to both foster and discourage the emergence of new ideas. The question is how to keep things balanced so that ideas ahead of their time survive long enough to find their time.

LITERATURE CITED

Alejandro, A. S. 2000. Regeneration in the metazoans: Why does it happen? BioEssays 22: 578–590.

Baker, S. M. 1909. On the causes of zoning of brown seaweeds on the seashore. New Phytologist 8: 196–202.

Baker, S. M. 1910. On the causes of zoning of brown seaweeds on the seashore. II. The effect of periodic exposure on the expulsion of gametes and on the gemination of the oospore. New Phytologist 9: 54–67.

Baker, S. M. 1913. Quantitative experiments on the effect of formaldehyde on living plants. Annals of Botany 27: 411–442.

Barlow, C., and T. Volk. 1992. Gaia and evolutionary biology. BioScience 42: 686–693.

Bell, G. 2000. The distribution of abundance in neutral communities. American Naturalist 155: 606–617.

Briggs, J. C. 1966. Zoogeography and evolution. Evolution 20: 282–289.

Camerini, J. R. 1996. Wallace in the field. Osiris 11: 44–65.

Clark, W. C. 2007. Sustainability science: A room of its own. Proceedings of the National Academy of Sciences USA 104: 1737–1738.

Cobb, M. 2006. Generation: The Seventeenth-Century Scientists Who Unraveled the Secrets of Sex, Life, and Growth. Bloomsbury Publishing, New York.

Connell, J H. 1961. Effects of competition, predation by *Thais lapillus*, and other factors on natural populations of the barnacle *Balanus balanoides*. Ecological Monographs 31: [illegible]–104.

Dall, W. H. 1877. On a provisional hypothesis of saltatory evolution. American Naturalist 11: 135–137.
Egerton, F. N. 1968. Leeuwenhoek as a founder of animal demography. Journal of the History of Biology 1: 1–22.
Erezyilmaz, D. F. 2006. Imperfect eggs and oviform nymphs: A history of ideas about the origins of insect metamorphosis. Integrative and Comparative Biology 46: 795–807.
Fischer-Piette, E. 1935. Histoire d'une mouliere: Observations sur une phase de desequilibrium faunique. Bulletin Biologique de la France et de la Belgique 2: 1–25.
Golley, F. B. 1993. A History of the Ecosystem Concept in Ecology. Yale University Press, New Haven.
Gorbachev, M. 1990. Mikhail Gorbachev on the environment. Population and Development Review 16: 198–203.
Grant, P. R., and B. R. Grant. 2007. How and Why Species Multiply: The Radiation of Darwin's Finches. Princeton University Press, Princeton.
Hamlin, W. M. 1996. Imagined apotheoses: Drake, Hariot, and Raleigh in the Americas. Journal of the History of Ideas 57: 405–428.
Hatton, H. 1938. Essais de bionomie explicative sur quelques especes intercotidales d'algues et d'animaux. Annales de l'Institut Océanographique de Monaco 17: 241–348.
Hatton, H., and E. Fischer-Piette. 1932. Observations et expériences sur le peuplement des côtes rocheuses par les Cirripèdes. Bulletin de L'Institut Océanographique, Monaco (no. 592, February), 1–15.
Hubbell, S. P. 2001. The Unified Neutral Theory of Biodiversity and Biogeography. Princeton University Press, Princeton.
Huggett, R. J. 1999. Ecosphere, biosphere, or Gaia? What to call the global ecosystem. Global Ecology and Biogeography 8: 425–431.
Hutchinson, G. E. 1943. The biogeochemistry of aluminum and of certain related elements. Quarterly Review of Biology 18: 1–29.
Jackson, J. B. C. 1981. Interspecific competition and species' distributions: The ghosts of theories and data past. American Zoologist 21: 889–901.
Kareiva, P., S. Watts, R. McDonald, and T. Boucher. 2007. Domesticated nature: Shaping landscapes and ecosystems for human welfare. Science 316: 1866–1869.
Lenhoff, H. M., and S. G. Lenhoff. 1984. Tissue grafting in animals: Its discovery in 1742 by Abraham Trembley as he experimented with Hydra. Biological Bulletin 166: 1–10.
Lenhoff, H. M., and S. G. Lenhoff. 1989. Challenge to the specialist: Abraham Trembley's approach to research on the organism—1744 and today. American Zoologist 29: 1105–1117.
Lenhoff, S. G., and H. M. Lenhoff. 1986. Hydra and the Birth of Experimental Biology, 1744: Abraham Trembley's Memoires Concerning the Polyps. Boxwood Press, Pacific Grove, California.
Lindberg, D. R. 1998. William Healey Dall: A neo-Lamarckian view of molluscan evolution. Veliger 41: 227–238.
Lubchenco, J., A. M. Olson, L. B. Brubaker, S. R. Carpenter, M. M. Holland, S. P. Hubbell, S. A. Levin, J. A. MacMahon, P. A. Matson, J. M. Melillo, H. A. Mooney, C. H. Peterson, H. R. Pulliam, L. A. Real, P. J. Regal, and P. G. Risser. 1991. The sustainable biosphere initiative: An ecological research agenda. Ecology 72: 371–412.
Margulis, L. 1968. Evolutionary criteria in Thallophytes: A radical alternative. Science 161: 1020–1022.

Marmont, D., and J. M. Ponder. 2004. Obituary: Dr. Paul Fisher, 1898–2003. Molluscan Research 24: 131–133.
Martin, W., and M. Müller. 1998. The hydrogen hypothesis for the first eukaryote. Nature 393: 37–41.
McCoy, E. D., and K. L. Heck, Jr. 1983. Centers of origin revisited. Paleobiology 9: 17–19.
Nee, S., and G. Stone. 2003. The end of the beginning for neutral theory. Trends in Ecology and Evolution 18: 433–434.
Parker, G. H. 1933. Anthony Van Leeuwenhoek and his microscopes. Scientific Monthly 37: 434–441.
Robles, C., and R. Desharnais. 2002. History and current development of a paradigm of predation in rocky intertidal communities. Ecology 83: 1521–1536.
Rotramel, G. L. 1973. The development and application of the area concept in biogeography. Systematic Zoology 22: 227–232.
Rupp, J. C. C. 1990. Matters of life and death: The social and cultural conditions of the rise of anatomical theatres, with special reference to seventeenth century Holland. History of Science 28: 263–287.
Schiller, J. 1974. Queries, answers, and unsolved problems in eighteenth century biology. History of Science 12: 184–199.
Schramm, M. 2001. Frederick II of Hohenstaufen and Arabic science. Science in Context 14: 289–312.
Smil, V. 2002. The Earth's Biosphere: Evolution, Dynamics, and Change. MIT Press, Cambridge, Massachusetts.
Stevens, P. F. 1994. The Development of Biological Systematics: Antoine-Laurent de Jussieu, Nature, and the Natural System. Columbia University Press, New York.
Swammerdam, J. 1758. The Book of Nature or The History of Insects. C. G. Seyffert, London.
Tansley, A. G. 1935. The use and abuse of vegetational concepts and terms. Ecology 16: 284–307.
Tansley, A. G., and R. S. Adamson. 1925. Studies of the vegetation of the English chalk. III. The chalk grasslands of Hampshire-Sussex border. Journal of Ecology 13: 177–223.
Trembley, A. 1742. Abstract of Part of a Letter from the Honourable William Bentinck, Esq; F. R. S. to Martin Folkes, Esq; Pr. R. S. Communicating the Following Paper from Mons. Trembley, of the Hague. Philosophical Transactions 42 (1683–1775): iii–xi.
van Leeuwenhoek, A. 1800. The select works of Antony van Leeuwenhoek: containing his microscopical discoveries in many of the works of nature. G. Sidney, London.
Vernadsky, W. J. 1945. The biosphere and the noösphere. American Scientist 33: 1–12.
Vitousek, P. M., H. A. Mooney, J. Lubchenco, and J. M. Melillo. 1997. Human domination of Earth's ecosystems. Science 277: 494–499.
Wallace, A. R. 1857. On the natural history of the Aru Islands. Annals and Magazine of Natural History 20: 355–376.
Whittaker, R. H., and L. Margulis. 1978. Protist classification and the kingdoms of organisms. BioSystems 10: 3–18.
Wiley, E. O. 1988. Vicariance biogeography. Annual Review of Ecology and Systematics 19: 513–542.
Wilkinson, D. M. 1999. Gaia and natural selection. Trends in Ecology and Evolution 14: 256–257.
Woese, C. R., and G. E. Fox. 1977. Phylogenetic structure of the prokaryotic domain:

The primary kingdoms. Proceedings of the National Academy of Sciences USA 74: 5088–5090.
Wood, C. A., and F. M. Fyfe. 1943. Frederick II of Hohenstaufen: The art of falconry. Stanford University Press, Stanford, California.

The Structure and Habits of Birds (1244–1250)

Frederick II of Hohenstaufen

OF THE DIVISION OF BIRDS INTO WATERFOWL, LAND BIRDS, AND NEUTRAL BIRDS

In this first section of our work we shall discuss those aspects of bird life it is necessary to understand before investigating the art of falconry. Birds, like all other animals, may be divided into various species in accordance with their sexual relations, their parturition, their methods of procuring food, their variations in residence at different seasons of the year (chiefly as a result of alterations in temperature), and their change in diet.

Consider, then, a general division of birds based on the different activities they employ in securing food, the great varieties of that food, the flights they make to localities both near and far in search of heat or cold, the variety of and support given them by their limbs, the peculiarities of their feathers, their art of flying, their contests, and their moulting. It is, therefore, quite obvious, in so far as our plan permits, that we, as practitioners of falconry who hunt with birds of prey, should discuss these avian traits so that we can better understand where, when, and how hunting birds can be taught to catch their prey. All that we do not include on the nature of birds can be found in Aristotle's book *On Animals*.

All birds may be divided as follows:

Waterfowl are birds that habitually live in, or near aquatic areas and whose organs are so fashioned that they may remain for indefinite periods immersed in water.

Land birds are those that prefer a continuous life on land, an existence for which their bodies are especially constructed.

Neutral birds are those that may change from one habitat to another, from earth to water and vice versa, as shown both by their preferences and by their bodily structure.

Although Aristotle declares that every creature may be classed either as a water resident or as a terrestrial animal, and that only fish are truly aquatic, and includes under the term of land animals all those that progress both by walking and by flying, yet he does not make the mistake of classifying all winged creatures as birds. We, however, following the usage of falconry experts and adopting its terms, divide birds (in whose mingled constituents the lightest of elements predominate and who are the lightest and most agile of the winged inhabitants of the air) into water, land, and neutral birds; and of all these we shall give examples. We find that they may also be divided into various genera and these again into a number of species.

Water birds rarely leave and prefer to remain in the water. They do not leave it in search of food or for other purposes except when they fly from one body of water to another or during their seasonal migrations. These species include mergansers, cormorants, swans, and those birds that Aristotle in his *Natural History* calls pelicans and which the Apulians dub "cofani." The last-named are as large and almost as white as swans; they have a long, broad beak that has a pouch attached that they open and shut like a fishnet. They also have a swimming web between their feet that extends to the hind toe; the latter condition is not found in other waterfowl with webbed feet.

These and many others rarely leave aquatic areas. Others occasionally do so for the reasons aforementioned, as well as in search of food, and then return to it according to their natural impulses. This is the rule with some species of geese, ducks, and similar birds. Certain land birds visit bodies of water, but only for drinking and bathing, among them quail, partridge, pheasants, common bustards, and lesser bustards (that are like the former only much smaller). To this class belong also peacocks and birds like them.

Certain birds live most of the time on land but resort to water not only for drinking and bathing but also to secure their food, like aquatic fowl, returning to the land for rest. These include the sea eagles, which dive into the sea, rivers, and swamps to catch fish, after which they return to the cliffs and trees where they dwell. It is proper to class them with land birds because they are birds of prey and as such ought not to be called waterfowl.

Of neutral birds one may distinguish three types. Some of them (like the curlew) prefer water to land, in response to the demands of their bodily structure. Then there are others that have a changeable residence but seem to prefer land to water, such as plover, lapwings, and snipe. Of these, plover love the water less than lapwings, and lapwings more than snipe. The latter more nearly approach the character of land birds than lapwings or plover, because, though both these birds often sleep on land, snipe slumber there more frequently. There are still other birds that remain as much in the water as on land, like the cranes, both large and small, also both kinds of storks, the white and the black. The latter are frequently seen wandering about, fishing in water and in swamps and other wet places, returning afterward to dry land.

Into these three classes (land, water, and neutral) are all birds divided.

It may be added that those neutral birds who spend the greater part of the time in the water are customarily called waterfowl by bird hunters and these as well as true aquatics are termed shore birds. Those neutral birds that follow their natural bent and live mostly on land are often called terrestrial; and both these and land birds may be styled field birds, or birds of the plain. Some of those species that live equally on land and water may be termed either land birds or water birds; but those neutrals that come twice a day to feed on dry land but return each time to an aqueous resort for safety and rest, although they frequent the meadows, are nevertheless to be classed as aquatic birds, since they most resemble true waterfowl in their habits and seek the water as a permanent place of refuge. Among such birds are the cranes. Those neutral birds, however, such as snipe, plover, and lapwings, that live in meadows but go to bathe and drink in the water and take refuge there when pursued by birds of prey, may well be called land birds; for, as will be observed in the chapter on bird defense, many land birds take to the water when chased by their raptorial foes, for example, the crows.

OF THE DIVISION OF BIRDS INTO RAPTORIAL AND NONRAPTORIAL SPECIES

Birds may be classified in still another manner—as raptorial and nonraptorial species. We call raptorial all those birds who employing their powerful flight and the special fitness of their members, prey upon any other bird or beast they are able to hold and whose sole sustenance is the flesh of such animals. These are the eagles, hawks, owls, falcons, and other similar genera. They feed only on their prey—never upon dead flesh or carrion (*carnibus cadaverum neque residuis*)—and are therefore called rapacious birds. Aristotle calls them "greedy-clawed" birds or sometimes "birds of the hooked claws"; but this nomenclature seems to us unsuitable, since it is erroneous in so far as birds such as jackdaws, the larger swallows, and vultures have hooked claws and yet may not properly be called raptores, as they do not feed upon their own quarry.

It was the habit of Aristotle and the philosophers to classify objects into positive and negative groups and to begin their discussions with the positive. Since it is our purpose to give special attention to raptorials, we shall first consider the nonrapacious (or negative) varieties; afterward we shall consider at length raptorial birds. Nonraptorial species are those (whether aquatic, land, or neutral birds) that do not live entirely by robbery; in fact they cannot be regarded as true birds of prey if they subsist in part by plunder and partly on grains and fruit, like some ravens, crows, and magpies less frequently certain species of vultures, and the so-called "bone-breakers" (lammergeiers)—also some ignoble eagles that never plunder other birds or quadrupeds but feed on dead bodies and scraps.

It is therefore, evident that all birds may be included under the two categories of rapacious and nonrapacious, and that birds of prey differ from harmless species not only in their method of securing food, as is herein described, but also in many other ways, as, for example, in the form of their members in their behavior, and in the thickness or sparseness of their plumage.

Among the characteristic forms of their organs may be mentioned: the beak, which in birds of prey is generally curved, strong, hard, and sharp; claws that are bent inward and are hard and needle-pointed; retracted eyes; a short neck, short legs, and the posterior toe of each foot very strong. The female is larger than the male. Not all of the foregoing is true of nonraptorial birds.

Functionally also they differ in that raptorials are more keen-sighted and have more acute hearing than other birds. They are strong in flight but walk badly. They dislike water and drink little, fly alone, and live long. They drive their young early from the nest and then abandon them; and this behavior is not that of nonrapacious birds.

As to plumage, it varies among raptores; the first year after hatching (when they are called sorehawks) they moult only once, while other birds (generally) shed their feathers twice. The large quill feathers of the wings and tail are limited to a definite number; this is not true of other birds.

In numbers also the two classes differ, for there are fewer rapacious birds than nonrapacious; and there are no raptores among aquatic and neutral birds, but only among land birds, and even here they are few in number; so that all water and neutral birds and the greater part of land birds are nonrapacious.

Rapacious birds (which are universally warmer and drier than aquatics and neutrals) dislike water for two reasons, one active and the other passive. Since they have not members and plumage of a suitable form, they do not live in the water, nor can they do so, because they cannot continue to stand in deep water, lacking long legs like those of herons and cranes, nor can they swim about with ease, as their feet and toes are not webbed like those of geese, ducks, coots, and nearly all aquatic birds. Were a raptorial bird overturned, or submerged, in water her feathers and quills would be more inclined than those of aquatic birds to become soaked, so that she could hardly fly, and her claws would become so softened that she would be unable to wound or hold her prey. For these reasons, birds of prey dread remaining in the water, since they are extremely feeble in that element. There are certain birds, however, similar to eagles but smaller, that perch above bodies of water (or on high banks) and, when they perceive fish in the water, suddenly drop on them, draw them out alive, and feed on them. They are, therefore, called fish eagles. Their members and plumage are better adapted for this purpose than are those of other raptores.

The genera into which raptores are divided, and the species in each genus, will be discussed more fully in another treatise and in other parts of this work. It is also to be noted that the same genera and species are given different names by diverse authors. Sometimes the same bird may have a variety of synonyms; and the same name applied to diverse birds that are so dissimilar that one cannot establish the true identity of a species simply by its name. In consequence of this multiplicity of terms, a description of the essential characters of individual birds is more difficult to furnish, whether they resemble or are different from another in the shape of the limbs, the movements they make, the way they feed, the care of their young, their mode of flight, and their style of defense. Let it, however, be remembered that, in general, their bodily conditions and their other peculiarities are due to definite causes.

Very different localities may possess the same species and genera not found elsewhere; or a single region may be the habitat of birds of a species found nowhere else; while, on the other hand, in one region may be seen a genus found elsewhere but of a different color, or varying in other respects, but which may be identified by the characters of its members, its feeding habits, and other essentials. Therefore when we give one example of a genus and speak of others as similar, it is not necessary to repeat all the identification marks, lest we be guilty of tiresome prolixity. Indeed there is a multitude of birds, aquatic, terrestrial, and neutral; and so many genera are in each class that it would take too long even to mention them. However, when birds presenting essentially the same (or entirely different) characteristics are encountered, they may be described as belonging to the same or to a different genus, and this will be true even of birds not seen elsewhere and not previously identified.

OF WATERFOWL: WHEN AND HOW THEY SEEK THEIR FOOD

In addition to the foregoing observations let us consider how aquatic birds secure their food. Their movements and methods are not alike in all instances because some of them swim well but fly with difficulty and, consequently, do not wander far from the water. This is specially true of cormorants but is not in agreement with the teaching of Aristotle, that those birds that are limited in flight are to an equal extent good pedestrians. Cormorants, it is true, do not fly with ease; yet they walk still worse. When applied to land birds this rule may in many instances hold good, but in the case of waterfowl with a limited flight range these swimmers have legs ill adapted to walking, whether they fly well or badly.

Certain water birds both swim and fly well, yet they rarely leave the water; for example, swans, pelicans (called in Italy, cofani), cormorants, and their relatives. Other birds, like water rails and their kin, do not swim or fly well, although they are true aquatic fowl. Others, e.g., the heron, cannot swim well but are good fliers, yet they do not wander far from wet localities.

Some waterfowl dive entirely under water to obtain their food; others immerse only the head and neck to the shoulders or to the middle of the body. Their food consists either entirely of aquatic produce, or it grows on land alone, or it may be derived from both these sources. Pelicans, cormorants, and mergansers live exclusively on fish.

Aquatic birds that subsist on terrestrial products only seek their food in or near the water, like swans who feed on herbs, grains, and fruit that, owing to the weakness of their bills, they are obliged first to soften in water before they can swallow them. Others secure their food near or far from water, wherever they can most easily obtain it, like geese that consume herbage in either planted fields or meadows. As regards those waterfowl that eat both aquatic and land products, some devour fish as well as terrestrial animals and aquatic reptiles. This is true of herons, who do not despise fish, frogs, snakes, or mice. Ducks may devour fish, herbage, fruit, and grains. From the foregoing it is clear that some swimming birds live on fish, like the divers, pelicans, and cormorants; others (the swans for example) reject these foods; while still other birds, who live on fish, like herons and their kindred, are not swimmers. Many others, such as the oyster catchers (or sea pie), are waterfowl that neither swim well nor eat fish.

OF THE EXODUS OF WATERFOWL TO THEIR FEEDING GROUNDS

Now that we have discussed the variety of food consumed by waterfowl, we must explain at what hours they set out to feed (that is, those that leave the water for that purpose); how often this departure occurs during the day; in what manner they fly to the feeding place; how they pass the intervening time and how they return home; in what order they fly back; what localities they visit; what decides them when to return to their aquatic resorts; and exactly how, meantime, they rest on the surface of the water itself.

With very few exceptions the exit of water birds from their usual resorts in search of food takes place during the daytime. Only those with moderate powers of movement and slight ability to fly, such as cormorants and coots, make this journey by night, and then solely because of their fear of birds of prey.

The return of aquatic birds from their hunting grounds is fairly definite; as a rule they leave home at sunrise and remain in their feeding resorts until the third hour, sometimes earlier, sometimes later, sooner if it is a hot day, later if it is cloudy and cool. In some instances they rest on the home water until about the ninth hour, going out again to feed until sunset warns them to leave for a night's rest. These hours may vary with changes in temperature. Some birds, geese for example, pass the night away from their home resort (especially during moonlight), when they consume large quantities of herbage, which is collected by the aid of the moon when the nights are long, short days being insufficient to allow gathering a sufficient food supply.

It sometimes happens that flocks of both large and small geese, as they fly to and from their feeding grounds, are accompanied by domesticated individuals. Wild geese rarely become tame. Only a few tame geese continue as such; the majority eventually revert to the wild state, as it is more natural for them to follow their normal instincts despite any advantages gained by domesticity.

Other waterfowl and neutral birds do not leave their usual resting places during the night. They are generally those birds that feed on grain and roots, which they cannot see at night even by the help of the clearest moonlight. Such are plover, lapwings, and cranes. Ducks, teal, and similar birds do not limit their hunt for food to any particular time but feed at all hours in and out of the water.

From the foregoing facts we may divide the bird's day into three spells, of which two are spent in their feeding grounds and the third in resting in their permanent resorts. They spend the entire night in repose upon the water; thence they make two journeys to and from the feeding grounds.

The manner of their departure from and their return to a habitat varies greatly. Many birds, when leaving a locality, join with others of the same species to form flocks that come back in the same order they left, that is, in two lines that form an angle. Seldom or never do we see among their number any individuals not of the same species; that is to say, geese always accompany geese, ducks flock with ducks, teal with teal, cranes with cranes, etc. If by any chance these birds happen to mingle, they do it because they see others about to fly and are taken with a desire to be on the move; but they soon separate and rejoin their fellows. There are, however, birds (lapwings, for example) that do not adopt any particular order in this flight but go and come as a medley, sometimes in groups, sometimes singly. Such birds, both neutrals and land species that are not raptorial, eat greedily, swallow their food quickly and as they find it. They are not obliged in any fashion to prepare it in advance, whether it be grains, fish, winged or wingless insects, or worms. They eat in this manner because, were other birds living on the same food to see it, the latter would have no time to approach before it was safely swallowed; nor do such birds like to eat alone, but prefer to fly in flocks and to feed in the company of their own kind.

OF THE ORDER IN WHICH BIRDS DEPART FOR THEIR FEEDING GROUNDS

Waterfowl and neutrals leave the water and fly to their feeding localities (as a rule) in the following order: First come the lesser geese, then the cranes, then the larger geese; last of all fly the bernacles and certain smaller kinds of geese that fly with them, called *blenectae*. After these geese come the other aquatic birds. Their return occurs in the following order: the first to leave and plunge into their home waters are the bernacles and teal (*blenectae*), followed by the cranes and geese, and finally the remaining flocks.

The localities they choose for feeding vary greatly according to the character of the food, the season of the year, and the ease (or difficulty) of escape from birds of prey. Ducks and related species prefer pasturage during the wet season and particularly during the rainy days of September, October, and November. This period is chosen mainly because at that time the rain dislodges the seeds of plants, fruits of trees and shrubs, which the waters collect and carry to rivulets, the shores of streams, and other shallow collecting grounds where the birds congregate. These are the occasions and places they prefer and most often take advantage of to feed. Moreover, not only ducks and other water birds frequent these areas but also those that live on worms, which they dig out of the earth or find on the ground. It is mostly during the rainy season that in such places worms abound. Abandoning their burrows most vermes come to the surface to escape water, which is noxious to them. Also at that time, because of the water-softened topsoil, it is easier for these creeping animals to come up than to dig deeper where the earth is hard.

OF THE RETURN OF AQUATIC BIRDS TO THEIR WATER OR SWAMP REFUGES

The home localities to which waterfowl return may be lakes, ponds, swamps, or some of the longer or smaller streams, and other water-covered areas sometimes called flats; but they prefer to swim about and plunge into some rocky and extensive body of water possessing islands and other advantages. Their homecoming takes place during the day that they may guard against such wild, animals as fish otters, foxes, and birds of prey. The last-named do not pursue waterfowl very much in the water, because swimmers can easily escape from them by diving. Even weak waterfowl more successfully defend themselves in this way than many other stronger birds that live away from water. Aquatic birds return to the water in the daytime for the purpose also of drinking and resting. This is especially true in summer. At night they remain standing in the water not only for protection against otters, foxes, wolves, and other wild animals that may harm them while sleeping but also that they may pass the night in peace, sleeping and resting.

OF THE POSITIONS ASSUMED BY BIRDS WHEN ASLEEP OR AWAKE ON OR IN THE WATER

During sleep swimming birds pass part of the time floating on the water, part of it near the shore with one or both feet on the bottom. Nonswimmers, aquatic birds or neutrals, keep either both feet on the ground under the water, or rest on one foot with the water up to the knee or above that point, depending upon its depth.

The larger number of water birds, like some land birds, when sleeping, turn round the head and rest it on the back between their shoulders. Indeed both swimmers and nonswimmers usually sleep on one foot, holding their heads on their backs below the shoulders so that the head as well as the cold and horny beak may be kept warm. This position not only provides warmth but prevents wetting the head (keeps it dry) and thus avoids possible freezing of the parts in very cold weather. Disease in the head, called *gipsus*, may easily set in from exposure to wet and cold.

Not only waterfowl but almost all other birds sleep on one foot in order that they may sleep lightly and be easily alarmed and readily awakened to meet approaching danger. Water birds asleep with one foot in the water readily sense any disturbance of the surrounding fluid and are thus warned of the approach of beasts of prey or other enemies. Since water is liquid and its limits are ill defined, it recedes from the point of disturbance with a circular motion that extends to the leg of the bird, who, feeling the motion, is instantly alarmed and put on guard. Moreover, a body that has several points of support stands more firmly and is more difficult to move than one resting less securely. Hence a bird resting on one foot only is easily roused. This, added to the fact that any person sleeping under threat of danger is easily disturbed, makes it doubly plausible that birds sleeping normally under these conditions will be quickly awakened.

Frederick II of Hohenstaufen (translated and edited by C. A. Wood and F. M. Fyfe, 1943). The Art of Falconry. I. The Structure and Habits of Birds; *chapters II–VIII, pp. 7–19. Stanford University Press, Stanford, California.*

Of the Spider (1800)

Antony van Leeuwenhoek

The following observations were made on those kinds of spiders, which are found in gardens, where they fix their webs to vines, herbs, and shrubs.

I have often seen these spiders, when dropping, or falling, as it seemed, from a tree, stop or support themselves in the midway, by means of their thread, and I found that this was done by the help of one of their hind feet, which they continually apply to the thread as they spin it. These feet are each of them furnished with three nails, or claws, standing separate, or apart from each other. Two of these claws are at the extremity of the foot and each of them is formed with teeth, or notches, like the cuts in a saw, growing narrower towards the bottom; and with these they are enabled to hold fast the thread, in like manner as the pulley or wheel, used by clock-makers, in their thirty-hour clocks, is contrived to lay hold of the clock-line, by means of the groove being also narrow at bottom.

The kind of spider I am now describing, has the hind part of its body much larger than is seen in other spiders; it is provided with eight longer and two shorter legs, which shorter ones are placed in the fore part of its body on each side of the head, and all furnished with an indented or notched claw as before described. Some will have it, that spiders have no more than eight legs, but this appears to be a mistaken opinion.

In these spiders I plainly perceived eight eyes, two of which are placed near to each other at the top of the head, and, in my judgment, designed to see those objects which are above the animal. Two others of them are situated a little lower down, in order to discover all objects in front; and on each side of the head are a pair of eyes close to each other, and of these, the two which stand forward, are to take in the view of all objects lying obliquely, or not straight in front; and the two which stand backward, are undoubtedly designed to behold all objects behind the animal. And if we consider that the pupils of these eyes are immoveable in the head, we may easily conclude, that this number is necessary, for enabling the spider to behold all circumjacent objects, and to go in search of its food.

I have often heard it said, that the spider has a sting, with which, it is also reported, it can kill the toad; but no one could tell me in what part of the body this sting was placed, therefore, I concluded that if there was one, it must be in the posterior or hind part as in other animals and insects; but on examination, I found this opinion to be groundless. The spider is, however, provided with two organs or weapons answering every purpose of a sting, which are placed in front of its head just below the eyes, and when not in use, they lie between the two shorter feet.

These weapons or instruments of offence, which are bent in the nature of claws, are very similar to the sting of the scorpion and the fangs of the Millepeda of India, and in each of these fangs (for so I will call them) is a small aperture, through which, in all probability, a liquid poison is emitted by the spider at the time it inflicts the wound.

I at several times included two or three large spiders in the same glass, and always found that when they approached each other, they would fight to that degree, as to be covered with the effusion of blood from their bodies, which was soon followed by the death of the wounded spider. I also observed, that the smaller spiders always avoided the larger, but when two of nearly equal size approached each other, neither would give way, but both of them grappled together furiously with their fangs, till one of them lay dead upon the spot, its body being as wet with the blood flowing from the wounds received; as if water had been poured upon it.

I at one time had a spider which was wounded by the bite of another in the thicker part of its leg, and from the wound there issued some blood, in quantity, about the size of a large grain of sand; this wounded leg, the spider held up, as unable to use it, and ṣoon afterwards the whole leg dropped from its body: whenever the breast or fore part of the spider was wounded, I always observed the wound to be mortal.

I had imagined, that when a spider applied its thread either to some foreign substance or to another thread, that the thread newly spun must be covered with some viscous or glutinous matter by which it became fastened, in like manner as we observe in silkworms threads. But I now found that the spider cannot fix its thread to any thing, without imprinting the hind part of its body on the place, by which pressure, it emits an incredible number of excessively small threads, diverging in every direction, from whence we may conclude, that as soon as the threads are exposed to the air, they lose their viscosity or gluey quality.

When I at first began the dissection of the spider, and endeavored to discover the viscous or gummy substance from whence their threads proceed, and could not satisfy myself in that particular, I was astonished, not being able to conceive how, from so moist a body as this creature's, there could in so short a time, be produced threads strong enough to bear the weight, not only of one, but of six spiders at a time, And upon endeavoring to discover the texture of these threads, I could at that time perceive no more, than that the same thread appeared in some places to be one and entire, and in others, to be composed of three, four, or more threads; and though I often endeavored to observe those threads immediately as they issued from the spider's body, I could not obtain a perfect view of them, notwithstanding which, I did not doubt, that what is commonly supported to be one thread, is, in fact, composed of many.

I determined therefore, so to fix a spider on its back, that it could not move the hind part of its body; and this being done, I contrived with a small pair of pincers to draw out from the body, that small part of the thread which projected from the organ or instrument from which the threads proceed, and then I perceived a great number of exceeding small threads issue forth, which, when at about one or two hairs breadth distance from the spider's body, united in one or two threads, and that in this manner the larger threads were composed.

Not content with these observations, I sat about devising means of keeping the threads separate, as they issue from the spider's body, so that I might be able to give some representation of their inconceivable fineness, and at three several times I succeeded herein to my wish. But yet, this fineness cannot by any efforts of the pen or pencil, be adequately described. For upon applying the utmost magnifying powers of the microscope, threads are discovered so exquisitely slender, as almost entirely to escape the sight. I have sometimes endeavored to count these threads as they issued forth, but always without success.

If we duly consider that the threads of spiders, which to the naked eye seem to be single, are composed of many smaller ones, and that they thence acquire the strength we observe them to have, we shall more than ever be assured, that no flexible bodies (except those made of metal, the component particles of which are, by the force of fire, most closely composed or knit together), can have any great strength or toughness, unless they are composed of oblong parts laid side by side, and that their strength or toughness will be greater where these oblong component parts are twisted together, or made to cohere by some glutinous matter, as are spun silk, linen garments, ropes and the like. And this is the reason why all the single threads of flax are very tough in proportion to their size, for each of them is composed of still smaller particles or fibres, which are not only joined together by a certain viscous or gummy matter, but are also surrounded with a coat or bark, as it may be called, whereby their inward component fibres are rendered still stronger.

Again, if we advert to the great number of excessively slender threads, proceeding, all at the same time from the body of the spider, we must acknowledge that this kind of formation is necessary, for were it a single thread which is spun by this creature with such celerity, the liquid matter of which it is formed, could not on its exposure to the air, become a solid substance so quickly as these lesser threads, an hundred or more of which, taken together, do not in my opinion equal the hundredth part of one of those hairs I can take from the back of my hand.

In a word, the inscrutable power and wisdom of the Almighty Creator, are manifestly displayed in the formation of such a thread as the spider's, the wonderful make of which is seldom observed, because, the fineness and delicacy of its texture are not discernible by the naked eye.

Upon beholding the exquisite tenderness, and also the multitude of these threads, I was struck with astonishment, upon considering how wonderful must be the organs in a spider's body to produce so many, and at the same time all distinct from each other. And although I never expected that I should be able to dive into this secret of nature, yet, upon dissecting the hind part of one of the largest spiders I could procure, and attentively examining it, I at length, with the greatest admiration, perceived a great number of excessively small organs, from each of which, one exquisitely fine thread proceeded, and these were so many, that I thought their number must at least exceed four hundred. They were not all placed close together, but in eight different spots or compartments, so that if the spider uses all these organs at the same time, eight several threads may be formed, each of which will consist of a great number of smaller ones. Again, these smaller threads differ in size, for one of the organs will be seen to spin a thread twice as large as the next adjoining to it.

If any person examines by the microscope that part towards the extremity of the spider's body, from whence its thread proceeds, he will observe the spot to be, as it were, surrounded by five several protuberances or risings, each ending in a point, and altogether forming a kind of enclosure; but from the anterior or forwardest of these five protuberances no threads proceed. The other four, on their outer sides are thick set with hairs, so that all the smaller organs designed to spin the threads, are situated towards the inside, the reason of which, I take to be, that they may be preserved uninjured, when the spider is creeping into holes, where it does not want to spin its web, or while running along the ground, or after its prey. When these last mentioned four protuberances are put aside from each other, there will be seen in the middle or space between them four smaller ones, each furnished with the like organs for spinning threads, but lesser in size and fewer in number.

These organs for spinning, being by this means all exposed to view, exhibit the appearance, as it were, of a field, thick set with an incredible number of pointed parts, each producing one thread; but these pointed parts are not made gradually tapering from the base to the point, they are formed, as if one were to imagine a small reed somewhat tapering, having a still smaller one joined to its taper end, and this latter terminating in a point, which point, in these organs I am now describing, is as fine as imagination can conceive.

Now if we lay it down as a fact, that a young spider which is several hundred times smaller than a full grown one, is furnished with the same organs as the larger, and that as the spider, so the organs do by degrees grow proportionally larger, the necessary conclusion is, that the threads spun by a young spider, are many hundred times finer than those spun by one full grown, which exquisite slenderness, it seems beyond the power of the human mind to form a true idea of.

After this, I took a small frog, whose body was about an inch and an half in length, which I put into a glass tube together with a large spider, in order to see the actions of these two animals when brought together; and I observed the spider pass over the frog without hurting it, though with its fangs displayed as if to attack the frog. Upon this, I caused the frog to fall against the spider, who, thereupon, stuck his fangs into the frog's back, making two wounds, one of which, exhibited a red mark, and the other a purple spot. I then brought the frog to the spider a second time, who, thereupon, stuck his fangs into one of the frog's fore feet, whereby some few of the blood vessels were wounded. And having provoked the spider a third time, he struck both fangs into the frog's nose, presently after which, I took the spider out of the glass. The frog, thus wounded, sat without motion, and in about the space of half an hour, it stretched out its hind legs and expired.

The next day I brought another frog, about the same size as the former, to the same spider, but though it was twice wounded, I did not perceive it to be injured thereby, perhaps because the spider's bite may not be so venomous in our climate as in warmer regions, or else, that the poison of this spider might have been exhausted by former attacks; the frog I threw back into the water whence I had taken it.

Towards the end of October, I took several of the largest spiders that could be got, and placed them in glasses apart by themselves, in order to wait for their laying eggs, which I purposed to open, and examine the contents. Two of these spiders, after being confined ten or twelve days, I found had laid their eggs, and enveloped them in so thick a web, that I was astonished to behold it, considering that it had been spun in a few hours space.

Some of these eggs I opened, and found the inside to be of a yellowish colour; the form of each egg was almost round, and nearly the thirtieth part of an inch in diameter, and the whole collection of eggs laid by one spider composed a rounding figure, almost spherical, nearly half an inch in diameter, from whence may be computed how great a number of eggs the spider lays. And one would almost think it impossible for so many to be contained within this creature's body; since upon viewing them with the naked eye, as they lie together in regular order, they occupy a larger space than the size of the animal itself. But it must be considered, and it is what I have often experienced in opening spiders, that the eggs while within their bodies are not of a globular figure, but being very soft they lie compressed together, and therefore are of divers shapes, but as soon as emitted from the spider they assume a spherical form, by reason of the equal pressure of the atmosphere on every part of them; and when of this round figure, being placed in exact order, side by side, and only touching each other in a point, they must necessarily, to our view, occupy more space than they did while in the animal's body.

I at first was not able to conceive by what means the spider could place its eggs exactly in the centre of the web, but now I was satisfied in that particular, for while I was observing a third spider which was fixing a web to the glass in order to lay her eggs in it, I saw that first she made a kind of thick layer of threads, and fastened them to the glass before she began to lay one egg; and it was most worthy of remark that this layer or stratum was not flat but curiously made with a roundish cavity. In about three quarters of an hour's space, began again observing the spider, I saw that this cavity was not only filled with eggs, but that eggs were piled up above the edges of it to the same height as the hollow of the cavity below, and the spider was then busied in spinning a web to enclose the eggs on every side. For this purpose she employed not only the hind part of her body from whence the threads were spun, but her two hinder feet, with which she placed the threads in due order. And now all the organs used in producing the threads appeared in view, each of them in the act of emitting its particular thread. I also observed the spider elevate the hind part of its body about the breadth of a straw, and then fix the thread which by the elevation had been drawn out to that length, to the web which was already spun about the eggs.

I was very desirous to see a spider in the act of laying its eggs, which at length I obtained a sight of, and observed that they were not emitted from the same part as is usual in all other minute animals; but from the fore part of its belly, not far from the hind legs, and near the place, I observed a kind of little hooked organ, handsomely shaped, which I had often before seen in this animal, and could not imagine for what purpose it was designed; but now I perceived, that it extended over that part whence the eggs issued and I therefore conjectured that its use was to deposit them in regular order within the web prepared to receive them.

On the first of January I put some spider's eggs into a glass tube which I constantly carried about me, in order to discover whether by the warmth I imparted to them, they would be hatched sooner than the usual time, which is in the spring; and on the 17th of January I saw above twenty five young spiders completely hatched, and as many more half way out of the eggs; and in the evening of the same day I counted above an hundred and fifty young ones. The next day, the number was not increased, for the remainder of the eggs, to the number of fifty, or thereabouts, were either barren, or the young spiders were dead within them.

Upon exposing the glass tube at this cold season, to the air for about a quarter of an hour, the young spiders lay without motion, but upon applying some warmth to it, they began to move, and the greater number of them crowded themselves together in an heap, after the manner of bees, within the web where the eggs had been. On the 21st of January I could discern eight eyes in each of them, which till then had not been visible, and on the 25th of January they began to spin webs in the same manner as full-grown spiders.

I had hitherto been at a loss to conceive how this great number of young spiders could be supplied with nourishment, considering that the natural food of this creature is the substance of other insects; but I now perceived that they had fed on the barren eggs which had been left in the glass, and they afterwards devoured one another till they were reduced to a very few in number.

I have often compared the size of the thread spun by full-grown spiders with a hair of my beard. For this purpose I placed the thickest part of the hair before the microscope, and from the most accurate judgment I could form, more than an hundred of such threads placed side by side could not equal the diameter of one such hair. If then we suppose such an hair to be of a round form, it follows that ten thousand of the threads spun by the full grown spiders when taken together will not be equal in substance to the size of a single hair. This is found by multiplying the number of spider's threads, constituting the diameter of the hair (which the Author computes to be one hundred) into itself, the contents of cylinders (which round threads may be called), being in the same proportion as the squares of their diameters, therefore 100 diameters of the thread multiplied by the same number 100, the square will be 10,000 the proportionate size of the hair, and this being multiplied by 400 the supposed bulk of a young spider compared with an old one, gives four millions (4,000,000), the proportion assigned by the Author to the young spiders threads.

To this if we add that four hundred young spiders at the time when they begin to spin their webs, are not larger than a full grown one, and that each of these minute spiders possesses the same organs as the larger ones, it follows, that the exceeding small threads spun by these little creatures, must be still four hundred times slenderer, and consequently that four millions of these minute spiders threads cannot equal in substance the size of a single hair. And if we farther consider of how many filaments or parts each of these threads consists, to compose the size we have been computing, we are compelled to cry out, O what incredible minuteness is here! And how little do we know of the works of nature!

I never could procure a sight of these animals when coupling together, either in the gardens or fields, nor when enclosed in glasses, for I always perceived the female to run away at the approach of the male, and having at one time enclosed three male spiders with a female in one glass, the female flew at the males with so much fury, and wounded them to such a degree, that blood issued from their legs and feet. Hereupon I killed the female, and the next day I saw two of the males lie dead, and the survivor employed in devouring the dead female.

These are the chief of my observations on the Spider, an animal held in such detestation by many, that they dread even the sight or approach of it, but in which we find as much perfection and beauty as in any other animal.

Van Leeuwenhoek, A. 1800. The select works of Antony van Leeuwenhoek: containing his microscopical discoveries in many of the works of nature. *G. Sidney, London; "Of the Spider," pp. 35–48.*

Observations Relating to the History of Bees (1758)

Jan Swammerdam

SOME PECULIAR OBSERVATIONS RELATING TO THE HISTORY OF BEES: A DESCRIPTION OF A HIVE OPENED THE TENTH OF MARCH, WITH AN ACCOUNT OF THE NUMBER OF CELLS IT CONTAINED.

On the tenth of March last I opened a hive, in which a young swarm of bees had been settled during the month of June, the preceding year, but they all died in the intervening February for want of honey. I examined the cells built from the month of June till the winter season, that is in the space of about four months, and counting them one by one, I found them to amount to 22,574; and the whole of this prodigious number was only of that kind of cells, in which the working bees are hatched and nursed, or the honey and bee-bread is stored up. Those in which bees had been already hatched, amounted to 7814; for it was very easy to distinguish them certainly from the others, by means of the skins and webs found in them, such things being always left behind by bees that have been hatched. All the other cells were formed for keeping honey, and the other cells are made to answer the same purpose, as soon as the young bees contained in them have acquired wings to fly abroad.

It appeared likewise that all these cells were contained in nine combs, as they are generally termed, or nine portions of the whole wooden structure, and these portions were large, oblong; of different forms, some diverging equally, others running out into two, three, or four angles. This variety in the figures of the combs is owing to some of them being built alone by themselves, and others close to each other; or to the necessity the bees were under of keeping clear of the sticks placed across the hive to support the wax, for this occasions them to make their combs sometimes of a triangular, and sometimes of other forms. Nor can we perceive, that in this business the bees observe any certain rule or order, since the figure of the cells themselves does not suffer by this liberty they give themselves.

Many of the little cells in which the honey was stored up, were twice as long as those intended for nests and nurseries, and were also irregularly built, crooked, and full of angles. Even the sides of the hexagonal cells did not every where exactly correspond with one another, but here and there might be seen a gap large enough to contain a pin's head, a thing never to be met with in a truly regular comb.

All the half combs of cells on one side of the perpendicular foundation, which runs through the middle of them, and against which the cells are horizontally placed, were built full one half as long again as those on the other side. There appeared here also many other irregularities, not to be seen in the cells that had served the purpose of hatching, such of them at least as had been quite finished.

From this prodigious number of cells, built between June and September, or October, we may entertain some idea of the great number of those that the bees construct from the month of March to the June or July of the following year. I believe they may amount to 50,000, as this is the time for supplying with cells the male, female, and working bees: but as yet I have not counted them.

A person fond of bees, and whose account I could credit, once told me that he had a hive placed upon the bare ground, and exceedingly well stocked with bees, insomuch that to make room for their combs, they had hollowed out the earth under their hive, extended their constructions very deep into this hollow, and thereby increased their numbers to a prodigious degree. But this is oftener practiced by wasps and hornets, as these insects naturally make their nests under ground.

A HIVE OPENED THE 14TH OF JUNE, THE NUMBER OF BEES AND NYMPHS FOUND IN IT, WITH A PARTICULAR DESCRIPTION OF MANY OTHER SINGULARITIES NOT AS YET KNOWN

In the beginning of June I bought a hive of bees, it produced a swarm the 14th of the same month. I received the young bees in another hive, and put this hive in water the day following, with all its new inhabitants. By this means I found the swarm consisted of one female, four males, and 2433 working bees, who had not made any wax since they swarmed.

The 16th of the same month, I likewise drowned in the same manner the bees that remained in the original hive, from which otherwise a second, and even a third, swarm might have been expected. In this hive I found one female, 693 males, and 8494, working bees. While I was employed in counting them, I let the water run off from the hive, that I might afterwards satisfy my curiosity in ascertaining the number of their cells, but I found the amount so great, especially that of the cells belonging to the working bees, that I thought proper to desist, for fear of losing the opportunity of making some other observations, that I imagined better deserved my attention.

I therefore reckoned with great care and exactness the little dwellings of the female bees, and found nineteen of them as yet building, but some a little more forward than others. There were besides fifteen more, in shape resembling a pear, and quite finished, which were all closed up with wax, and curiously disposed on the edges of the combs. Some of them stood by themselves, others lay close to each other, three, four, or five together. Others again were built quite close to the cells of the male bees; some were situated obliquely; others horizontally, so as to resemble a beerglass lying on its side; and, in the same manner, all the cells of both males and working bees. Some on the other hand were built in an inverse position, with their openings looking downwards, as the cells of hornets are generally found. Finally, I discovered the cell of a female eat through on the fore part, being that out of which the young queen bee had escaped that led the swarm of the 14th of June.

In nine of the cells belonging to the females, which I found closed up as just now mentioned, there were as many female bees arrived at their full size and furnished with wings ready expanded; and some of them were still alive. Some of these females were quite gray, and others of a somewhat darker colour, according to the time that had elapse since they changed their skins, and that which they were still to continue within the cell. Not one of them had as yet attempted to open itself a passage to fly abroad.

In the other five of the covered and closed up cells belonging to the females, I found as many nymphs as females. One of these five cells contained a nymph, which already begun to grow gray on the back, and was upon the point of throwing off its old skin; but in the other four nymphs there was no appearance of this colour, they being as it were still in their infancy, and for the most part resembling in whiteness the curds of milk. The eyes alone had by degrees acquired a watery purplish colour, and the same might be observed of the three distinct smaller eyes, which are more conspicuously perceivable in the insect, while it remains in this state, than afterwards when grown to its full perfection.

Under the belly and tail of these nymphs, I found the exuviæ and air tubes that had dropped from them, on their exchanging the form of worms for that of nymphs. I could also perceive the remains of their food, which on pouring water upon it looked like soft starch or gum tragacanth, beginning to swell; it was of the colour of pure amber, and of a somewhat subacid flavour.

In the upper part, under the wax with which these fourteen cells were closed up, I could discern the web which the nymphs spread in that part, while they continue in the form of worms. The upper web was very strong, and made of distinct thread, but in the lower part of the cell it looked like a membrane; for at the time these worms labour to shut up their cells with such webs, they are obliged to move their bodies in every direction, and thereby rub their food, and perhaps too their excrements also against their work, so as to fill the intervals between the threads that compose it with a kind of glue, and thereby reduce its surface to an evenness like that of a natural membrane.

I opened besides all the closed up and covered cells of the males; many of these cells were situated near those belonging to the working bees, and contained in a single comb, hanging at the bottom of the hive. The rest of the male's cells were built in the midst of those of the working bees, with common party walls or partitions. Of these closed up cells belonging to the males, I reckoned in all 858. In 234 of them I found as many worms, which had not as yet changed to nymphs, but some of them were nearer that period than others. In 146 cells there was the same numbers of milk-white nymphs, which had but just thrown off their skins. In 44 more cells the eyes of the nymphs were just beginning to acquire a watery and light purplish color. In 414 other cells I found as many nymphs, whose eyes were of a deep purple. And lastly, in the 20 remaining cells there were nymphs just upon the point of shedding their skins, and appearing in the form of male bees: the gray and hairy members of the young males appeared plainly in these, through the transparent membrane which still enclosed them.

After this I reckoned all the other male cells, and found them to amount to 1508, of which 720 were entirely empty, the male bees sometime before hatched and bred up in these, having taken their flight; 268 more were not as yet perfected, nor had been used for hatching; 520 of the same cells, in which also no worms had been yet hatched, were full of the purest virgin honey. I counted besides all these 1701 empty oblong cells, which, though considerably bigger than the male cells were not unlike them: neither had any hatching been performed in these, their form not being regular enough for that operation; therefore they could only serve to lay up honey. This circumstance likewise makes me imagine that these oblong cells are not to be looked upon as male cells, but to be reckoned amongst the store houses which the bees build for their winter provisions; for we find they make cells of the same oblong form, but like the cells of the working bees, to answer that purpose.

The number of closed up cells belonging to the working bees amounted to 6468, and in all of these I found nymphs under the same variety of circumstances with those which I had found in the male cells. It is therefore needless to waste words in explaining their different appearances, nor had I leisure to count the numbers at every period of growth and step towards their perfection: besides some of these nymphs began to have a very disagreeable smell.

I reckoned also 210 cells full of bees-bread, which was also heaped up here and there in the combs of the working bees, in particular cells disposed between those which had nymphs in them, or which were full of honey, but none of these bread-cells were closed up.

As to the remaining cells, those newly built, as well as the empty ones, in which bees had been hatched, and those constructed the year before as store-houses for honey, or nurseries, I had not time to count their prodigious numbers. Neither did I count the closed cells, which were disposed in the upper part of the hive, and were now ready to burst with honey. But my curiosity led me to weigh the honey itself, and I found it amounted to seven pounds.

In all this hive I did not meet with a single egg, nor with any worms, but such as were full grown; so that by this time the working bees must have got over the heaviest part at their yearly labour, for there was no longer any necessity for building cells, or nursing of young bees, nothing more remained but the agreeable task of gathering honey for the support of themselves, and of the males and females, and making preparation for the second, third, and fourth swarm, which I could easily see were to be produced from the different stages in which the nymphs of the future queens appeared, and from the different periods at which it was of consequence necessary these should make their appearance abroad. This induces me to believe, that the old females continue, even during the intervals of swarming, their labours for the propagation of the species, as I have already observed in describing the hive opened on the 22d of August.

Many of the working bees belonging to this original hive were still of a grayish colour which is a certain proof that they have not been long out of their cells. Nor did I observe one amongst them that had lost its wings; whereas such crippled bees are frequently seen in spring or autumn. This circumstance makes it probable, that most of the last year's male bees had been taken off by a violent or natural death, and succeeded by a new generation. Nor need this opinion appear improbable to any, for if on the 14th of June I could count 6468 nymphs in one hive, and 2433 bees in one swarm, we may easily guess what a prodigious number of bees must be produced in the interval of time between March and June, and that between June and September; no doubt a multitude sufficient to supply a hive with a number of new inhabitants, three times greater than that of those which had possessed it the preceding year, or summer months, supposing them all to have unfortunately perished: the queen alone survives a longer time, though I can scarce believe her life is of above two or three years.

As by what I have here observed, it plainly appears, that fifteen young female bees had been produced in one hive, and in the space of time required for one swarming; and as experience informs us that bees seldom swarm in this country above three or four times, and that after the last swarm they kill their queens, which are then no longer of any use, we may conclude that at this time the old and impotent queen undergoes this fate, and is succeeded by a young one, better able to propagate the species. This opinion indeed stands in need of more experiments to confirm it, and such experiments may be easily made by any one who is willing to sacrifice a few hives to his curiosity.

It is surprising how tenacious of life bees are; after the hive and all its inhabitants had, in consequence of my orders, been kept under water for a considerable time, and I had begun to count them, as if they were perfectly suffocated, they began by degrees to recover life, as it were, and fly about the hive, so that I found myself under the necessity of causing them to be again put under water, and though I had reason to think none of them could outlive this second submersion, yet there appeared many after it with signs of life, and some of them recovered themselves so well as to live after this three days and two nights without eating.

ALL ACCOUNT OF SEVERAL WONDERFUL PARTICULARITIES DISCOVERED ON OPENING A HIVE THAT HAD A FEW DAYS BEFORE RECEIVED A YOUNG SWARM

Happening to be in the country on the 25th of July, I observed a great swarm of bees, which, on its hanging to an elm, I ordered to be received into a hive; but in a little time they all left this new habitation, and fled back to the elm, where they hung entangled by each others legs. The female bee had not dropped into the hive with the others: I was therefore obliged to have recourse to another shaking; when having brought the female into the hive, all the rest soon followed.

On the 26th of July the weather was tolerably good, with a bright sunshine; the 27th cloudy; the 28th and 29th rainy: on the 30th on examining the hive, I found at the bottom of it upon the ground where it stood, a piece of a honey-comb, which had fallen thither, either because it had not been strongly enough fastened to the top of the hive, or because too many bees had lighted upon it at one time. This piece of a comb contained 418 cells for the working bees, some were building, and others were finished, and there were also ten eggs sticking to the wax by one of their ends. All the forenoon of the 31st it was rainy and about midday very cloudy and windy, with some rain. In the evening I ordered the hive to be taken into my chamber, in order to examine what the bees had done in the space of these six days.

But as I was afraid of being stung in this enterprise, I resolved to have all the bees killed before I went to handle or inspect them, for this reason I fumigated them with a bundle of lighted matches rolled up in linen rags, to such a thickness, that it would just fit in the upper opening of the hive. All my endeavors to kill these bees this way were however to no purpose; for after plying them with this fume, from eight o' clock to eleven, lighting the matches from time to time, as they went out, the bees continued alive; but they seemed grievously complaining of, and resenting the injury offered them, with the most horrid noise and loudest buzzings.

The next morning all was quiet again, so I removed the hive, at the bottom of which I found some hundreds of bees lying dead upon the ground; but the greatest part of them were still alive, and some of them were beginning to fly away. I therefore resolved to fumigate the hive a second time, and I gave its inhabitants liberty to escape while it was doing. For fear of being stung on this occasion, I took a half pint bottle, and having rolled some soft paper about the neck of it, thrust it into the opening of the hive, taking care afterwards to stop all gaps between the door or opening of the hive, and the neck of the bottle with more paper of the same kind. As soon as the sulphureous vapour began to fill the hive, the bees in the greatest hurry and confusion and with the most dreadful buzzing, rushed to the number of 1893 in a manner all at once into the bottle, which I then removed to substitute another in its place; and by repeating the operation in this manner, I at last so thoroughly accomplished my purpose, that not the least noise could be heard in the hive.

Having then turned the hive upside down, I found the queen lying dead, in appearance, upon the ground, and some of the others which had fallen upon the ground, killed downright and wet all over; whilst some other bees that had remained in the upper part of the hive, were quite dry, and when put into the bottles flew about as briskly as if they had not received the least harm.

I next poured some water upon the prisoners I had in the bottle; by this means they were all drowned in a very short time. I then made my examination, and found the swarm consisted of 5669 bees, and was therefore a very good one, according to the judgment I had formed of it on its first appearance. Nevertheless, as the season was very far advanced; and the spot the bees lighted upon very ill furnished with materials for making honey, I thought it worth while to sacrifice them to the curiosity I had of knowing what work such a number could perform in so short a time, and withal in so unfavourable weather.

Among this great multitude, there was but one female bee. The greatest number of them were working bees, which are neither males nor females; and there were besides these and the female bee already mentioned, only 33 male bees, preposterously called by the vulgar hatching bees; for the young bees are hatched by the mere heat of the summer, and that which is caused by the perpetual hurry and motion of the old bees flying about, or working in the hive. It is very remarkable that the bottle into which the first 1898 bees driven out of the hive had been received, was thoroughly heated by the perpetual motion of these imprisoned creatures, and the warm vapours which exhaled from their bodies.

The number of waxen cells begun and finished, including those of the comb I had found on the ground on my first examining the hive, amounted to 3392: they were all of the same size and form, and were intended only for nests to hatch the working bees. In 236 of the cells some honey had been stored up, but it had been afterwards made use of; as very little could be then gathered abroad. It was no difficult matter to distinguish the cells thus made use of from the others, for they had received a yellow tincture from the honey deposited in them; whereas those which had not as yet been employed this way were of a shining white.

There were also 62 of these cells, in which the bees had already begun to lay up their ordinary food or bread called erithace. This substance was of a changeable colour, between a yellow and a purplish red; but perhaps this tinge might be owing to the fumigation: the whiteness of the unemployed wax was in some parts also impaired by the same means; coloured and covered besides with black spots.

In 35 cells I found as many eggs fixed in them at one end, so that including the eggs found in the comb, which had fallen to the ground as already mentioned, there were 45 eggs in all. There were besides in 150 of the cells so many new-hatched worms, but these lay almost insensible and motionless. They were of different sizes. All these worms were surrounded with that kind of food, which the most expert observers of bees think is honey thrown up by the old ones, out of their stomachs. This kind of honey is white, like a solution of gum tragacanth, or starch dissolved in water, and is almost insipid: it shows nothing remarkable on being viewed with the microscope. In the worms themselves I could perceive pulmonary tubes of a silver whiteness running most beautifully on each side through their little transparent bodies.

I examined attentively the wax cemented by way of foundation to the top of the hive, but I could find no difference between that and the other wax of which the cells consist. They appear both to have the same nature and properties. I could not, however, but admire this strong union or fastening; this substance being just spread upon the hive like a crust, and consequently fastened to it by a very small portion of its surface; whereas the rest of the wax hung perpendicularly from this foundation, without any lateral or other support whatsoever, as if a wooden bowl were fixed to a plain ceiling by a small part of its circumference.

This hive contained the rudiments of a great many more such combs of wax, of an oval form, and full of cells on each side: the empty spaces left between the combs, for the bees to pass and repass, did not exceed half an inch in breadth, so that it is plain the comb I found open upon the ground, and in which I reckoned 418 cells, had been torn from its foundation by its own weight, and that of the bees walking upon it. Hence it appears, with what good reason those who keep bees, place sticks crossways in their hives, that the combs may have the more support; and accordingly we observe that in these hives, the bees themselves on each side suspend their combs to these sticks.

Considering the great multitude of bees employed in building the waxen cells, which I have been just examining, there is no great reason to be surprised at their having done so much work that way, though the time they had to do it in was so short, and the weather so unfavourable.

But it is really astonishing to think how a single female could lay so many eggs in the same small interval, and withal deposit every egg in a separate cell, and there firmly fasten it. We must also allow some time for laying the perpendicular foundations. It is, moreover, very surprising how these eggs should so speedily turn to worms, and how those worms should grow so very suddenly to their state of change. But I must now conclude, and I shall do it with the following account of what the hive I have been describing contained.

Swammerdam, J. 1758. The Book of Nature or The History of Insects. *C. G. Seyffert, London; "Some Peculiar Observations Relating to the History of Bees," pp. 232–235.*

History of a mussel bed: Observations on a phase of faunal disequilibrium (1935)

Edouard Fischer-Piette

HISTORICAL INTRODUCTION

At the end of the last century, Fauvel devoted himself to a study of variations in the marine fauna around Saint-Vaast-la-Hougue. He was the first to practice observations of this type in an ample and continuous way, and in spite of the interest in facts obtained and the fact that one hears many naturalists wish this kind of research to be pursued, the course by Fauvel has not been followed since then except in brief and timid attempts.

Since 1929 I have regularly observed variations in the marine fauna of Saint-Malo, by visiting the same stations each year at the same seasons of the year. My study, like that of Fauvel, will apply in principle to all the species; but, here as in Saint-Vaast, the most evident facts, the most easily followed and also the most interesting, relate to variations observed in the abundance of mussels (*Mytilus edulis*). The study of these variations permitted me to recognize the interactions of several species, mussels, purpurids and barnacles, following the disruption of the faunal equilibrium, interactions which have the effect of progressively bringing back the initial state of equilibrium. I present these facts in a special study, detaching them from the whole work related to the variations of all species, for which I continue to collect data each year.

Before presenting the data, we must recall the principal observations made up to now on the variation in mussel beds. It has been recognized for a very long time that the abundance of mussels is liable to vary, and, as Fauvel recalls in 1901 "the periodicities of the mussel beds, in some places, is well recognized by the marine administration." Fauvel has studied these variations and their effects from a biological point of view, in the region of Saint-Vaast-la-Hougue. He has reported that mussels of the bay of Isigny would rapidly colonize the region of St. Vaast from time to time, then disappear under the attacks of their enemies, man and especially starfish (*Astercanthion rubens*). Let us not fear to cite largely his work to demonstrate the complexity of the biological consequences of these variations:

> When the mussel beds extend, the starfish arrive from all surrounding areas and, finding abundant food, grow and multiply with astonishing rapidity. Their voracity added to man's destructive activities does not take long to get the better of the mussels, but then when the prey begins to decline, the starfish, victims of famine, tends itself to disappear or at least diminish, soon leaving a field open to a new invasion of mussels. This is, without a doubt, one of the important causes of the periodicities of mussel beds on our coasts.
>
> The development of mussels on a rocky coast has other consequences. The slime which collects between their shells and byssal threads does not take long to suppress the animals needing clean water, like many sponges, bryozoans, hydrozoans and colonial ascidians, and with them their associates. The invasion of the mussels is a disaster for the biologist. After the death of the mussels, the soft slime disappears little by little and a coarse sand formed from remains of their shells settles between the rocks. This is a new, rather poor habitat, but sometimes containing special species. A certain length of time elapses before the rock surfaces, well cleared, are covered anew with a cloak of algae, sponges, ascidians and bryozoans, and that the coarse sand becomes a prairie of seagrass. Then, one fine day, a new invasion of mussels occurs, and the cycle begins again.
>
> Does it not seem that we have here a kind of cycle, if one may express it thus? We see the development of one species bring about the loss of a certain number of others at the same time as the multiplication of animals which live at its expense and bring about its destruction in turn. In many periodic appearances of certain species should we not discover thus

a cycle of successive associations furnishing us with the explanation of these variations in the fauna of a locality?

This work of Fauvel forms the essence of our biological understanding of the variations of mussel beds. I can see citing next only the work of Joubin (1910).

Joubin notes the weak development of mussel beds in the bay of Saint-Malo. He adds: "one attributes this fact to the great abundance of *Octopus vulgaris* which has been swarming in the bay for several years." Likewise, at Chausey, he remarks (1910) that the mussels "have diminished a lot this year because of the abundance of octopi." One can see that the case cited by Joubin is very different from that observed by Fauvel.

One will see likewise, that the case I have observed is very different from the two we have just recalled. Let us say right away that men, starfish and octopi, have not intervened in the present case to limit the extension of the mussel beds, and that this role has been fulfilled by the mollusc *Purpura lapillus* L.; but, here, as in the case observed by Fauvel, there is a cycle of variations which concerns several species.

These observations concern a long stretch of coast from Bréhat to Barfleur. But observations have not been made with continuity except on the rocks situated in front of Saint-Enogat and Saint-Lunaire, and in particular on the Décollé Point at Saint-Lunaire. This point has been chosen for these observations because its topography and its fauna were, for a long time (since 1925), known to me in all their details, for I had made them a basis of comparison for the population descriptions that appeared in my thesis work (1929). From 1925 to 1929 the state of the fauna of this station appeared stable. In 1929 began the expansion of the mussel beds, which is the object of this work.

We must therefore first recall the state of equilibrium existing until 1929. Then we will describe the observations since 1929.

DESCRIPTION OF DÉCOLLÉ POINT, AND OF ITS FAUNA IN A PERIOD OF EQUILIBRIUM

Décollé Point, situated at Saint-Lunaire, is elongated from south to north. The two flanks, east and west, rather steep, are very similar from a topographical point of view: compact granular gneiss forms the banks, rather regular overall because erosion of this gneiss gives relatively smooth and little tormented surfaces. The rocky crest is separated into three unequal sections by two benches, which I will call proximal bench and distal bench.

From the population point of view, the two flanks, east and west, present certain differences, for the west flank is very battered, and the east flank is partially sheltered. These population differences have been described in my work of 1929: algae less rare on the east than west; *Balanus perforatus* (barnacles of large size) occurring on the west flank and in the benches, but not on the east flank; among the small barnacles, *Balanus balanoides* is very abundant on the two flanks, whereas *Chthamalus stellatus*, very abundant on the west flank at all levels, does not exist on the east flank except above the level of the high tide at dead water.

But the difference that interests us the most concerns *Mytilus edulis*. In this period (1925–1929), *Mytilus edulis* was entirely absent on the east flank (except in an isolated bed close to the beach of Saint-Lunaire). On the west flank, the mussels exhibit a certain abundance and principally from the level of mid-tide to high tide at dead water. These are individuals of a mediocre size (around 40 mm), strongly curved, with a stout shell ornamented with numerous growth lines, which fill the crevices and depressions but do not overflow them. As Joubin has already remarked (1910), the mussels of the region of Saint-Malo are scabby, that is to say covered with small barnacles. The small barnacles are called "galls of the sea" by fishermen. Here these barnacles are principally *Balanus balanoides*, with some *Chthamalus stellatus*. These mussels, thus hidden in the crevices and clothed by barnacles, are not visible from far. I regret not having taken any photograph during the first years of these observations. But photograph number 10, taken in 1933, reproduces well the appearance that this bed of mussels used to present, from 1925 to 1929. Over the course of these five years 1925–1929, this aspect did not change in any fashion. This state existed moreover probably since 1917; indeed, the fisherman of the region all recall that in 1916 the mussels were more abundant than normal, but they have not observed other phases of abundance since that time.

I will recall finally, that at this time the presence of *Purpura lapillus* was noted on the two flanks, east and west, with the notation "abundant."

ACCOUNTS OF THE MUSSELS, PURPURIDS AND BARNACLES DURING THE PERIOD FROM 1925–1929

The barnacles *Balanus balanoides* and *Chthamalus stellatus* clothed the rock with an extremely crowded population, and formed thus the very base of the population, as they do moreover at all the unprotected stations of the region; they clothed equally the larger organisms attached to the rocks, principally limpets and mussels. These small barnacles served as nourishment for the purpurids [*drilling snails*].

To feed on a barnacle, the purpurid spreads the opercular plates with the aid of its proboscis. The proboscis is introduced through the orifice thus opened and consumes the flesh of the barnacle. The purpurids disregard the very small barnacles, and do not attack them until they are 6 months to 1 year old.

Purpura lapillus had no apparent relation with the mussels. Never did I see a mussel drilled by a purpurid, although my attention had been drawn to this point by a work of P. H. Fischer (1922). It was not rare to see the purpurids attached to mussels, but on separating them one saw barnacles interposed: the purpurid fed on balanids sitting on the mussel, and not on the mussel itself.

The state of the animal populations that we have just described, according to all appearance constituted a state of equilibrium, which already existed in 1925 (the beginning of my observations) and continued up to 1929. In 1930, I found that this state of equilibrium had come to an end, due to the settlement of a large quantity of juvenile *Mytilus edulis*. We shall report with some detail the stages of the expansion of the mussel beds and the various facts that accompanied it.

EXTENSION OF THE MUSSEL BEDS FROM 1929

The 16th of March 1930, I noticed at Décollé, that the mussels formed much thicker groups on the west flank than before, overflowing the crags, extending themselves progressively, and succeeding in covering rather extensive portions of the rocks; moreover, the mussels had invaded the benches which traverse the point, and reached thus up to the east flank, without, however, spreading themselves out on this flank itself. The juveniles must have begun to settle in summer 1929.

At the end of summer 1930, a very abundant juvenile set reinforced the settlement of the already conquered surfaces, and carried out new conquests on all fronts.

In 1931, the juveniles having grown, one noticed that the mussels succeeded in covering extensive rock surfaces, being able in certain places, and particularly in the benches, to form carpets of many square meters without interruptions. Moreover, the colonization had spread equally onto the east flank, up to that point completely devoid of mussels, and still constituted there, in April, only a few groups (from 10–30) localized in the crevices; but in August, the appearance of the east flank succeeded at some points in reproducing the appearance which was observed from 1925–1929 on the west flank, and, moreover, the mussels began to appear on the south portion of this east flank.

From 1932 on, the facts become more complex. The progression of the mussels continues on the east flank, particularly at the north extremity of this flank where very extensive populations are formed. But at other points a regression manifests itself that we must now look at, and which is due to the intervention of the purpurids, which begin to drill the mussels. The effect of this regression is first felt at those points of the station which were the first colonized, that is to say on the west flank and in the benches.

Over the course of the study of these facts of regression, we will have to cite certain observations made at other stations on the rocks of the Grand Vidé and the Petit Vidé in front of Saint-Enogat. We must then, at first, mention the fact that these rocks, nearly devoid of mussels from 1925 to 1929, were the site of an invasion of mussels parallel to that produced at Décollé, but which was delayed one year: the juveniles arrived there in the summer of 1930; in 1931 vast territories were covered with mussels; in 1932 this movement continued in considerable proportions; in 1933 one noted again an extension at one point, the north-west coast of Petit Vidé; since then no extension has occurred.

Let us see thus what makes up the action of the purpurids on the mussels; but, to understand it well, it is necessary first to know that the mussels made the barnacles disappear, and then to understand the circumstances in which this disappearance induced the release of the action of the purpurids.

ACTION OF MUSSELS ON THE BARNACLES

Before the arrival of the mussels, the rocks were, as we have said, almost completely covered with small barnacles, *Chthamalus stellatus* and especially *Balanus balanoides*. The mussels settled themselves on this carpet of barnacles. After awhile, under this covering of mussels and in the middle of the tangle of byssal filaments, the barnacles die. Most of them, pulverized by the growth of the mussels, disappear entirely. There remain meanwhile many empty shells, from which the inhabitants had perished probably from asphyxiation or by starvation. Here, on these very battered rocks, it does not seem necessary to consider the action of the slime observed by Fauvel at Saint-Vaast on the other organisms.

One would think that, as compensation for the disappearance of this coat of barnacles covering the surface of the rocks, another coat would form on the surface of the carpet of mussels. Now this is not the case: the newly arrived mussels did not become "scabby," they remained "clean."

Explaining this cleanness moreover is not perfectly easy. Many factors come into play. First, there may be a question of incompatibility between the speed of growth of a barnacle and that of the mollusc upon which it is inserted. This incompatibility does not exist in normal times because the mussels grow slowly, but during the period of expansion of the mussel beds the mussels grew a great deal more rapidly. Second, this rapid growth is at the same time continuous: the growth lines are very little marks. Now it is along the growth striations that the cypris larvae used to fix themselves in normal times: on the very smooth surfaces these larvae are unable to hold. Finally and above all, the principal reason seemed to me to be the following. Over the course of the flourishing of the mussels, groups of them become so thick that the most external mussels do not have much connection with the rock, being driven out by the growth of others, and are peeled off by waves. This phenomenon is the most evident. They carry away with them the newly attached barnacles; now, the attachment of the larval barnacles taking place only twice a year (spring for *Balanus balanoides*, autumn for *Chthamalus stellatus*, after Hatton and Fischer-Piette, 1932), and uniquely on the most external mussels, the young barnacles are rapidly swept away with these mussels; and the mussels left behind remain clean up to the following settlement of cypris larvae.

ACTION OF PURPURIDS ON THE MUSSELS

From the point of view of the relations between the purpurids and the mussels, nothing changed until the autumn of 1931, that is to say during the first two years of the extension of the mussel beds: the two species continued to coexist without a single drilling observed, the purpurids feeding themselves always on balanids without touching the mussels. At the end of 1931 some drill holes were seen, and from this moment on the purpurids are content with themselves to drill the *Mytilus* more and more frequently. This phenomenon took on such amplitude, that the movement of extension of the mussels gave place, we have said, to a brutal regression, but beforehand we must recall of what consists the act of drilling, and relate the observations that we have made on the drilling instinct.

To feed on a mussel, the purpurid applies its foot on the closed shell of the mussel, and places the extreme end of its proboscis in contact with the shell. At this place a perfectly regular excavation is produced, which is deepened little by little and becomes a cylindrical space traversing the entire thickness of the shell, that one could compare to a hole made by a punch, if it were a little contracted at its distal extremity. It seems almost certain that it is the radula of the purpurid which is the essential agent of this boring. Other hypotheses have been put forth, in particular the intervention of an acid secretion. For all these questions I refer to the works of P. H. Fischer (1922) and of Pelseneer (1924). The drill hole once completed, the proboscis enters by this orifice and the purpurid devours the tissues of the mussel. It leaves nothing from there that is for its support: the gonads when they are developed, and the other tissues, are always consumed by it.

THE CIRCUMSTANCES WHERE THE PROCESS OF DRILLING IS INITIATED BY THE PURPURIDS

How was the implementation of the drilling process initiated? Why did it not appear for such a long time after the beginning of the mussel bed extensions? It was easy for me to establish it without ambiguity. I observed, in effect, that, among the first times where the drill holes appeared, the purpurids which drilled the mussels were invariably purpurids entirely surrounded by mussels, and having no possibility of attaining a single barnacle. As long as barnacles remained in the proximity of a purpurid, it was able to feed itself, without attacking the nearby mussels.

One can conceive how the purpurids occupied with feeding on balanids, can be progressively surrounded by mussels whose groups, overflowing the nearby crevices, come into conflict, recovering and destroying the barnacles as it has been said. At some point, the purpurids, having relentlessly devoured the last barnacles in the limited domain that remains at their disposition, see this domain reduced to nothingness and find themselves in contact with the mussels. Their habitual food is entirely lacking because the mussels themselves do not carry any barnacles, contrary to that which exists normally. It is from this moment on that the purpurids attack the mussels.

ERRORS AND CORRECTION OF THE DRILLING INSTINCT

It is rather curious to see thus developing at a given point, a phenomenon so complex as the act of drilling, in animals that had not exhibited this instinct until then, and would have been able besides not to exhibit it all. Because, it is necessary to insist, the purpurids whose generations succeeded one another at Décollé and in all of the region from 1925 did not drill at any moment of their existence. It is necessary to specify that the purpurids appeared to nourish themselves exclusively on barnacles, and that I have never seen them drilling any animal species whatever, before 1931. Moreover, after 1931, they drilled only mussels, and never limpets or other species. The capacity for drilling has therefore been transmitted through generations without being necessary.

Another curious fact is that, when the instinct of drilling begins to appear, it does not present its definitive perfection at first: errors are very frequent at the start, and one witnesses next a progressive correction. Here are some examples of their correction.

During the first months where the drill holes were able to be observed, it was not rare to see purpurids attacking empty and gaping mussels, and effecting there a complete drill hole, even though, according to all appearance, not trace of any flesh remained in the inside of that shell. At the same time it was not rare to find purpurids entered in the interior of the gaping valves and drilling a valve from the inside towards the outside. It is not a question of the purpurids simply taking refuge in these shells: on lifting the purpurid one saw it retract its proboscis and one could observe the presence of the drilling attempt. The observation of these imperfections at the beginning is interesting, for it has often been assumed that the instinct of drilling was quite certain, and that they did not drill except well knowingly (see Pelseneer, 1924).

After the first year (after the end of 1932), such errors appeared no more. I saw nothing except the purpurids drilling, from the exterior to the interior, of living mussels, or at least containing flesh. Later the "instinct" continued to perfect itself: starting in autumn 1933, one could observe that, in the presence of a gaping mussel still containing flesh, the purpurids no longer practiced drilling: they only introduced their proboscis through the gape between the valves and consumed the flesh directly. At the same time, from this moment on, one no longer saw the purpurids drilling very young mussels: they contented themselves with forcing apart the valves, with the aid of their proboscis, as they had done previously for the opercular plates of barnacles. I observed to the same degree this mode of feeding in England, the 12th of May 1933, on the rocks of Pennance Point near Falmouth: in one whole region of shore, the heads of rocks, due to an uncemented crust, were entirely devoid of barnacles and carried nothing except mussels of very small size. On these mussels were numerous *Purpura lapillus* occupied at devouring them after having forced the valves.

We would have therefore before our eyes a sort of apprenticeship of the purpurids in their method of feeding on mussels. Pelseneer (1924, p. 41) speaks of a case where *Natica* (penetrating *Donax* and *Tellina*) took advantage of the experience acquired at the end of a series of frustrations, which "is the criterion, generally accepted, of the existence of intelligence." But the notion of individual education does not suffice in the present case. It could not apply except to the first generations of purpurids which attempted to drill (1931–1932). In the following generations, the individuals knew right away to drill without errors, without passing through the stages of apprenticeship of which we have spoken above (the existence of these following generations makes no doubt: I will speak at length of the intense multiplication of the purpurids starting in 1932). Is it necessary to believe, over the course of the generations of purpurids that followed on the rocks of Décollé since 1931, that the transmission of the drilling instinct by a hereditary means had involved the transmission of acquired perfections? We do not maintain it strongly, for this would be to touch on the very difficult question of the inheritance of acquired characters (or, more precisely, it would concern here re-acquired characters, after having been lost during the numerous years where the generations followed one another without having an opportunity to drill. This loss of characters is no less curious than their re-acquisition).

What other hypotheses can we invoke? On the off chance I will suggest the following of which I am not certain: is it not possible that the instinct may be susceptible to varying in certain years more than others, and that the first drillings became apparent in our region in one year, or, by coincidence, the instinct finds itself particularly variable from the point of view of the choice of shells to drill and of the side, internal or external, to start drilling at. In the following years, simple coincidence again, the variation could carry itself to other things: the instinct not only allows drilling, it also permits the direct introduction of the proboscis between the valves. In other words, the apprenticeship mentioned above will be the simple appearance of one of many possibilities. Once again, I am not certain of this.

MULTIPLICATION OF THE PURPURIDS

Some months after the purpurids had begun to drill, as far back as the spring of 1932, one was able to observe that their number had increased in enormous proportions. I had qualified the purpurids found on the Décollé from 1925 to 1929 as abundant. This abundance was relative, it consisted of seeing on the average many purpurids per square meter. But, from 1932 on, their number increased to the point that one easily counted hundreds and thousands of them per square meter. Must we think, by comparison with the case of *Asterias rubens* cited by Fauvel, that this large number was due to a migration of individuals run up from all sides, attracted by the mussels? Certainly not. As a matter of fact, after 1932 as before, the purpurids, far from being attracted by the mussels, did not attack them except when constrained or forced, when they had no barnacles at their disposition. Moreover, one observed with one's own eyes, starting in 1931, an enormous increase in the amount of egg laying by purpurids. It seems thus certain that the purpurids were increasing in number by a very active multiplication.

One could consider this case as commonplace, because it is almost the rule that predators multiply when their prey have multiplied. It is nevertheless interesting to observe that this diet to which the animal has not had recourse except when constrained and forced, is more profitable to it than the preceding regime to which it remained still faithful up to the limit of possibility. Feeding on barnacles, the purpurids lived an inactive life. Why did they not profit as soon as possible from the new, infinitely more profitable regime that found itself in their immediate reach, given that their species is precisely with a specialized means of attack for the regime in question? There is here, at least in appearance, a singular failure of instinct.

But things go much further than that: we will see soon that following the destruction of the mussels by purpurids, the barnacles reappear on the locations stripped of mussels: now we will state that from this moment on, the purpurids cease for the most part to feed on mussels, and return to their first diet. There would be then a true preference in this species for the diet, apparently, the least profitable for the species. To explain this preference, one could invoke, at least humorously the "Law of Least Effort," the individual having greater ease sucking barnacles than drilling mussels. But the scientific explanation remains to be found.

DESTRUCTION OF THE MUSSELS BY PURPURIDS REGRESSION OF THE MUSSEL BEDS

The purpurids, surrounded by mussels, began their work at the underpinning: they drilled the mussels situated basally against the rock surface, in preference to the mussels situated superficially. When some mussels thus situated in contact with the rock were dead, they washed away in their decline (under cover of the first blows of the sea) all of the mussels situated exterior to them, so that it is by large pieces that the mussels found themselves removed. One understands that under these conditions the mussel bed regressed very rapidly.

The work of the purpurids was manifested principally in the cracks of the rocks, then on the flat parts. After a certain time these regions were stripped, and no mussels subsisted except on protruding parts of the rocks. Thus an inverse distribution was achieved to that which existed at the beginning, where the mussels were contained in the cracks.

The mussels progressively diminishing in number, the purpurids succeed to be in certain beds, as numerous then later more numerous than the mussels. It was not rare, as far back as then, to see many purpurids simultaneously attacking a single mussel. The fact that one purpurid will have already achieved a perforation and commenced to devour the flesh of the mussel, does not hinder the other purpurids from achieving the perforations they have undertaken, and of taking their part of the remaining flesh. A bed having reached this point disappears with the greatest rapidity. But the destruction was not complete on all the beds making up the mussel bed: we will see further on under these circumstances an important number of mussels will escape from the massacre.

DESTRUCTION OF MUSSELS BY AGENTS OTHER THAN THE PURPURIDS

Were the purpurids the sole agent of destruction of the mussels? What was the role of the mussel enemies cited by Fauvel and Joubin: starfish, octopi, and man?

At no time have I seen starfish on the mussel beds studied from Bréhat to Contentin. In the effected region, starfish were rare (except for the inoffensive *Asterina gibbosa*), and remained this way during the extension phase of the mussel beds.

At no time have I seen octopi on the mussel bed, nor on any of the exposed rocks of the region: they keep themselves in the more sheltered regions; at no time can the coastal natives be blamed for the destruction of the mussels in these last years. Finally, I will state precisely that, during the year 1932, where the destruction of the mussels was very active, the octopi were nearly non-existent in the region. I have pointed out before (Bull. St. Servan, vol 10, p 28) that over the course of the year 1932 we were able to get only eight octopi total for the aquarium at Saint-Servan, in spite of frequent fishing with a trawl net and with a seine, and in spite of the promise of recompense which we made there to the sailors and children; this rarity moreover had been observed simultaneously in Bretagne, in Contentin and in Calvados.

These facts would not be understood to express the least doubt about the observations of Fauvel and Joubin. They show only that we have dealt with a different case from those that these authors have observed.

On the intervention of man here is what I can say.

The mussels of the region have never been the objects of regular exploitations, except on the rocky reefs of cape Fréhel and at certain localities in the bay of Saint-Brieue. In the effected region properly speaking, it was only during the periods of abundance (1916; 1931–32), that profitable harvesting was done, and the only points where these harvestings were "profitable" were the rocky reefs situated in front of Paramé, where the mussels are particularly nice. But on the coastal rocks (among those which are studied), the mussels have always been neglected because of their small size and of their poor quality. Even at the time of the greatest abundance people carried out there only insignificant appropriations. The unemployed, the tourists, from time to time collected a basket of mussels; but the effects of these appropriations passed unnoticed. To prove this, I chose a place where people made particularly frequent collections (because of easy access) and where they were the only enemy of the mussels, the purpurids being diminished on this point. I made there, at regular intervals, successive photographs of a single bed: the appearance remained completely the same, which proves that the cleared spaces were rapidly filled up. One sees thus, that the intervention of people was negligible.

But the purpurids were nevertheless not the only agent of destruction of the mussels. Many mussels, as a matter of fact, were destroyed mechanically by the very fact of their abundance: I have already indicated strongly that the mussels situated superficially on a layer of their congeners, were eliminated in a regular fashion. But, moreover, it was common to see the entire bed of mussels detached from a rather large surface under the shock of a wave, because the very thickness of this bed presented an excellent grasp to the wave. It was very easy to observe this fact in regions deficient in purpurids, as well as in the regions provided with purpurids whose work at the underpinnings often only facilitated (in enormous numbers it is true) the action of waves. We must state precisely that this action of the waves was not capable by itself of wiping out a bed, but only of creating some clearings by removing the most heavily developed and most exposed groups. In summary, it acts there only to eliminate excess mussels.

PHASES OF ACTIVITY AND INACTIVITY IN THE PURPURIDS – DESTRUCTION OF PURPURIDS BY MUSSELS

Purpurids are not always active. Under the influence of severe cold, and more again under the influence of heat, they become inactive for considerable lengths of time. Their repose resembles "aestivation": they are hidden in the cracks and depressions of the rocks, or else at the base of patches of mussels, often in very numerous groups. Those that are found on a smooth wall and without refuge, retract and fall to the base of the rocks, where the sea removes them.

On the days that represent the passage from inactive to active periods, the purpurids, active when the sea leaves them, remain that way as long as they are in the shade, and become inactive when the sun strikes them. The inactivity that overtakes the purpurids is not known to overtake the mussels: the attachment of their young, and their growth, continues. The young mussels that attach around an immobile purpurid, succeed to clasp this purpurid in a lacework of byssal filaments, and the purpurid is bound shortly thereafter: when it revives, it is incapable of breaking the byssal filaments. The byssal threads are sufficient to immobilize a purpurid.

The purpurid thus bound, dies of starvation. However, in the case where its aperture is against a mussel, it drills this mussel and devours it. Often it repeats the drilling act many times on the same valve, without benefit. It finishes always by dying.

The number of purpurids so sequestered by the mussels is considerable, one can count as many of them as one wants during the time of a low tide. But, relative to the prodigious number of purpurids existing in the mussel bed, it is unimportant, and one can say that this destruction of purpurids by mussels does not impede in a meaningful way the ravages made on the mussels by purpurids.

CESSATION OF MUSSEL DESTRUCTION BY THE PURPURIDS – THE ROLE OF BARNACLES

The ravages carried out by the purpurids returned to nakedness, in a multitude of places, the rock previously covered by mussels. These cleared surfaces became populated with barnacles when the swarming of the latter comes about: spring for *Balanus balanoides*, autumn for *Chthamalus stellatus*.

The purpurids then cease to drill mussels and return to feeding on barnacles, when the latter have attained a sufficient size (after 6 months to 1 yr. of growth). The remaining mussels even then, are protected from destruction, the mussel bed is thus not totally annihilated.

The barnacles do not attach solely to the rock. They also attach, we have said, to mussels. We have stated how, at the time of the greatest extension of the mussel beds, the superficially situated mussels fell off and were carried away with the barnacles that covered them, leaving uncovered a layer of "clean" mussels. But then, at the time when the mussel bed is in great regression, the thickness of the bed of mussels returns to being slight and the elimination of which we have just spoken does not occur. In such a way the barnacles cover the mussels again in a durable fashion, and these latter merit anew the epithet "scabby." Even then the purpurids finding themselves on the mussels, as well as the purpurids finding themselves on the rock, have barnacles at their disposition, and they generally cease drilling mussels to content themselves with sucking barnacles. I say "generally" for nothing is absolute. For many months one could still see purpurids drilling the mussels even though barnacles of a sufficient size were at their disposition. But, overall, the drillings become more and more rare, and one can say that the purpurids return to their initial diet. Consequently everything passes as if the reappearance of the barnacles had protected the mussels from total destruction.

Is this exactly the explanation of the disdain of the purpurids for the last mussels? One asks oneself if it was not necessary rather to see the proof that the mussels have lost their quality. Indeed, the active reproduction of the mussels (the production of a large quantity of juveniles), and their rapid growth, seems to have been only once. From 1933 on both of them slackened. The slackening of reproduction could only be due to an extrinsic cause, namely the progressive destruction of the mussel bed by purpurids. But the slackening of growth cannot be explained except by a lesser vitality of the mussels (that this vitality was or not under the influence of external conditions), involving possibly a lesser nutritive quality. The mussels would not have in that case as much attraction for the purpurids.

I am not certain that this is the best explanation. Even if the mussels had less quality, the purpurids would truly complete the destruction of them if they had not had another food at their disposition. On the other hand, even at the height of the prosperity of the mussels, there existed numerous mussels of poor appearance, of gibbous shape and with slow growth marked by strong striations, and they were more often drilled than the others; inversely, during the later times, there were moreover numerous mussels with a good appearance, smooth and of large size, and the purpurids abandoned them as well as the mussels of size growth.

I admit thus that it is indeed the presence of barnacles which preserved the rest of the mussel beds from total destruction. Let us see now of what consists the rest.

ACTUAL STATE OF THE BED (JULY 1934) AFTER THE UNFOLDING OF VARIOUS PHENOMENA MENTIONED ABOVE

Presently, we observe the following state, which is variable depending on the stations since the date of mussel appearance has varied depending on the stations.

(1°) On the west side of Décollé, the only site that has had primitively a bed of mussels, and from which started the extension of these mussels, the actual state represents almost exactly the primitive state. We can say that the cycle is finished. The purpurids have all returned to their primitive food diet (barnacles); the mussels, again low in numbers, found in crevices. We have said that the purpurids have first cleaned the crevices, and then attacked the mussels found on projecting site. It is likely that the crevices have been restocked during that time, favored to other regions, on which the newborns would have had difficulty to hang, since the rock would be too smooth because of the absence of Cirripedes. However, I have not watched closely the mechanisms of restocking of these crevices, and scabby again; the barnacles have again colonized the site that was theirs. This state was found already during the summer of 1933, and since then, this aspect has not been modified: the stability has come back.

(2°) In the benches and east side, the destruction goes on in certain places that are particularly well stocked, and lately stocked with mussels; elsewhere, in places where a low number of mussels were established, they have grown without being attacked; it will be interesting to see if the initial state (total absence of mussels on the east side) happens again. For that to occur, there should not be establishment of newborn mussels that would replace dying ones.

(3°) On the rocks of Saint-Enogat (Grand Vidé and Petit Vidé), the initial state has come back, or is being reconstituted, where the purpurids have intervened. However, there are places, on these reefs that have never borne many purpurids. In those places, mussels have developed freely, and still constitute discontinuous carpets. However, they have the tendency to become "scabby," which proves that either their growth is being slowed or there are fewer newborns.

SUMMARY OF THE INTERACTIONS OF THE VARIOUS SPECIES

We observe that the studied species act on one another in a complex manner.

Purpura lapillus on *Balanus balanoides* (and *Chthamalus stellatus*): they feed on them normally (before multiplication of mussels, and after their destruction).

Purpura lapillus on *Mytilus edulis*: they feed on it during great abundance of mussels.

Mytilus edulis on *Balanus balanoides* (and *Chthamalus stellatus*): destroy them by covering them up during expansion of the mussel bed.

Mytilus edulis on *Purpura lapillus*: the mussels confine and destroy a certain number of purpurids during the inactive time of the latter.

Balanus balanoides (and *Chthamalus stellatus*) on *Mytilus edulis*: indirect action: stop the destruction of mussels and prevent them from disappearing completely.

CONCLUSIONS. REFLECTIONS ON FAUNAL EQUILIBRIUM

In the preceding pages, we have mentioned the various facts on fauna variations, as they relate to a mussel bed whose history has been followed for 9 consecutive years.

Now, we should separate these various facts, as far as interpretation is possible, in two distinct categories: those whose cause is not apparent to us, and those that can be explained. The fact whose explanation is not apparent to us is the initial fact from which all others derive. It is the fixation of an abnormal quality of newborn mussels during several consecutive years.

The other facts derive from this one, and we can without simplifying too much, summarize in the following manner, the succession of facts.

(a.) The arrival of a large number of newborn mussels induce death of small barnacles *Balanus balanoides* and *Chthamalus stellatus* on large surfaces where these barnacles are swamped by the encrustation of mussels.

(b.) The *Purpura lapillus*, which were feeding on these barnacles, can no longer feed on them: they change diet and start perforating the mussels. At first, the perforating instinct is variable and produces interesting mistakes. An "apprenticeship" occurs. Everything happens as though the following generations benefit right away from this "apprenticeship."

(c.) The *Purpura lapillus* multiply in large proportions: although they change their diet only when absolutely needed, the new diet is a lot more profitable than the old diet.

(d.) From this multiplication of purpurids, a rapid decline of the bed of mussels, eaten by the purpurids, occurs.

(e.) Surfaces stripped of mussels by the action of purpurids become colonized again with barnacles when the season of larval fixation of the latter occurs.

(f.) As soon as those barnacles are of a sufficient size, purpurids start again to feed on them, thereby coming back to their original diet.

(g.) The mussels, thus left alone by the purpurids, are no longer destroyed.

At the level of fauna equilibrium, this is, in a coarse manner, how we can represent the facts of this station where the mussels are generally limited in numbers: If the mussels get to multiply in an abnormal way, thereby creating a strong disequilibrium, a direct consequence of the desequilibrium (suppression of barnacles) is to trigger an antagonistic factor, the purpurids, which start destroying the mussels by perforating them, whereas normally they would not touch them. This destruction of mussels could lead to total disappearance if they did not trigger (by "liberation" of rock surfaces) another factor that inhibits the action of the first: the coming back of the barnacles that divert the purpurids from their prey. Therefore, mussels survive in a limited number, and the various species that accompanied them have returned back to their respective places and normal interactions.

The final state is equivalent to the initial one.

The initial state was a state of equilibrium that had lasted for several years. It will be interesting to see if the final state also constitutes a state of persistent equilibrium.

LITERATURE CITED

Fauvel, P. Les variations de la faune marine. *Feuille des jeunes Naturalistes*, XXXI, 1901.

Fischer, P. H. Sur les Gasté ropodes perceurs. *Journ. de Conchyliologie*, LXVII, 1922.

Fischer-Piette, E. Recherches de Bionomie et d'Océanographic littorales sur la Rance et le littorales de la Manche. *Ann. Inst. Océanogr.*, V, 1929.

Fischer-Piette, E. Répartition des principales expèces fixées sur les rocher battus des côtes et des îles de la Manche, de Lannion à Fé camp. *Ann Inst. Océanogr.*, XII, 1932.

Fischer-Piette, E. Etudes sur les variations de la faune marine, parues dans le *Bull. du Laboratoire de St-Servan*, fasc. V, 1930; VIII, 1932; XI, 1933.

Hatton, H. et Fischer-Piette, E. Observations et expériences sur le peuplement des côtes rocheuses par les Cirripè des. *Bull. Inst. Océanogr.* n° 592, 1932.

Joubin, L. Etudes sur les gisements de Mollusques comestibles des côtes de France. La baie de Saint-Malo. *Bull. Inst. Océanogr.* n° 172, [1909?] [year not printed in original. *Trans.*]

Joubin, L. Etudes sur les gisements de Mollusques comestibles des côtes de France. La baie de Cancale. *Bull. Inst. Océanogr.* n° 174, 1910.

Pelseneer, P. Comment mangent divers Gastéropodes aquatiques. *Ann. Soc. royale zool. de Belgique*, LV, 1924.

Fischer-Piette, E. 1935. "Histoire d'une mouliere" (History of a mussel bed). "Observations sur une phase de disequilibrium faunique" (Observations on a phase of faunal disequilibrium). Bulletin Biologique de la France et de la Belgique *2:1–25. Figures*

1–4 have not been included. Translated by A. R. Palmer, Department of Biological Sciences, University of Alberta, Edmonton, AB T6G 2E9. Revised February 23, 2007.

On a provisional hypothesis of saltatory evolution (1877)

William H. Dall

It has long been brought forward, as against the Evolutionary Theory, that there were missing links in the chain of development, which could not fairly be charged to the account of deficiencies in the palæontological record. This is the chief weapon of all opponents to the doctrines so generally received by modern naturalists. The number of instances in which the objection is well founded has been much exaggerated, but that there are cases of the kind will not, I think, be denied by any impartial student, though some imprudent partisans of the new faith have rather scoffed at the idea.

Having confidence that evolution when fully understood in all its modes will prove amply sufficient to account for all phases of organization, and realizing that leaps, gaps, saltations, or whatever they may be called, do occur, I have for some years made this branch of the subject a matter of reflection in the hope of arriving at some clue to the mode.

I have had my attention more especially called to the matter in studying a phase of the kind of evolution I have here termed saltatory, which is especially referred to in Cope's paper on the Origin of Genera, where, if I recollect rightly, it finds expression in the paradox that "the same species may belong to two different genera." That is, more explicitly, that species which are abundantly proved to be distinct from each other by generic characteristics may be, so far as their specific characters are concerned, not distinguishable from one another. Such cases are mentioned by Cope in the paper alluded to, and there are other well-known instances of the paradox among birds, Crustacea, and Brachiopoda.

(1.) As an illustration of how the apparent leaps, which nature occasionally exhibits, may still be perfectly in accordance with the view that all change is by minute differences gradually accumulated in response to the environment, I would offer the following example:

In any sloping gutter of a paved street not too cleanly swept, everyone will have noticed during a sudden shower how small particles of earth and other materials will sometimes act as a dam, producing a puddle which, relieved by partial drainage, may for a time appear to remain in status quo. A time comes, however, when the gradually accumulated pressure suddenly sweeps the dam before it for a short distance. Then another similar pool is formed, and so on indefinitely.

(2.) The modern idea of a species may be stated to be a greater or lesser number of similar individual organisms in which for the time being the majority of characters are in a condition of more or less stable equilibrium; and which have the power to transmit these characters to their progeny with a tendency to maintain this equilibrium.

(3.) This tendency may be in some cases sufficiently strong to resist for a considerable period the changes which a gradual modification of the environment may tend to bring about. When the latter has reached a pitch which renders the resistance no longer effectual, it is conceivable that a sudden change may take place in the constitution of the organism, rapidly adapting it once more to its surroundings, upon which the tendency to equilibrium may reassert itself in the minor characteristics, and these may, as it were, crystallize once more in a manner not dissimilar in its results to the form which was recognizable in the earlier generic type.

(4.) If among a certain assemblage of individuals forming a species the tendency to maintain the specific equilibrium is (as it should be, *a priori*) transmitted to individual offspring in different degrees of intensity, a gradual separation may take place between those with the stronger tendency to equilibrium, and those with less.

(5.) Those yielding to the pressure of the environment (let us say in the manner indicated in paragraph 3) must by the law of natural selection become better adapted to it, and with their changed generic structure may be able to persist.

(6.) On the other hand, those with the broader base, so to speak, with an inherited tendency to remain unshaken by the modifications of the environment, may be conceived as being and remaining, through this tendency, less injuriously affected by adverse circumstances and consequently might still endure.

(7.) In short, natural selection in the one case might find its fulcrum in the easy adjustment of characters; and in the other case in the inherited persistency in equilibrium, by which the organism would be rendered more or less indifferent to the injurious elements of the environment as well as to its favorable phases.

(8.) The intermediate individuals, by the hypothesis, would be those least fitted to persist in any case, and therefore would be few in number and rapidly eliminated. Then we should have a parallel series of species in two or even more genera, existing simultaneously.

(9.) The above hypothesis would account for the special class coming under the paradox quoted, and has an important bearing on the interpretation of certain embryological changes. For other forms of Saltatory Evolution attention should be directed to the inherited tendency to equilibrium which is the converse of the inherited tendency to vary, but which has hardly been granted the place in the history of evolution to which its importance entitles it. Mr. Darwin, whom nothing escapes, has apparently recognized it in his testimony to the "remarkably inflexible organization" of the goose. Other writers seem to have been chiefly attracted by the converse of this tendency, as, under the circumstances, is most natural.

It seems as if the preceding reasoning might serve as a key to many puzzling facts in nature, and perhaps deprive the catastrophists of their most serviceable weapon.

Dall, W. H. 1877. "On a provisional hypothesis of saltatory evolution." American Naturalist *11: 135–137.*

On the Natural History of the Aru Islands (1857)

Alfred Russel Wallace

In December 1856, I left Macassar in one of the trading prows which make an annual voyage to these islands. On January 1st, 1857, we arrived at the Ké Islands. Here we remained six days, while the natives, who are clever boat-builders, finished two small vessels our captain purchased for the Aru trade. During this time I made daily excursions in the forests, collecting birds and insects; but the weather was showery, and the coralline-limestone rocks, which everywhere protrude through the thin soil, are weatherworn into such sharp-edged, honeycombed, irregular surfaces, as to make any distant excursions almost impossible. The great

Fruit Pigeon of the Moluccas (*Carpophagaænea*) was abundant, its loud, hoarse cooings constantly resounding through the forest. Crimson Lories of two or three species were also plentiful, but were so wary that we could not obtain any. All other birds were scarce, and I only obtained thirteen species in all, many of which will, however, I think, prove new, viz.:

TABLE 1

Megapodius, sp., same species at Aru	*Tropidorhynchus*, n. sp.?
Carpophaga ænea	*Cinnyris*, D. sp.
Ptilonopus, n. sp.	*Zosterops citrinella*, Müll.
Macropygia, sp., same at Aru	*Rhipidura*, two species
Dicrurus, sp.	*Muscicapidæ aliæ*, two species
	Psittacus (Geoffroyus?), sp.

Among the birds offered for sale, *Eclectus linnæi* and *Psittacodis magnus* were the most abundant. Of Mammalia I saw none, and could only learn that a wild pig and a species of *Cuscus* inhabited the island. The only reptiles I saw were lizards of two or three kinds, one of which, a very long and slender species, with a finely-pointed tail of a most brilliant blue, swarmed everywhere on the low herbage, gliding among leaves and twigs in the most rapid and elegant manner. Of insects I made a nice little collection, the natives bringing me several very fine Coleoptera. A considerable proportion appears to be quite new; those known being a mixture of New Guinea and Molucca species. It would occupy too much space to enter into any details on this extensive class; I shall therefore give only the results of my six days work, as follows:

TABLE 2

Coleoptera	70 species
Lepidoptera	50 species
Diptera	19 species
Hymenoptera	24 species
Hemiptera and others	31 species
	194 species of insects

It was here I first made acquaintance with the Papuan race in their native country, and it was with the greatest interest I studied their physical and moral peculiarities, and noted the very striking differences that exist between them and the Malays, not only in outward features, but in their character and habits.

A day and a half's sail brought us to the trading settlement of Dobbo, situated on a sandy spit running out in a northerly direction from the island of Wamma, which here approaches to within a mile of the great island of Aru. Having obtained the use of one of the palm-thatched sheds here dignified with the name of houses, arranged my boxes and table, and put up a drying-shelf indoors and out, protected by water-insulation from the attacks of ants, I was ready to commence my exploration of the unknown fauna of Aru. I had brought with me two boys, whose sole business was to shoot and skin birds, while I attended entirely to insects, and to the observation and registry of the habits of the birds and animals I met with in my walks in the forest. The first fortnight was very unpropitious, violent gusts of wind and driving rain allowing us to do very little out of doors, and making the drying of the little we obtained a matter of great difficulty. It soon became apparent that in this small island there was a very limited number of birds, and I determined to go as soon as possible to the large island; but that was not an easy matter, and I now found that I should have brought from Macassar three men accustomed to the islands, and who could take me wherever I wanted to go. As it was, I had to get natives, and there was, as usual, all sorts of delay, and then there was an alarm of pirates, and unfortunately it was not a false alarm. A fleet of the celebrated Ilanun pirates, from the island of Maguidanao, had really arrived; they attacked a small vessel not far from Dobbo, which, escaping from them with one man wounded, brought the news. Then came messengers from one of the northern islands, telling how they had been attacked, and many taken prisoners, and the rest of the population had all fled to the mainland. Now for some time there was no more hope of my getting boats and men. Guards were set in Dobbo, and prows were got ready to go after the pirates. A few days more, and the crew of one of our captain's small vessels which had gone trading among the islands, returned stripped of everything. They had got on shore, while the pirates plundered the prow, taking everything, even to the men's boxes and clothes. They reported that the pirates were all at the east side of the islands, where the merchants send their small vessels to buy pearl-shell and tripang, and there was no danger of their returning again to this side, where they had more to fear and less to get. Now, too, I received a letter I had been expecting from the Governor of Amboyna, with orders to the Aru chief to give me assistance; and, after two months residence in Dobbo, I succeeded in getting a boat and two natives, and set off for the great island of Aru.

I visited several localities, and at length, finding a good one near the centre of the island, I stayed there six weeks, and got, on the whole, a very fine collection of birds. Returning to Dobbo, I intended to make another short excursion; but lameness, produced by the constant irritation of insect-bites on my legs, kept me in the house for several weeks, and the east wind became so strong, and the weather so wet and boisterous, as to render traveling by sea in a small boat out of the question. A little later, one of my bird-skinners left me, and the other was laid up with intermit tent fever, so I was compelled to make the best of it, and get what I could in the mall island till the commencement of July, when we returned direct to Macassar.

Having thus given an outline of my journey, I shall proceed to give some account of the ornithology and general natural history of the Aru Islands, and a summary of the collections I have made there. The very first bird likely to attract one's attention at Dobbo is a most beautiful brush-tongued parroquet, closely allied to *Trichoglossus cyanogrammus*. It frequents in flocks the Casuarina trees, which line the beach, and its crimson under wings and orange breast make it a most conspicuous and brilliant object. Its twittering whistle may be heard almost constantly in the vicinity of the trees it frequents. Almost the only other birds which approach the village are a swallow (*Hirundo nigricans*), found also in New Guinea and Australia, and an *Artamus*, probably *A. papuensis*, which perches occasionally on the house-tops, or on dead trees in the neighbourhood. A little black-and-white wagtail flycatcher (*Rhipidura* sp.) may also often be seen among bushes, and on the sea-beach, chirping musically, and waving laterally its expanded tail whenever it alights.

In the forest which everywhere covers the islands, somber and lofty as on the banks of the Amazon, a different set of birds is met with, the two most abundant being both New Guinea species, *Cracticus varius* and *Phonygama viridis*. The former has a loud and very varied note; sometimes a fine musical whistle; at others (principally when alarmed), a harsh, toad-like croak. It is very active, flying about from tree to tree and from bush to bush, seeking after insects, or feeding on small fruits. It is a long time before one can recognize its various cries for those of one and the same bird. The *Phonygama* is a very powerful and active bird; its legs are particularly strong, and it clings suspended to the smaller branches, while devouring the fruits on which alone it appears to feed. Its affinities seem to be with the Paradiseas rather than with the Garrulidæ. Next to these, two species of *Dacelo* are most frequently met with, and their loud monotonous cry, very much resembling the bark of a dog, most frequently heard. A large crow, with a fine sky-blue iris, and hoarse cawing cry, is also not uncommon; and now I have mentioned all the birds, except parrots and pigeons, that are common enough to be at all characteristic of the forest near Dobbo. For noise, however, the Psittacidæ surpass all others, and the Yellow-crested Cockatoo (*Plyctolophus galerita*) is an absolute nuisance. Instead of flying away when alarmed, as other birds do, it circuits round and round from one tree to another, keeping up such a grating, creaking, tympanum-splitting scream, as to oblige one to retire as soon as possible to a distance. Far more agreeable is the low cooing of the pigeons, several fine species of which are not uncommon. *Carpophaga pinon*, is plentiful, and another, which seems to be *C. zoë*, rather scarce; while *C. alba*, L., is common everywhere. Of the smaller and more beautiful species there are also three, *Ptilonopus perlatus*, *P. pulchellus*, and *P. purpuratus*. These birds are all very difficult to obtain in good condition, because their feathers fall so readily; but they are always acceptable, as their flesh (especially that of the smaller species) is perhaps equal in delicacy and flavour to that of any birds whatever.

In one or two excursions which I made to the mainland, immediately opposite Dobbo, I obtained the two beautiful flycatchers, *Arses telescophthalma*, and *A. chrysomela*, as well as some species of *Ptilotis* and other small birds new to me. It was not, however, until I was regularly established in the central forests of the large island that I obtained a true insight into the ornithological fauna of Aru. Then a host of new species burst upon me, revealing the richness of the country, and its intimate connexion with New Guinea. *Paradisea apoda*, *P. regia*, *Microglossus aterrimus*, *Brachyurus macklotti*, *B. novæ-guineæ*, *Tanysiptera*, *Eurystomus gularis*, *Carpophaga*, with several small flycatchers, thrushes and shrikes, and that most magnificent of the swallow-tribe, *Macropteryx mystaceus* were what I now obtained—almost all New Guinea species, or new, and none of them found on the smaller islands. Of the beautiful little "King Bird of Paradise," I obtained several specimens in perfect plumage and excellent condition. It feeds, I believe, entirely on fruit, frequenting lofty trees in the deep forest, where it is very active, flying from branch to branch, shaking its wings, and expanding its beautiful fan-shaped breast-plumes. When quite at rest, or feeding, these plumes are closed and concealed beneath the wing. Of the "Great Bird of Paradise," I have recorded my observations in a separate paper. The Black Cockatoo is a very curious bird, of most disproportionate form and dimensions. Its huge head certainly weighs as much as its whole body. The legs are very long and slender for the tribe, while its wings are large and powerful. Its cry is a shrill whistle, very different from that of most other cockatoos. The bill of the male is larger, and the apex more produced, than in the female; but the crest-plumes are equally long in both. The *Tanysiptera* is a Kinghunter, feeding on insects, worms, etc., which it picks up from the ground in the damp forest. Its coral-red bill is always dirty from this cause, and sometimes so incrusted with mud that the bird seems to have been actually digging for its food. The *Syma torotoro*, also occurs, but much more rarely, and seems to have very similar habits. Two species of *Megapodius* are plentiful, and the immense mounds of earth and leaves formed by them are scattered all over the forest. These mounds are generally from 5 to 8 feet high, and from 15 to 30 feet in di ameter. But the giant of the Aru forests is the Cassowary (*Casuarius galeatus*); it is by no means uncommon, and the young are brought in numbers to Dobbo, where they soon become tame, running about the streets, and picking up all sorts of refuse food. When very young, they are striped with broad lines of rich brown and pale buff. This gradually fades into a dull pale brown, and in the old bird changes to black. They sit down to rest on their tibiæ, and lie down on their breast to sleep; they are very frolicsome, having mock fights, rolling on their backs, and leaping in a most ridiculous manner with all the antics of a kitten. The same species is said to be found in Ceram, and also in the small island of Goram, as well as in New Guinea. The following list shows the number of species in each of the principal tribes and families which I have observed in Aru:

TABLE 3

Grallæ and Natatores	12	One duck, near *Ana radjah*, *Less.*
Gallinaceæ	15	Twelve pidgeons, *Alecthelia urvillei*
Accipitres (Falconidæ)	4	
Psittaci	10	
Paradiseidæ	2	New Guinea species
Cinnyridæ	5	Three New Guinea species
Meliphagidæ	9	Six *Ptilotis*
Sturnidæ	2	Both New Guinea species
Corvidæ	1	
Garrulidæ	1	
Laniidæ	3	Two New Guinea species
Turdidæ	6	
Pittidæ	2	Both New Guinea species
Maluridæ	2	
Oriolidæ	2	
Artamidæ	1	New Guinea species
Muscicapidæ	13	Four or five New Guinea species
Edollidæ	6	
Coraciadæ	1	*Eurystomus*, same at Macassar and Lombock
Hirundinidæ	3	one New Guines, one Australian
Caprimulgidæ	2	*Podargus* and *Caprimulgus*
Alcedinidæ	11	Four or five New Guinea species
Cuculidæ	3	*Centropus* and *Chrysococcyx*
Total species	116	

From this list, and the preceding observations, it will be seen that many Australian genera and some species occur in Aru; while, considering the very small number of species known from New Guinea, and the necessarily very imperfect exploration of Aru in such a short time, the number of identical species is very remarkable. I believe that nearly one-half of the hitherto-described species of passerine birds from New Guinea will be found in my Aru collections, a proportion which we could only expect if all the species of the latter country inhabit also the former. Such an identity occurs, I believe, in no other countries separated by so wide an interval of sea, for the average distance of the coast of Aru from that of New Guinea is at least 150 miles, and the points of nearest approach upwards of 100. Ceylon is nearer to India; Van Diemen's Land is not farther from Australia, nor Sardinia from Italy; yet all these countries present differences more or less marked in their faunas; they possess each their peculiar species, and sometimes even peculiar genera. Almost the only islands possessing a rich fauna, but identical with that of the adjacent continent, are Great Britain and Sicily, and that circumstance is held to prove that they have been once a portion of such continents, and geological evidence shows that the separation had taken place at no distant period. We must, therefore, suppose Aru to have once formed a part of New Guinea, in order to account for its peculiar fauna, and this view is supported by the physical geography of the islands; for, while the fathomless Molucca sea extends to within a few miles of them on the west, the whole space eastward to New Guinea, and southward to Australia, is occupied by a bank of soundings at a uniform depth of about 30 or 40 fathoms. But there is another circumstance still more strongly proving this connexion: the great island of Aru, 80 miles in length from north to south, is traversed by three winding channels of such uniform width and depth, though passing through an irregular, undulating, rocky country, that they seem portions of true rivers, though now occupied by salt water, and open at each end to the entrance of the tides. This phenomenon is unique, and we can account for their formation in no other way than by supposing them to have been once true rivers, having their source in the mountains of New Guinea, and reduced to their present condition by the subsidence of the intervening land.

This view of the origin of the Aru fauna is further confirmed by considering what it is not, as well as what it is; its deficiencies teach as much as what it possesses. There are certain families of birds highly characteristic of the Indian Archipelago in its western and better-known portion. In the Peninsula of Malacca, Sumatra, Java, Borneo and the Philippine Islands, the following families are abundant in species and in individuals.

They are everywhere *common birds*. They are the *Bucerida*, *Picida*, *Bucconida*, *Trogonida*, *Meropida*, and *Eurylaimida*; but not one species of all these families is found in Aru, nor, with two doubtful exceptions, in New Guinea. The whole are also absent from Australia. To complete our view of the subject, it is necessary also to consider the Mammalia, which present peculiarities and deficiencies even yet more striking. Not one species found in the great islands westward inhabits Aru or New Guinea. With the exception only of pigs and bats, not a genus, not a family, not even an order of mammals is found in common. No Quadrumana, no Sciuridæ, no Carnivora, Rodentia, or Ungulata inhabit these depopulated forests. With the two exceptions above mentioned, all the mammalia are *Marsupials*; in the great western islands there is not a single marsupial! A kangaroo inhabits Aru (and several New Guinea), and this, with three or four species of *Cuscus*, two or three little rat-like marsupials, a wild pig and several bats, are all the mammalia I have been able either to obtain or hear of.

It is to the full development of such interesting details that the collector and the systematist contribute so largely. In this point of view the discovery of every new species is important, and their correct description and accurate identification absolutely necessary. The most obscure and minute species are for this purpose of equal value with the largest and most brilliant, and a correct knowledge of the distribution and variations of a beetle or a butterfly as important as those of the eagle or the elephant. It is to the elucidation of these apparent anomalies that the efforts of the philosophic naturalist are directed; and we think, that if this highest branch of our science were more frequently alluded to by writers on natural history, its connexion with geography and geology discussed, and the various interesting problems thence arising explained, the too prevalent idea—that Natural History is at best but an amusement, a trivial and aimless pursuit, a useless accumulating of barren facts—would give place to more correct views of a study, which presents problems as vast, as intricate, and as interesting as any to which the human mind can be directed, whose objects are as infinite as the stars of heaven and infinitely diversified, and whose field of research extends over the whole earth, not only as it now exists, but also during the countless changes it has undergone from the earliest geological epochs.

Let us now examine if the theories of modern naturalists will explain the phenomena of the Aru and New Guinea fauna. We know with a degree of knowledge approaching to certainty, that at a comparatively recent geological period, not one single species of the present organic world was in existence; while all the Vertebrata now existing have had their origin still more recently. How do we account for the places where they came into existence? Why are not the same species found in the same climates all over the world? The general explanation given is, that as the ancient species became extinct, new ones were created in each country or district, adapted to the physical conditions of that district. Sir C. Lyell, who has written more fully, and with more ability, on this subject than most naturalists, adopts this view. He illustrates it by speculating on the vast physical changes that might be effected in North Africa by the upheaval of a chain of mountains in the Sahara. "Then," he says, "the animals and plants of Northern Africa would disappear, and the region would gradually become fitted for the reception of a population of species *perfectly dissimilar in their forms, habits, and organization.*" Now this theory implies, that we shall find a general similarity in the productions of countries, which resemble each other in climate and general aspect, while there shall be a complete dissimilarity between those which are totally opposed in these respects. And if this is the general law which has determined the distribution of the existing organic world, there must be no exceptions, no striking contradictions. Now we have seen how totally the productions of New Guinea differ from those of the Western Islands of the Archipelago, say Borneo, as the type of the rest, and as almost exactly equal in area to New Guinea. This difference, it must be well remarked, is not one of species, but of genera, families, and whole orders. Yet it would be difficult to point out two countries more exactly resembling each other in climate and physical features. In neither is there any marked dry season, rain falling more or less all the year round; both are near the equator, both subject to the east and west monsoons, both everywhere covered with lofty forest; both have a great extent of flat, swampy coast and a mountainous interior; both are rich in Palms and Pandanaceæ. If, on the other hand, we compare Australia with New Guinea, we can scarcely find a stronger contrast than in their physical conditions: the one near the equator, the other near and beyond the tropics; the one enjoying perpetual moisture, the other with alternations of excessive drought; the one a vast ever-verdant forest, the other dry open woods, downs, or deserts. Yet the faunas of the two, though mostly distinct in species, are strikingly similar in character. Every family of birds (except *Menuridæ*) found in Australia also inhabits New Guinea, while all those striking deficiencies of the latter exist equally in the former. But a considerable proportion of the characteristic Australian genera are also found in New Guinea, and, when that country is better known, it is to be supposed that the number will be increased. In the Mammalia it is the same. Marsupials are almost the only quadrupeds in the one as in the other. If kangaroos are especially adapted to the dry plains and open woods of Australia, there must be some other reason for their introduction into the dense damp forests of New Guinea, and we can hardly imagine that the great variety of monkeys, of squirrels, of Insectivora, and of Felidæ, were created in Borneo because the country was adapted to them, and not one single species given to another country exactly similar, and at no great distance. If there is any reason in the hardness of the woods or the scarcity of wood-boring insects, why woodpeckers should be absent from Australia, there is none why they should not swarm in the forests of New Guinea as well as in those of Borneo and Malacca. We can hardly help concluding, therefore, that some other law has regulated the distribution of existing species than the physical conditions of the countries in which they are found, or we should not see countries the most opposite in character with similar productions, while others almost exactly alike as respects climatic and general aspect, yet differ totally in their forms of organic life.

In a former number of this periodical we endeavored to show that the simple law, of every new creation being closely allied to some species already existing in the same country, would explain all these anomalies, if taken in conjunction with the changes of surface and the gradual extinction and introduction of species, which are facts proved by geology. At the period when New Guinea and North Australia were united, it is probable that their physical features and climate were more similar, and that a considerable proportion of the species inhabiting each portion of the country were found over the whole. After the separation took place, we can easily understand how the climate of both might be considerably modified, and this might perhaps lead to the extinction of certain species. During the period that has since elapsed, new species have been gradually introduced into each, but in each closely allied to the pre-existing species, many of which were at first common to the two countries. This process would evidently produce the present condition of the two faunas, in which there are many allied species—few identical. The great well-marked groups absent from the one would necessarily be so from the other also, for however much they might be adapted to the country, the law of close affinity would not allow of their appearance, except by a long succession of steps occupying an immense geological interval. The species, which at the time of separation were found only in one country, would, by the gradual introduction of species allied to them, give rise to groups peculiar to that country. This separation of New Guinea from Australia no doubt took place while Aru yet formed part of the former island. Its separation must have occurred at a very recent period, the number of species common to the two showing that scarcely any extinctions have since taken place, and probably as few introductions of new species.

If we now suppose the Aru Islands to remain undisturbed during a period equal to about one division of the Tertiary epoch of geologists, we have reason to believe that the change of species of Vertebrata will become complete, an entirely new race having gradually been introduced, but all more or less closely allied to those now existing. During the same period a new fauna will also have arisen in New Guinea, and then the two will present the same comparative features that North Australia and New Guinea do now. Let the process of gradual change still go on for another period regulated by the same laws. Some species will then have become extinct in the one country, and unreplaced, while in the other a numerous series of modified species may have been introduced. Then the faunas will come to differ not in species only, but in generic groups. There would be then the resemblance between them that there is between West India Islands and Mexico. During another geological period, let us suppose Aru to be elevated, and become a mountainous country, and extended by alluvial plains, while New Guinea was depressed, reduced in area, and thus many of its species perhaps extinguished. New species might then be more rapidly introduced into the modified and enlarged country; some groups, which had been nearly extinct in the other, might thus become very rich in species, and then we should have an exact counterpart of what we see now in Madagascar, where the families and some of the genera are African, but where there are many extensive groups of species forming peculiar genera, or even families, but still with a general resemblance to African forms. In this manner, it is believed, we may account for the facts of the present distribution of animals, without supposing any changes but what we know have been constantly going on. It is quite unnecessary to suppose that new species have ever been created "perfectly dissimilar in forms, habits, and organization" from those which have preceded them; neither do "centres of creation," which have been advocated by some, appear either necessary or accordant with facts, unless we suppose a "centre" in every island and in every district which possesses a peculiar species.

It is evident that, for the complete elucidation of the present state of the fauna of each island and each country, we require a knowledge of its geological history, its elevations and subsidences, and all the changes it has undergone since it last rose above the ocean. This can very seldom be obtained; but a knowledge of the fauna and its relation to that of the neighbouring countries will often throw great light upon the geology, and enable us to trace out with tolerable certainty its past history. A consideration of the birds of Aru has led us at some length into this subject, both on account of the interest attached to it, and because we are not aware of any attempt to explain in detail how the existing distribution of species has arisen, or strictly to connect it with those changes of surface which all countries have undergone. The Birds and Mammalia only have been used for illustration, because they are much better known than any other groups. The Insects, however, of which I have made a very extensive collection, furnish exactly similar results, and were these, particularly the Coleoptera, well known, they would perhaps be preferable to any group for such an inquiry, from the great number of their genera and species, and the very limited range which many of them attain. In imperfectly explored countries, however, birds are almost always better known than any other group, as a larger proportion of the whole number of species may be obtained in a limited time. I think it probable that I have collected more than half the birds inhabiting Aru, while I do not imagine I have obtained one-fifth part of the Insects. The following is a brief summary of my collections in this class:

TABLE 4

Coleoptera	572 species
Lepidoptera	229 species
Hymenoptera	214 species
Diptera	185 species
Hemiptera and Homoptera	130 species
Othroptera and Neuroptera, etc.	34 species
Making a total of	1364 species

Lest the conchologists should think I have quite neglected their interests, I may mention, that I have collected all the land-shells I could find or procure from the natives. I have only obtained, however, 25 species. Almost all are Helices (20 species), some pretty and some of curious forms, but I am not sufficiently acquainted with shells to say how much novelty they present. It is remarkable that I have not found a single *Bulimus*, which in Celebes was the most abundant group; the few *Cyclostomata* are also small and obscure. Reptiles are scarce. I did not see a snake six times in as many months. There are, however, on the shores many sea snakes, whose bite is very deadly. The natives spear and eat them. Lizards are rather plentiful in species and individuals; they are almost all plant-dwellers, and run on the leaves and twigs with great agility. The coasts swarm with fish in immense variety, and mollusca innumerable. A shell-collector would obtain a fine harvest, but I have been too fully occupied myself to attend to any of these last-mentioned groups; having often found the greatest difficulty in properly drying and securing any bird and insect collections in the rude houses, boats, and sheds I have been compelled to occupy. Damp, mites, ants, rats and dogs, are all enemies which must be guarded against with ever-watchful vigilance, and from all of them I have suffered more or less severely. Bird and animal skins require daily exposure to air and sun for weeks before they are dry enough to pack away. In this time they accumulate to such an extent, that it is a constant puzzle and difficulty to find places to put them in, so as to keep them free from ants, which establish colonies inside the skin, whence they sally out to gnaw the eyelids, the base of the bill and the feet; arsenic they laugh to scorn; and there is absolutely nothing that will keep them away but water-isolation, which again requires space and constant care to keep perfect. When to these are added insect specimens by thousands, requiring still greater care, the mere labour of watching the collections during the time they must remain exposed to the air, to sunshine, and often to artificial heat, is greater than a collector in a temperate climate, and residing in weather-tight roomy houses, can have any conception of. These remarks are merely my apology for not collecting everything, which stay-at-home naturalists often imagine may be as easily done anywhere else as in England.

Wallace, A. R. 1857. "On the natural history of the Aru Islands." Annals and Magazine of Natural History *20: 355–376.*

Evolutionary criteria in Thallophytes: A radical alternative (1968)

Lynn Margulis

Abstract — The classical assumptions, upon which all previous phylogenies for the lower plants (Thallophytes) have been based, are claimed to be erroneous. An alternative view, that the eukaryotic cell arose in the late Precambrian from prokaryotic ancestors by a specific series of symbioses, is referred to here. Mutually consistent phylogenies, one for the prokaryotes, another for the lower eukaryotes, can be constructed on the basis of the symbiotic theory. The resulting prokaryote phylogeny is presented here; it is claimed to be more consistent with cytological data, measured DNA base ratios, and the fossil record than the several classical partial phylogenies for Thallophytes recently published.

Klein and Cronquist have recently assembled data relevant to the possible phylogenetic relationships among the lower organisms (1). It is evident from their presentation of at least 14 different, and often mutually exclusive, "partial phylogenies" (1, figs. 20 and 22, a and b; scheme A, B, and C, p. 26, for example) that these new data do not clarify evolutionary relationships in the group as a whole. Taxonomic schemes should help us to make predictions. When we are told that a giraffe is a mammal, we infer that the female suckles her young. Without knowing anything else about *Acer pseudoplatanus* except that it is dicotyledonous, one can deduce that it photosynthesizes and that it has true leaves, roots and stems, flowers, and many other traits. These concepts, so obvious to the great evolutionists such as Simpson (2), have been often ignored by many new "biochemical evolutionists" (for example (3)) who tend to disregard whole organisms—the objects upon whose populations selection in the natural environment acts.

A new approach to phylogeny of the Thallophytes is obviously needed, and one such approach is suggested here. In Table 1 the principles upon which it is based are compared with the conventional ones of Klein and Cronquist (1). They have been discussed in much greater detail elsewhere (4). On the basis of these alternative principles a single, unified phylogenetic tree for almost all prokaryotic and eukaryotic organisms can be devised. The basic concept of the origin of eukaryotes from prokaryotes by a series of specific symbioses is outlined in Fig. 1. Details of possible derivations of various well-known and presumably natural prokaryotic groups are shown in Fig. 2. The scheme illustrating evolution of the various eukaryotic lines has already been published. For the details of the right side of Fig. 1, see fig. 1, p. 228 of (4). No attempt has been made here to use any but common names. Although, no doubt, there are errors in the details of the scheme, there is no datum known to the author that contradicts the idea. This is true of both the geological record (5) and modern biochemical data. For example, see (6) for relationships between plastids and blue-green algae; see (7) for relationships between bacteria and blue-green algae, and (8) for a possible phylogenetic status of the mitochondrion. The fact that the DNA base ratios (6, 9) can be easily superimposed on the chart (Fig. 2) lends credence to the idea that the concept is correct.

TABLE 1. Evolutionary criteria in Thallophytes.

ASSUMPTIONS OF KLEIN AND CRONQUIST (*1*)	ALTERNATIVE ASSUMPTIONS (*4*)
1. The basic dichotomy between organisms of the present-day world is between Animals and Plants.	The basic dichotomy between organisms of the present-day world is between Prokaryotes and Eukaryotes.
2. Photosynthetic eukaryotes (higher plants) evolved from photosynthetic prokaryotes (blue-green algae, "uralgae").	Photosynthetic eukaryotes (higher algae, green plants) and non-photosynthetic eukaryotes (animals, fungi, protozoans) evolved from a common nonphotosynthetic (amoebo-flagellate) ancestor. There is not now, nor was there ever, an "uralga."
3. The evolution of plants and the photosynthetic pathways occurred monophyletically on the ancient earth.	The evolution of photosynthesis occurred on the ancient earth in bacteria and blue-green algae; higher plants evolved abruptly from prokaryotes when the heterotrophic ancestor (2 above) acquired plastids by symbiosis.
4. Animals and fungi evovled from plants by loss of plastids.	Animals and most eukaryotic fungi evolved directly from protozoans.
5. Mitochondria differentiated in the primitive plant ancestor.	Mitochondria were present in the primitive eukaryote ancestor when plastids were first acquired by symbiosis.
6. The primitive plant differentiated the complex flagellum, the mitotic system, and all of the other eukaryote organelles.	Mitosis evolved in heterotrophic eukaryotic protozoans by differentiation of the complex flagellar system.
7. All organisms evolved from a primitive ancestor monophyletically by single steps.	All prokaryotes evolved from a primitive ancestor by single mutational steps; all eukaryotes evolved from a primitive eukaryote ancestor by single mutational steps. Eukaryotes evolved from prokaryotes by a specific series of symbioses.
8. Morphological, biochemical, and physiological characters are useful in classification of the Thallophytes.	Only total gene-based biochemical pathways resulting in the productin of some selectively advantageous markers are reliable "characters" in classification; morphology is useless in prokaryotes (Fig. 2).
Result of foregoing assumptions	
Nothing predicted; no consistent phylogeny possible, many predicted organisms not found, for example "uralgae"; no correlation with fossil record possible; no presentation of phylogeny as a function of time elapsed is possible.	Major biochemical pathways predicted; consistent phylogeny constructed; biological discontinuity at Precambrian boundary predicted.

If the genus *Cyanidium* (1, p. 219) had in fact neither mitochondria nor endoplasmic reticulum, it might have represented an inexplicable "uralgan" contradiction, for the theory (4) predicts that no plastid-containing organisms without mitochondria ever evolved. However, recent electron micrographs (10) shows that *Cyanidium* is in these respects a typical eukaryote. That some eukaryotic algae may have lost their originally symbiotic plastids, and later reestablished new symbioses with somewhat

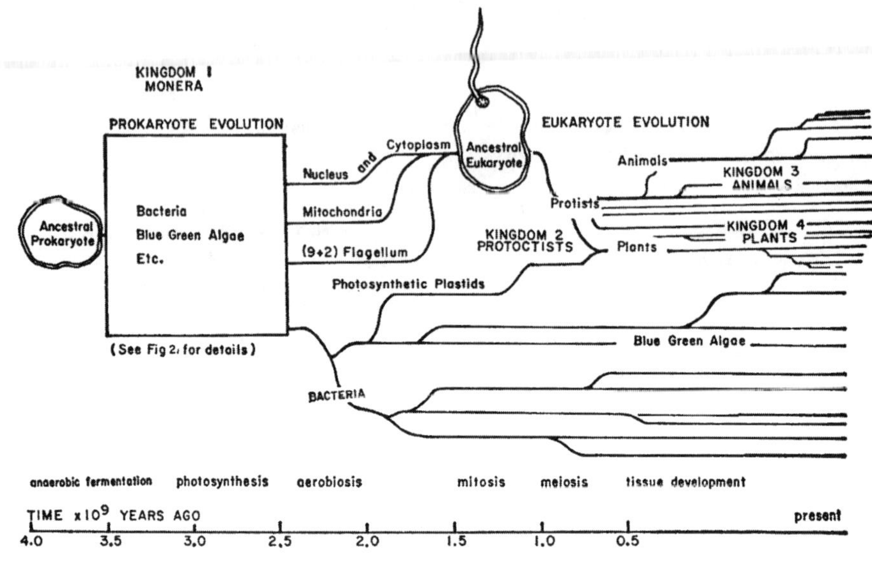
KINGDOM I
MONERA
PROKARYOTE EVOLUTION
Ancestral Prokaryote
Bacteria
Blue Green Algae
Etc.
(See Fig 2. for details)
Nucleus and Cytoplasm
Mitochondria
(9+2) Flagellum
Photosynthetic Plastids
Ancestral Eukaryote
EUKARYOTE EVOLUTION
Animals
KINGDOM 3
ANIMALS
Protists
KINGDOM 4
PLANTS
KINGDOM 2
PROTOCTISTS
Plants
Blue Green Algae
BACTERIA
anaerobic fermentation
photosynthesis
aerobiosis
mitosis
meiosis
tissue development
TIME x10⁹ YEARS AGO
present
4.0
3.5
3.0
2.5
2.0
1.5
1.0
0.5

FIG. I.

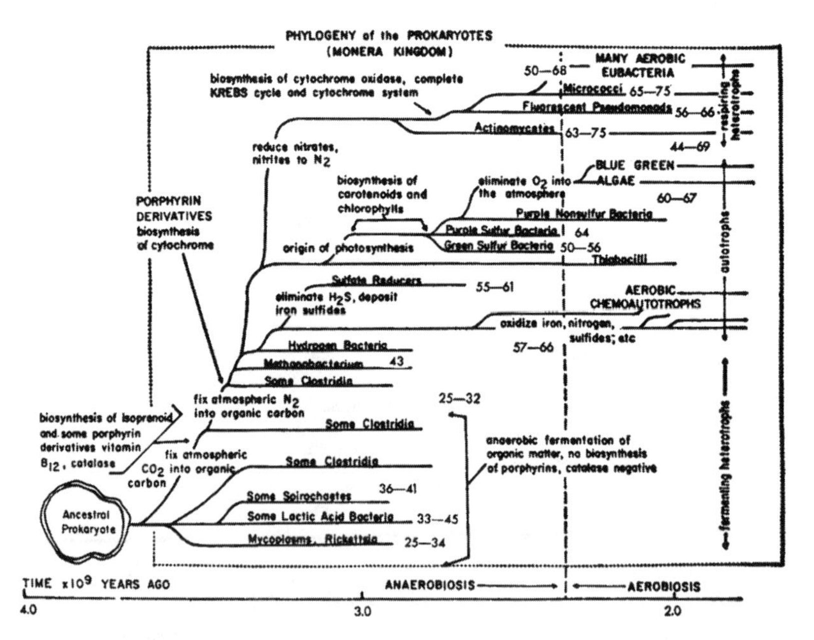
PHYLOGENY of the PROKARYOTES
(MONERA KINGDOM)
MANY AEROBIC EUBACTERIA
50—68
biosynthesis of cytochrome oxidase, complete KREBS cycle and cytochrome system
Micrococci 65—75
Fluorescent Pseudomonads 56—66
Actinomycetes 63—75
44—69
respiring heterotrophs
reduce nitrates, nitrites to N2
BLUE GREEN ALGAE
biosynthesis of carotenoids and chlorophylls
eliminate O2 into the atmosphere
60—67
PORPHYRIN DERIVATIVES biosynthesis of cytochrome
Purple Nonsulfur Bacteria
Purple Sulfur Bacteria 64
Green Sulfur Bacteria 50—56
origin of photosynthesis
Thiobacilli
autotrophs
Sulfate Reducers
55—61
eliminate H2S, deposit iron sulfides
AEROBIC CHEMOAUTOTROPHS
oxidize iron, nitrogen, sulfides; etc
57—66
Hydrogen Bacteria
Methanobacterium 43
Some Clostridia
fix atmospheric N2 into organic carbon
25—32
biosynthesis of isoprenoid and some porphyrin derivatives vitamin B12, catalase
Some Clostridia
fix atmospheric CO2 into organic carbon
Some Clostridia
anaerobic fermentation of organic matter, no biosynthesis of porphyrins, catalase negative
Some Spirochaetes 36—41
Some Lactic Acid Bacteria 33—45
Mycoplasms, Rickettsia 25—34
Ancestral Prokaryote
fermenting heterotrophs
TIME x10⁹ YEARS AGO
ANAEROBIOSIS
AEROBIOSIS
4.0
3.0
2.0

FIG. 2.

different forms of bluegreen algae, is indeed to be expected. Such anomalous symbioses could include not only *Cyanidium* but also *Cyanophora paradoxa* and *Glaucocystis nostochinearum* (11).

The amassed data fit the proposed phylogeny as well as they do any of the several schemes presented by Klein and Cronquist (1). Furthermore, the symbiotic theory enables one to make many predictions. For example, that the usual pathway of the Krebs cycle and cytochrome electron transport oxidations will be found lacking in detail in prokaryotic chemo- and photoautotrophs, although present in all eukaryotic photoautotrophs; that all cells containing chloroplasts also have membrane-bound nuclei and can produce steroid derivatives; that all cells with the complex (9 + 2) flagellum of eukaryotic cells (12) must also contain cytochrome oxidase and other mitochondrial enzymes (13); that all cells with the "higher chromosomes" seen in classical mitosis (for example, red algae and ascomycetes)

TABLE 2. The Four Kingdom Classification modified after Copeland (*16*).

Kingdom	EXAMPLES OF ORGANISMS	APPROXIMATE TIME OF EVOLUTION (MILLIONS OF YEARS AGO)	MAJOR TRAITS THAT ENVIRONMENTAL SELECTION PRESSURES ACTED ON TO PRODUCE	MOST SIGNIFICANT SELECTIVE FACTOR
Monera	All prokaryotes; bacteria, blue-green algae, actinomycetous fungi, and so forth	Early-Middle Precambrian (3000–1000)	Photosynthesis and aerobiosis	Solar radiation, increasing atmospheric oxygen concentrations
Protoctista	All "higher" (eukaryotic) algae: green, yellow-green, red and brown, and so forth; all protozoans, phycomycetous fungi, ascomycetes, basidiomycetes and so forth	Late Precambrian Early Paleozoic (1500–500)	Classical mitosis and meiosis: obligate recombination each gerneration; more efficient nutrition	Depletion of organic nutrients
Animalia	Metazoa: all animals developing from zygotes	Paleozoic (600 on)	Tissue development for heterotrophic specializations	Transitions from aquatic to terrestrial and aerial environments
Plantae	Metaphyta: all green plants (above green algae)	Paleozoic (600 on)	Tissue development for autotrophic specializations	Transitions from aquatic to terrestrial environments

had a (9 + 2) flagellated ancestor and retain the relevant DNA of that "protoflagellum" (4) even if they lack visible (9 + 0) centrioles and basal bodies (14); that all eukaryotes potentially form the colchicine-sensitive protein of the microtubules (15); that all eukaryotic plant cells contain at least three different nonnuclear ("satellite") DNA's; and that steroid and flavonoid derivatives will be found only in relatively young sediments-much younger than those which first contain photosynthetically reduced carbon. By challenging the students of the enormously diverse Thallophytes to find contradictions to the theory proposed here, perhaps some appropriately focused research will be stimulated. If it proves generally acceptable, the division of living organisms into four kingdoms proposed by Copeland (16) logically follows (Fig. 1 and Table 2).

REFERENCES AND NOTES

1. R. Klein and A. Cronquist, Quart. Rev. Biol. 42, 105 (1967).
2. G. G. Simpson, Major Features of Evolution (Columbia Univ. Press, New York, 1953).
3. S. Aaronson and S. H. Hutner, Quart. Rev. Biol. 41, 13 (1966).
4. L. Sagan, Theoret. Biol. 14, 225 (1967). This paper was written by the present author under a former name.
5. J. W. Schopf, in McGraw-Hill Yearbook of Science and Technology (McGraw-Hill, New York, 1967), p. 46.
6. M. Edelman, D. Swinton, S. Schiff, B. Zeldin, H. Spstein, Bact. Rev. 31, 315 (1967).
7. P. Echlin and I. Morris, Biol. Rev. 40, 193 (1965).
8. P. Borst, A. M. Kroon, C. J. C. M. Ruttenberg, Genetic Elements Properties and Function, D. Shugar, Ed. (Academic Press, London, 1967), p. 81.
9. L. R. Hill, J. Gen. Microbiol. 44, 419 (1966); J. Marmur, S. Falkow, M. Mandel, Ann. Rev. Microbiol. 17, 329 (1963).
10. R. Troxler, personal communication.
11. W. T. Hall and G. Claus, J. Phycol. 3, 37 (1967).
12. I. R. Gibbons, in Formation and Fate of Cell Organelles, K. B. Warren, Ed. (Academic Press, New York, 1967).
13. Subsequent dedifferentiation or secondary loss of mitochondria, for example, in yeast and trypanosomes may of course occur (4).
14. E. Stubblefield and B. Brinkley, in Formation and Fate of Cell Organelles, K. B. Warren, Ed. (Academic Press, New York, 1967); D. Mazia, ibid.; A. V. Grimstone, ibid.
15. G. Borisy and E. W. Taylor, J. Cell. Biol. 34, 535 (1967).
16. H. F. Copeland, Classification of the Lower Organisms (Pacific Books, Palo Alto, Calif., 1956).

17. I thank Profs. R. Lewin and R. Estes for advice, and the Boston University Graduate School for support.

Margulis, L. 1968. "Evolutionary criteria in Thallophytes: A radical alternative." Science *161: 1020–1022. Reprinted with permission from AAAS.*

Observations and Experiments upon the Freshwater Polypus (1742)

Abraham Trembley

The animal in question is an aquatic insect, of which mention is made in the *Philosophical Transactions* for the Year 1703. N°. 283. Art. IV. page 1307 and N°. 288. Art. I. pag. 1494. Its body, which is pretty slender, has on its anterior extremity, several horns which serve it instead of legs and arms, and which are yet slenderer than the body. The mouth of the *Polypus* is in that anterior extremity; it opens into the stomach, which takes up the whole length of the body. This whole body forms but one pipe; a sort of a gut, which can be opened at both ends. The length of the body of a *Polypus* varies according to its different species, and according to many other circumstances, to be mentioned hereafter.

I know two species, of which I have seen some individuals extend their bodies to the length of an inch and a half; but this is uncommon. Few are generally found above 9 or 10 lines long; and even these are of the larger kind. The body of the *Polypus* can contract itself, so as not to be above a line, or thereabouts, in length. Both in contracting and extending itself, it can stop at any degree imaginable, between that of the greatest extension, and of the greatest contraction.

The length of the arms of the *Polypus* differs also according to the several species: those of one of the species that I know can be extended to the length of seven inches at least. The number of legs or arms is not always the same in the same species. One seldom sees in a *Polypus*, come to its full growth, fewer than six. The same may be said of the extension, and of the contraction of the arms, which I have said concerning the body.

The body and the arms admit of inflexion in all their parts, and that in all manner of ways. From the different degrees of extension, contraction, and inflexion, which the body and the arms of the *Polypus* admit of, results a great variety of figures, which they can form themselves into.

The insects do not swim; they crawl upon all the bodies they meet with in the water, upon the ground, upon plants, upon pieces of wood, etc. Their most common position is, to fix themselves by their posterior end to something, and so to stretch their body and arms forwards into the water. They make use of their progressive motion, to place themselves conveniently, so as to catch their prey. They are voracious animals: their arms extended into the water, are so many snares, which they set for a number of small insects that are swimming there. As soon as any of them touches one of the arms, it is caught.

The *Polypus* being seized of a prey, conveys it to his mouth, by contracting or bending his arm. If the prey be strong enough to make resistance, he makes use of several arms. A *Polypus* can master a worm twice or thrice as long as himself. He seizes it, he draws it to his mouth, and what is more, swallows it whole. If the worm comes endways to the mouth, he swallows it by that end; if not, he makes it enter double into his stomach, and the skin of the *Polypus* gives way. The size of the stomach extends itself so as to take in a much larger bulk than that of the *Polypus* itself, before it swallowed that worm. The worm is forced to make several windings and folds in the stomach, but does not keep there long alive; the *Polypus* sucks it, and after having drawn from it what serves for his nourishment, he voids the remainder by his mouth, and these are his excrements. According as the weather is more or less hot, the *Polypus* eats more or less, oftener or less often. They grow in proportion to what they eat; they can bear to be whole months without eating, but then they waste in proportion to their fasting.

The observations related in the *Philosophical Transactions*, principally concern the manner in which these insects multiply. What is said there of them, is true and exact. The more one searches into the manner how a *Polypus* comes from the body of its parent, the more evidently is one persuaded, that it is done by a true vegetation. There is not on the body of a *Polypus* any distinguished place, by which they bring forth their young. I have some of them, that have greatly multiplied under my eyes, and of which I might almost say, that they have produced young ones, from all exterior parts of their body. A *Polypus* does not always put forth a single young one at a time; it is a common thing to find those which produce five or six: I have kept some which have put forth nine or ten at the same time, which when one dropped off another came into its place.

These insects seem so many stems from which issue many branches. I have learned by a continual attention to two species of them, that all the individuals of these species produce young ones. I have for two years had under my eye thousands of them; and though I have observed them constantly, and with attention, I never observed any thing like copulation. Upon supposition, that this copulation is performed in some secret manner: I tried at first to be sure it had not taken place between two of them, after they were severed from the body of their parent. To this end,

I took young ones, the moment they came from the parent, which was alone in a glass; or I even parted them with scissors: each of these young ones I put into perfect solitude, I fed them every one separately in a glass; they all multiplied, not only themselves, but also their offsprings, which from generation to generation, as far as the seventh, were all confined to solitude with the same precaution.

Another fact, which I have observed, has proved to me, that they have the faculty of multiplying, before they are severed from their parent. I have seen a *Polypus*, still adhering, bring forth young ones; and those young ones themselves have also brought forth others. Upon supposition, that perhaps there was some copulation between the parent and young ones, whilst they were yet united; or between the young ones coming from the body of the same parent; I made diverse experiments, to be sure of the fact; but not one of those experiments ever led me to anything that could give the idea of a copulation.

The *Polypus* multiplies more or less, as he is more or less fed, and as the weather is more or less warm. If plenty of food, and a sufficient degree of warmth concur, they multiply prodigiously.

I now proceed to the singularities resulting from the operations I have tried upon them.

If the body of a *Polypus* is cut into two parts transversely, each of those parts become a complete *Polypus*. On the very day of the operation, the first part, or anterior end of the *Polypus*, that is, the head, the mouth, and the arms; this part, I say, lengthens itself, it creeps, and eats. The second part, which has no head, gets one: a mouth forms itself, at the anterior end; and shoots forth arms. This reproduction comes about more or less quickly, according as the weather is more or less warm. In summer, I have seen arms begin to sprout out 24 hours after the operation, and the new head perfected in every respect in a few days. Each of those parts, thus become a perfect *Polypus*, performs absolutely all its functions. It creeps, it eats, it grows, and it multiples; and all that, as much as a *Polypus* which never had been cut.

In whatever place the body of a *Polypus* is cut, whether in the middle, or more or less near the head, or the posterior part, the experiment has always the same success. If a *Polypus* is cut transversely, at the same moment, into three or four parts, they all equally become so many complete ones. The animal is too small to be cut at the same time into a great number of parts; I therefore did it successively. I first cut a *Polypus* into four parts, and let them grow; next I cut those quarters again; and at this rate I proceeded, till I had made 50 out of one single one: and here I stopped, for there would have been no end to the experiment. I have now actually by me several parts of the same *Polypus*, cut into pieces above a year ago; since which time, they have produced a great number of young ones.

A *Polypus* may also be cut in two, lengthways. Beginning by the head, one first splits the said head, and afterwards the stomach: the *Polypus* being in the form of a pipe, each half of what is thus cut lengthways forms a half-pipe; the anterior extremity of which is terminated by the half of the head, the half of the mouth, and part of the arms. It is not long before the two edges of those half-pipes close, after the operation: they generally begin at the posterior part, and close up by degrees to the anterior part. Then, each half-pipe becomes a whole-one, complete: A stomach is formed, in which nothing is wanting; and out of each half-mouth a whole-one is also formed. I have seen all of this done in less than an hour; and that the *Polypus*, produced from each of those halves, at the end of that time did not differ from the whole-ones, except that it had fewer arms; but in a few days grew more. I have cut a *Polypus*, lengthways, between seven and eight in the morning; and between two and three in the afternoon, each of the parts has been able to eat a worm as long as itself.

If a *Polypus* is cut lengthways, beginning at the head, and the section is not carried quite through; the result is, a *Polypus* with two bodies, two heads, and one tail. Some of those bodies and heads may again be cut, lengthways, soon after. In this manner I have produced a *Polypus* that had seven bodies, as many heads, and one tail. I afterwards, at once, cut off the seven heads of this new *Hydra*: seven others grew again; and the heads, that were cut off, became each a complete *Polypus*.

I cut a *Polypus*, transversely, into two parts: I put these two parts close to each other again, and they reunited where they had been cut. The *Polypus*, thus reunited, eat the day after it has undergone this operation: it is since grown, and has multiplied. I took the posterior part of one *Polypus*, and the anterior of another, and I have brought them to reunite in the same manner as the foregoing: next day, the *Polypus* that resulted, eat: it has continued well these two months since operation: it is grown, and has put forth young ones, from each of its parts of which it was formed. The two foregoing experiments do not always succeed; it often happens, that the two parts will not join again.

In order to comprehend the experiment I am now going to speak of, one should recollect, that the whole body of a *Polypus* forms only one pipe, a sort of gut, or pouch.

I have been able to turn that pouch, the body of the *Polypus*, inside-outwards; as one may turn a stocking. I have several by me, that have remained turned in this manner; their inside is become their outside, and their outside their inside: they eat, they grow, and they multiply, as if they had never been turned.

Facts like these I have related, to be admitted, require the most convincing proofs. I venture to say, I am able to produce such proofs. They arise from the detail of my experiments, from the precautions I have taken to avoid all uncertainties, from the care I have used to repeat the same experiment several times, from the assiduity and attention with which I have observed them. All this would require a discussion too long to be related here. I might also appeal to the quantity and number of persons who have been witness to these facts; as well of those who have seen me observe, as of those who have observed themselves. For brevity-sake, I have omitted several curious and material facts.

If any persons in England shall be desirous to make observations on the *Polypus*, and to repeat my experiment; I hope I shall be able to send some over, in case they shall not be found there. They are to be looked for in such ditches whose water is stocked full with small insects. Pieces of wood, leaves, aquatic plants, in short, everything is to be taken out of the water, that is met with at the bottom, or on the surface of the water, on the edges, and in the middle of the ditches. What is taken out, must be put into a glass of clear water, and these insects, if there are any, will soon discover themselves; especially if the glass is let stand a little, without moving it; for thus the insects, which contract themselves when they are first taken out, will again extend themselves when they are at rest, and become thereby so much more the remarkable. In order to feed them, one must know how to provide oneself with insects fit for their food. If that is thought necessary, I will point out the means I make use of for that purpose.

I am ready to impart to every one who shall desire to make observations on these animals, all the means and contrivances I have used; to enable them to practice the same, and to judge of them. I shall set forth all these means and contrivances in the history of the *Polypus*, which I am now at work upon. But if, before its publication, any information should be desired, I again repeat, that I shall be ready and willing to furnish them.

Trembley, A. 1742. "Abstract of Part of a Letter from the Honourable William Bentinck, Esq; F. R. S. to Martin Folkes, Esq; Pr. R. S. Communicating the Following Paper from Mons. Trembley, of the Hague." Philosophical Transactions *42 (1683–1775):iii–xi. Reprinted with permission of the Royal Society.*

On the causes of zoning of brown seaweeds on the seashore (1909)

Sarah M. Baker

The conditions which determine the zonal growth of algæ between the tidal limits appear likely to be very simple, and thus afford a very promising field for experimental study. The present paper is a contribution in this direction. The actual zonal distribution of algæ on the shore has been very thoroughly studied by Börgesen, and others. But in order to have a clear understanding of the conditions under which the seaweeds were growing at the particular locality chosen for the present work, a set of measurements was taken to find their vertical distribution in relation to the tides. The experiments were carried out on the eastern side of White Cliff Bay, near Bembridge, Isle of Wight, where there is a continuous stretch of gently inclined limestone and marl rocks upon which the algæ grow profusely, for a distance of about two miles along the coast.

The level of high water at the spring tide was found by marking from a boat the highest point to which the tide rose on a given rock. This point was used as the standard point of reference. At low tide an observer was stationed at this rock, while two others went down and found the seaweeds. The vertical distance below high water was found by sighting across from the reference point to the horizon, and adjusting a slider on a graduated vertical scale into line with this. Sighting to the horizon involves an error of not more than half-an-inch (1.2 cm.) in the more distant readings taken (*i.e.* about 100 yards or metres). By this method the vertical height of any rock could be measured to within an inch or so. Several readings were taken at different points on the rocks for each species of seaweed. The zoning seemed to be similar at all points on the rocks measured. The zones merged into one another, but were on the whole very well defined. The readings were taken as far as possible on gently inclined rocks, and no account was taken of seaweeds growing in rock pools, or even in hollows which might form temporary rock pools, because the seaweeds in the pools were often entirely different from those on the rocks nearby. Thus in a rock pool high up on the shore, in the zone of *Fucus ceranoides*, there were flourishing specimens of *Fucus serratus* and also of *Halidrys siliquosa*, which are usually found below the low water of the neap tides.

The following Table shows the extreme readings, that is those taken on the extreme edges of each zone, and also mean readings taken from the middle of the zones, where each species was at its thickest.

TABLE I

Vertical Distances Below H. W. S. T.						
				FEET		METRES
Spring Tide	...	High Water	...	0	=	0
"	...	Low Water	...	13	=	4.0
Neap Tide	...	High Water	...	3	=	0.9
"	...	Low Water	...	9	=	2.7

	Upper Limit				*Lower Limit*				*Mean Reading*			
Species of Seaweed	FT	INS		MTRS	FT	INS		MTRS	FT	INS		MTRS
Fucus ceranoides	1	9	=	0.5	3	0	=	0.9	2	6	=	0.8
Ascophyllum nodosum	2	3	=	0.7	7	3	=	2.2	3	6	=	1.0
Fucus vesiculosus	4	9	=	1.4	8	6	=	2.6	5	6	=	1.7
Fucus serratus	5	9	=	1.7	12	0	=	3.7	9	0	=	2.7
Halidrys siliquosa	11	6	=	3.5								
Laminarias	12	9	=	3.9								

These readings have been plotted graphically on the diagram (Figure 27). It appears that the time during which the seaweeds were covered by the sea was the primary factor in determining the zoning. The rock-pool vegetation confirms this. The seaweeds growing high on the shore were never, or very rarely, found in pools, whereas those which were only uncovered at the lowest tides were found in a great majority of rock-pools.

There are three phases of the plant's life history which may be influenced by drying, *viz*:

1. Germination of the zygote.
2. Vegetative growth.
3. Reproduction and dispersal of gametes of zygotes.

The first and last of these phases offer some difficulties to experimental study, owing to the small size of the gametes. To see whether it would be possible to grow the seaweeds under different conditions with regard to desiccation from those to which they were accustomed, a considerable number of very small plants were collected off the rocks, each species being taken from the very centre of its zone; these were divided as equally as possible between three jars. The sea-water in the jars was changed once every twelve hours; but during the course of the twelve hours one

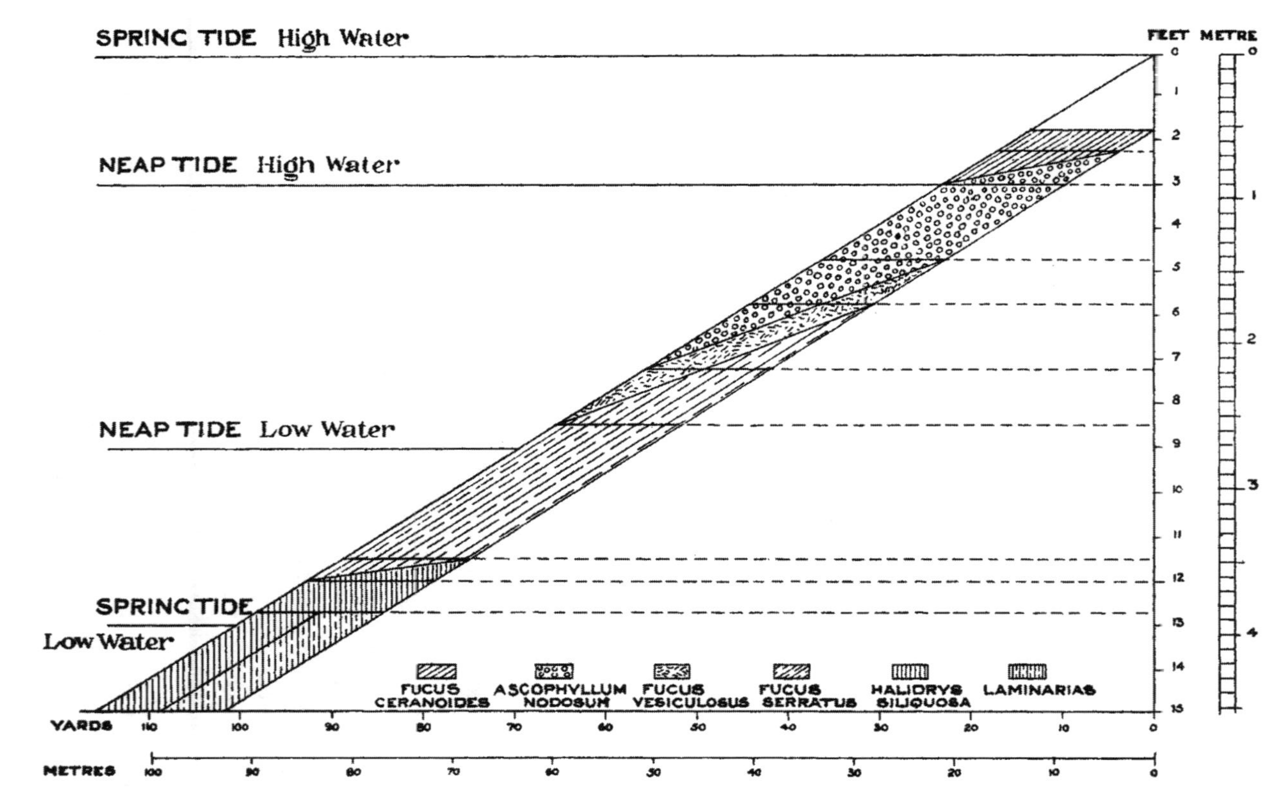

SPRINC TIDE High Water
NEAP TIDE High Water
NEAP TIDE Low Water
SPRINC TIDE
Low Water
FEET
METRE
FUCUS CERANOIDES
ASCOPHYLLUM NODOSUM
FUCUS VESICULOSUS
FUCUS SERRATUS
HALIDRYS SILIQUOSA
LAMINARIAS
YARDS
METRES

FIG. 27.

jar (A) was left dry for one hour and underwater for eleven hours; the second (B) was dry for six hours and underwater for six hours; the third (C) was dry for eleven hours and underwater for one hour. By this means a rough imitation of the periodic exposure by the tides was obtained. The plants were treated in this way for twenty-four days, and even in this comparatively short period they showed marked differences according to the conditions.

The specimens of *Fucus serratus* died in two or three days in jar C. In jar B most of the plants flagged in ten to fifteen days, but one was still alive at the end of the experiment. In jar A they grew very quickly, and after the first six or seven days were far ahead of the other seaweeds in the jar; but after this they flagged and one or two of the specimens finally died.

The specimens of *Fucus vesiculosus* grew rapidly in jar A, and none of the plants died. In jar B this seaweed grew quickly, but some of the specimens became rather shriveled towards the end of the time. In jar C only two specimens just survived, grew very little and shriveled at the base.

Ascophyllum nodosum grew well in jars A and B, and one specimen survived in C, though much stunted.

Fucus ceranoides was very slow growing in all the jars, but it grew best in B, where there was exposure for half the period. None of the specimens died.

From the way in which, the seaweed grew, one would draw the following conclusions:

1. That the species of seaweeds growing high up on the shore have a power of resisting desiccation, which is not possessed by those growing lower down, and that this power decreases regularly in those species growing towards the lower levels.

2. That the seaweeds which can best resist desiccation grow most slowly, and those that grow most quickly are the least tolerant of desiccation.

The species which resist desiccation most must have most protection from loss of water, and hence will absorb water most slowly; and since these plants get their nourishment from general absorption of water by the thallus, the best-protected plants must grow most slowly. In the case of *Fucus ceranoides* there seems to be a definite physiological adaptation against desiccation, for it grows best when it is dry for a large proportion of the time. It is probable that, growing on the rocks, the seaweeds would never become so dry as in the glass jars, which were left exposed to all the weather conditions; but the results obtained give an indication of the reasons for the zoning observed.

In the lower zones the primary factor is probably rate of growth; the quickest growing forms would supersede the others. In the upper zones the determining factor is tolerance of desiccation. It is easy to see that these factors would cause fairly sharp zoning of the forms with these varying properties. There may be also other factors depending on the reproductive systems of the adult plants on the dispersal of the gametes, and on the power of germination of the zygotes under different conditions.

I have to thank my brothers for kind assistance in carrying out this work.

Baker, S. M. 1909. "On the causes of zoning of brown seaweeds on the seashore." New Phytologist *8:196–202. Reprinted with permission of Wiley-Blackwell.*

On the causes of zoning of brown seaweeds on the seashore. II. The effect of periodic exposure on the expulsion of gametes and on the germination of the oospore (1910)

Sarah M. Baker

As was stated in a previous paper on this subject (New Phytologist, Vol. VIII., p. 196, 1909), it seems probable that the most important physical factor governing the zoning of Brown Seaweeds is the periodic exposure by the tides. In order to find out, as far as possible, why some species grow high on the shore and others lower down, it is necessary to work out in detail the effect of exposure on the different species. Some experiments on the effect of periodic dessication on vegetative growth have already been described (*loc. cit.*); the work described in the present paper was carried out with a view to determining the influence of this factor on the germination, and also on the dehiscence mechanism of the four species under observation. Some details are added on the mode of expulsion of the gametes from the conceptacles.

Except where otherwise stated, all these experiments were carried out at White Cliff Bay, Bembridge, Isle of Wight, during July and August; and were made, as far as possible, on the seashore.

The order of zoning observed on the shore was as follows:—(1) *Fucus spiralis.* (2) *Ascophyllum nodosum.* (3) *Fucus vesiculosus.* (4) *Fucus serratus*; beginning from the high water line and going downwards.

I. EFFECT OF PERIODIC DESSICATION ON GERMINATION.

To obtain ripe gametes, receptacles of both kinds were left dry for a few hours, and then put into filtered seawater. The water containing the eggs and sperms was mixed and left standing twelve hours to ensure fertilisation; it was then divided equally between three straight open glass tubes, the lower ends of which were covered with filter paper protected by muslin on each side. The tubes were provided with wire hooks by means of which they could be hung inside glass jars of filtered seawater, and so be filled with water. As soon as the tubes were removed from the water it ran out through the filter paper, leaving the young plants dry; the last few drops of water were expelled by blowing into the tops of the tubes. The plants were left dry, and in water, alternately for definite intervals of time; thus imitating the periodic exposure by the tides.

The results of four weeks' treatment may be summarised as follows:

In water 11 hours, dry 1 hour, in every twelve. The plants of all four species were at about the same stage of development, most of them had a much divided rounded thallus with a well-developed branched pseudopodium, and a few hairs at the top of the thallus. In all the species certain specimens had reached only about the eighth division, and occasionally the thallus was perfectly round, without any development of a pseudopodium.

In water 6 hours, dry 6 hours, in every twelve. The specimens of *Fucus spiralis* and *Fucus vesiculosus* were at nearly the same stage as in the former case; most of them having the much divided round thallus, occasionally with hairs, and attached by a well-developed pseudopodium. There were, however, specimens which had only divided about eight times and even only once or twice. The plants of *Ascophyllum nodosum* were at an earlier stage, the most advanced specimens showing seven divisions; while in the tube containing *Fucus serratus* no plant was found showing more than four divisions, and many had only divided twice. The pseudopodia were well developed in all these.

In water 1 hour, dry 11 hours, in every twelve. The specimens of *Fucus spiralis* had many of them divided three times, though others showed only one division. The most advanced specimen of *Fucus vesiculosus* had three divisions, another had divided once, and several were not divided. The pseudopodia were well-marked in both these species. No germinated specimens were found in the tubes containing *Fucus serratus* and *Ascophyllum nodosum*, though several undivided oospores were found.

It appears from these results that the species occupying the highest zones can germinate, and become attached under all conditions; but those growing in the lower zones will not germinate under the tidal conditions obtaining in the highest zones. It is noteworthy that *Fucus vesiculosus* behaves in this connection, as though it occupied a higher zone than *Ascophyllum nodosum*, which is the reverse of the order found on the shore in this locality.

II. EFFECT OF EXPOSURE TO THE AIR ON THE EXPULSION OF GAMETES FROM THE CONCEPTACLES.

The first essential is to have a means of measuring the number of reproductive bodies ejected from a definite number of conceptacles under various conditions. The method employed was as follows. The specimens were collected as the tide ebbed, whenever possible, and were left in a large rock pool until required. They were then carefully washed, and exposed to the open air, by leaving them lying on flat stones on the shore, for definite periods of time. The dried receptacles were covered with a measured quantity of water, after trimming them so that there were 200 conceptacles in each receptacle. After twelve hours the receptacles were removed and the number of bundles of gametes in the water was counted. The bundles of oospores were sufficiently large to be fairly easily counted with the naked eye, or a small lens, when the water was contained in a white glazed vessel. The number of bundles of antherozoids was estimated by taking out, with a pipette, several drops of approximately the same size (after violently stirring the liquid) and counting the number of bundles in each drop under the microscope; the mean value of three readings has been used in every experiment. The weights of ten consecutive drops of pure water, delivered from the same pipette, were found to vary from 0.0197 gram to 0.0229 gram per drop, the mean value being 0.0213 gram; so the variation in volume of drops is not more than about 16%. The pipette used to measure out the water was a short one with a fine capillary just above the bulb, so that the liquid adjusted itself mechanically to a definite level in the capillary, by means of surface tension. This enabled several small quantities of water to be accurately measured out in a very short time. The volume of the pipette used was found to be 3.348 ccs. which is about 160 times the volume of one drop. This method of counting the numbers was employed in all the experiments referred to in this paper except where otherwise stated.

The methods are open to several errors:

(*a*). Errors in counting; these are extremely easy to make when the numbers involved are large, and can only be avoided by care and system.

(*b*). Errors due to breaking up of the bundles of gametes can be obviated to a certain extent in the case of the oospores, which can be roughly counted; and the number of bundles broken up is then estimated dividing by eight for the *Fuci* and by four for *Ascophyllum*.

(*c*). Errors due to insufficient stirring and variation in size of drops in the measurement of the dehiscence of the male conceptacles.

Besides these a still more important source of error was found to be the variation in ripeness of the individual specimens, however carefully they were chosen. The only possible method of getting over this was to take a number of readings with different specimens for each experiment, and use only the mean value. The number of readings taken was, however, not sufficient to prevent this source of error from being a serious one.

The results of a series of experiments carried out to ascertain the effect of different durations of exposure on the four species is given below in tabular form. The numbers for the oospores indicate the number of bundles ejected from 200 conceptacles. The numbers for the bundles of

TABLE 1. *Fucus serratus* (male)

Hours Dry	4	3	2	1	1/2	0
Experiments	78	21	118	45	29	43
	27	54	13	14	8	12
	21	23	7	–	12	77
	24	–	23	–	10	22
	–	–	–	–	54	–
	–	–	–	–	56	–
Means	37	33	38	29	28	38

Fucus serratus (female)

Hours Dry	8	3	2	1
Experiments	187	67	64	28
	151	140	72	52
	171	240	–	18
Means	169	169	68	33

sperms indicate the mean numbers found in one drop, *i.e.*, 1/160th the number ejected from 200 conceptacles, or approximately the number from one conceptacle. The values under the headings "o" were obtained by leaving the seaweeds in water for ten hours, then transferring them to fresh seawater and leaving them twelve hours, and then counting the number of bundles ejected.

The experimental methods were varied in the case of this alga. Six or seven receptacles were covered with two measures of water, after counting the number of conceptacles, and the results reduced to the values for 200 conceptacles and one measure.

The values for the bundles of eggs and sperms were found from the same specimens in this species. The experiments have been so arranged that a given specimen occupies corresponding positions in both tables. It is to be noted that the specimens giving many oospores give also, as a rule,

TABLE 2. *Fucus vesiculosus* (male)

Hours Dry	9	6	1 1/2	1/2
Experiments	8	8	8	4
	–	–	–	3
Means	8	8	8	4

Fucus vesiculosus (female)

Hours Dry	9	6	3	1 1/2	1/2
Experiments	59	133	64	76	27
	90	97	56	58	33
	230	103	68	39	41
	54	–	124	64	–
Means	108	111	78	59	34

TABLE 3. *Ascophyllum nodosum* (male)

Hours Dry	6	5	4	3	2
Experiments	10	8	4	4	1

Ascophyllum nodosum (female)

Hours Dry	4	3	2	1/2
Experiments	61	59	21	0

TABLE 4. *Fucus spiralis* (male)

Hours Dry	24	14	9	6	3	1	0
Experiments	12	20	18	1	7	3	9
	19	20	11	3	2	4	3
	13	10	11	6	4	2	4
	5	36	14	6	4	1	7
Means	12	23	12	4	4	3	6

Fucus spiralis (female)

Hours Dry	24	14	9	6	3	1	0
Experiments	138	264	207	35	90	80	61
	203	174	158	92	27	64	67
	312	147	89	73	76	74	29
	127	420	323	99	118	33	62
Means	195	251	194	74	78	63	54

many sperms; apparently both kinds of gametes ripen at about the same time. The mean values have been plotted graphically against the time of exposure (Fig. 1). It will be noticed that the curves are all very similar in shape. At a certain point there is a rapid increase in the number of bundles liberated with time of exposure, followed by an approximately constant value varying very little with the time. It looks as though a certain length of time of exposure was required to make the mechanism efficient, and this appears to vary with each species. In the diagram the points at which the mechanism may be said to become efficient have been indicated by arrows for each species; in the case of *Ascophyllum nodosum* the experiments were not carried far enough for this point, if it exists, to be reached. The length of time required for efficiency increases progressively as the seaweeds grow higher and higher on the shore; the order being the same as that of the zoning observed. It is noteworthy that *Ascophyllum nodosum* is the only species in which exposure to the air appears essential to the expulsion of gametes; in all the Fuci considerable dehiscence takes place even when they are not exposed, though the maximum efficiency may not have been reached. The experimental methods are, however, not sufficiently trustworthy to do more than to indicate the probabilities for the intermediate species, and to contrast the behaviour of the two extreme species. It will be remembered that the receptacles of *Fucus serratus* are only very slightly swollen, while those of *Fucus spiralis* and the other two

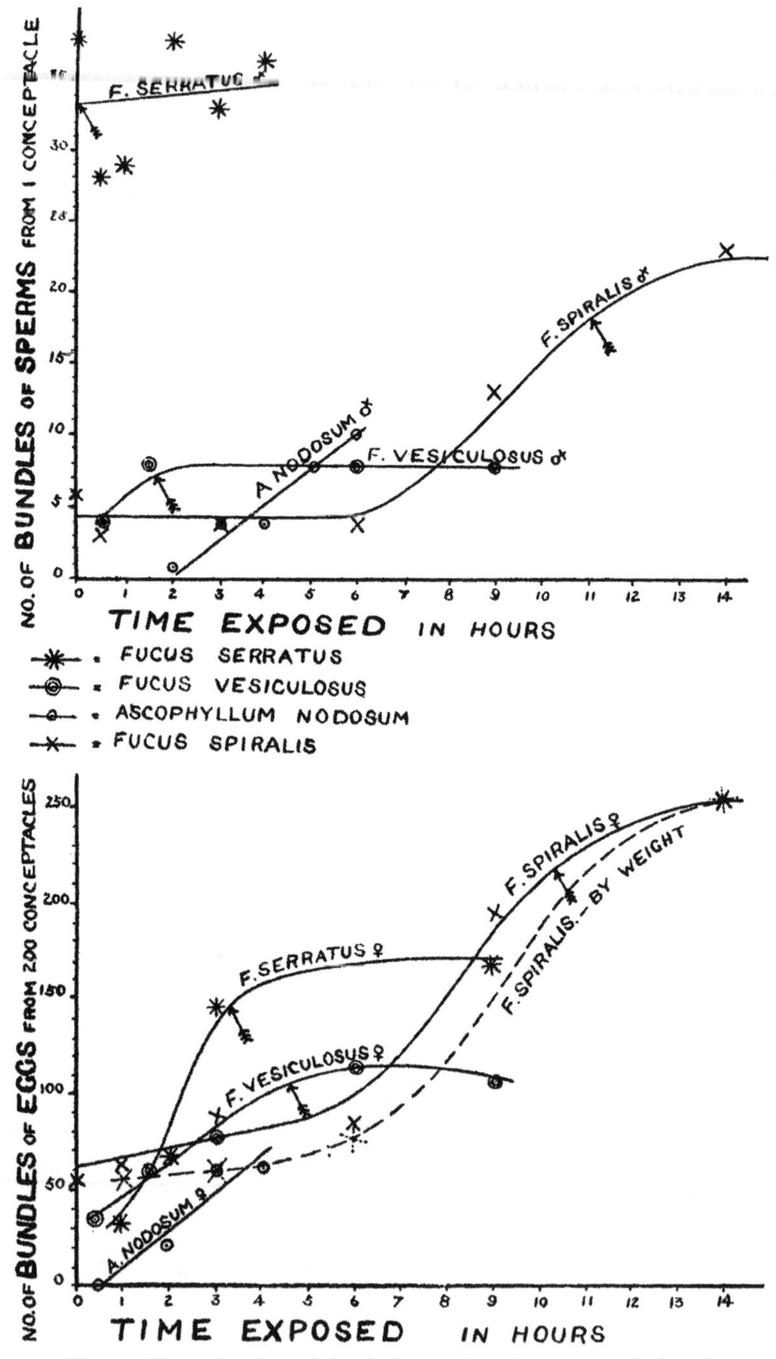

FIG. 1. Diagram illustrating the variation in the number of gametes expelled, with time of exposure. The arrows indicate the points at which the dehiscence mechanism of each species may be considered to have become efficient. The dotted curve represents the proportional values found for *Fucus spiralis* by weighing the gametes.

species are enlarged and filled with mucilage, so that for *Fucus serratus* a short time of exposure would be as effective in drying as a much longer exposure for the other species.

III. OBSERVATIONS CONNECTED WITH THE METHOD OF EXPULSION OF THE BUNDLES OF GAMETES.

Farmer and Williams (Phil. Trans. B. (1898), 190, p. 629–631), Oltmanns (Beiträge zur Kenntniss der Fucaceen. Bibl. Botanica 3, 14,1–19), Thuret (Ann. d. Sc. Nat., Bot. IV., 2, 197–214), and others have shown that the bundles of gametes are extruded by means of the mucilage which collects in drops at the openings of the conceptacles, when they are exposed to the air. It should be noticed that this effect appears to be much more marked when the receptacles are exposed to a very moist atmosphere, than when they are left lying in the sunshine and wind, as was the case in the present experiments. On the shore, as Peierce has pointed out (Torreya, Vol. 2, No.9, p. 134–136) the plants tend to lie over one another and so, keep the air round the receptacles fairly humid; this makes it difficult to imitate the real conditions of the shore.

Farmer (loc. cit.) has also suggested that the long paraphyses occurring in the conceptacles play a part in the dehiscence mechanism. A chance observation led to the suspicion that this effect had, if anything, been underrated. A section was cut through the ostiole of a very ripe male conceptacle of *Ascophyllum nodosum* which had been exposed overnight. After covering the section with seawater and a glass, the ejection of about 20 bundles of spermatozoids through the ostiole was observed during a period of about half-an-hour. It was noticed that each bundle moved, from its point of attachment, between two of the paraphyses; also that its movement was by a series of irregular jumps, and further that beyond the region of paraphyses, which project a little way out of the ostiole, no further motion occurred. It appeared as though the ejection was caused by the swelling up of the adjacent paraphyses on being put into water, the bundles moving in the direction of least resistance away from the point of attachment of the paraphyses. Although several hundreds of sections of receptacles in various stages of dryness have been cut, this phenomenon has not since been observed. One bundle of sperms has been seen moving by jerks in *Fucus vesiculosus*, and two on different occasions in *Fucus serratus*, but not more.

It was thought advisable to try the effect on the dehiscence of removing some of the paraphyses. The hairs were removed by putting the receptacle into a liquid, allowing the liquid to solidify (thus embedding the projecting paraphyses) and then pulling off the solid mass which contained the hairs. *Fucus spiralis* was used for the experiment, since in this species the palaphyses project in huge tufts from the conceptacles. The first liquid used was paraffin wax. The receptacle was dipped into molten wax, then at once put into cold water, and the loosened shell of wax removed with a penknife. Some of the paraphyses which were only dragged partly out of the conceptacles, and were still attached to small pieces of wax, could be easily cutoff with a sharp knife. In order to eliminate, as much as possible, the error due to a different degree of ripeness in the receptacles, the twin receptacles formed by the dichotomy of the thallus were used; one being waxed and the other being similarly treated, only without the waxing. It was found that the waxing was only efficient in removing the projecting paraphyses when the receptacles were comparatively dry; this introduced a difficulty, because the waxing had to be done after exposing the receptacles, and so any gametes which had collected in drops of mucilage outside the ostioles of the conceptacles were removed with the paraphyses. As far as could be judged by rough counting on the dull wax, the number of bundles of eggs thus removed was, as a rule, comparatively small. In addition to this there is the possibility that small pieces of wax may be left in the ostioles and so prevent the expulsion of gametes.

A set of exactly similar experiments was made with *Fucus serratus* (in which the paraphyses do not project from the conceptacles) as a check to the above. If the actual process of waxing has any effect on the seaweeds, it should be shown in these experiments where no paraphyses are removed by it. The results were very variable, though the mean values seem to show that, apart from the removal of the paraphyses, waxing has very little effect on the dehiscence.

The means of sixteen experiments with *Fucus spiralis* (in which waxing removes some of the paraphyses) were:

Male Waxed = 7 bundles liberated. Check = 13 bundles liberated.
Female Waxed = 63 bundles liberated. Check = 166 bundles liberated.

With *Fucus serratus* (in which waxing does not remove the paraphyses) the results were:

Male (means of 10 experiments) waxed = 16 bundles liberated. Check = 13.
Female (means of 4 experiments) waxed = 119 bundles liberated. Check = 115.

The second set of experiments was carried out in London in November and December, on material which was obtained from Mersea Island,

Essex. The liquid used was water, and this was selected for two reasons. First, it is not solid at ordinary temperatures, and so cannot block up the conceptacles; and, secondly, the hairs can be removed before exposing the receptacles. Two twin receptacles were put into two test-tubes; one was washed and covered with distilled water, the other similarly treated with seawater. The test-tubes were put into larger tubes and the tops padded with cotton wool. These air jacketed tubes were put into the same freezing mixture, so that whatever effect the cold had upon the seaweeds should be shared equally by both. The distilled water froze into a solid mass; but the seawater, owing to its higher osmotic pressure, did not freeze so readily; and when it did freeze the ice formed in crystals and not in a solid mass. It was found advisable, however, to melt any crystals when they were formed; as the mere solidification of the water round them is sufficient to remove some of the delicate hairs. The block of ice was broken off the one receptacle, and then both were exposed for a definite time, put into seawater, and the number of bundles of gametes expelled from each counted. The results of a series of ten experiments with *Fucus spiralis* are shown in the Table below.]

TABLE 5

	Male		*Female*			*Male*		*Female*	
No.	ICED	CHECK	ICED	CHECK	*No.*	ICED	CHECK	ICED	CHECK
I.	4	14	102	168	VI.	5	15	34	51
II.	16	22	70	106	VII.	*9*	*9*	*54*	*48*
III.	3	7	55	81	VIII.	8	13	114	299
IV.	4	7	61	89	IX.	7	11	44	63
V.	3	5	15	31	X.	9	14	53	69

The mean values are:

Male Iced = 7 bundles expelled. Check = 12 bundles expelled.
Female Iced = 60 bundles expelled. Check = 100 bundles expelled.

The values *in italics* are the only cases in which the iced receptacle does not show less dehiscence than the check. The mean values do not exhibit so large a difference as that given by the waxing method, though this may be partly due to the fact that they were made with material from a different locality, and at a different time of year. It appears, therefore, that the paraphyses play an important part in the extrusion of the gametes;

TABLE 6

	Male		Female			Male		Female	
No.	F.W.	S.W.	F.W.	S.W.	*No.*	F.W.	S.W.	F.W.	S.W.
I.	7	1	91	35	VII.	16	4	48	29
II.	8	3	171	92	VIII.	13	7	86	62
III.	13	6	109	73	IX.	*3*	*3*	*78*	*80*
IV.	15	6	137	99	X.	*4*	*4*	96	64
V.	16	9	138	61	XI.	3	2	*66*	*74*
VI.	18	3	92	67	XII.	2	1	54	33

the removal of some of them, and the consequent disarrangement, being sufficient to affect the mechanism considerably.

Another series of experiments was made at White Cliff Bay in which the receptacles, dried in the usual way, were allowed to extrude their reproductive bodies into fresh water instead of seawater. If the swelling of the paraphyses plays a part in the mechanism, the smaller osmotic pressure of the fresh water should cause them to swell up more and hence to expel more gametes. As it must often rain when the plants are exposed by the tides, they must be accustomed to the action of fresh water. The values found using *Fucus spiralis* are shown in the Table above. The check values with seawater were done with the twin receptacle to that used for the fresh water, as in the former experiments.

The mean values are:

Male F.W. = 10 bundles expelled. S.W. = 4 bundles expelled.
Female F.W. = 97 bundles expelled. S.W. = 64 bundles expelled.

The values *in italics* are those in which the number in fresh water is, not greater than in seawater. The majority of these experiments were made with only a short time of exposure, for the sake of convenience.

A similar set of experiments with *Fucus serratus* gave mean values:

Male (means of 7 experiments) F.W. = 31 bundles expelled. S.W. = 11
Female (means of 2 experiments) F.W. = 113 bundles expelled. S.W. = 51

Mr. B. J. Salisbury has suggested to me that the concentration of the film of seawater covering the plants when they are exposed by the tides is an important factor in the extrusion of gametes. In view of the function of the paraphyses in this mechanism I think it is extremely probable that osmotic effects of this kind are utilized by the plant. At any rate the mechanism is by no means a simple one, and is probably made up of several factors; among these the swelling of the paraphyses on any lowering of osmotic pressure may, I think, be reckoned.

IV. CHECK EXPERIMENTS BY WEIGHING THE GAMETES.

As the methods employed throughout all these experiments are open to so many objections, it was thought advisable to repeat one series of experiments, using an entirely different method, as a check on the values obtained. These experiments were performed in the Laboratories at University College, London, on account of the apparatus required. The material was obtained from Mersea Island, Essex. The method employed was as follows. Fifty receptacles of *Fucus spiralis* were very carefully washed with filtered seawater, and then dried in an oven for a definite time at 20°C (as these experiments were carried out in November and December, and the others in August, the temperature of 20°C was chosen to imitate summer conditions). They were then put into filtered seawater. After twenty hours the receptacles were removed one by one from the liquid, washing each with distilled water into the vessel. The remaining liquid was filtered through a tared asbestos gooch crucible, with the help of a suction pump. After thorough washing with distilled water, the crucible containing the oospores was dried at 55°–60°C for some hours, cooled in a dessicator and weighed; the process of drying and weighing being repeated until the weight was constant.

The advantages of this method are: (*a*) the use of a large number of receptacles, (*i.e.*, 50 receptacles estimated to represent about 4,200 conceptacles), and (*b*) the definite temperature used for drying the specimens, which it was not possible to obtain for the former experiments. Its disadvantages are: (*a*) the very small weight of the gametes, which makes errors in weighing or the possible entrance of dust of considerable importance, and (*b*) the sliminess of the liquid that has contained the receptacles, which makes the whole process extremely tedious.

The results obtained are shown in the Table below. They have been compared with those found by counting the bundles of oospores (as the weight of the bundles of sperms may be considered negligeable), by working out the proportions between the results found by the two methods; and, on the whole, there is a tolerable agreement between them.

The general order obtained by the two methods is the same. The weighing-waxing experiment was done at the same time, and with the

TABLE 7

Time Dry (in hours)	WEIGHTS IN GRAMS	PROPORTION BY WEIGHT	PROPORTION BY COUNTING
24	0.0121	100	100
17	0.0141	118	129
6	0.0043	36	38
3	0.0034	28	40
1	0.0032	26	32
Waxed	0.0093	65	37
Checked	0.0141	100	100
Iced	–	–	59
Freshwater	0.0042	125	145
Check	0.0034	100	100

same material, as the icing experiments; so it is possibly more comparable with these than with the waxing experiments done by the counting method. The results have been plotted with a dotted line on the same diagrama as the values obtained by counting to show the similarity between the two curves.

V. CONCLUSIONS.

We are now in a position to estimate, to a certain extent, some of the effects of dessication on these four species of algae. The results have all been obtained under more or less artificial conditions; but they may serve to indicate some of the causes of the zoning observed on the shore. These effects may be briefly summarised.

Those seaweeds which grow in the upper zones are capable of resisting dessication both during germination and vegetative growth. Also their receptacles are protected by being filled with mucilage, and this seems to make their dehiscence mechanism most efficient when they are dry for a considerable length of time. Also, in both *Fucus spiralis* and *Ascophyllum nodosum* the paraphyses project considerably from the ostioles of the conceptacles; and this may be an adaptation to ensure the gametes being very rapidly expelled during the short time they are covered by water. In a few experiments with *Fucus spiralis* it was found that during the first hour, after putting a dry receptacle into water, half the number of bundles of gametes were given off that were expelled during twenty-four hours.

The seaweeds growing in the lower zones have become adapted to very rapid growth; they are thus able to supersede the more slow-growing and protected forms, in their own zones, but at the same time they have not the power of resisting dessication, so that they cannot grow in the upper zones. Also their dehiscence mechanism has become efficient for very short times of exposure; and they are able to choke out any stray member of a higher zone, growing lower down, which cannot compete with them in the number of reproductive bodies given off. At the same time in *Fucus serratus*, the lowest form, the receptacles are very little protected.

On the whole it seems as though the greatest competition had been called into play in the lowest zones, the dry and uncongenial regions of the upper shore being left to the most tolerant forms, which, if left to themselves, are able to grow anywhere on the shore.

I wish to express my warmest thanks to Professor F. W. Olivet for his kind interest and encouragement; also to Mr. A. D. Cotton for his advice and assistance, and to Miss M. K. Bohling for assistance in some of the experiments.

Baker, S. M. 1910. "On the causes of zoning of brown seaweeds on the seashore. II. The effect of periodic exposure on the expulsion of gametes and on the germination of the oospore." New Phytologist *9:54–67. Reprinted with permission of Wiley-Blackwell.*

The chalk grasslands of Hampshire-Sussex border: The effects of rabbits (1925)

Alfred G. Tansley and Robert S. Adamson

THE EFFECTS OF RABBITS

It appears that the excessive multiplication of rabbits in certain places on the Downs [*east coast of southern England*] has only taken place within comparatively recent years, and that it is mainly the result of definite preservation or the formation of rabbit warrens for purposes of sport. The occurrence of rabbits in great numbers is in fact quite local, and the animals do not normally travel far in search of food even when it has become exceedingly scarce in the immediate neighbourhood of their burrows. Thus at a distance of 150 to 200 yards from a large aggregation of burrows, round which for some yards the vegetation is completely destroyed, there is little sign of rabbit attack. As they multiply, of course, the animals have to go further for their food, but they never go further than they need.

The incidence of rabbit pressure on the vegetation thus varies from place to place, and also from time to time, as the rabbits are preserved or are systematically trapped. Shooting alone does not effect more than a temporary reduction in their numbers.

The herbage of a good sheep pasture, not too heavily grazed, varies according to soil and water content from about 2 to 4 inches (5 to 10 cm.) in height. Anything like heavy rabbit attack reduces the herbage to a height of ½"–1" (or say 1 to 2.5 cm.), and ruins the pasturage for sheep. Sheep and cattle of course eat off the seedlings of woody plants, but they do not commonly graze right up to bushes, so that in the near neighbourhood of these the herbage grows taller and woody seedlings may establish themselves.

This is apparently often the cause of the frequent alternation on common grazing land of clumps of bushes with closely grazed pasture. If some bushes have established themselves at a time when pasturage is in abeyance or is insignificant, they will continue to grow and increase by the formation of clumps round the original bushes in spite of grazing, though they cannot colonize the area generally. Thus the sharp alternation between two different types of vegetation originates and is maintained. An example of this has been observed adjoining an old neglected plantation on Cissbury Hill in the central Downs of Sussex. In 1885 an area here was covered with rough herbage among which were numerous young plants of *Ulex europaeus* and *Ligustrum vulgare* with *Crataegus monogyna*, *Prunus spinosa* and *Ilex aquifolium*. Thirty-five years later this area bore dense clumps of bushes 2 or 3 metres high, consisting of these species with *Viburnum lantana*, *Sambucus nigra*, *Rubus* spp. and *Lonicera periclymenum*. Between the clumps were tracts of pure, fairly closely grazed grassland. Close round each clump was taller herbage with some woody seedlings. This is a good example of local differentiation through the action of the biotic factor. It is noteworthy that the bulk of the woody vegetation was spinose, and all but one species had fleshy fruits, i.e. were probably bird sown.

But in a heavily rabbit-attacked area the rabbits not only shave the herbage close to the ground but nibble the bushes themselves up to the height they can reach. The edge of scrub or woodland in such an area has a characteristic clean cut appearance with a sharp lower limit to the foliage of the bushes, about 18 inches (45 cm.) above the ground, and bare soil under and immediately round the shrubs, a glance at which at once betrays the presence of rabbits in great numbers.

It does not appear that moderately heavy rabbit attack has much effect on the floristic composition of chalk pasture, though it makes the establishment of woody plants completely impossible. A sample of a rabbit infested area shows much the same list of species as one in which rabbits are scarce or absent. A comparison of the lists outside and inside the Downley enclosures will show the plants which do not exist, or are at least very hard to find on a rabbit attacked area, but which appear within a few years when rabbits (and other grazing) are excluded. The competition of the new comers, however, soon alters the balance of the vegetation, apart altogether from the arrival of woody plants.

On steep northern exposures and other places shaded from direct insolation the existence of heavy rabbit attack does however alter the vegetation very markedly, as already described. Here mosses are generally especially prominent in the herbage, and these the rabbits appear to neglect altogether, while nibbling down the flowering plants. The result is that the mosses, especially the Hylocomia, become dominant, though a variety of the common chalk grasses and herbs maintain themselves, protected by the tufts and carpets of moss, through which their shoots grow. This is well seen in the list from the steep northern escarpment of War Down, and it is also seen on the Downley "A" area close to the edge of the wood, as well as in other places in Sussex on the main escarpment of the Lower Chalk.

There are also a certain number of flowering plants normally refused by rabbits, but none of these appears to be a constant constituent of chalk grassland, though some occur here and there. On the flat or gently sloping summit of Holt Down for instance, where neither soil nor flora is that of a chalk down, there is a large area of rabbit warren, and in this most of the herbage is very severely eaten down. The following species were found to be untouched: *Senecio jacobaea*, *Glechoma hederacea*, *Teucrium scorodonia* (these three were the commonest untouched plants); *Urtica dioica*, *Verbascum thapsus*, *Erythraea centaurium*, *Cynoglossum officinale*, *Cirsium arvense*, *C. lanceolatum*. Of the chalk grassland "constants" *Thymus serpyllum* seems to be relatively immune, and probably *Cirsium acaule*.

When rabbit attack is very heavy indeed, however, the flora is impoverished, and, especially on fairly steep slopes, the soil is opened to erosion, the vegetation being completely destroyed. This normally happens immediately around the burrow complexes on a hillside, but more extensive examples were seen on the east slopes of War Down, in the eastern coombes of the Butser Hill *massif*, and on the north side of Windmill Hill (near Chalton). The last case was examined in some detail.

Windmill Hill. Even on the upper part of the north slope of this hill, where the declivity is not more than 12° or 15°, a great deal of soil is thus exposed, more than half the area in many places being completely bared. Moles (*Talpa europea*) add to the destruction by turning up loose soil mixed with lumps of chalk from the subsoil, and the rain washes the earthy constituents away and disintegrates the chalk, so that the surface becomes covered, largely or entirely, with small chalk fragments. The bare white patches thus formed show up on the hillside from a great distance away.

Lower down the hillside, where the slope is steeper (20°–25°), an area 30 m. × 15 m. round a collection of burrows is entirely covered with this loose chalk. Scattered in this were plants of *Urtica dioica*, *Senecio jacobaea* and *Atropa belladonna*, standing up tall and conspicuous from the bare chalk, and nearly all quite untouched by rabbits. A few plants of the Deadly Nightshade were badly attacked, some of the stems being half gnawed through, and a few of the Ragworts were also nibbled. Here and there were a few scattered plants of *Sedum acre*, *Arenaria serpyllifolia*, *Cerastium vulgatum* and *Myosotis arvensis*, all quite untouched. A little to the east of this completely bare area is a steepsided coombe with the degenerate remnant of an old beechwood. Here also, the soil is practically naked over an acre or so of the valley side, though with a more numerous flora of scattered plants—all rabbit attacked except *Urtica dioica* and *Solanum dulcamara*.

The following lists illustrate the degeneration of the vegetation as a result of the rabbit attack. ["R," "A," "F," and "O" mean rare, abundant, frequent, and original.] The first gives the species occurring on a typical area of two square metres near the top of the hill, with a very slight slope to the north, at a distance of perhaps 80 metres from the nearest rabbit burrow complex. The turf is very heavily rabbit-eaten, the herbage not exceeding 2 cm. in height, but still continuous. Only eight species of flowering plants occurred on this small area:

TABLE I

Agrostis tenuis	R	Prunella vulgaris	F
Euphrasia nemorosa	R	Ranunculus bulbosus	O
Festuca ovina	A	Taraxacum erythrospermum	O
Lotus corniculatus	A	Thymus serpyllum	F

In the immediate neighbourhood there were, in addition, *Senecio jacobaea*, *Helianthemum vulgare* (known to be usually refused by rabbits),

Galium verum and *G. erectum* (believed to be relatively unpalatable). This is of course an exceedingly poor flora for chalk grassland even when rabbit eaten. For instance there were 30 species of flowering plants on one square metre of the Downley "B" area in 1914, and 24 on two square metres of the north-east slope of War Down in the same year—both areas considerably rabbit eaten, while on the most severely rabbit-eaten slope of Butser Hill (herbage 1.5–2.5 cm.) there were 22 species of flowering plants on 12 square metres, which would certainly give more than eight species on two square metres, though it shows the nearest approach to the Windmill Hill conditions.

The soil was comparable in depth with that of the "primitive" chalk grassland near Buriton Limeworks but was different in nature. The top 1½ inches (3.75 cm.) was a brown loam, densely matted with roots. Below this was 2 inches (5 cm.) of similar loam mixed with small chalk lumps, resting on partly disintegrated chalk *in situ* with brown soil in the fissures.

Though still a highly calcareous soil, with a distinctly high lime content in the top 1½ inches, this is decidedly more leached and contains more humus than the Buriton Limeworks "*a*" soil, and it is noteworthy that its humus content is much greater in the lower layers. It is probably an older soil, belonging to a later stage in the succession than the Buriton Limeworks "*a*" soil, and we can only suppose that its surface layers have been impoverished in humus by persistent severe rabbit-attack diminishing the supply of humus-forming material. The presence of *Agrostis tenuis* and *Prunella vulgaris*, in spite of the very small number of species, probably mark the difference of age. The diminution in size of the plants and their parts, and the shallowness of the general rooting are seen to be very marked.

Passing now to the semi-bare areas nearer the rabbit burrow complexes, we find a sharp differentiation between the flora of the islands of turf and that of the loose soil bared by the action of the rabbits and moles.

(*1*) *Species occurring in turf still remaining*

(2) *Bryophytes in disintegrated turf on margins of turf "islands."*

(*3*) *Species on bare soil with chalk lumps partially covering loam*

Of these three distinct communities (1) the turf plants show an impoverished list of a fairly advanced chalk pasture, but with some open soil species, and a marked poverty in grasses, (2) the disintegrated margins of the turf islands bear a collection of mosses which include a proportion of calciphilous species, but of which *Polytrichum piliferum* usually at least occurs on non-calcareous soil, while (3) the bare soil shows a scattered collection of plants none of which is a constant constituent of chalk grassland, though a few occasionally occur in it. Most of these plants which had colonised the bare soil were quite untouched by the rabbits. The conspicuous feature was the abundance of *Arenaria serpyllifolia* and *Sedum acre*.

On the steep (east-facing) slope of the adjoining coombe there was no turf, and large tracts of soil were covered with chalk fragments quite bare of plants. The following flowering plants occurred: no bryophytes were met with. Here, in addition to a few chalk pasture species, there are a majority of open soil plants, ten of which occur in the last list, the relative abundance of *Arenaria serpyllifolia* and *Sedum acre* again being noticeable. *Urtica dioica* and *Solanum dulcamara* were the only two species absolutely untouched by rabbits.

The occurrence of *Teucrium scorodonia* is interesting. This plant has rather the reputation of being a calcifuge and doubtless the immense majority of individuals of the species in this country occur in light soils very poor in lime, but here it grows on soil containing an enormous proportion of calcium carbonate. Nor is this an isolated instance. In a similar heavily rabbit-attacked locality on Butser Hill it grows in sheets rooted in practically pure chalk, and it has been observed in the same habitat elsewhere on the South Downs. It is also a characteristic species on dry rocky limestone slopes in the ashwoods of Derbyshire and West Yorks.

CONCLUSIONS

The general conclusions are very much what might have been expected from the more or less vague knowledge we already possessed.

The first significant colonization of bare chalk is by "chomophytes," not "lithophytes." A wide variety of species can colonize chalk talus and particularly finely divided chalk debris, and which do so depends largely on the surrounding vegetation. When chalk grasslands surround the bare area *Festuca ovina* comes at once to the fore, and remains, in this region, the typical dominant or co-dominant of pastured chalk grassland. Mosses play a fairly prominent part from an early stage, but the first colonists are different (largely calcicole) species from the later ones (largely *Brachythecium purum*, and in shaded places species of Hylocomia), which remain constant constituents of mature grassland. In "primitive" grassland, i.e. when the soil is shallow and contains a very high percentage of carbonates, and little humus even in the superficial layer, the variety of species is small and these, both flowering plants and mosses, are of xerophilous type.

With increasing depth of soil and increasing humus and water-content a much greater variety of herbaceous species appears, and the typical richness of the chalk grassland community is developed. *Festuca ovina* remains dominant or co-dominant with *Carex flacca*, or at least very abundant, equaling or exceeding any other species in quantity. Of other grasses *Avena pratensis* is the commonest, with *Trisetum flavescens* and *Avena pubescens* usually less common. *Bromus erectus* is rare and local in this region. *Briza media* is occasional but pretty widely distributed. This condition is maintained so long as the land is moderately grazed. The herbage varies from 2–4 inches (5–10 cm.) in height according to grazing and water content.

Moderately heavy rabbit attack reduces the general height of the herbage to one inch (2.5 cm.) or less, but does not effect any great change in the list of species present, though it injures some species more than others and thus relatively favours the resistant species. But on the whole the species of chalk grassland, being hemicryptophytes, are able to withstand rabbit attack as well as sheep grazing. On steep northern exposures, and where the vegetation is otherwise shaded, mosses, especially the Hylocomia, are very abundant, and here heavy rabbit attack at once gives the moss vegetation dominance, reducing the phanerogamic vegetation to a subordinate position.

Very heavy rabbit attack which stops short of baring the soil does however appear greatly to reduce the number of species. Exactly how this happens has not been determined. To obtain an answer to the question it would be necessary to study the behaviour of an area of grassland rich in species from the time when it was first exposed to the attack.

Except immediately around burrows, where scratching is an additional factor preventing the maintenance of vegetation, it is especially though not exclusively on fairly steep slopes that the ground is extensively bared by rabbits, especially when aided by moles. Loose soil together with chalk lumps is brought up and the soil is washed down the slope by rain, leaving the surface covered, partially or entirely, by loose fragments of chalk. Such areas are in part completely bare of plants, but are tenanted here and there by species which rabbits do not touch or only occasionally bite. The root relations of these are interesting.

In the absence of rabbit pressure and with not too heavy grazing there is a tendency to the invasion of chalk grassland by plants not characteristic of it. This is the result of increasing depth of soil, both by dissolution of chalk below and by the accumulation of humus above, together with the washing out of carbonates from the surface layers. Some of these plants are species which have a distributional mode on acid soils, such for instance as *Potentilla erecta* and *Calluna vulgaris.* Of the grasses there are shallow rooted forms like *Agrostis vulgaris,* which cannot be called characteristic of typical chalk grassland, though it is often present and sometimes abundant, and *Holcus lanatus,* which is very local, though where this grass appears it often becomes locally dominant. More important are the deeper-rooted grasses which increase in, or invade chalk pasture whose soil is deep. The Avenae, which appear when the soil is still shallow, increase notably, often at the expense of *Festuca ovina,* when the soil deepens and grazing is not severe, but in this region *Avena pratensis* rarely approaches dominance, as it does for instance in enclosure A (the one with the most chalky soil) at Downley. On the central and eastern downs, where the plague of rabbits is not nearly so widespread and severe the dominance or co-dominance of *Avena pratensis* is much commoner, though of the taller growing grasses, *Bromus erectus* (very local on the western downs) is the typical dominant in the east. *Trisetum flavescens* is widespread but rarely very abundant and never dominant. *Dactylis glomerata,* while by no means a characteristic chalk grassland species, is a fairly constant member of the herbage except on the shallowest soils. The other meadow grasses play little part in this region, and no areas have been met with such as those which often occur on the central and eastern downs, especially on the north-facing escarpment, where a considerable number of species of tall grasses, those already mentioned, *Festuca rubra, Poa pratensis, Arrhenatherum elatius, Deschampsia caespitosa,* and even *Festuca elatior,* alternate or compete for dominance, accompanied by diminution or disappearance of many of the lower growing species of the typical sheep's fescue chalk pasture. This is doubtless to be correlated with a much more restricted occurrence of rabbits in the central and eastern regions. The western area, part of which is dealt with in this paper, is a region of large estates with extensive woodlands, maintained primarily for the amenities of their owners, and with relatively little agricultural land. The centre and east are regions of farms, where rabbits are not encouraged, and the extensive grasslands are treated as an integral part of the farm system.

Tansley, A. G., and R. S. Adamson. 1925. "Studies of the vegetation of the English chalk: III. The chalk grasslands of Hampshire-Sussex border." Journal of Ecology *13: 177–223. Reprinted with permission of Wiley-Blackwell.*

The biosphere and the noösphere (1945)

Vladimir J. Vernadsky

THE BIOSPHERE

Living matter is the totality of all organisms present on the earth at any one time. It is usually such a totality that is important, though in dealing with the effect of man on the processes of this planet, a single individual may be of importance. The living matter of the earth may be regarded as the sum of the average living matter of all the taxonomically recognizable groups. Each of these groups is said to consist of homogeneous living matter.

Living matter exists only in the biosphere. This includes the whole atmospheric troposphere, the oceans, and a thin layer in the continental regions, extending down three kilometers or more. Man tends to increase the size of the biosphere. The biosphere is distinguished as the domain of life, but also, and more fundamentally, as the region where changes due to incoming radiation can occur.

Within the biosphere, matter is markedly heterogeneous and may be distinguished as inert matter or living matter. The inert matter greatly predominates in mass or volume. There is a continual migration of atoms from the inert matter to living matter and back again. All the objects of study in the biosphere are to be regarded as the *natural bodies* of the biosphere. They may be of varying complexity, inert, living, or bio-inert as in the case of soil or lake water. The study of all phenomena has a unity, leading to the production of a body of systematized knowledge, the *corpus scientiarum*, which tends to grow like a snowball; this *corpus* includes all systematized knowledge, and is contrasted to the results of philosophy, religion, and art where truth may be revealed intuitively; the systematized history of these activities belongs to the corpus.

Two concepts have been inadequately stressed in the past: (a) Pasteur was correct in regarding the preponderance of optically active compounds as the most characteristic general property of living matter and its products; this idea is of immense importance; (b) the functions of living organisms in the energetics of the biosphere have been seriously neglected. Biogeochemical energy may be expressed in the velocity with which the biosphere could be colonized by a given species. For certain bacteria, the limiting velocity of extension of a dividing chain of cells tending to embrace the whole circumference of the earth would tend to approach the velocity of sound.

Bearing in mind these introductory principles, the difference between living matter and inert matter of the biosphere is expressed in a table, here given in condensed form. The differences in this table are not merely differences with regard to energetics and chemical properties. They also involve a fundamental difference in the spacio-temporal manifestations of living and inert matter. It is suggested that the geometry appropriate to the bodies of living organisms may be different from that appropriate to inert bodies.

I. A. Living natural bodies exist only in the biosphere and only as discrete bodies. They have never been observed to arise except from other living bodies. Their entry into the biosphere from cosmic space is hypothetical and has never been proved.

B. Discrete inert forms are concentrated in the biosphere but are also found much deeper in the earth's crust. They are created in the biosphere, but also enter it from below in volcanic phenomena, and from cosmic space as meteorites and dust.

II. A. Living natural bodies, in their cellular morphology, protoplasmic nature and reproductive capacity have a unity, which must be connected with their genetic connection with each other in the course of geological time.

B. Inert natural bodies are extremely diverse and have no common structural or genetic connections.

III. A. Chemical differences between right and left enantiomorphs characterize the state of physical space occupied by living organisms. Either left or right enantiomorphs predominate.

B. The left and right enantiomorphs of the same chemical compounds have the same chemical properties in inert bodies. The numbers of dextral and sinistral crystals formed in an inert medium are the same.

IV. A. New living natural bodies are born only from pre-existing ones. From time to time new generations arise differing from the preceding ones. The rise of the central nervous system has increased the geological role of living matter, notably since the end of the Pliocene.

B. New inert bodies are created in the biosphere irrespective of the natural bodies that previously existed. In general the same kind of natural bodies are formed by inert processes, as were formed two billion years ago. New kinds of inert bodies appear only under the influence of living matter, notably man.

V. A. There are no liquid or gaseous living bodies, though liquids and gases are present in the mesomorphous or solid living bodies. Spontaneous, largely self-regulated movement is characteristic of living bodies. This may be passive as in reproduction, but the effect of reproduction is for the biosphere to be colonized, by a process comparable to the expansion of a gas.

B. Liquid and gaseous inert bodies take the form of the receptacles in which they are contained. Solid or mesomorphous inert bodies in general show no motion peculiar to the body as a whole.

VI. A. There is a continual stream of atoms passing to and from living organisms from and into the biosphere. Within the organisms a vast and changing number of molecules are produced by processes not otherwise known in the biosphere.

B. Inert natural bodies change only from outside causes, with the exception of radio-active materials.

VII. A. The number of living natural bodies is quantitatively related to the size of the biosphere.

B. The number of inert natural bodies is defined by the general properties of matter and energy, and is independent of the size of the planet.

VIII. A. The mass of living matter has remained fairly constant, being determined by radiant solar energy and the biogeochemical energy of colonization, but apparently the mass increases towards a limit, and the process is not yet complete.

B. The area of manifestation of inert natural bodies in the biosphere is limited by the size of the latter and increases only as the biosphere is expanded by the motion of living matter.

IX. A. The minimum size of a living natural body is determined by respiration, and is of the order of 10^{-6} cms. The maximum size has never exceeded $n \cdot 10^4$. The range, 10^{10}, is not great.

B. The minimum size of an inert natural body in the biosphere is determined by the degree of dispersion of matter and energy, i.e. the size of the ultimate particles of physics. The maximum size is determined by the size of the biosphere. The range is 10^{40} or more.

X. A. The chemical composition of living bodies is a function of their own properties.
B. The chemical composition of inert bodies is a function of the properties of the medium in which they are formed.

XI. A. The number of kinds of chemical compounds in living bodies is connected with the kinds of individual organisms and probably reaches many millions.
B. The number of different kinds of chemical compounds in inert bodies is limited to a few thousands.

XII. A. The processes in living matter tend to increase the free energy of the biosphere.
B. All inert processes, save radio-active disintegration, decrease the free energy of the biosphere.

XIII. A. Living natural bodies are always mesomorphous, and except in latent conditions H and O as water predominate, with an extremely complicated mixture of other compounds. The chemical composition of any one kind of living matter, though not exhibiting stoichiometric relationships, is definitely determined and more constant than the isomorphous mixtures constituting natural minerals.
B. The chemical composition of inert natural bodies may correspond to nearly pure chemical compounds with precise stoichiometric relations between the elements. In minerals solid solutions predominate.

XIV. A. Isotopic ratios may be markedly changed by the processes in living matter.
B. The isotopic ratios do not change markedly in the inert natural bodies of the biosphere, though outside the biosphere, deep in the crust, such changes may occur.

XV. A. The vast majority of living natural bodies change their forms in the process of evolution. The rates at which these changes occur are, however, widely divergent.
B. The majority of inert bodies in the biosphere are stable, and so lack variety.

XVI. A. The processes of living natural bodies are not reversible in time.
B. All physiochemical processes in inert natural bodies are reversible in time.

In everyday life one used to speak of man as an individual, living and moving freely about our planet, freely building up his history. Until recently the historians and the students of the humanities, and to a certain extent even the biologists, consciously failed to reckon with the natural laws of the biosphere, the only terrestrial envelope where life can exist. Basically man cannot be separated from it; it is only now that this indissolubility begins to appear clearly and in precise terms before us. He is geologically connected with its material and energetic structure. Actually no living organism exists on earth in a state of freedom. All organisms are connected indissolubly and uninterruptedly, first of all through nutrition and respiration, with the circumambient material and energetic medium. Outside it they cannot exist in a natural condition.

In our century the biosphere has acquired an entirely new meaning: it is being revealed as a planetary phenomenon of cosmic character. In biogeochemistry we have to reckon with the fact that living organisms actually exist not on our planet alone, and not in the terrestrial biosphere only. It seems to me that so far this has been established beyond doubt only for all so-called "terrestrial planets," that is, for Venus, the Earth, and Mars.

The thought of life as a cosmic phenomenon was alive long ago, as evidenced by the archives of science, including Russian science. At the end of the seventeenth century, the Dutch scientist Christian Huygens (1629–1695) put forward that problem in his last work, "Cosmotheoros" which was published after his death. This book, upon the initiative of Peter the Great, was twice published in Russian in the first quarter of the eighteenth century, under the title, "The Book of Contemplation of the World". In it Huygens established the scientific generalization that "life is a cosmic phenomenon somehow sharply distinct from inert matter." I have recently called this generalization the "Huygens principle".

Living matter, by weight, constitutes an insignificant part of our planet. Presumably, this is observed in the whole course of geological time; in other words this relation is geologically eternal. Living matter is concentrated in a thin but more or less continuous film on the surface of land in the troposphere, in the forests and fields, and permeates the whole ocean. Its quantity is calculated to be of the order of 0.25 per cent of the weight of the biosphere. On land it descends under the surface in non-continuous accumulations, probably down to an average depth of less than 3 kilometers.

THE NOÖSPHERE

We are approaching the climax in the Second World War. In Europe war was resumed in 1939 after an intermission of twenty-one years; it has lasted five years in Western Europe, and is in its third year in our parts in Eastern Europe. As for the Far East, the war was resumed there much earlier, in 1931, and is already in its twelfth year. A war of such power, duration and strength is a phenomenon unparalleled in the history of mankind and of the biosphere at large. Moreover, it was preceded by the First World War which, although of lesser power, has a causal connection with the present war.

In our country that First World War resulted in a new, historically unprecedented, form of statehood, not only in the realm of economics, but likewise in that of the aspirations of nationalities. From the point of view of the naturalist (and, I think, likewise from that of the historian) an historical phenomenon of such power may and should be examined as a part of a single great terrestrial geological process, and not merely as a historical process.

In my own scientific work the First World War was reflected in a most decisive way. It radically changed my geological conception of the world. It is in the atmosphere of that war that I have approached a conception of nature, at that time forgotten and thus new for myself and for others, a geochemical and biogeochemical conception embracing both inert and living nature from the same point of view. I spent the years of the First World War in my uninterrupted scientific creative work, which I have so far continued steadily in the same direction.

Twenty-eight years ago, in 1915, a "Commission for the Study of the Productive Forces" of our country, the so-called KEPS, was formed at the Academy of Sciences. That commission, of which I was elected president, played a noticeable role in the critical period of the First World War. Entirely unexpectedly, in the midst of the war, it became clear to the Academy of Sciences that in Tsarist Russia there were no precise data concerning the now so-called strategic raw materials, and we had to collect and digest dispersed data rapidly to make up for the lacunae in our knowledge. Unfortunately by the time of the beginning of the Second World War, only the most bureaucratic part of that commission, the so-called Council of the Productive Forces, was preserved, and it became necessary to restore its other parts in a hurry.

By approaching the study of geological phenomena from a geochemical and biogeochemical point of view, we may comprehend the whole of the circumambient nature in the same atomic aspect. Unconsciously such an approach coincides for me with what characterizes the science of the twentieth century and distinguishes it from that of past centuries. The twentieth century is the century of scientific atomism.

At that time, in 1917–1918, I happened to be, entirely by chance, in the Ukraine, and was unable to return to Petrograd until 1921. During all those years, wherever I resided, my thoughts were directed toward the geochemical and biogeochemical manifestations in the circumambient nature, the biosphere. While observing them, I simultaneously directed both my reading and my reflection toward this subject in an intensive and systematic way. I expounded the conclusions arrived at gradually, as they were formed, through lectures and reports delivered in whatever city I happened to stay, in Ialta, Poltava, Kiev, Simferopol, Novorossiisk, Rostov, and so on. Besides, in almost every city I stayed, I used to read everything available in regard to the problem in its broadest sense. I left aside as much as I could all philosophical aspirations and tried to rest only on firmly established scientific and empiric facts and generalizations, occasionally allowing myself to resort to working scientific hypotheses. Instead of the concept of "life," I introduced that of "living matter," which now seems to be firmly established in science. "Living matter" is the totality of living organisms. It is but a scientific empirical generalization of empirically indisputable facts known to all, observable easily and with precision. The concept of "life" always steps outside the boundaries of the concept of "living matter"; it enters the realm of philosophy, folklore, religion, and the arts. All that is left outside the notion of "living matter."

In the course of geological time living matter morphologically changes according to the laws of nature. The history of living matter expresses itself as a slow modification of the forms of living organisms which genetically are uninterruptedly connected among themselves from generation to generation. This idea had been rising in scientific research through the ages, until, in 1859, it received a solid foundation in the great achievements of Charles Darwin (1809–1882) and Wallace (1822–1913). It was cast in the doctrine of the evolution of species of plants and animals, including man. The evolutionary process is a characteristic only of living matter. There are no manifestations of it in the inert matter of our planet. In the cryptozoic era the same minerals and rocks were being formed which are being formed now. The only exceptions are the bio-inert natural bodies connected in one way or another with living matter.

The change in the morphological structure of living matter observed in the process of evolution unavoidably leads to a change in its chemical composition.

While the quantity of living matter is negligible in relation to the inert and bio-inert mass of the biosphere, the biogenic rocks constitute a large part of its mass, and go far beyond the boundaries of the biosphere. Subject to the phenomena of metamorphism, they are converted, losing all traces of life, into the granitic envelope, and are no longer part of the biosphere. The granitic envelope of the earth is the area of former biospheres. In Lamarck's book, "Hydrogeologie" (1802), containing many remarkable ideas, living matter, as I understand it, was revealed as the creator of the main rocks of our planet. Lamarck never accepted Lavoisier's (1743–1794) discovery. But that other great chemist, J. B. Dumas (1800–1884), Lamarck's younger contemporary, who did accept Lavoisier's discovery, and who intensively studied the chemistry of living matter, likewise adhered for a long time to the notion of the quantitative importance of living matter in the structure of the rocks of the biosphere.

The younger contemporaries of Darwin, J. D. Dana (1813–1895) and J. Le Conte (1823–1901), both great American geologists (and Dana a mineralogist and biologist as well) expounded, even prior to 1859, the empirical generalization that the evolution of living matter is proceeding in a definite direction. This phenomenon was called by Dana "cephalization," and by Le Conte the "psychozoic era." Dana, like Darwin, adopted this idea at the time of his journey around the world, which he started in 1838, two years after Darwin's return to London, and which lasted until 1842.

Empiric notions of a definite direction of the evolutionary process, without, however, any attempt theoretically to ground them, go deeper into the eighteenth century. Buffon (1707–1788) spoke of the "realm of man," because of the geological importance of man. The idea of evolution was alien to him. It was likewise alien to Agassiz (1807–1873), who introduced the idea of the glacial period into science. Agassiz lived in a period of an impetuous blossoming of geology. He admitted that geologically the realm of man had come, but, because of his theological tenets, opposed the theory of evolution. Le Conte points out that Dana, formerly having a point of view close to that of Agassiz, in the last years of his life accepted the idea of evolution in its then usual Darwinian interpretation. The difference between Le Conte's "psychozoic era" and Dana's "cephalization" thus disappeared. It is to be regretted that, especially in our country, this important empirical generalization still remains outside the horizon of our biologists.

The soundness of Dana's principle, which happens to be outside the horizon of our palaeontologists, may easily be verified by anyone willing to do so on the basis of any modern treatise on palaeontology. The principle not only embraces the whole animal kingdom, but likewise reveals itself clearly in individual types of animals. Dana pointed out that in the course of geological time, at least two billion years and probably much more, there occurs an irregular process of growth and perfection of the central nervous system, beginning with the crustacea (whose study Dana used to establish his principle), the molluscs (cephalopoda), and ending with man. It is this phenomenon that he called cephalization. The brain, which has once achieved a certain level in the process of evolution, is not subject to retrogression, but only can progress further.

Proceeding from the notion of the geological role of man, the geologist A. P. Pavlov (1854–1929) in the last years of his life used to speak of the anthropogenic era in which we now live. While he did not take into the account the possibility of the destruction of spiritual and material values we now witness in the barbaric invasion of the Germans and their allies, slightly more than ten years after his death, he rightly emphasized that man, under our very eyes, is becoming a mighty and ever-growing geological force. This geological force was formed quite imperceptibly over a long period of time. A change in man's position on our planet (his material position first of all) coincided with it. In the twentieth century, man, for the first time in the history of the earth, knew and embraced the whole biosphere, completed the geographic map of the planet Earth, and colonized its whole surface. Mankind became a single totality in the life of the earth. There is no spot on earth where man cannot live if he so desires. Our people's sojourn on the floating ice of the North Pole in 1937–1938 has proved this clearly. At the same time, owing to the mighty techniques and successes of scientific thought, radio and television, man is able to speak instantly to anyone he wishes at any point on our planet. Transportation by air has reached a speed of several hundred kilometers per hour, and has not reached its maximum. All this is the result of "cephalization," the growth of man's brain and the work directed by his brain.

The economist, L. Brentano, illuminated the planetary significance of this phenomenon with the following striking computation: if a square meter was assigned to each man, and if all men were put close to one another they would not occupy the area of even the small Lake of Constance between the borders of Bavaria and Switzerland. The remainder of the earth's surface would remain empty of man. Thus the whole of mankind put together represents an insignificant mass of the planet's matter. Its strength is derived not from its matter, but from its brain. If man understands this and does not use his brain and his work for self-destruction, an immense future is open before him in the geological history of the biosphere.

The geological evolutionary process shows the biological unity and equality of all men, *Homo sapiens* and his ancestors, *Sinanthropus* and others; their progeny in the mixed white, red, yellow, and black races evolves ceaselessly in innumerable generations. This is a law of nature. In a historical contest, as for instance in a war of such magnitude as the present one, he finally wins who follows that law. One cannot oppose with impunity the principle of the unity of all men as a law of nature. I use here the phrase "law of nature" as this term is used more and more in the physical and chemical sciences, in the sense of and empirical generalization established with precision.

The historical process is being radically changed under our very eyes. For the first time in the history of mankind the interests of the masses on the one hand, and the free thought of individuals on the other, determine the course of life of mankind and provide standards for men's ideas of justice. Mankind taken as a whole is becoming a mighty geological force. There arises the problem of the reconstruction of the biosphere in the interests of freely thinking humanity as a single totality. This new state of the biosphere, which we approach without our noticing is the noösphere.

In my lecture at the Sorbonne in Paris in 1922–23, I accepted biogeochemical phenomena as the basis of the biosphere. The contents of part of these lectures were published in my book, "Studies in Geochemistry," which appeared first in French, in 1924, and then in a Russian translation, in 1927. The French mathematician Le Roy, a philosopher, accepted the biogeochemical foundation of the biosphere as a starting point, and in his lectures at the College de France in Paris, introduced in 1927 the concept of the noösphere as the stage through which the biosphere is now passing geologically. He emphasized that he arrived at such a notion in collaboration with his friend de Chardin, a great geologist and palaeontologist, now working in China.

The noösphere is a new geological phenomenon on our planet. In it for the first time man becomes a large-scale geological force. He can and must rebuild the province of his life by his work and thought, rebuild it radically in comparison with the past. Wider and wider creative possibilities open before him. It may be that the generation of our grandchildren will approach their blossoming.

Here a new riddle has arisen before us. Thought is not a form of energy. How then can it change material processes? That question has not as yet been solved. As far as I know, it was first posed by an American scientist born in Lvov, the mathematician and biophysicist Alfred Lotka. But he was unable to solve it. As Goethe (1740–1832), not only a great poet but a great scientist as well, once rightly remarked, in science we only can know *how* something occurred, but we cannot know why occurred.

As for the coming of the noösphere, we see around us at every step the empirical results of that "incomprehensible" process. That mineralogical rarity, native iron, is now being produced by the billions of tons. Native aluminum, which never before existed on our planet, is now produced in any quantity. The same is true with regard to the countless number of artificial chemical combinations (biogenic "cultural" minerals) newly created on our planet. The number of such artificial minerals is constantly increasing. All of the strategic raw materials belong here. Chemically, the face of our planet, the biosphere, is being sharply changed by man, consciously, and even more so, unconsciously. The aerial envelope of the land as well as all its natural waters are changed both physically and chemically by man. In the twentieth century, as a result of the growth of human civilization, the seas and the parts of the ocean closest to shore become changed more and more markedly. Man now must take more and more measures to preserve for future generations the wealth of the seas which so far have belonged to nobody. Besides this, new species and races of animals and plants are being created by man. Fairy tale dreams appear possible in the future: man is striving to emerge beyond the boundaries of his planet into cosmic space. And he probably will do so.

At present we cannot afford not to realize that, in the great historical tragedy through which we live, we have elementally chosen the right path leading into the noösphere. I say elementally, as the whole history of mankind is proceeding in this direction. The historians and political leaders only begin to approach a comprehension of the phenomena of nature from this point of view. The approach of Winston Churchill (1932) to the problem, from the angle of a historian and political leader, is very interesting.

The noösphere is the last of many stages in the evolution of the biosphere in geological history. The course of this evolution only begins to become clear to us through a study of some of the aspects of its biosphere's geological past. Let me cite a few examples. Five hundred million years ago, in the Cambrian geological era, skeletal formations of animals, rich in calcium, appeared for the first time in the biosphere, those of plants appeared over two billion years ago. That calcium function of living matter, now powerfully developed, was one of the most important evolutionary factors in the geological change of the biosphere. A no less important change in the biosphere occurred from seventy to one hundred and ten million years ago, at the time of the Cretaceous system, and especially during the Tertiary. It was in this epoch that our green forests, which we cherish so much, were formed for the first time. This is another great evolutionary stadium, analogous to the noösphere. It was probably in these forests that man appeared around fifteen or twenty million years ago.

Now we live in the period of a new geological evolutionary change in the biosphere. We are entering the noösphere. This new elemental geological process is taking place at a stormy time, in the epoch of a destructive world war. But the important fact is that our democratic ideals are in tune with the elemental geological processes, with the laws of nature, and with the noösphere. Therefore we may face the future with confidence. It is in our hands. We will not let it go.

Vernadsky, W. J. 1945. The biosphere and the noösphere. American Scientist *33: 1–12. Reprinted with permission of Sigma Xi, The Scientific Research Society.*

Unification

Peter R. Grant

> It is the intertwined and interacting mechanisms of evolution and ecology, each of which is at the same time a product and a process, that are responsible for life as we see it, and as it has been. (Valentine 1973, 58)

Marston Bates, preeminent spokesman for the mid-twentieth century Western worldview of natural history, set forth a prescription of the field this way:

> The diversity of life is extraordinary. There is said to be a million or so different kinds of living animals, and hundreds of thousands of kinds of plants. But we don't need to think of the world at large. It is amazing enough to stop and look at a forest or at a meadow—at the grass and trees and caterpillars and hawks and deer. How did all these different kinds of things come about; what forces governed their evolution; what forces maintain their numbers and determine their survival or extinction; what are their relations to each other and to the physical environment in which they live? These are the problems of natural history, problems that concern ourselves as animals and that concern us even more as originators of this thing we call civilization—which is, after all, merely a rather special sort of animal community. (Bates 1950, 8)

Bates (1950) abhorred the specializations that were transforming biology then into ecology and evolution, entomology and ornithology, and attempted to hold a large part of biology together as natural history. The same has been attempted in this section of papers and chapters. The theme is the confluence of ecology and evolution in a modern view of natural history. The section begins where the confluence began, with two primary roots, one ecological and the other evolutionary. They fuse together in the trunk of a metaphorical tree of natural history.

First, the ecological root. The modern naturalist goes forth with guidebook, net, or binoculars and attempts to identify what is seen or caught. This tradition goes back to the late seventeenth and early eighteenth centuries when John Ray in England and Carl von Linné (Linnaeus) in Sweden systematically described, named, and ordered living things. Their books were the forerunners of field guides, home guides in effect. They gave focus and help to inquisitive observers who wanted to attach a name to what they had seen. One such observer was the Reverend Gilbert White (1720–1793). He lived in the village of Selbourne in southern England. The first selection comes from two of his letters (White 1813).

Lacking neighbors interested in natural history, White embarked on an extensive correspondence with two like-minded naturalists, Thomas Pennant and Daines Barrington, from 1767 to 1786. His correspondents praised him for his meticulous observations, and in the case of Barrington even published some of them. The first letter reprinted here is the most famous, because in it he uses observations on the appearance, song, and feeding station of warblers to identify a new and cryptic species. It is a tentative step towards the development of a new concept, the concept of an ecological niche.

Natural history before then has been referred to as the science of describing (Ogilvie 2006). The second letter shows that White was more than just a describer and cataloguer of facts. He attempted to interpret what he saw, and if his interpretations were unsatisfactory he would ask his correspondents for help, as in the case of an apparently new migration of ring ouzels (a kind of blackbird) in southern England. However, like a scientist he disciplined his conjectural thinking. Consider this passage:

> The question that you put with regard to those genera of animals that are peculiar to America, viz. how they came there, and whence? Is too puzzling for me to answer; and yet so obvious as often to have struck me with wonder. If one looks into the writers of the subject little satisfaction is to be found. Ingenious men will readily advance plausible arguments to support whatever theory they shall chuse to maintain; but then the misfortune is, every one's hypothesis is each as good as another's, since they are all founded on conjecture. (White 1813, 68)

Gilbert White is a significant figure in the history of natural history for another reason. His writings exemplify the value of making and recording observations in the same place for many years, learning what in modern language has been called the ecology of place (Billick and Price 2010). White lived in the same village for nearly fifty years, at the end and most intense part of the Little Ice Age (Lamb 1977). He was thus aware of both short-term and long-term changes, the occurrence of rare events such as especially cold winters, and their effects; for example, migrant thrushes died whereas residents survived. He was able to detect the gradual decline in species such as the wood pigeon, and its cause, a reduction in beech forests. This has an ominously modern ring, especially when coupled with the realization that several of the bird species he observed, some fairly common, can no longer be seen anywhere in England (corncrake) or even in the whole of the British Isles (wryneck).

The second root to the tree is evolutionary, and it is illustrated here with the writings not of a naturalist but of a mathematician; an idiosyncratic choice of a brilliant, if idiosyncratic, scientist! The world of biologists knows that Mendel's experiments with peas established the particulate nature of inheritance, a key to understanding evolution that had escaped Darwin. But had it escaped all others? Historians, especially Bentley Glass (1959), have given credit to the Frenchman Pierre Louis Moreau de Maupertuis (1698–1759) for discovering the basis of what we now call genetics. He was president of Frederick the Great's Academy of Sciences in Berlin. Amongst other activities, he visited Lapland to demonstrate that the earth was flatter at the poles than elsewhere.

Part of a chapter written by Glass (1959) has been chosen in place of the original writing of Maupertuis (1745) because it is more sharply focused on his discoveries, as well as being in clear English. It also better serves the purpose of revealing how the discovery was made, in other words the logical steps by which an explanation was extracted from a set of observations. This makes it comparable to the naturalist tradition, even though at face value his inquiries appear almost as remote as possible from the naturalists'.

Maupertuis investigated four generations of polydactyly in humans, and the occurrence of albinism in black people, and applied the mathematical theory of probability to rule out chance as an explanation of the patterns he documented. He undertook experiments in breeding dogs to throw light on his theories, and formulated a particulate theory of heredity. This theory involved the mutual attraction of analogous particles provided by each parent, and implied segregation, dominance, and independent assortment. He appears to have had a rudimentary theory of evolution in terms of errors in transmission equivalent to what we now call mutations, and speculated this could be the basis for the origin of new species. In short, he was a prescient forerunner of Mendel and deVries.

Particulate inheritance was an idea a century ahead of its time (see part 4). Like the mountain stream that is far removed from the lowland lake, one needs to see the whole chain of intermediate effects to appreciate Maupertuis's trickle-down influence on the theory of evolution; how he influenced the writings of others such as Diderot and Buffon, and through them, ultimately, Darwin. It is possible to imagine a different history, in which Maupertuis, thinking solely of the cellular determinants of adult form, corresponded with White, thinking solely about the whole organism and its relation to the external environment. If this had happened the ecological and evolutionary roots of natural history might have come together and nourished each other so much earlier!

These roots came together in the mind of Charles Darwin (1809–1882). Darwin drew upon vast experience as field naturalist, observer, and collector in devising a theory of evolution to explain the diversity of the living world. This part of the history of natural history is so well known it needs little discussion. His book *On the Origin of Species* (1859) was actually not so much an account of speciation as an explanation of adaptation of organisms to their environment by natural selection. After 1859 he addressed a multitude of interesting questions in this new framework. Darwin is fascinatingly revealed in the role of observer-naturalist in a chapter reproduced here from a book he wrote on pollination of orchids (Darwin 1862). Like his work on the movement of plants and insectivorous plants, they show him concerned to understand how plants work in order to explain their diversity; in this case the mechanisms of donating pollen to insects in such a way that the pollen is transferred unerringly to the stigmas of the next flower they visit—"one of the most wonderful cases of adaptation which has ever been recorded" (Darwin 1877, 30).

In 1847, Henry Walter Bates (1825–1892) accepted an invitation from his friend Alfred Russel Wallace to accompany him on an expedition to the Amazon to collect animals and plants. They knew nothing of Darwin's revolutionary ideas when they set out but they did have an interest in the question of how species were formed, and there is reason to believe that each of them, like Darwin, found the answer. After collecting together, Wallace and Bates parted company and followed different branches of the Amazon. Wallace returned to England after four years. A few years later he sent Bates a draft of a paper (Wallace 1855) that outlined the first steps towards a theory of evolution by natural selection, and Bates replied "the theory I quite assent to, and, you know, was conceived by me also, but I profess I could not have propounded it with so much force and completeness" (Shoumatoff 1988, xii).

Bates stayed in the Amazon region for the best eleven years of his life. On his return to England he wrote a hugely successful book on his travels (Bates 1863) and one scientific masterpiece (Bates 1861). The latter, as can be seen in this section, illustrates the early confluence of ecological and evolutionary strands of natural history. It is about mimicry.

Camouflage and crypsis is everywhere, writes Shoumatoff (1988, xiv) in the introduction to a reissue of Bates's book: "It is as if the rain forest were a dangerous, riotous carnival, and everyone who would dance in it must wear a costume." The costume that Bates attempted to understand was often bright and conspicuous and shared by unrelated species. Why?

In this paper Bates leads us through the intricacies of butterfly wing colors and patterns along a path called logic. First he performs a systematic analysis to determine affinities, and corrects an error of combining separate families through superficial similarities. This is important for subsequent elucidation of mimicry. Then he asks a series of questions, and supplies answers, as he gets closer and closer to an explanation for the amazing resemblances. How do we know that resemblance is due to "counterfeiting" and not to shared inheritance or shared environments? Which are the mimics, the copiers, and which are the models that are copied? Having answered these questions and discarded alternative explanations, he zeroes in on the key question, the cause of mimicry, by asking what advantages the ones who are copied possess. Nothing in habitat, structure, or behavior makes them safe from insectivorous animals, and yet Bates never saw them persecuted by birds, dragonflies, lizards, or Asilid flies, therefore the probable answer is they are unpalatable. This is a critical insight. From this it follows that mimics must share the protection that models possess. How is the mimicry brought about? Natural selection: poor imitators are selected against and better imitators are favored. An admiring and grateful Darwin complimented him on providing the best, most convincing, example of his theory of evolution. We now call it Batesian mimicry. In the flowering of modern natural history it continues to be visited and revisited (e.g., Jiggins et al. 2006).

If Henry Walter Bates is underappreciated as a naturalist of adaptation and natural selection, Robert Perkins (1866–1955) is the unsung hero of adaptive radiation (Grant 2000a, 2001). "Nothing can be more improving to a young naturalist," Darwin (1839, 607) had written, "than a journey in a distant country." Darwin was twenty-three when he first set foot in Brazil, so was Bates, Wallace was twenty-five, and so was Perkins when he arrived in his distant country, Hawaii, in 1892. He had been sent from England as a collector. After two and a half years he returned to England to supervise the preparations of his specimens, then returned to Hawaii twice more. Altogether he was there for six years, often alone for long periods of time in remote parts of rain-soaked, mountainous wilderness. Like Bates he was more than a collector, and like him he had plenty of time to reflect on his collections. The two naturalists shared an impressive ability to reason cautiously from observations to unknown causes; to reconstruct a probable, dynamic, past from which has arisen the known present.

Putting his knowledge of beetles, snails, and birds together, Perkins (1913) asked how often they had arrived in Hawaii, and how they had diversified. In answer to the first question, colonization of the archipelago must be a rare event, he argued, because whole families with a wide distribution elsewhere are missing from the native fauna. Their absence cannot be explained by failed attempts because when humans introduced several of the missing species they flourished. Furthermore, since the most differentiated and distinctive species are found on Kauai, this is likely to be the island where the original immigrants arrived. They then spread eastwards, and speciated. How had they speciated? He compared closely related species, and concluded, "Reviewing the whole fauna, the widespread effect of segregation in species-formation becomes evident,—isolation appears the most important factor" (Perkins 1913, lxx).

This is the backdrop to a selection of his writing; a synthesis of his work with birds (Perkins 1903). It exemplifies his interpretative skills, and shows how he arrived at new insights on competition between species and an evolutionary trend from generalization to specialization in feeding habits and diets. One of the gems explains the evolution of the curved beaks of nectar-feeding honeycreeper finches as coupled to the evolution of curved corollas of endemic lobelias (beautifully illustrated by Pratt 2005). It is an example of coevolution, and a link with my final selection by Janzen (discussed below). There is another extraordinary value in his accounts of Hawaiian birds, stemming from the fact that one can no longer walk in the forests he explored as they have been destroyed, nor can one see some of the bird species he observed because they are now extinct (Scott et al. 2001). In some cases he was the last observer.

California was for Joseph Grinnell (1877–1939) what Hawaii was for Perkins: a world full of animals and plants calling for an explanation. What gripped Grinnell perhaps most was the geographical distribution of vertebrate animals, especially mammals. Like Perkins he developed his own ideas of speciation in geographically isolated populations: allopatric speciation (Grinnell 1924). He is better known for articulating a modern concept of the ecological niche (Whittaker et al. 1973). These two ideas are connected: adaptation to local environmental conditions (niche evolution) results in speciation in populations that are geographically isolated.

The starting point of the paper reprinted here is the simple observation that pocket gophers spend 99 percent of the time below ground, and must be vulnerable to predators in the remaining 1 percent. How, then, can one explain the patchy distribution of these ecologically specialized animals on such a large geographical scale? There are lowland groups of pocket gophers and there are high-mountain-dwelling groups, and there is wide separation within and between groups, and yet over time they have colonized all of California and more. Grinnell takes the reader step by step through an analysis of their morphological variation and the differences in their affinities to particular habitats. Current barriers to dispersal are too large to cross; therefore the spatial array of habitats must have changed in the past. He sketches out an explanation for one particular situation in the neighborhood of Salton Sea, and is finally led to a far-reaching conclusion: the history of the habitats has determined the adaptations and geographical distribution of modern populations of pocket gophers. Modern genetic studies have filled in some important details in this broad picture of pocket gopher natural history in showing, for example, that the populations exchange individuals more than one would think from their apparent low vagility (Patton and Yang 1977).

The last reading brings us to the modern era of making sense of observations of organisms in nature by reasoning from ecological and evolutionary principles, leavened by a healthy dose of common sense. Daniel H. Janzen (1939–) took up the challenge of explaining how ants and the acacias they inhabit could have evolved in such a way that both benefit from the association (Janzen 1966). The swollen thorns of acacias are used as a home by ants. Ants protect the acacias from herbivores. Thus the two participants of a mutualism have undergone coevolution, and Janzen worked out in exquisite detail how it probably happened. This required him to think like an ant, imagine he was an acacia. While strongly reminiscent of the work of Henry Bates in the Amazon its true antecedent was the acute observations of Thomas Belt, a nineteenth-century naturalist in Nicaragua. Amongst other things Belt (1874) described how the tips of the leaflets of the acacias inhabited by ants were modified in structure and composition and fed upon by the ants. These Beltian bodies are both encouragement to the ants and a reward to them for defending the tree against herbivores.

Modern naturalists look back to the Janzen paper as the pioneer of a more comprehensive field of coevolutionary studies (Thompson 1994) that attempts to determine how sets of ecologically interdependent species affect each other's evolution.

To conclude, the theme of these selections is the confluence of ecology and evolution in a modern view of natural history. Darwin foretold it in a buoyant letter to the botanist J. D. Hooker in 1856 (Darwin 1887, 194): "And what a science Natural History will be, when we are in our graves, when all the laws of change are thought one of the most important parts of Natural History." This idea became a prophesy in the *Origin of Species* (Darwin 1859, 455): "When the views entertained in this volume on the origin of species, or when advantageous views are generally admitted, we can dimly foresee that there will be a considerable revolution in natural history." There has been; and its influence has spread far beyond what is typically thought to be natural history, into all of biology, from molecules to ecosystems (Grant 2000b).

LITERATURE CITED

Bates, H. W. 1861. Contributions to an insect fauna of the Amazon Valley. Transactions of the Linnean Society 32: 495–566.

Bates, H. W. 1863. The Naturalist on the River Amazons. J. Murray, London.

Bates, M. 1950. The Nature of Natural History. Scribner, New York.

Belt, T. 1874. The naturalist in Nicaragua. E. Bumpas, London.

Billick, I. J., and M. V. Price, eds. 2010. Ecology of Place. University of Chicago Press, Chicago.

Darwin, C. 1839. Journal of Researches into the Natural History and Geology of the Countries Visited During the Voyage of H.M.S. Beagle Round the World. Henry Colborn, London, U.K.

Darwin, C. 1859. On The Origin of Species by Means of Natural Selection. J. Murray, London, U.K.

Darwin, C. 1862. The Various Contrivances by which Orchids are Fertilized by Insects. J. Murray, London, U.K.

Darwin, C. 1877. The Various Contrivances by which Orchids are Fertilized by Insects, 2nd ed. J. Murray, London, U.K.

Darwin, F., ed. 1919. The Life and Letters of Charles Darwin: Including an Autobiographical Chapter. Vol. 1. D. Appleton and Co., New York and London. Originally published in 1887.

Glass, B. 1959. Maupertuis, pioneer of genetics and evolution. In Forerunners of Darwin: 1745–1859, B. Glass, O. Temkin, and W. L. Straus, Jr., eds., pp. 51–83. Johns Hopkins University Press, Baltimore.

Grant, P. R. 2000a. R. C. L. Perkins and evolutionary radiations on islands. Oikos 89: 195–201.

Grant, P. R. 2000b. What does it mean to be a naturalist at the end of the twentieth century? American Naturalist 155: 1–12.

Grant, P. R. 2001. Reconstructing the evolution of birds on islands: 100 years of research. Oikos 92: 385–403.

Grinnell, J. 1924. Geography and evolution. Ecology 5: 225–229.

Grinnell, J. 1927. Geography and evolution in the pocket gopher. University of California Chronicle 28: 247–262.

Janzen. D. H. 1966. Coevolution of mutualism between ants and acacias in Central America. Evolution 20: 249–275.

Jiggins, C. D., R. Mallarino, K. R. Willmott, and E. Bermingham. 2006. The phylogenetic pattern of speciation and wing pattern change in neotropical *Ithomia* butterflies (Lepidoptera: Nymphalidae). Evolution 60:1454–1466.

Lamb, H. H. 1977. Climate: Present, Past, and Future. Vol. 2. Climatic History and the Future. Methuen, London.

Maupertuis, P. L. M. de. 1745. Vénus Physique, contenant deux dissertations, l'une sur l'origine des hommes et des animaux; et l'autre sur l'origine des noirs. La Haye, France.

Ogilvie, B. W. 2006. The Science of Describing: Natural History in Renaissance Europe. University of Chicago Press, Chicago.
Patton, J. L., and S. Y. Yang. 1977. Genetic variation in *Thomomys bottae* pocket gophers: Macrogeographic patterns. Evolution 31: 697–720.
Perkins, R. C. L. 1903. Vertebrata. In Fauna Hawaiiensis, D. Sharp, ed., pp. 365–466. Cambridge University Press, Cambridge.
Perkins, R. C. L. 1913. Introduction to Fauna Hawaiiensis, D. Sharp, ed., pp. xv–lxxii. Cambridge University Press, Cambridge.
Pratt, H. D. 2005. The Hawaiian Honeycreepers Drepanidinae. Oxford University Press, Oxford.
Scott, J. M., S. Conant, and C. Van Riper III (eds.). 2001. Evolution, ecology, conservation, and management of Hawaiian birds: A vanishing avifauna. Studies in Avian Biology, no. 22.
Shoumatoff, A. 1988. Introduction to The Naturalist on the River Amazons, by H. W. Bates (1863). Penguin Books, London.
Thompson, J. N. 1994. The Coevolutionary Process. University of Chicago Press, Chicago.
Valentine, J. W. 1973. Evolutionary Paleoecology of the Marine Biosphere. Prentice-Hall, Englewood Cliffs, N.J.
Wallace, A. R. 1855. On the law which has regulated the introduction of new species. Annals and Magazine of Natural History 16: 184–196.
White, G. 1813. Natural History of Selborne. White, Cochrane & Co., London. Originally published in 1789.
Whittaker, R. H., S. A. Levin, and R. B. Root. 1973. Niche, habitat, and ecotope. American Naturalist 107: 321–338.

The Natural History of Selborne (1813)

Gilbert White

LETTER XIX. TO THE SAME.
SELBORNE, AUG. 17, 1768.

[The addressee is Thomas Pennant, the leading British zoologist at the time.]

Dear Sir—I have now, past dispute, made out three distinct species of the willow-wrens (*Motacillae trochili*), which constantly and invariably use distinct notes. But, at the same time I am obliged to confess that I know nothing of your willow-lark. In my letter of April the 18th, I had told you peremptorily that I knew your willow-lark, but had not seen it then; but, when I came to procure it, it proved in all respects a very *Motacilla trochilus*; only that it is a size larger than the two other, and the yellow-green of the whole upper part of the body is more vivid, and the belly of a clearer white. I have specimens of the three sorts now lying before me; and can discern that there are three gradations of sizes, and that the least has black legs, and the other two flesh-coloured ones. The yellowest bird is considerably the largest, and has its quill feathers and secondary feathers tipped with white, which the others have not. This last haunts only the tops of trees in high beechen woods, and makes a sibilous grasshopper-like noise now and then, at short intervals, shivering a little with its wings when it sings; and is, I make no doubt now, the *regulus non cristatus* of Ray; which he says, "*cantat voce stridula locustae*". Yet this great ornithologist never suspected that there were three species.

LETTER XX. TO THE SAME. SELBORNE, OCT. 8, 1768.

It is, I find, in zoology as it is in botany; all nature is so full, that that district produces the greatest variety which is the most examined. Several birds, which are said to belong to the north only, are, it seems, often in the south. I have discovered this summer three species of birds with us which writers mention as only to be seen in the northern counties. The first that was brought me (on the 14th of May) was the sandpiper (*Tringa hypoleucus*): it was a cock bird, and haunted the banks of some ponds near the village; and, as it had a companion, doubtless intended to have bred near that water. Besides, the owner has told me since, that on recollection he has seen some of the same birds round his ponds in former summers.

The next bird that I procured (on the 21st of May) was a male red-backed butcher-bird (*Lanius collurio*). My neighbour, who shot it, says that it might easily have escaped his notice had not the outcries and chattering of the whitethroats and other small birds drawn his attention to the bush where it was: its craw was filled with the legs and wings of beetles.

The next rare birds (which were procured for me last week) were some ring-ousels (*Turdi torquati*).

This week twelvemonths a gentleman from London being with us, was amusing himself with a gun, and found, he told us on an old yew hedge where there were berries, some birds like blackbirds, with ring of white round their necks; a neighbouring farmer also at the same time observed the same; but, as no specimens were procured, little notice was taken. I mentioned this circumstance to you in my letter of November the 4th, 1767 (you, however, paid but small regard to what I said, as I had not seen these birds myself): but last week the aforesaid farmer, seeing a large flock, twenty or thirty, of these birds, shot two cocks and two hens; and says, on recollection, that he remembers to have observed these birds again last spring, about Ladyday, as it were on their return to the north. Now, perhaps these ousels are not the ousels of the north of England, but belong to the more northern parts of Europe; and may retire before the excessive rigour of the frosts in those parts; and return to breed in spring when the cold abates. If this be the case, here is discovered a new bird of winter passage, concerning whose migrations the writers are silent; but if these birds should prove the ousels of the north of England, then here is a migration disclosed within our own kingdom, never before remarked. It does not yet appear whether they retire beyond the bounds of our island to the south; but it is most probable that they usually do, or else one cannot suppose that they would have continued so long unnoticed in the southern counties. The ousel is larger than a blackbird, and feeds on haws; but last autumn (when there were no haws) it fed on yew-berries: in the spring it feeds on ivy-berries, which ripen only at that season, in March and April.

I must not omit to tell you (as you have been so lately on the study of reptiles) that my people, every now and then of late, draw up, with a bucket of water from my well, which is sixty-three feet deep, a large black warty lizard, with a fin tail and yellow belly. How they first came down at that depth, and how the were ever to have got out thence without help, is more than I am able to say.

Many thanks are due to you for your trouble and care in the examination of a buck's head. A far as your discoveries reach at present, they seem much to corroborate my suspicions; and I hope Mr. may find reason to give his decision in my favour; and then, I think, we may advance this extraordinary provision of nature as a new instance of the wisdom of God in the creation.

As yet I have not quite done with my history of the *aedienemus*, or stone curlew; for I shall desire a gentleman in Sussex (near whose house these birds congregate in vast flock in the autumn) to observe nicely when they leave him (if they do leave him), and when they return again in the spring: I was with this gentleman lately, and saw several single birds.

White, G. 1813. The Natural History of Selborne. *White, Cochrane & Co., London (originally published in 1789); Letters XIX and XX, pp. 76–79.*

Maupertuis, Pioneer of Genetics and Evolution (1959)

Bentley Glass

MAUPERTUIS AS EPIGENETICIST

It is in his biological ideas that Maupertuis was most clearly gifted with prevision. Here he must be reckoned as fully a century or a century and a half before his time. His biological ideas may be considered under the three heads of the formation of the individual, the nature of heredity, and the evolution of species, although obviously these are so closely interrelated that the division is largely artificial.

He began with an interest in the formation of the embryo, and quickly put his finger on the soundest argument against the preformationists, who believed the embryo to be fully preformed before conception, and present either in the sperm or the egg. This argument led him to a study of heredity, and he may be justly claimed as the first person to record and interpret the inheritance of a human trait through several generations. He was also the first to apply the laws of probability to the study of heredity. He was led by the facts he had uncovered to develop a theory of heredity that astonishingly forecast the theory of the genes. He believed that heredity must be due to particles derived both from the mother and from the father, that similar particles have an affinity for each other that makes them pair, and that for each such pair either the particle from the mother or the one from the father may dominate over the other, so that a trait may seemingly be inherited from distant ancestors by passing through parents who are unaffected. From an accidental deficiency of certain particles there might arise embryos with certain parts missing, and from an excess of certain particles could come embryos with extra parts, like the six-fingered persons or the giant with an extra lumbar vertebra whom Maupertuis studied. There might even be complete alterations of particles—what today we would call "mutations"—and these fortuitous changes might be the beginning of new species, if acted upon by a survival of the fittest, and if geographically isolated so as to prevent their intermingling with the original forms. In short, virtually every idea of the Mendelian mechanism of heredity and the classical Darwinian reasoning from natural selection and geographic isolation is here combined, together with De Vries' theory of mutations as the origin of species, to a synthesis of such genius that it is not surprising that no contemporary of its author had a true appreciation of it.

Even in his earliest biological essays, "Observations et experiences sur une des especes de salamandre" (1727), and "Experiences sur les scorpions" (1731), Maupertuis evinced an absorbing interest in reproduction and those problems that it inevitably generates. Having observed both eggs in the ovaries and live young to the number of 54 in the oviducts of single female salamanders, Maupertuis remarked that "these animals appear very suitable for enlightening the mystery of generation." In the study of scorpions it was again the presence of living young (up to 65 in number) within the female, and the fantastic courtship of the sexes which excited his interest, as it did that of Fabre a century and a half later.

The prevailing ideas regarding the origin of the individual in Maupertuis' time were those of preformation. William Harvey, in 1651, had plainly declared his conviction, based on the study of the development of the embryo chick, in favor of epigenesis, that is, of the view that the parts of the embryo are formed in succession out of unorganized material; but the application of the microscope to the reinvestigation of the matter, by Malpighi, led to the common opinion that Harvey had been wrong. Malpighi was perhaps led astray by his inability to find any but a fairly well developed embryo in freshly laid eggs, not recognizing that the embryo might have begun its development while the egg was descending the oviduct. Swammerdam's investigations of insect metamorphosis—in particular, his demonstration that a perfectly formed butterfly is to be found within the chrysalis—seemed convincing demonstrations of a like sort, although the idea was pushed far beyond demonstrable limits. The caterpillar must mask a perfect butterfly, within which there would be eggs containing, *multum in parvo*, caterpillars, chrysalids, and butterflies of a generation still to come; and so on.

Another notable scientific doctrine was involved in the common acceptance of the preformation theory in the seventeenth and eighteenth centuries. This was the belief in spontaneous generation. As long as it was believed that living creatures could arise, spontaneously from mud, filth, and putrescent matter, either epigenesis or some preformation of germs disseminated throughout nature seemed inescapable, at least for such organisms. The demonstration by Redi, in 1668, that maggots hatch only from flies' eggs, and do not appear in putrefying meat except when the flies have access to it, cleared the way for the encasement ("emboitement") theory of preformation.

The predominance of the preformationist doctrine led to heated claims in behalf of the ovum by the followers of Malpighi and Swammerdam, on the one hand, and of the spermatozoa by the disciples of Leeuwenhoek, on the other; since, if, the embryo is preformed, it could not logically be preformed in both parents alike. The discovery of the mammalian egg (or, in reality, of the ovarian follicle) by Regnier de Graaf, in 1672, was a great victory for the ovists, who claimed that the spermatic animalcules possessed only the function of stirring and mixing the two seminal fluids. The animalculists countered with purported observations of homunculi within the heads of the spermatozoa. But this is not the place to trace the exaggerated and frequently ridiculous claims and counterclaims of these two schools of thought. Suffice it to say that by 1740 there remained very few epigenesists indeed, and the encasement theory of preformation prevailed almost universally.

In 1745 a small anonymous volume appeared from the press entitled *Venus Physique, contenant deux dissertations, l'une sur l'origine des hommes et des animaux: et l'autre sur l'origine des noirs.* This book was written in a popular style for gentlemen and ladies of the court. The author, who later turned out to be Maupertuis, dared to adopt the discredited epigenetic theory of development. He was led to this by a consideration of the plain facts of biparental heredity, which had been pointed out by Aristotle and again by Harvey, but which the ovists and animalculists attempted to explain away. According to them, heredity might be regarded as a spiritual essence, transferable in the seminal fluids, and impressing itself upon the preformed embryo provided by one or the other of the parents. This essentially vitalistic idea was abhorrent to Maupertuis. A mathematician trained in Cartesian views, an early convert to Newton's teachings, Maupertuis belies the remark made by Driesch that all epigenesists were vitalists, for Maupertuis was a consistent mechanist. In criticizing the views of the preformationists, Maupertuis would have nothing to do with essences and spiritual "virtues." He argued against the possibility that maternal impressions can affect the foetus, and to him biparental heredity necessarily implied corporeal contributions from each of the two parents.

MAUPERTUIS' INVESTIGATION OF POLYDACTYLY

One may well wonder what led Maupertuis to these views. The answer is sufficiently clear. He had already, prior to 1745, conducted a careful study in human genetics that is matched by only one other of its time. It seems that a young albino negro had been brought to Paris and naturally enough created a great stir. This started Maupertuis to thinking about heredity. Surely the albino condition must be hereditary, for although it seemed to appear sporadically among Negroes, it was reportedly not rare in Senegal, where whole families were said to be "white." Of the famous albino Indians of Panama he had also heard, and discoursed about them with no little sentimentality. "The black color," he said, "is just as hereditary in crows and blackbirds, as it is in Negroes: I have nevertheless seen white blackbirds and white crows a number of times."

Shortly after coming to Berlin, Maupertuis began to hunt for a case he could study at first-hand. He was fortunate in his search: he ran across the polydactylous Ruhe family. The investigation had been completed some time before the *Venus Physique* was written, and from the internal evidence seems to have been the main directive to the thoughts and arguments presented in that book. When it is remembered that this was Maupertuis' first extensive venture into biological investigation, that the collection of human pedigrees was a novel enterprise, and that even in the present century the necessity of making a complete record of all normal as well as all abnormal members of a family has frequently been neglected, one can only marvel at his perspicacity. Maupertuis described his study of this hereditary trait no less than three separate times, in the *Venus Physique*, in the *Systeme de la Nature*, and in the *Lettres*. It is in the latter (Letter XIV) that we find the fullest account, which follows:

> A great physician proposes in a useful and inquiring work (L'art de faire eclorre des oiseaux domestique, par M. de Reaumur, t. I.I., memo 4) to perform some experiments on this question of biparental inheritance. In the race (genre) of fowls it is not rare to see types (races), which bear five toes on each foot; it is hardly more to see those which are born without rumps. M. de Reaumur proposes to mate a hen with five toes with a four-toed cock, a four-toed hen with a five-toed cock; to do the same experiment with the rumpless cocks and hens: and he regards these experiments as able to decide whether the foetus is the product solely of the father, solely of the mother, or of the one and the other together.

I am surprised that that skilful naturalist, who has without doubt carried out these experiments, does not inform us of the result. But an experiment surer and more decisive has already been entirely completed. That peculiarity of the supernumerary digits is found in the human species extends to entire breeds (races); and there one sees that it is equally transmitted by the fathers and by the mothers.

Jacob Ruhe, surgeon of Berlin, is one of these types. Born with six digits on each hand and each foot, he inherited this peculiarity from his mother Elisabeth Ruhen, who inherited it from her mother Elisabeth Horstmann, of Rostock. Elisabeth Ruhen transmitted it to four children of eight she had by Jean Christian Ruhe, who had nothing extraordinary about his feet or hands. Jacob Ruhe, one of these six-digited children, espoused, at Dantzig in 1733, Sophie Louise de Thüngen, who had no extraordinary trait: he had by her six children; two boys were six-digited. One of them, Jacob Ernest, had six digits on the left foot and five on the right: he had on the right hand a sixth finger, which was amputated; on the left he had in the place of the sixth digit only a stump.

One sees from this genealogy, which I have followed with exactitude, that polydactyly (sex-digitisme) is transmitted equally by the father and by the mother; one sees that it is altered through the mating with five-digited persons. Through these repeated matings it must probably disappear; and must be perpetuated through matings in which it is carried in common by both sexes.

Maupertuis was quite struck by this apparent weakening of the trait with time, and it led him to the conclusion that through repeated matings with normal individuals the trait might in time disappear. In other words, the deviations of nature tend to fade out and "her works always tend to resume the upper hand." Maupertuis had evidently a clear idea that most if not all abnormalities are disadvantageous, and as we shall see later, this became an important part of his evolutionary thought.

It is truly astonishing to discover that in the second edition of that very book by Reaumur which served as inspiration to Maupertuis, there is a celebrated account of the inheritance of human polydactyly that is as circumstantial and complete as that of the Ruhe family. Charles Bonnet, in 1762, struggled vainly to deal with its implications, so embarrassing to the theory of preformation and to ovists in particular. A Maltese couple, named Kelleia, whose hands and feet were constructed upon the ordinary human model, had a son, Gratio, who possessed six perfectly moveable fingers on each hand and six toes, deformed and crownlike, on each foot. This man married a normal woman and had by her four children. The eldest, Salvator, a boy, had six fingers and six toes; the second, George, had five fingers and toes, but his hands and feet were slightly deformed; the third, Andre, was normal; the fourth, a girl, Marie, had five fingers and toes, but her thumbs were slightly deformed. All of these children grew up and married normal persons. The eldest had four children, two boys and a girl polydactylous, and one normal boy. The second, George, had three girls, all polydactylous, and a normal son. Two of these girls had six fingers and six toes on each side, but one had only five toes on his left foot. The fourth of Gratio Kelleia's children, his daughter Marie, had one boy with six toes and three normal children (one boy; two girls). The normal son of Gratio had only normal children.

How did Maupertuis come to overlook this account by Reaumur, which so closely resembles his own investigation? An examination of Reaumur's rather rare book shows that the edition of 1749 does not contain the account. It was first added in the second edition, of 1751, in an appended section in square brackets at the very end of the book. It must be presumed that Maupertuis never saw it.

This well-known pedigree was cited by T. H. Huxley in his essay of April 1860, entitled "(Darwin on) the Origin of Species," but with one egregious addition to the original. Huxley stated: "A Maltese couple, named Kelleia, whose hands and feet were constructed upon the ordinary human model, had born to them a son, Gratio. . . . " Nowhere in Reaumur's account however is it said that both the parents of Gratio Kelleia definitely had normal hands and feet. I was consequently misled in my original essay on Maupertuis when I relied on Huxley's account for this point, which would instantly be recognized by a geneticist as highly significant. For if both parents had been carefully examined and had been found to have normal digits, one might suppose that a mutation had occurred in one of the germ cells from which Gratio issued, although the very mild expression of the deformity in his daughter Marie is sufficient warning of a highly variable manifestation of the condition and a possibly reduced penetrance, such that occasionally a seemingly normal person might transmit the deformity. It would be necessary, therefore, to establish that Gratio Kelleia's ancestry for more than one generation back had been free of the trait before the assumption of a mutation would be warranted. However, Maupertuis might well have seized upon the sudden appearance in the son of normal parents of a hereditary condition that thereafter descended in unbroken lineage as a striking example of one of those fortuitous and tenacious alterations of the hereditary particles which he thought might constitute the beginning of a new species.

MAUPERTUIS' THEORY OF HEREDITY AND THE ORIGIN OF THE FOETUS

This investigation of polydactyly was the work upon which Maupertuis founded his theory of the formation of the foetus and the nature of heredity, a theory that was more than a century before its time and that brilliantly anticipated the discoveries of Mendel and de Vries. Maupertuis started out with the idea of chemical attraction, that attraction between the particles of different elements which results in the formation of compounds between them. This new chemical concept of Geoffroy and other French chemists of the time was much discussed, and Maupertuis was predisposed to it on account of his early devotion to the concept of gravitational attraction, which seemed to him to be an analogous phenomenon. As Maupertuis conceived it, two substances possess a tendency to unite by virtue of their chemical affinity; but if a third appears on the scene with a greater affinity for one of the two, it "unites with it while making it take leave of the other." Maupertuis boldly applies these laws of chemistry to living beings. "Why," he asks, "if this force exists in Nature, would it not operate in the formation of the body of animals? Let us suppose that there are in each of the semens particles destined to form the heart, the head, the intestines, the arms, the legs; and these particles may each have a greater uniting power with that one which, in order to form the animal, has to be its neighbor, than with any other; the foetus will form, and were it yet a thousand times more organized than it is, it would form." Several years later, when Maupertuis, under the pseudonym of Doctor Baumann, wrote the *Systeme de La Nature* (1751), his ideas regarding these hereditary particles had apparently been influenced both by Leibniz's famous monads and by Buffon's theory of organic particles disseminated throughout nature, the latter theory having appeared in the meantime; and Maupertuis attributed to these postulated particles a property "akin to that which in us we term desire, aversion, memory." Moreover, although in the *Venus Physique* he proposed pangenesis with cautious reservation and only as a hypothesis worthy of investigation, in the *Systeme de la Nature*, no doubt encouraged in this direction by Buffon's embracement of the idea, he spoke without reserve. In the *Venus Physique* the 1746 edition reads simply: "As to the manner whereby there form in the semen of each animal particles analogous to those of that animal, I do not at all examine it here." The same passage in the later version included in the *Oeuvres* (1756) is amplified as follows: "As to the matter of which, in the semen of each animal, particles like that animal are formed, it would be a very bold conjecture, but one perhaps not destitute of all truth, to think that each part furnishes its own germs. Experiment could perhaps clear up this point, if one tried over along period to mutilate certain animals generation after generation; perhaps one would see the parts cut off diminish little by little; perhaps in the end one would see them annihilated." In the *Systeme de la Nature* we find: "The elements suitable for forming the foetus swim in the semens of the father and mother animals; but each, extracted from the part like that which it is to form, retains a sort of recollection of its old situation; and will resume it whenever it can, to form in the foetus the same part."

This particulate theory of heredity and the formation of the foetus is logically analogous to Mendelian theory, and far in advance of those ideas that heredity is determined by indivisible entities, fluids or vapors, subject to irrevocable blending through intermating, that were current until late in the nineteenth century. If in one direction Maupertuis' hereditary particles therefore look back to the monads of Leibniz, in the other they look forward to the concept of the genes; and they have really much more in common with the latter than the former. Of course, Maupertuis confused the hereditary particles with the effects they produce and with the parts whose development they control, but that was hardly avoidable at the time. After all, this was nearly a century before the formulation of the Cell Theory. To see that at bottom heredity must depend on a sort of organic, chemical memory, and to attribute this to separable particles that maintain their nature in combination is extraordinary enough. And what can one say of the perspicacity that proposed, as a test of pangenesis, the very experiment that Weismann was to perform nearly 150 years later?

Next to the particulate nature of the hereditary material, the most important of Mendel's principles is that of segregation. The hereditary units, or genes, as we now call them, are present in pairs as a consequence of fertilization, and the members of each pair segregate from one another in the reproductive cells prior to the production of another generation of offspring. This principle, too, was foreshadowed by Maupertuis. However the particles might previously be combined in each one of the parents, the particles from the two semens, he supposed, would unite separately in accordance with their affinities and, since corresponding particles from the mother and the father would be most alike, they would unite two by two and exclude other combinations. He says:

> One ought not to believe that in the two semens there are only precisely the particles which are needful to form one foetus, or the number of foetuses that the female is to bear: each of the two sexes without doubt furnishes a great many more than are necessary. But the two particles, which are to be adjacent once being united, any third, which could have made the same union, would no longer find its place, and would remain useless. It is this, it is by these repeated operations, that the child is formed from the particles of the father and the mother, and often bears visible marks that it partakes of the one and of the other.

In the seminal fluid of each individual the particles suitable for forming traits like those of that individual are the ones which are ordinarily most numerous, and which have the greatest combining power; although there are a great many others for different traits—the particles analogous to those of the father and the mother being the most numerous, and having the most combining power, will be those which most commonly unite; and they will ordinarily form animals like those from which they are come. This is further clarified by Maupertuis' consideration of hybrids:

> For as soon as there is a mixing of species, experience teaches us that the child resembles both the one and the other. If the elements come from animals of different species, but in which there still remains sufficient affinity between the elements; these more attached to the form of the father, those to the form of the mother, they will produce hybrid animals. Finally if the elements come from animals who no longer have between them sufficient analogy, the elements not being able to assume, or not being able to retain a suitable arrangement, generation becomes impossible.

Thus, Maupertuis goes on, one can explain the sterility of hybrids such as the mule.

> One of the most singular phenomena, and one of the most difficult to explain, is the sterility of hybrids. Experiment has shown that any animal born of the coupling of different species cannot reproduce. Could one not say that in the parts of the hinny and of the mule, the elements having taken a particular arrangement which was neither that which they had had in the ass, nor that which they had had in the mare; when these elements pass into the semens of the hinny and of the mule, the habitude of this last arrangement being most recent, and the habitude of the arrangement which they had had in the ancestors being stronger, because contracted over a greater number of generations, the elements remain in a certain equilibrium, and unite neither in one manner nor in the other

To anyone familiar with the causes of hybrid sterility this last paragraph has a startlingly modern ring. For Maupertuis was essentially correct. A set of chromosomes is indeed an arrangement of hereditary elements, and very often in hybrids the chromosomes derived from the two pure species are incapable of normal segregation, because of differences in the arrangements of their genes. The sterile hybrid between the radish and the cabbage furnishes an example. Radish chromosomes cannot pair with cabbage chromosomes. Segregation produces no effective germ cells, for neither an array wholly of radish elements nor one wholly of cabbage elements is likely to recur, and random mixtures of radish and cabbage chromosomes are physiologically ill assorted. In spite of the crudity of Maupertuis' idea, it is impossible not to admire his insight.

Mendel is often credited with having discovered genetic dominance, although actually it was known long before his day, having been described by Knight in 1823. Maupertuis also arrived, though vaguely, at the idea of dominance. This came about from two considerations. In the first place, as we know today, polydactyly is a dominant trait. In the second place, the other genetic character in which Maupertuis became interested, and to which he devoted the second part of the *Venus Physique*, was albinism in negroes, and this is a recessive. Maupertuis was therefore aware that whereas polydactyly descends regularly from affected persons, married to normals, to some but not all of their offspring, albinism, on the other hand, seemed to appear sporadically among negroes, albino negroes being born of parents both of whom were black. Yet Maupertuis was convinced that albinism was hereditary. Maupertuis was thus faced with the problem of accounting in his theory for the different modes of inheritance, and concluded: "There could be, on the other hand, arrangements so tenacious that from the first generation they dominate over all the previous arrangements, and efface the habitude of these."

Had Maupertuis evolved a metaphysical system of heredity and embryogenesis, and done nothing more, he could hardly be ranked above his contemporary biologists. Metaphysical systems, in the eighteenth century, were "a dime a dozen." Georges Herve, the first person to have considered the work of Maupertuis in the light of Mendelian genetics, has said: "His right to figure in a gallery of pre-Mendelian genetics would in that case have been disputable; if he holds a place there, it is because he knew enough to carry his researches into experimental realms." He diligently pursued the collection of further evidence on polydactyly, and with some success, since he could write: "I have found in Berlin two six-digited persons, and I have given the genealogy of one. I have not been able to follow with sufficient exactitude the genealogy of the other, who is a foreigner, and who has concealed himself from me: but he had six-digited children, and I have been assured that this polydactyly has been hereditary in his family for a long time. A scientist illustrious in Germany and Minister to the Duke of Würtemberg, M. de Bulfinger, was of such a family, and born with a sixth finger that his parents had cut off as a monstrosity."

Like Mendel, Maupertuis applied mathematics to genetic investigation, by calculating the probability that the polydactyly observed might be only a sporadic accident and not really inherited. He wrote:

> But if one wished to regard the continuation of polydactyly as an effect of pure chance, it would be necessary to see what the probability is that this accidental variation in a first parent would be repeated in his descendants. After a search, which I have made in a city which has one hundred thousand inhabitants, I have found two men who had this singularity. Let

us suppose, which is difficult, that three others have escaped me; and that in 20,000 men one can reckon on one six-digited: the probability that his son or daughter will not be born with polydactyly at all is 20,000 to 1; and that his son and his grandson will not be six-digited at all is 20,000 × 20,000 or 400,000,000 to 1: finally the probability that this singularity will not continue during three generations would be 8,000,000,000,000 to 1; a number so great that the certainty of the best demonstrated things of physics does not approach these probabilities.

This is not only an excellent example of scientific caution, but also represents what is without doubt the first application to genetics of one of the most important of the principles of the mathematics of probability, that of the probability of coincidence of independent items. It was this very principle that Mendel applied so effectively in his analysis of segregation, random recombination, and independent assortment.

BREEDING EXPERIMENTS

But Maupertuis was not content to make only this analysis. He undertook actual breeding experiments with animals to test out his theories, although of the results of these he has unfortunately left us only the account of a single one. It is related that he "adored animals and lived surrounded by them." "You are more pleased with Mme. d'Aiguillon than with me," wrote Mme. du Deffand to him one day, "she sends you cats." And Frederick wrote, too: "I know that at Paris just as at Berlin you are enjoying the delights of good company—I am only afraid that Mme. la duchesse d'Aiguillon is spoiling you. She loves parrots and cats, which is a prodigious merit in your eyes." Maupertuis had established himself in the outskirts of Berlin, in a spacious house adjacent to the royal park, near the present Tiergarten; and this house he had converted into a virtual Noah's ark. Samuel Formey, permanent secretary of the Berlin Academy, has left us the following description:

The house of M. de Maupertuis was a veritable menagerie, filled with animals of every species, who failed to maintain the proprieties. In the living rooms troops of dogs and cats, parrots and parakeets, etc. In the forecourt all sorts of strange birds. He once had sent from Hamburg a shipment of rare hens with a cock. It was sometimes dangerous to pass by the run of these animals, by whom some had been attacked, I was especially afraid of the Iceland dogs. M. de Maupertuis amused himself above all by creating new species by mating different races together; and he showed with complaisance the products of these matings, who partook of the qualities of the males and of the females who had engendered them. I loved better to see the birds, and especially the parakeets, which were charming.

It was of the Iceland dogs that Maupertuis has left us true account of his breeding experiment:

> Chance led me to meet, with a very singular bitch, of that breed that is called in Berlin the Iceland Dogs: she had her whole body the color of slate, and her head entirely yellow; a singularity which those who observe the manner in which the colors are distributed in this sort of animals will find perhaps rarer than that of supernumeraral digits. I wished to perpetuate it; and after three litters of dogs by different fathers which did not yield anything of the sort, at the fourth litter she gave birth to one who possessed it. The mother died; and from that dog, after several matings with different bitches, there was born another who was exactly like him. I actually have them both.

His breeding of dogs led him to wonder particularly about the supernumerary fifth digit which is not uncommon on the hind foot: "There are no animals at all upon whom supernumerary digits appear more frequently than upon dogs. It is a remarkable thing that they ordinarily have one digit less on the hind feet than on those in front, where they have five. However, it is not at all rare to find dogs who have a fifth digit on the hind feet, although most often detached from the bone and without articulation. Is this fifth digit of the hind feet then a supernumerary? Or is it, in the regular course, only a digit lost from breed to breed throughout the entire species, and which tends from time to time to reappear? For mutilations can become hereditary just as much as superfluities." Were all dogs, in other words, once five-toed on both front and back feet? Have we here a remnant, a vestige of a once functional structure? These observations might well have been made by Charles Darwin.

HIS THEORY OF EVOLUTION

Maupertuis' studies thus led him to evolution. Here with certainty he must be ranked above all the precursors of Darwin. To begin with, he was faced with the problem of accounting for supernumerary digits, albinism, and other hereditary anomalies on the basis of his theory of generation. This he solved ingeniously. "If each particle is united to those that are to be its neighbors, and only to those, the child is born perfect. If some particles are too distant, or of a form too little suitable, or too weak in affinity to unite with those with which they should be united, there is born a monster with deficiency. But if it happens that superfluous particles nevertheless find their place, and unite with the particles whose union was already sufficient, there is a monster with extra parts." Even Mendel did not foresee that deficiencies and duplications of the hereditary material might constitute a basis of abnormal development, a sort of mutation! Maupertuis comments on the remarkable fact that in monsters with extra parts, these are always to be found in the same locations as the corresponding normal parts: two heads are always on the neck: extra fingers are always on the hand, extra toes on the foot. This is very difficult to explain on the basis of the theory that monsters come from the union of two foetuses or eggs, which was the explanation forced upon the preformationists by the nature of their views; but it was not at all difficult to explain on the basis of Maupertuis' concepts. He described the skeleton of a giant man, preserved in the Hall of Anatomy of the Academy in Berlin with an extra vertebra in the lumbar region, inserted in a regular fashion between the ordinary vertebrae. How could this be the remains of a second foetus fused with the first?, he asked.

But on Maupertuis' particulate theory, "chance, or the scarcity of family traits, will sometimes make rarer assemblages; and one will see born of black parents a white child, or perhaps even a black of white parents. There are elements so susceptible of arrangement, or in which recollection is so confused, that they become arranged with the greatest facility;" elements which represent the condition in an ancestor rather than that in the immediate parent may enter into union in forming the embryo, producing resemblance to the ancestor rather than to the parent, but also "a total forgetfulness of the previous situation" may occur.

Maupertuis thus came to the conclusion that hereditary variants are sudden, accidental products—mutations, to use the modern term. Moreover, since negroes could by mutation produce "whites" (i.e. albinos), it was clear that racial, or species, differences—the distinction was not too clear in the eighteenth century—are produced by mutations. To Maupertuis, exactly as to Hugo de Vries a century and a half later, a species was merely a mutant form that had become established in nature. The evidence for this was clear from the artificial breeds of domestic animals. As in the case of Charles Darwin a century later, it was in particular the pigeons that clinched the argument.

> Nature contains the basis of all these variations: but chance or art brings them out. It is thus that those whose industry is applied to satisfying the

> taste of the curious are, so to say, creators of new species. We see appearing races of dogs, pigeons, canaries, which did not at all exist in Nature before. These were to begin with only fortuitous individuals; art and the repeated generations have made species of them. The famous Lyonnes every year created some new species, and destroyed that which was no longer in fashion. He corrects the forms and varies the colors: he has invented the species of the harlequin, the mopse, etc.

And then Maupertuis wonders seriously why this art should be restricted to animals. Might sultans in their seraglios practice a similar art? Had not Frederick William of Prussia built an armed force of giant soldiers, and thereby, thought Maupertuis, singularly increased the stature of the Prussian people?

If the ingenuity of man can produce species, why not nature, either by "fortuitous combinations of the particles of the seminal fluids, or effects of combining powers too potent or too weak among the particles" or by the action of the environment, such as the effect of climate or nutrition, on the hereditary particles. It is worth emphasizing, for it has been misunderstood, that Maupertuis raises the latter possibility only as one worthy of investigation; but clearly at this point he anticipated both Erasmus Darwin and Lamarck in suggesting the possibility of evolution through an inheritance of environmentally modified characters. Even so, it is the direct mutational action of heat or other factors on the hereditary material itself that Maupertuis seems most to have had in mind.

> For the rest, although I suppose here that the basis of all these variations is to be found in the seminal fluids themselves, I do not exclude the influence that climate and foods might have. It seems that the heat of the torrid zone is more likely to foment the particles which render the skin black than those which render it white: and I do not know to what point this influence of climate or of foods might extend, after long centuries of time.

It is likewise clear that Maupertuis understood that most mutant forms are deleterious and at a disadvantage in comparison with the normal or wild types. "What is certain is that all the varieties which can characterize new species of animals and plants, tend to become extinguished: they are the deviations of Nature in which she perseveres only through art or system. Her works always tend to resume the upper hand."

How, then, account for the distribution of different races and species? The "thousand" human varieties are insuperable difficulties for the preformationist; but by mutation, migration and isolation they are readily accounted for by Maupertuis. Perhaps, he suggested, in the tropics all the peoples are dark of skin in spite of the interruptions caused by the sea, because of the heat of the torrid zone over a long period of time. The geographical isolation of Nature's deviations must play a part here, for:

> In traveling away from the equator, the color of the people grows lighter by shades. It is still very brown just outside the tropics; and one does not find complete whiteness until one has reached the temperate zone. It is at the limits of the zone that one finds the whitest peoples. Well, men of excessive stature, and others of excessive littleness, are species of monsters; but monsters, which can become peoples, were one to apply himself to multiplying them. Are there not races of giants and dwarfs? These have become established, either by the suitability of climates, or rather because, in the time when they commenced to appear, they would have been chased into these regions by other men, who would have been afraid of the Colossi, or disdain the Pygmies. However many giants, however many dwarfs, however many blacks, may have been born among other men; pride or fear would have armed against them the greater part of mankind; and the more numerous species would have relegated these deformed races to the least habitable climates of the Earth. The Dwarfs will have retired toward the arctic pole: the Giants will have inhabited the Magellanic lands: the Blacks will have peopled the torrid zone.

However naïve these anthropological conceptions may be—and they were an easy target for the sharp gibes of Voltaire—there is nevertheless a groping here for a truth that was only to be captured fully by Charles Darwin and Alfred Russel Wallace in a later day. There is no naïvety, only pure genius, in these final words:

> Could one not explain by that means how from two individuals alone the multiplication of the most dissimilar species could have followed? They could have owed their first origination only to certain fortuitous productions, in which the elementary particles failed to retain the order they possessed in the father and mother animals; each degree of error would have produced a new species; and by reason of repeated deviations would have arrived at the infinite diversity of animals that we see today; which will perhaps still increase with time, but to which perhaps the passage of centuries will bring only imperceptible increases.

Glass, B. 1959. In Forerunners of Darwin: 1745–1859, *B. Glass, O. Temkin, and W. L. Straus, Jr., eds. The Johns Hopkins University Press, Baltimore, MD; "Mau-*

pertuis, Pioneer of Genetics and Evolution," pp. 51–83. This excerpt is a revision of Glass, "Maupertuis and the beginning of genetics," Quarterly Review of Biology *22 (1947):196–210, and Glass, Maupertuis, "A Forgotten Genius,"* Scientific American *193 (October 1955):100–110. Reprinted with permission of The Johns Hopkins University Press.*

On the Various Contrivances by which Orchids are Fertilized by Insects (1862)

Charles Darwin

For my purpose British Orchids may be divided into three groups, and the arrangement is, for the most part, a natural one. But I leave out of consideration the British species of *Cypripedium*, with its two anthers, of which I know nothing. Of these three groups the first consists of the Ophreæ, which have pollinia furnished at their lower ends with a caudicle, congenitally attached to a viscid disc. The anther stands above the rostellum. The Ophreæ include most of our common Orchids.

First, for the genus *Orchis*. The reader may find the following details rather difficult to understand; but I can assure him, if he will have patience to make out this first case, the succeeding cases will be easily intelligible.

The accompanying diagrams (Fig. I) show the relative position of the more important organs in the flower of the Early *Orchis* (*O. mascula*). The sepals and the petals have been removed, excepting the labellum with its nectary. The nectary is shown only in the side view (*n* Fig. I. A); for its enlarged orifice is almost hidden in shade in (Fig. I. B) the front view. The stigma (*s*) is bilobed, and consists of two almost confluent stigmas; it lies under the pouch-formed (*r*) rostellum. The anther (*a* in Fig. I. A and B) consists of two rather widely separated cells, which are longitudinally open in front: each cell includes a pollen-mass or pollinium.

A pollinium removed out of one of the two anther-cells is represented by Fig. I. C; it consists of a number of wedge-formed packets of pollen-grains (see Fig. I. F, in which the packets are forcibly separated), united together by excessively elastic, thin threads. These threads become confluent at the lower end of each pollen-mass, and compose the (*c* in

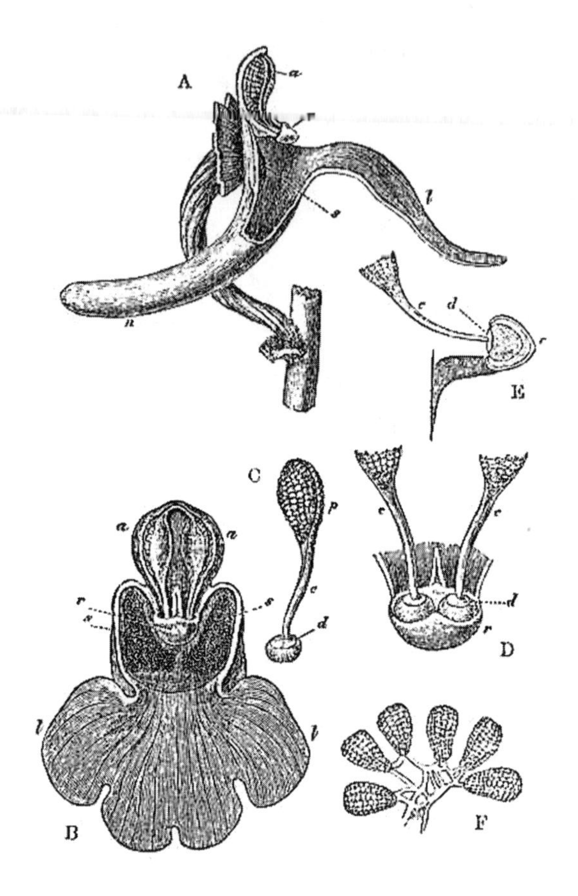

FIG. I.

Fig. I. C) straight elastic caudicle. The end of the caudicle is firmly attached to the viscid disc (*d* in Fig. I. C), which consists (as may have been seen in the section, Fig. I. E) of a minute oval piece of membrane, with a ball of viscid matter on it's under side. Each pollinium has its separate disc; and the two balls of viscid matter lie enclosed together (Fig. I. D) within the rostellum.

The rostellum is a nearly spherical, somewhat pointed projection (*r* in Fig. I. A and B) overhanging the two almost confluent stigmas, and must be fully described, as every detail of its structure is full of signification. A section through one of the discs and balls of viscid matter is given (Fig. I. E); and a front view of both viscid discs within the rostellum (Fig. I. D) is likewise given. This latter figure (Fig. I. D) probably best serves to explain the structure of the rostellum; but it must be understood that the front lip is here considerably depressed. The lowest part of the anther is united to the back of the rostellum, as may be seen in Fig. I. B. At an early period of growth the rostellum consists of a mass of polygonal cells, full of brownish matter, which cells soon resolve themselves into two balls of an extremely viscid semi-fluid substance, void of structure. These viscid masses are slightly elongated, almost flat on the top, and convex below. They lie quite free within the rostellum (being surrounded by fluid), except at the back, where each viscid ball firmly adheres to a small portion or disc of the exterior membrane of the rostellum. The ends of the two caudicles are strongly attached to these two little discs of membrane.

The membrane forming the whole exterior surface of the rostellum is at first continuous; but as soon as the flower opens the slightest touch causes it to rupture transversely in a sinuous line, in front of the anther-cells and of the little crest or fold of membrane (see Fig. I. D) between them. This act of rupturing makes no difference in the shape of the rostellum, but converts the front part into a lip, which can easily be depressed. This lip is represented considerably depressed in Fig. I. D, and its edge is seen, Fig. I. B, in the front view. When the lip is thoroughly depressed, the two balls of viscid matter are exposed. Owing to the elasticity of the hinder part or hinge, the lip or pouch, when not pressed down, springs up and again encloses the two viscid balls.

I will not affirm that the rupturing of the exterior membrane of the rostellum never takes place spontaneously; and no doubt the membrane is prepared for the rupture by having become very weak along defined lines; but several times I saw the act ensue from an excessively slight touch—so slight that I conclude that the action is not simply mechanical, but, for the want of a better term, may be called vital. We shall hereafter meet with other cases, in which the slightest touch or the vapour of chloroform causes the exterior membrane of the rostellum to rupture along certain defined lines.

At the same time that the rostellum becomes transversely ruptured in front, it probably (for it was impossible to ascertain this fact from the position of the parts) ruptures behind in two oval lines, thus separating and freeing from the rest of the exterior surface of the rostellum the two little discs of membrane, to which externally the two caudicles are attached, and to which internally the two balls of viscid matter adhere. The line of rupture is thus very complex, but strictly defined.

As the two anther-cells open longitudinally in front from top to bottom, even before the flower expands, as soon as the rostellum is properly ruptured from the effects of a slight touch, its lip can be easily depressed, and, the two little discs of membrane being already separate, the two pollinia now lie absolutely free, but are still embedded in their proper places. So that the packets of pollen and the caudicles lie within the anther-cells; the discs still form part of the posterior surface of the rostellum, but are separate: and the balls of viscid matter still lie concealed within the rostellum.

Now let us see how this complex mechanism acts. Let us suppose an insect to alight on the labellum, which forms a good landing-place, and to push its head into the chamber at the back of which lies the stigma, in order to reach with its proboscis the end of the nectary; or, which does equally well to show the action, push a sharply-pointed common pencil into the nectary. Owing to the pouch-formed rostellum projecting into the gangway of the nectary, it is scarcely possible that any object can be pushed into it without the rostellum being touched. The exterior membrane of the rostellum then ruptures in the proper lines, and the lip or pouch is most easily depressed. When this is effected, one or both of the viscid balls will almost infallibly touch the intruding body. So viscid are these balls that whatever they touch they firmly stick to. Moreover the viscid matter has the peculiar chemical quality of setting, like a cement, hard and dry in a few minutes' time. As the anther-cells are open in front, when the insect withdraws its head, or when the pencil is withdrawn, one pollinium, or both, will be withdrawn, firmly cemented to the object, projecting up like horns.

The firmness of the attachment of the cement is very necessary, as we shall immediately see; for if the pollinia were to fall sideways or backwards they could never fertilize the flower. From the position in which the two pollinia lie in their cells, they diverge a little when attached to any object. Now let us suppose our insect to fly to another flower, or insert the pencil, with the attached pollinium, into the same or into another nectary: it will be evident that the firmly attached pollinium will be simply pushed against or into its old position, namely, into its anther-cell. How then can the flower be fertilized? This is effected by a beautiful contrivance: though the viscid surface remains immovably affixed, the apparently insignificant and minute disc of membrane to which the caudicle adheres is endowed with a remarkable power of contraction (as will hereafter be more minutely described), which causes the pollinium to sweep through about 90 degrees, always in one direction, viz., towards the apex of the proboscis or pencil, in the course, on an average, of thirty seconds. Now after this movement and interval of time (which would allow the insect to fly to another flower), it will be seen, that, if the pencil is inserted into the nectary, the thick end of the pollinium will exactly strike the stigmatic surface.

Here again comes into play another pretty adaptation, long ago noticed by Robert Brown ('Transactions of the Linnæan Society,' vol. xvi.

p. 731.). The stigma is very viscid, but not so viscid as when touched to pull the whole pollinium off the insect's head or off the pencil, yet sufficiently viscid to break the elastic threads by which the packets of pollen grains are tied together, and leave some of them on the stigma. Hence a pollinium attached to an insect or to the pencil can be applied to many stigmas, and will fertilize all. I have seen the pollinia of *Orchis pyramidalis* adhering to the proboscis of a moth, with the stump-like caudicle alone left, all the packets of pollen having been left glued to the stigmas of the flowers successively visited.

One or two little points must still be noticed. The balls of viscid matter within the pouch-formed rostellum are surrounded with fluid; and this is very important, for, as already mentioned, the viscid matter sets hard when exposed to the air for a very short time. I have pulled the balls out of their pouches and have found that in a few minutes they entirely lost their power of adhesion. Again, the little discs of membrane, the movement of which, as causing the movement of the pollinium, is so absolutely indispensable for the fertilization of the flower, lie at the upper and back surface of the rostellum, and are closely enfolded and thus kept damp within the bases of the anther-cells; and this is very necessary, as an exposure of about thirty seconds causes the movement of depression to take place; but as long as the disc is kept damp the pollinium remains ready for action whenever removed by an insect.

Lastly, the pouch, after being depressed, springs up to its former position; and this is of great service; for if this action did not take place, and an insect after depressing the lip failed to remove either viscid ball, or if it removed one alone, in the first case both, and in the second case one of the viscid balls would be left exposed to the air; consequently they would quickly lose all adhesiveness, and the pollinia would be rendered absolutely useless. That insects often remove one alone of the two pollinia at a time in many kinds of *Orchis* is certain; it is even probable that they generally remove only one at a time, for the lower and older flowers almost always have both pollinia removed, and the younger flowers close beneath the buds, which will have been seldomer visited, have frequently only one pollinium removed. In a spike of *Orchis maculata* I found as many as ten flowers, chiefly the upper ones, which had only one pollinium removed; the other pollinium being in place, with the lip of the rostellum well closed up, and all the mechanism perfect for its subsequent removal by some insect.

The description now given of the action of the organs in *Orchis mascula* applies to *O. morio*, *fusca*, *maculata*, and *latifolia*, and to *Aceras anthropomorpha*. These species present slight and apparently coordinated differences in the length of the caudicle, in the direction of the nectary, in the shape and position of the stigma, but they are not worth detailing. In all, the pollinia undergo, after removal from the anther-cells, the curious movement of depression, which is so necessary to place them in a right position on the insects head to strike the stigmatic surface of another flower. In *Aceras* the caudicle is unusually short; the nectary consists of two minute rounded depressions; the stigma is transversely elongated; the two viscid discs lie so close together within the rostellum that they affect each others outline; this is worth notice, as a step towards the two becoming absolutely confluent, as in *O. pyramidalis*. Nevertheless, in *Aceras* a single pollinium is sometimes removed by insects, though more rarely than with the other species.

We now come to *Orchis pyramidalis*, one of the most highly organized species that I have examined, and which is ranked by several botanists in a distinct genus. The relative position of the parts is here considerably different from what it is in *O. mascula* and its allies. There are two quite distinct rounded stigmatic surfaces placed on each side of the pouch-formed rostellum. This latter organ, instead of standing some height above the nectary, is brought down so as to overhang and partially to close its orifice. The antechamber to the nectary formed by the union of the edges of the labellum to the column, which is large in *O. mascula* and its allies, is here small. The pouch-formed rostellum is hollowed out on the under side in the middle: it is filled with fluid. The viscid disc is single, of the shape of a saddle, carrying on its nearly flattop or seat the two caudicles of the pollinia; of which the two truncated ends firmly adhere to its upper surface. Before the membrane of the rostellum ruptures, it can be clearly seen that the saddle-formed disc forms part of the continuous surface of the rostellum. The disc is partially hidden and kept damp (which is of great importance) by the largely over-folded basal membranes of the two anther-cells. The upper membrane of the disc consists of several layers of minute cells, and is therefore rather thick; it is lined beneath with a layer of highly adhesive matter, which is formed within the rostellum. The single saddle-formed disc strictly answers to the two minute, oval, separate discs of membrane to which the two caudicles of *O. mascula* and its allies are attached: two separate discs have here become completely confluent. When the flower opens and the rostellum has become symmetrically ruptured, either from a touch or spontaneously (I know not which), the slightest touch depresses the lip, that is, the lower and bilobed portion of the exterior membrane of the rostellum, which projects into the mouth of the nectary. When the lip is depressed, the under and viscid surface of the disc, still remaining in its proper place, is uncovered, and is almost certain to adhere to the touching object. Even a human hair, when pushed into the nectary, is stiff enough to depress the lip or pouch; and the viscid surface of the saddle adheres to it. If, however, the lip be touched too slightly, it springs back and re-covers the under side of the saddle.

The perfect adaptation of the parts is well shown by cutting off the end of the nectary and inserting a bristle at that end; consequently in a reversed direction to that in which nature intended moths to insert their probosces, and it will be found that the rostellum may easily be torn or penetrated, but that the saddle is rarely or never caught. When the saddle sticking to a bristle together with its pollinia is removed, the under lip instantly curls closely inwards, and leaves the orifice of the nectary more open than it was before; but whether this is of any real use to moths which so frequently visit the flowers, and consequently to the plant, I will not pretend to decide.

Lastly, the labellum is furnished with two prominent ridges, sloping down to the middle and expanding outwards like the mouth of a decoy; these ridges perfectly serve to guide any flexible body, like a fine bristle of hair, into the minute and rounded orifice of the nectary, which, small as it is, is partly choked up by the rostellum. This contrivance of the guiding ridges may be compared to the little instrument sometimes used for guiding a thread into the fine eye of a needle.

Now let us see how these parts act. Let a moth insert its proboscis (and we shall presently see how frequently the flowers are visited by Lepidoptera) between the guiding ridges of the labellum, or insert a fine bristle, and it is surely conducted to the minute orifice of the nectary, and can hardly fail to depress the lip of the rostellum; this being effected, the bristle comes into contact with the now naked and sticky under surface of the suspended saddle-formed disc. When the bristle is removed, the saddle with the attached pollinia is removed. Almost instantly, as soon as the saddle is exposed to the air, a rapid movement takes place, and the two flaps curl inwards and embrace the bristle. When the pollinia are pulled out by their caudicles, by a pair of pincers, so that the saddle has nothing to clasp, I observed that the tips curled inwards so as to touch each other in nine seconds, and in nine more seconds the saddle was converted by curling still more inwards into an apparently solid ball. The probosces of the many moths which I have examined, with the pollinia of this *Orchis* attached to them, were so thin that the tips of the saddle just met on the under side. Hence a naturalist, who sent me a moth with several saddles attached to its proboscis, and who did not know of this movement, very naturally came to the extraordinary conclusion that the moth had cleverly bored through the exact centres of the so-called sticky glands of some Orchid.

Of course this rapid clasping movement helps to fix the saddle with its pollinia upright on the proboscis, which is very important; but the viscid matter rapidly setting hard would probably suffice for this end, and the real object gained is the divergence of the pollinia. The pollinia, being attached to the flat top or seat of the saddle, project at first straight up and are nearly parallel to each other; but as the flat top curls round the cylindrical and thin proboscis, or round a bristle, the pollinia necessarily diverge. As soon as the saddle has clasped the bristle and the pollinia have diverged, a second movement commences, which action, like the last, is exclusively due to the contraction of the saddle-shaped disc of membrane. This second movement is the same as that in *O. mascula* and its allies, and causes the divergent pollinia, which at first projected at right angles to the needle or bristle, to sweep through nearly 90 degrees towards the tip of the needle, so as to become depressed and finally to lie in the same plane with the needle. In three specimens this second movement was effected in from 30 to 34 seconds after the removal of the pollinia from the anther-cells, and therefore in about 15 seconds after the saddle had clasped the bristle.

The use of this double movement becomes evident if a bristle with pollinia attached to it, which have diverged and become depressed, be pushed between the guiding ridges of the labellum into the nectary of the same or another flower; for the two ends of the pollinia will be found to have acquired exactly such a position that the end of the one strikes against the stigma on the one side, and the end of the other at the same moment strikes against the stigma on the opposite side. These stigmas are so viscid that they rupture the elastic threads by which the packets of pollen are bound together; and some dark-green grains will be seen, even by the naked eye, remaining on the two white stigmatic surfaces. I have shown this little experiment to several persons, and all have expressed the liveliest admiration at the perfection of the contrivance by which this Orchid is fertilized.

As in no other plant, or indeed in hardly any animal, can adaptations of one part to another, and of the whole to other organized beings widely remote in the scale of nature, be named more perfect than those presented by this *Orchis*, it may be worth while briefly to sum them up. As the flowers are visited both by day and night-flying Lepidoptera, I do not think that it is fanciful to believe that the bright-purple tint (whether or not specially developed for this purpose) attracts the day-fliers, and the strong foxy odour the night-fliers. The upper sepal and two upper petals form a hood protecting the anther and stigmatic surfaces from the weather. The labellum is developed into a long nectary in order to attract Lepidoptera, and we shall presently give reasons for suspecting that the nectar is purposely so lodged that it can be sucked only slowly (very differently from in most flowers of other families), in order to give time for the curious chemical quality of the viscid matter on the under side of the saddle setting hard and dry. He who will insert a fine and flexible bristle into the expanded mouth of the sloping ridges on the labellum will not doubt that they serve as guides; and that they effectually prevent the bristle or proboscis from being inserted obliquely into the nectary. This circumstance is of manifest importance, for, if the proboscis were inserted obliquely, the saddle-formed disc would become attached obliquely, and after the compounded movement of the pollinia they could not strike the two lateral stigmatic surfaces.

Then we have the rostellum partially closing the mouth of the nectary, like a trap placed in a run for game; and the trap so complex and perfect, with its symmetrical lines of rupture forming the saddle-shaped disc above, and the lip of the pouch below; and, lastly, this lip so easily depressed that the proboscis of a moth could hardly fail to uncover the viscid disc and adhere to it. But if this did fail to occur, the elastic lip would rise again and re-cover and keep damp the viscid surface. We see the viscid matter within the rostellum attached to the saddle-shaped disc alone, and surrounded by fluid, so that the viscid matter does not set hard till the disc is withdrawn. Then we have the upper surface of the saddle, with its attached caudicles, also kept damp within the bases of the anther-cells, until withdrawn, when the curious clasping movement instantly commences, causing the pollinia to diverge, followed by the movement of depression, which compounded movements together are exactly fitted to cause the ends of the two pollinia to strike the two stigmatic surfaces. These stigmatic surfaces are sticky enough not to tear off the whole pollinium from the proboscis of the moth, but by rupturing the elastic threads to secure a few packets of pollen, leaving plenty for other flowers.

But let it be observed that, although the moth probably takes a considerable time to suck the nectar of any one flower, yet the movement of depression in the pollinia does not commence (as I know by trial) until the pollinia are fairly withdrawn out of their cells; nor will the movement be completed, and the pollinia be fitted to strike the stigmatic surfaces, until about half a minute has elapsed, which will give ample time for the moth to fly to another plant, and thus effect a union between two distinct individuals. Lastly, we have the wonderful growth of the pollen tubes and their penetration of the stigma, as well as the mysteries of germination, though these are common to all phanerogamic plants.

Orchis ustulata resembles *O. pyramidalis* in some important respects, and differs in others. The labellum is deeply channeled; this channel, which replaces the guiding ridges of *O. pyramidalis*, leads to the small triangular orifice of the short nectary. The upper angle of the triangle is overhung by the rostellum, the pouch of which is rather pointed below. Owing to this position of the rostellum, close to the mouth of the nectary, the stigma is necessarily double and lateral; but we here have an interesting gradation, showing how easily the single and slightly lobed medial stigma of *O. maculata* would pass through the bilobed stigma of *O. mascula* into that of *O. ustulata*, and thence into the truly double stigma of *O. pyramidalis*; for in *O. ustulata*, directly under the rostellum, there is a narrow rim, in direct continuity with the two lateral stigmas, and which itself has the character of a true stigma, as it is formed of utriculi, or true stigmatic tissue, exactly like that of the lateral stigmas. The viscid discs are somewhat elongated. The pollinia undergo the usual movement of depression, and in acquiring this position the two diverge slightly, so as to be ready to strike the lateral stigmas.

The divergence seemed due to the manner or direction in which the membrane forming the top of the disc contracted obliquely; but I am not sure of this observation.

I have now described the structure, as seen in fresh specimens, of most of the British species of the genus *Orchis*. All these species absolutely require the aid of insects for their fertilization. This is obvious from the fact that the pollinia are so closely embedded in their anther-cells, and the disc with its ball of viscid matter in the pouch-formed rostellum, that they cannot be shaken out by violence. We have also seen numerous contrivances by which the pollinia assume, after an interval of time, a position adapted to strike the stigmatic surface; and this indicates that the pollinia are habitually carried from one flower to another. But to prove that insects are necessary I covered up a plant of *Orchis morio* under a bell-glass, before any of its pollinia had been removed, leaving three adjoining plants uncovered; I looked at the latter every morning, and daily found some of the pollinia removed, till all were removed with the exception of the pollinia in one flower low down on one spike, and with the exception of those in one or two flowers at the apex of each spike, which were never removed. I then looked at the perfectly healthy plant under the bell-glass, and it had, of course, all its pollinia in their cells. I tried an analogous experiment with specimens of *O. mascula* with exactly the same result. It deserves notice that the spikes which had been covered up, when subsequently left uncovered, had not their pollinia removed, and did not, of course, set any seed, whereas the adjoining plants produced plenty of seed; and from this fact I infer that probably there is a proper season for each kind of *Orchis*, and that insects cease their visits after the proper season has passed, and the regular secretion of nectar has ceased.

I have been in the habit for twenty years of watching Orchids, and have never seen an insect visit a flower, excepting butterflies twice sucking *O. pyramidalis* and *Gymnadenia conopsea*. M. Ménière, first name is unknown (in Bull. Bot. Soc. de France, tom. i. 1854, p. 370) says he saw, in Dr. Guépins collection, bees collected at Saumur with the pollinia of Orchids attached to their heads; and he states that a person who kept bees near the Jardin de la Faculté (at Toulouse) complained that his bees returned from the garden with their heads charged with yellow bodies, of which they could not free themselves. This is good evidence how firmly the pollinia become attached. There is nothing to show whether the pollinia in these cases belonged to the genus *Orchis* or to other genera of the family, some of which I know are visited by bees. I have evidence in a bumble and hive bee sent me by Professor Westwood, with pollinia attached to them; and Mr. F. Bond informs me that he has seen pollinia attached to other species of bees; but I feel almost certain that bees do not habitually visit the common British species of *Orchis*. On the other hand, I have met with several accounts in entomological works of pollinia having been observed attached to moths. Mr. F. Bond was so kind as to send me a large number of moths in this condition, with permission, at the risk of the destruction of the specimens, to remove the pollinia; and this is quite necessary, in order to ascertain to what species the pollinia belong. Singularly all the pollinia (with the exception of a few from Orchids of the genus *Habenaria*, presently to be mentioned) belonged to *O. pyramidalis*. I here give the list of twenty-three species of Lepidoptera, with the pollinia of *O. pyramidalis* attached to their proboscis:

Polyommatus alexis, *Lycæna phlæas*, *Arge galathea*, *Hesperia sylvanus*, *Hesperia linea*, *Syrichthus alveolus*, *Anthrocera filipendulæ*, *Anthrocera trifolii*, *Lithosia complanata*, *Leucania lithargyria* (two specimens), *Caradrina blanda*, *Caradrina alsines*, *Agrotis cataleuca*, *Eubolia mensuraria* (two specimens), *Hadenia dentina*, *Heliothis marginata* (two specimens), *Xylophasia sublustris* (two specimens), *Euclidia glyphica*, *Toxocampa pastinum*, *Melanippe rivaria*, *Spilodes palealis*, *Spilodes cinctalis*, *Acontia luctuosa*.

A large majority of these moths and butterflies had two or three pairs of pollinia attached to them, and invariably to the proboscis. The *Acontia* had seven pair, and the *Caradrina* no less than eleven pair! The probosces of these two latter moths presented an extraordinary arborescent appearance. The saddle-formed discs adhered to the proboscis, one before the other, with perfect symmetry (as necessarily follows from its insertion having been guided by the ridges on the labellum), each saddle bearing its pair of pollinia. The unfortunate Caradrina, with its proboscis thus encumbered, could hardly have reached the extremity of the nectary, and would soon have been starved to death. These two moths must have sucked many more than the seven and eleven flowers, of which they bore the trophies, for the earlier attached pollinia had lost much of their pollen, showing that they had touched many viscid stigmas.

The list shows, also, how many species of Lepidoptera visit the same kind of *Orchis*. The *Hadena* also frequents *Habenaria*. Probably all the Orchids provided with spur-like nectaries are visited indifferently by many kinds of moths. I have twice observed *Gymnadenia conopsea*, transplanted many miles from its native home, with nearly all its pollinia removed. Mr. Marshall in Ely has made the same observation on transplanted specimens of *O. maculata* ('Gardeners Chronicle,' 1861, p. 73). Mr. Marshall's communication was in answer to some remarks of mine on the subject previously published in the 'Gardeners Chronicle,' 1860, p. 528. I have not sufficient evidence, but I suspect that the Neotteæ and Malaxeæ, which have not tubular nectaries, are frequented by other orders of insects. *Listera* is generally fertilized by small Hymenoptera; *Spiranthes* by bumblebees. Mr. Marshall found that fifteen plants of *Ophrys muscifera*, transplanted to Ely, had not one pollen-mass removed; so it was during the first summer with *Epipactis latifolia* planted in my own garden; during the following summer six flowers out of ten had their pollinia removed by some insect. These facts possibly indicate that certain Orchids require special insects for their fertilization. On the other hand, *Malaxis paludosa*, placed in a bog about two miles from that in which it grew, had most of its pollinia immediately removed.

The list that follows serves to show that in most cases moths perform the work of fertilization effectually. But the list by no means gives a fair idea how effectually it is done; for I have often found nearly all the pollinia removed; but generally I kept an exact record in exceptional cases only, as may be seen by the appended remarks. Moreover, in most cases, the pollinia that had not been removed were in the upper flowers beneath the buds, and many of these would probably have been subsequently removed. I have often found abundance of pollen on the stigmas of flowers which had not their own pollinia removed, showing that they had been visited by insects; in many other flowers the pollinia had been removed, but no pollen had, as yet, been left on their stigmas.

In the second lot of *O. morio*, given in the list, we see the injurious ef-

TABLE I

Species	NUMBER OF FLOWERS WITH BOTH OR ONE POLLINIUM REMOVED. FLOWERS LATELY OPEN EXCLUDED.	NUMBER OF FLOWERS WITH ONLY ONE POLLINIUM REMOVED. THESE FLOWERS ARE INCLUDED IN THE COLUMN TO THE LEFT.	NUMBER OF FLOWERS WITH NEITHER POLLINIUM REMOVED.
Orchis morio[1]	22	2	6
Orchis morio[2]	110	23	193
Orchis pyramidalis[3]	39		8
Orchis pyramidalis[4]	102		66
Orchis pyramidalis[5]	57		166
Orchis maculata[6]	32	6	12
Orchis maculata[7]	21	5	7
Orchis maculata[8]	28	17	50
Orchis latifolia[9]	50	27	119
Orchis fusca[10]	8	5	54
Aceras anthropomorpha[11]	63	6	34

1. Three small plants. N. Kent

2. Thirty-eight plants. N. Kent. These plants were examined after nearly four weeks of extraordinary cold and wet weather in 1860; and therefore under the most unfavourable circumstances

3. Two plants. N. Kent and Devonshire

4. Six plants from two protected valleys. Devon

5. Six plants from a much-exposed bank. Devon

6. One plant. Staffordshire. Of the twelve flowers which had not their pollinia removed, the greater number were young flowers under the buds

7. One plant. Surrey

8. Two plants. N. and S. Kent

9. Nine plants from S. Kent, sent me by the Rev. B. S. Malden. The flowers were all mature

10. Two plants. S. Kent. Flowers quite mature, and even withered

11. Four plants. S. Kent

fects of the extraordinary cold and wet season of 1860 in the infrequency of the visits of insects, and, consequently, on the fertilization of this Orchid. Very few seed-capsules were produced this year.

In *O. pyramidalis* I have examined spikes in which every single expanded flower had its pollinia removed. The forty-nine lower flowers of a spike from Folkestone (sent me by Sir Charles Lyell) actually produced forty-eight fine seed-capsules; and of the sixty-nine flowers in three other spikes, seven alone had failed to produce capsules. These facts show conclusively how well moths had performed their office of marriage-priests.

In the list, the third lot of *O. pyramidalis* grew on a steep grassy bank, overhanging the sea near Torquay, and where there were no bushes or other shelter for moths; being surprised how few pollinia had been removed, though the spikes were old, and very many of the lower flowers had withered, I gathered, for comparison, six other spikes from two bushy and sheltered valleys, half a mile on each side of the exposed bank; these spikes were certainly younger, and would probably have had several more of their pollinia removed; but in their present condition we see how much more frequently they had been visited by moths, and consequently fertilized, than those growing on the much exposed bank. The Bee Ophrys and *O. pyramidalis*, in many parts of England, grow mingled together; and they did so here, but the Bee Ophrys, instead of being, as usual, the rarer species, was here much more abundant than *O. pyramidalis*; no one would readily have suspected that probably one chief reason of this difference was, that the exposed situation was unfavourable to moths, and therefore to the seeding of *O. pyramidalis*; whereas, as we shall hereafter see, the Bee Ophrys is independent of insects.

I counted many spikes of *O. latifolia*, because, being familiar with the usual state of the closely-allied *O. maculata*, I was surprised to observe in nine nearly withered spikes how few pollinia had been removed. In one instance, however, I found *O. maculata* even worse fertilized; for seven spikes, which had borne 315 flowers, produced only forty-nine seed-capsules—that is, each plant on an average produced only seven capsules: in this case the plants had grown in greater numbers close together, forming large beds, than I had ever before observed; and I imagined that there were too many plants for the moths to suck and fertilize. On some other plants, growing at no great distance, I found above thirty capsules on each spike.

Orchis fusca offers a more curious case of imperfect fertilization. I examined ten fine spikes from two localities in South Kent, sent to me by Mr. Oxenden and Mr. Malden: most of the flowers on these spikes were partly withered, with the pollen mouldy even in the uppermost flowers; hence we may safely infer that no more pollinia would have been removed. I examined all the flowers only in two spikes, on account of the trouble from their withered condition, and the result may be seen in the list, namely, fifty-four flowers with both pollinia in place, and only eight with one or both removed. We see in this Orchid, and in *O. latifolia*, neither of which had been sufficiently visited by moths, that there are more flowers with one pollinium than with both removed. I casually examined many flowers in the other spikes of *O. fusca*, and the proportion of pollinia removed was evidently not greater than in the two given in the list. The ten spikes had borne 358 flowers, but, in accordance with the few pollinia removed, only eleven capsules had been formed: five of the ten spikes bore not a single capsule; two spikes had only one, and one bore as many as four capsules. As corroborating what I have previously said on pollen being often found on the stigmas of flowers which have their own pollinia in place, I may add that, of the eleven flowers which had produced capsules, five had both pollinia still within their now withered anther-cells.

From these facts the suspicion naturally arises that *O. fusca* is so rare a species in Britain from not being sufficiently attractive to our moths, and consequently not producing a sufficiency of seed. C. K. Sprengel ('Das entdeckte Geheimniss,' etc. s. 404) noticed, that in Germany *O. militaris* (ranked by Bentham as the same species with *O. fusca*) is likewise imperfectly fertilized, but more perfectly than our *O. fusca*; for he found five old spikes bearing 138 flowers, which had set thirty-one capsules; and he contrasts the state of these flowers with those of *Gymnadenia conopsea*, in which almost every flower produces a capsule.

An allied curious subject remains to be discussed. The existence of a well-developed spur-like nectary seems almost to imply the secretion of nectar. But Sprengel, a most careful observer, thoroughly searched many flowers of *O. latifolia* and of *O. morio*, and could never find a drop of nectar; nor could Krünitz (Quoted by J. G. Kurr in his 'Untersuchungen über die Bedeutung der Nektarien,' 1833, s. 28. See also 'Das entdeckte Geheimnis,' s. 403.) find nectar either in the nectary or on the labellum of *O. morio*, *fusca*, *militaris*, *maculata*, and *latifolia*. I have looked to all the species hitherto mentioned in this work, and could find no signs of nectar; I examined, for instance, eleven flowers of *O. maculata*, taken from different plants growing in different districts, and taken from the most favourable position on each spike, and could not find under the microscope the smallest bead of nectar. Sprengel calls these flowers "Scheinsaftblumen," or sham-nectar-producers; that is, he believes, for he well knew that the visits of insects were indispensable for their fertilization, that these plants exist by an organized system of deception. But when we reflect on the incalculable number of plants which have existed for enormous periods of time, all absolutely requiring for each generation insect-agency; when we think of the special contrivances clearly showing that, after an insect has visited one flower and has been cheated, it must almost immediately go to a second flower, in order that impregnation may be effected (of which fact we have the plainest evidence in the large number of pollinia attached to the probosces of those moths which had visited *O. pyramidalis*), we cannot believe in so gigantic an imposture. He who believes in this doctrine must rank very low the instinctive knowledge of many kinds of moths.

To test the intellect of moths I tried the following little experiment, which ought to have been tried on a larger scale. I removed a few already-opened flowers on a spike of *O. pyramidalis*, and then cut off about half the length of the nectaries of the six next not-expanded flowers. When all the flowers were nearly withered, I found that thirteen of the fifteen upper flowers with perfect nectaries had their pollinia removed, and two alone had their pollinia still in their anther-cells; of the six flowers with their nectaries cut off, three had their pollinia removed, and three were still in place; and this seems to indicate that moths do not go to work in a quite senseless manner.

Nature may be said to have tried, but not quite fairly, this same experiment; for *Orchis pyramidalis*, as shown by Mr. Bentham ('Handbook of the British Flora,' 1858, p. 501.) often produces monstrous flowers without a nectary, or with a short and imperfect one. Sir C. Lyell sent me several spikes from Folkestone with many flowers in this condition: I found six without a vestige of a nectary, and their pollinia had not been removed. In about a dozen other flowers, having either short nectaries, or with the labellum imperfect, with the guiding ridges either absent or developed in excess and rendered foliaceous, the pollinia in one alone had been removed, and the ovarium of another flower was swelling. Yet I found that the saddle-formed discs in the first six and in the dozen other flowers were perfect, and that they readily clasped a needle when inserted in the proper place. Moths had removed the pollinia, and had thoroughly well fertilized the perfect flowers on the same spike; so that they must have neglected the monstrous flowers, or, if visiting them, the derangement in the complex mechanism had hindered the removement of the pollinia, and prevented their fertilization.

From these several facts I still suspected that nectar must be secreted by our common Orchids, and I determined to examine *O. morio* rigorously. As soon as many flowers were open, I began to examine them for twenty-three consecutive days: I looked at them after hot sunshine, after rain, and at all hours: I kept the spikes in water, and examined them at midnight, and early the next morning: I irritated the nectaries with a bristle, and exposed them to irritating vapours: I took flowers which had quite lately had their pollinia removed by insects, of which I had independent proof on one occasion by finding within the nectary grains of some foreign pollen (I may mention that on soaking and separating the laminæ of the proboscis of a moth, which had the pollinia of a Habenaria attached to its head, a surprising number of pollen-grains of some other plant were seen in the water); and I took other flowers which from their position on the spike would soon have had their pollinia removed; but the nectary was invariably quite dry.

I still thought that the secretion might perhaps take place at the earliest dawn, as I have found that the secretion of nectar in flowers of other orders ceases and commences in the most rapid manner. Consequently, as *O. pyramidalis* is visited (as may be seen in the foregoing list) by butterflies and by several day-flying moths (such as *Anthrocera* and *Acontia*), I carefully examined its nectary, taking plants from several localities and the most likely flowers, as just explained; but the glittering points within the nectary were absolutely dry. Hence we may safely conclude that the nectaries of the above-named Orchids neither in this country nor in Germany ever contain nectar.

In examining the nectaries of *O. morio* and *maculata*, and especially of *O. pyramidalis*, I was surprised at the degree to which the inner and outer membranes forming the tube or spur were separated from each other,—also at the delicate nature of the inner membrane, which could be most easily penetrated,—and, lastly, at the quantity of fluid contained between these two membranes. So copious is this fluid, that, having at first merely cut off the ends of the nectaries of *O. pyramidalis*, and gently squeezing them on glass under the microscope, such large drops of fluid exuded from the cut ends that I concluded that the nectaries certainly did contain nectar; but when I carefully made, without any pressure, a slit along the upper surface, and looked into the tube, I found that the inner surface was quite dry.

I then examined the nectaries of *Gymnadenia conopsea* (a plant ranked by some botanists as a true *Orchis*) and of *Habenaria bifolia*, which are always one-third or two-thirds full of nectar: the inner membrane presented the same structure in being covered with papillæ, but there was a plain difference in the inner and outer membranes being closely united, instead of being, as with the above-named species of *Orchis*, in some degree separated from each other and charged with fluid. Hence I am led to suspect that moths penetrate the lax inner membrane of the nectaries of these Orchids, and suck the copious fluid between the two membranes. I am aware that this is a bold hypothesis; for no case is recorded of nectar being contained between the two membranes of a nectary (The nearest approach to this supposed case, yet really distinct, is the secretion of nectar in several monocotyledonous plants (as described by Ad. Brongniart in Bull. Soc. Bot. de France, tom. i. 1854, p. 75) from between the two walls (feuillets) which form the divisions of the ovarium. But the nectar in this case is conducted to the outside by a channel; and the secreting surface is homologically an exterior surface.) or of Lepidoptera penetrating with their delicate probosces even the laxest membrane.

We have seen how numerous and beautifully adapted the contrivances are for the fertilization of Orchids. We know that it is of the highest importance that the pollinia, when attached to the head or proboscis of an insect, should not fall sideways or backwards. We know that the ball of viscid matter at the extremity of the pollinium rapidly becomes more and more viscid, and sets hard in a few minutes' time: therefore we can see that it would be an advantage to the plant if the moth were delayed in sucking the nectar, so as to give time for the viscid disc to become immovably affixed. Assuredly moths would be delayed if they had to bore through several points of the inner membrane of the nectary, and to suck the nectar from the intercellular spaces. This explanation of the good thus gained in some degree corroborates the hypothesis that the nectaries of the above-named species of *Orchis* do not secrete nectar externally, but into internal cavities.

The following singular relation supports this view more strongly. I have found nectar within the nectaries of only five British species of Ophreæ, namely, in *Gymnadenia conopsea* and *albida*, in *Habenaria bifolia* and *chlorantha*, and in *Peristylus* (or *Habenaria*) *viridis*. The first four of these species have the viscid surface of the discs of their pollinia, not enclosed within the pouch, but naked, which by itself shows that the viscid matter has a different chemical nature from that in the species of true *Orchis*, and does not rapidly set hard when exposed to the air. But to make sure of this I removed the pollinia from their anther-cells, so that the upper as well as the under surfaces of the viscid discs were freely exposed to the air; in *Gymnadenia conopsea* the disc remained sticky for two hours, and in *Habenaria chlorantha* for more than twenty-four hours. In *Peristylus viridis* the viscid disc is covered by a pouch-formed membrane, but this is so minute that botanists have overlooked it. I did not, when examining this species, see the importance of exactly ascertaining how rapidly the viscid matter set hard; but I copy from my notes the words written at the time: "Disc remains sticky for some time when removed from its little pouch."

Now the bearing of these facts is clear: if, as is certainly the case, the viscid matter of the discs of these five latter species is so viscid as to serve at once for the firm attachment of the pollinia to insects, and does not quickly become more and more viscid and set hard, there could be no use in moths being delayed in sucking the nectar by having to bore through the inner membrane of the nectaries at several points; and in these five species, and in these alone, we find copious nectar ready stored for their use in the open tubular nectaries. If this relation, on the one hand, between the viscid matter requiring some little time to set hard, and the nectar being so lodged that moths are delayed in getting it; and, on the other hand, between the viscid matter being at first as viscid as ever it will become, and the nectar lying all ready for rapid suction, be accidental, it is a fortunate accident for the plant. If not accidental, and I cannot believe it to be accidental, what a singular case of adaptation!

Darwin, C. R. 1862. On the Various Contrivances by which Orchids are Fertilized by Insects. *J. Murray, London, U.K; Chapter I, pp. 9–53.*

Lepidoptera: *Heliconidæ* (1861)

Henry W. Bates

The family *Heliconidæ* was established by Mr. E. Doubleday in 1847, in Doubleday and Hewitson's 'Genera of Diurnal Lepidoptera.' It was founded on a number of butterflies, remarkable for the elongated shape of their wings, and peculiar (with the exception of one genus, *Hamadryas*, which the author placed provisionally in the family) to the intertropical and subtropical zones of America. Many of them had been described by the older authors under *Heliconia*, *Mechanitis*, and several other ill-defined genera. They had been previously (in 1836) united in a tribe, *Heliconides*, by Dr. Boisduval in his 'Species General des Lepidopteres;' but this comprehended also the group *Acræidæ*, which Doubleday excluded from the family. Linnæus treated them as a section of the genus *Papilio*, under the name of *Heliconii*. The nearest allies of the *Heliconidæ* are the *Acræidæ* just mentioned and the *Danaidæ*: all are distinguished from the true *Nymphalidæ* by the discoidal cell of the hind wings being always closed by perfect tubular nervules. Mr. Doubleday, placing more reliance on the shape of the antennæ and the abdominal border of the hind wings than on the far more important character above named, was led to exclude the genus *Eueides* from the family: this rendered the definition of the two groups very difficult, if not impossible, *Eueides* having the wing-cells closed in the same way as the *Heliconidæ*. Excepting that I re-admit *Eueides*, and exclude *Hamadryas*, which does not enter into the series of the American

Heliconidæ, the family will be treated in the present memoir as defined in the work above quoted.

The position of the *Heliconidæ* in the order Lepidoptera may be understood when I state that in a natural system the group would stand at the head of the whole series of families of which the order is composed. At least, this should be its place according to the view now taken of the order by many systematists, who arrange the families of *Rhopalocera*, or butterflies, according to their degree of dissimilarity to the *Heterocera*, or moths—in other words, according as their structure shows a lower or a higher stage in an ascending scale of organization. For, as the lower families of moths are allied to other orders of insects, the further a group recedes from them in structure, the higher is the grade of perfection of the Lepidopterous type which it exhibits. The families show their degree of affinity to moths by many characters, the principal of which is the structure of the anterior legs in the adult state of the insects. The *Heterocera* have always six perfect legs: most of the families of *Rhopalocera* have the anterior pair in a more or less rudimentary condition; and as the atrophy seems to have reached its furthest stage in the *Heliconidæ*, this group must be considered as occupying the highest rank in the order. Other characters accompany the one derived from the structure of the legs, which it is unnecessary here to enumerate. It will be seen from these remarks that the order Lepidoptera is one of those groups in the Animal Kingdom which show, beyond the many collateral branches of development that always exist, a clear linear advancement of organization.

The most interesting part of the natural history of the *Heliconidæ* is the mimetic analogies of which a great many of the species are the objects. Mimetic analogies, it is scarcely necessary to observe, are resemblances in external appearance, shape, and colours between members of widely distinct families: an idea of what is meant may be formed by supposing a Pigeon to exist with the general figure and plumage of a Hawk. Most modern authors who have written on the group have mentioned the striking instances of this kind of resemblances exhibited with reference to the *Heliconidæ;* but no attempt has been made to describe them fully, nor to explain them. I will give a short account of the leading facts, and then mention some circumstances which seem to throw light on their true nature and origin.

A large number of the species are accompanied in the districts they inhabit by other species which counterfeit them in the way described. The imitators belong to the following groups: *Papilio*, *Pieris*, *Euterpe*, and

Leptalis (fam. *Papilionidæ*), *Protogonius* (*Nymphalidæ*), *Ithomeis* (*Eryjainidæ*), *Castnia* (*Castniadæ*), *Dioptis*, *Pericopis*, *Hyelosia*, and other genera (*Bombyaidæ* Moths). I conclude that the *Heliconidæ* are the objects imitated, because they all have the same family facies, whilst the analogous species are dissimilar to their nearest allies—perverted, as it were, to produce the resemblance, from the normal facies of the genus or family to which they severally belong. The resemblance is so close, that it is only after long practice that the true can be distinguished from the counterfeit, when on the wing in their native forests. I was never able to distinguish the *Leptalides* from the species they imitated, although they belong to a family totally different in structure and metamorphosis from the *Heliconidæ*, without examining them closely after capture. They fly in the same parts of the forest, and generally in company with the species they mimic.

I have already given an account of the local modifications to which the *Heliconidæ* are subject. It is a most curious circumstance, that corresponding races or species of counterfeiting groups accompany these local forms. In some cases I found proof that such species are modified from place to place to suit the peculiar forms of *Heliconidæ* there stationed. As this is an important point, and one which throws light on the origin of mimetic species, I must ask the reader's careful attention to the details, referring to the plates. Plate LV. fig. 1 *a* (*Ithomia Flora*) and fig. 1 (*Leptalis Theonoë*) represent a Heliconide and its imitator, both of which inhabit the banks of the Cuparí, a river belonging to the Amazon system, in 55° W. long. Neither of these is found on the Upper Amazons (60° to 70° W. long.), where I made the remaining part of my observations on these insects. At Ega, on this upper river, in 65° W. long., two species of *Ithomia* occurred, which I consider to be local varieties or races of *I. Flora*, namely, *I. Onega* (Pl. LV. fig. 2 *a*) and *I. Illinissa* (Pl. LV. fig. 6 *a*). It is immaterial to the question in hand whether these be considered absolutely distinct species or races; the *Leptalis* which was found in their company was the form called *L. Lysinoë* (Pl. LV. fig. 3), with its admitted varieties (figs. 4, 5, 6, and 8). Only one of these varieties of *Leptalis* mimics an *Ithomia;* this is our fig. 6, which evidently counterfeits *Ithomia Illinissa* (fig. 6 *a*). The prevailing form of *Leptalis*, the *L. Lysinoë* (fig. 3), has no resemblance to any *Ithomia* of Ega, but is, when flying, a wonderful imitation of the *Stalachtis Duvalii* (Pl. LV. fig. 3 *a*), a common insect belonging to a genus (family *Erycinidæ*) equally flourishing and abundant in individuals with the members of the family *Heliconidæ*. I think there will be no doubt in the mind of anyone that the Ega *Leptalides* are local varieties of the Cuparí *L. Theonoë* (fig. 1), when all the connecting links between them are studied in the figures given on our two plates. It is highly probable, therefore, that this species

PLATE LV.

has been by some means modified with especial reference to the changed *Ithomiæ*, or other insects, of the locality. The varieties, figs. 4, 5, and 8, were excessively rare: they have the appearance of sports, and show how variable the species has been in this district.

The same takes place at St. Paulo, in 69° W. long. Here we find the *Ithomiæ* again changed. Neither the *I. Flora* of the Cuparí and Lower Amazons nor the *I. Illinissa* of Ega occurs; but the second Ega species, *I. Onega*, inhabits the district, and several other species not found in other places, amongst them *I. Ilerdina* (Pl. LVI. fig. 4 *a*), *I. Chrysodonia* (Pl. LVI. fig. 3 *a*), and *I. Virginia* (Pl. LVI. fig. 6 *a*). The prevailing species of *Ithomia* of the locality being thus changed, how stands it with the *Leptalides*? They are changed also, and again with close reference to the *Ithomiæ*. I found a number of different varieties, which I could not doubt were local forms of the same species as that found on the Cuparí and at Ega. Thus, there was one (Pl. LV. fig. 2) closely resembling *L. Theonoë* (fig. 1), but modified to produce a nearer imitation of the *Ithomia Onega* (Pl. LV. fig. 2 *a*), which I believe to be a local form of *I. Flora*. Another (Pl. LVI. fig. 3) resembled *Ithomia Chrysodonia* (Pl. LVI. fig. 3 *a*); but the imitation is not fixed or

PLATE LVI.

exact in all the specimens taken, as may be seen by comparing figs. 1, 2, 3, of the same Plate. We here detect nature, as it were, *striving* after a correct imitation: the explanation of this will be attempted further on. A third form of *Leptalis* found at St. Paulo is the one figured Pl. LVI. fig. 6, which mimics the *Ithomia Virginia* (Pl. LVI. fig. *6 a*). Besides these, a few varieties occurred which did not closely counterfeit any *Ithomia*; they were very much rarer than the others. I figure two of these (Pl. LV. figs. 7, 9), to show how they connect the other more strongly modified varieties with the Ega forms.

The *Ithomiæ* concerned in these imitations have the character of true species, being distinct and constant, with the exception of *I. Chrysodonia*, whose varieties are detailed under the head of the species, which is variable, and throws light on the origin of the rest. They are all excessively numerous in individuals, swarms of each kind being found in the localities they inhabit. The *Leptalides* are exceedingly rare; they cannot be more than as 1 to 1000 with regard to the *Ithomiæ*. It may be asked, how can we know they are all varieties (using the term as meaning forms descended from others) of one species? I must refer to the figures given, which, although they do not include all the connecting varieties that were collected, show how nearly all the forms are linked together. The most distinct amongst them are those figured Pl. LVI. figs. 4 and 6. The feature which distinguishes fig. 4 is the white colour of the disk of the hind wings, and the veins which traverse it. This character is shown to be due to variation, from the facts that *Ithomia Oncidia*, an undoubted variety of *I. Chrysodonia* (or *Orolina*), exhibits a commencement of this milky shade of the wings, and that many individuals of *I. Ilerdina* (Pl. LVI. fig. 4 *a*) display steps of modification in the colours of the veins. The variety figured (Pl. LVI. fig. 6) appears distinct, from the single pale spot near the tips of the wings; an approximation to this is seen in the variety figured Pl. LV. fig. 9, which is an undoubted modification of *L. Lysinoë* (Pl. LV. fig. 3). The remarkable variety figured (Pl. LV. fig. 4) has been described by the only author who has treated these insects (Mr. Hewitson) as a variety of *L. Lysinoë*. In a polymorphic form, like this *Leptalis*, none of the varieties can be taken from the rest and denominated species, (using the term as meaning forms which cannot have descended from other closely allied ones), without exercising the art of species-making in the most arbitrary manner. For if we allow so great a latitude to variation as that from figs. 3 to 4, 5, 6, 7, 8, and 9, Pl. LV., how can we venture to say that natural modification, having gone so far, was incompetent to go further, so as to produce figs. 4 and 6, Pl. LVI., and that those forms must have arisen by some unknown agency? It is true, they have not arisen by simple variation, or sports, in one generation, but, as we shall presently see, by an external agency accumulating the modifications of many generations in two diverging directions. As the connecting links have not all been found, they may be called species: the word is of little importance. The habits of all are the same. When I had collected only two or three of the most distinct, I considered them separate species; but intermediate forms successively occurred, every capture tending to link the whole more closely together. The explanation that the whole are the result of hybridization from a few originally distinct species cannot at all apply in this case, because the distinct forms whose intercrossing would be required to produce the hybrids are confined to districts situated many hundred miles apart.

None of these *Leptalides* have been found in any other district or country than those inhabited by the *Ithomiæ* which they counterfeit. A species very closely allied to *L. Lysinoë*, var. *Argochloë* (Pl. LVI. fig. 6), has been received from Mexico (*L. Antherize*); but an *Ithomia*, of nearly the same colours (*I. Nero*) also inhabits Mexico. Many other species of *Leptalis*, of much larger size than the one here discussed, also mimic *Heliconidæ*, the objects of imitation not being *Ithomiæ*, but other genera of the family. Two of these are figured on Pl. LVI. *L. Orise* (Pl. LVI. fig. 8) is a remarkably exact counterfeit of *Methona Psidii* (fig. 8 *a*), the resemblance being carried to minutiæ, such as the colour of the antennæ and the spotting of the abdomen. *L. Amphione*, var. *Egaëna* (Pl. LVI. fig; 7), is very curious, as being a satellite of *Mechanitis Polymnia*, var. *Egaëna* (fig. 7 *a*), both peculiar to the district of Ega—the typical *L. Amphione* being found at Surinam, in company with the typical *M. Polymnia*, which it resembles—local varieties or sister species of *Leptalis Amphione* accompanying local varieties of *Mechanitis Polymnia* in other parts of tropical America.

Several species of *Dioptis*, a genus of moths, and *Ithomeis*, a genus of *Erycinidæ*, also accompany these species or distinct local forms of *Ithomia*. A few of the moths are figured on Pl. LV. figs. 10, 11, 12, 13. The imitations may not appear very exact from the figures; but when the insects are seen on the wing in their native woods, they deceive the most experienced eye.

A similar series of mimetic analogies occurs in the Old World, between the Asiatic and African *Danaidæ*, or representatives of the *Heliconidæ*, and species of other families of butterflies and moths. No instance is known in these families of a tropical species of one hemisphere counterfeiting a form belonging to the other. A most remarkable case of mimicry has been recorded by Mr. Trimen in a *Papilio* of Southern Africa, *P. Cenea*, whose male wears to deception the livery of one species of *Danais*, namely, *D. Echeria*, whilst the female resembles a quite different one, *D. Chrysippus*,—both

African. Mimetic analogies, however, are not confined to the Lepidoptera; most orders of insects supply them; but they are displayed only by certain families. Many instances are known where parasitic bees and two winged flies mimic in dress various industrious or nestbuilding bees, at whose expense they live in the manner of the cuckoo. I found on the banks of the Amazons many of these cuckoo bees and flies, which all wore the livery of working bees peculiar to the country.

The instances of this kind of analogy most familiar to European entomologists are those of the European species of *Trochilium* (a genus of moths), which strangely mimic various bees, wasps, and other Hymenopterous and Dipterous insects. The parallelism between these several forms and their geographical relations have not yet, I believe, been investigated. The resemblances seem to be more closely specific in tropical countries than in Europe; and I think it likely that the counterfeits in high latitudes may not always be found in company with their models. It is possible the geographical relations between the species concerned may have been disturbed by the great climatal and geological changes which have occurred in this part of the world since the date when they first came into existence.

Not only, however, are *Heliconidæ* the objects selected for imitation; some of them are themselves the imitators; in other words, they counterfeit each other, and this to a considerable extent. Species belonging to distinct genera have been confounded, owing to their being almost identical in colours and markings; in fact, many of them can scarcely be distinguished except by their generic characters. It is a most strange circumstance connected with this family, that its two sections, or subfamilies, have been mingled together by all authors, owing to the very close resemblance of many of their species. Analogies between the two subfamilies have been mistaken for affinities. It is sometimes difficult to understand in these cases which is the imitator and which the imitated. We have, however, generally a sure test in the one set exhibiting a departure from the normal style of colouring of their congeners, whilst the other are conformable to their generic types. The species of *Napeogenes* are, by this criterion, evidently all imitators of *Ithomiæ;* they are also rare insects, like the *Leptalides.* The mimetic species of *Heliconius* must be, for the same reason, imitators.

These imitative resemblances, of which hundreds of instances could be cited, are full of interest, and fill us with the greater astonishment the closer we investigate them; for some show a minute and palpably intentional likeness which is perfectly staggering. I have found that those features of the portrait are most attended to by nature which produce the most effective deception when the insects are seen in nature. The faithfulness of the resemblance, in many cases, is not so striking when they are seen in the cabinet. Although I had daily practice in insect collecting for many years, and was always on my guard, I was constantly being deceived by them when in the woods. It may be asked, why are mimetic analogies so numerous and amazingly exact in insects, whilst so rare and vague in the higher animals? The only answer that I can suggest is, that insects have perhaps attained a higher degree of specialization, after their type, than most other classes: this seems to be shown by the perfection of their adaptive structures and instincts. Their being more numerous and striking in tropical than in temperate countries is perhaps attributable to the more active competitive life, and the more rapid succession of their generations, in hot than in cold countries.

It is not difficult to divine the meaning or final cause of these analogies. When we see a species of moth which frequents flowers in the daytime wearing the appearance of a wasp, we feel compelled to infer that the imitation is intended to protect the otherwise defenseless insect by deceiving insectivorous animals, which persecute the moth, but avoid the wasp. May not the Heliconide dress serve the same purpose to the *Leptalis?* Is it not probable, seeing the excessive abundance of the one species and the fewness of individuals of the other, that the Heliconide is free from the persecution to which the *Leptalis* is subjected?

I think it clear that the mutual resemblance in this and other cases cannot be entirely due to similarity of habits or the coincident adaptation of the two analogues to similar physical conditions. This is a very abstruse part of our subject; for I think the facts of similar variation in two already nearly allied forms do sometimes show that they have been affected in a similar way by physical conditions. A great number of insects are modified in one direction by a seaside habitat. I found, also, the general colours of many widely different species affected in a uniform way in the interior of the South American continent. But this does not produce the specific imitation of one species by another; it only prepares the way for it.

It is perhaps true that the causes (to be discussed presently), which produce a close or mimetic analogy, cannot operate on forms which have not already a general resemblance, owing to similarity of habits, external conditions, or accidental coincidence. Species or groups which have this kind of resemblance to each other have been called by Dr. Collingwood recurrent animal forms. The English bee moths owe the narrow and pointed shapes of their wings, which already approximate them to bees, to their blood relationship to the Hawk-Moth family. Their bee-like size, form, and flight doubtless arise from their bee-like habits. A close specific analogy between anyone of these and a bee, such as exists between the insects discussed in this memoir, could scarcely be due to an accidental resemblance like that between the Hawk-Moth and a bee, or to similarity of habits. It would mean an adaptation of the Moth with especial reference to the bee.

I believe, therefore, that the specific mimetic analogies exhibited in connexion with the *Heliconidæ* are adaptations—phenomena of precisely the same nature as those in which insects and other beings are assimilated in superficial appearance to the vegetable or inorganic substance on which, or amongst which, they live. The likeness of a beetle or a lizard to the bark of the tree on which it crawls cannot be explained as an identical result produced by a common cause acting on the tree and the animal.

Some of the imitations by insects of inanimate and living objects are very singular, and may be mentioned in this place. Many caterpillars of moths, but sometimes the cases only which are manufactured and inhabited by the caterpillars; have a most deceptive likeness to dry twigs and other objects. Moths themselves very frequently resemble the bark on which they are found, or have wings coloured and veined like the fallen leaves on which they lie motionless. The accidental general resemblance between the shape of moths' wings and leaves here gives nature the groundwork for much mimetic analogy. It has been pointed out by Rossler that the buff-tip moth, when at rest, is intended to represent a broken piece of lichen-covered branch—the coloured tips of these wings, when they are closed, resembling a section of the wood. Other moths are deceptively like the excrement of birds on leaves. I met with a species of phytophagous beetle (*Chlamys Pilula*) on the Amazons, which was undistinguishable by the eye from the dung of Caterpillars on foliage. These two latter cases of imitation should be carefully considered by those who would be inclined to think that the object of mimetic analogies in nature was simply variety, beauty, or ornament: nevertheless these are certainly attendants on the phenomena; some South-American *Cassidæ* resemble glittering drops of dew on the tips of leaves, owing to their burnished pearly gold colour. Some species of Longicorn Coleoptera (*Onychocerus Scorpio* and *Conoentricus*) have precisely the colour and sculpture of the bark of the particular species of tree on which each is found. It is remarkable that other, species of the same small group of *Longicornes* (*Phacellocera Buquetii*, *Cyclopeplus Batesii*) counterfeit, not inanimate objects, like their near kindred just cited, but other insects, in the same way as the *Leptalides* do the *Heliconidæ*.

Amongst the living objects mimicked by insects are the predacious species from which it is the interest of the mimickers to be concealed. Thus, the species of *Scaphura* (a genus of Crickets) in South America resemble in a wonderful manner different sand wasps of large size, which are constantly on the search for crickets to provision their nests with. Another pretty Cricket, which I observed, was a good imitation of a tiger beetle, and was always found on trees frequented by the beetles (*Odontocheilæ*). There are endless instances of predacious insects being disguised by having similar shapes and colours to those of their prey; many spiders are thus endowed: but some hunting spiders mimic flower-buds, and station themselves motionless in the axils of leaves and other parts of plants to wait for their victims.

The most extraordinary instance of imitation I ever met with was that of a very large caterpillar, which stretched itself from amidst the foliage of a tree which I was one day examining, and startled me by its resemblance to a small snake. The first three segments behind the head were dilatable at the will of the insect, and had on each side a large black pupillated spot, which resembled the eye of the reptile: it was a poisonous or viperine species mimicked, and not an innocuous or colubrine snake; this was proved by the imitation of keeled scales on the crown, which was produced by the recumbent feet, as the caterpillar threw itself backwards. The Rev. Joseph Greene, to whom I gave a description, supposes the insect to have belonged to the family *Notodontidæ*, many of which have the habit of thus bending themselves. I carried off the caterpillar, and alarmed every one in the village where I was then living, to whom I showed it. It unfortunately died before reaching the adult state.

I think it will be conceded that all these various kinds of imitative resemblances belong to the same class of phenomena, and are subject to the same explanation. The fact of one species mimicking an inanimate object, and another of an allied genus a living insect of another family, sufficiently proves this. I do not see how they differ from the adaptations of organs or instincts to the functions or objects they relate to. All are adaptations, either of the whole outward dress or of special parts, having in view the welfare of the creatures that possess them.

Every species in nature may be looked upon as maintaining its existence by virtue of some endowment enabling it to withstand the host of adverse circumstances by which it is surrounded. The means are of endless diversity. Some are provided with special organs of offence, others have passive means of holding their own in the battle of life. Great fecundity is generally of much avail, added to capabilities, active or passive, of wide dispersion; so that when the species is extirpated in one part of its area of distribution, the place is refilled by migration of individuals from another part. A great number have means of concealment from their enemies, of one sort or other. Many are enabled to escape extermination, or obtain subsistence, by disguises of various kinds: amongst these must be reckoned the adaptive resemblance of an otherwise defenseless species to one whose flourishing race shows that it enjoys peculiar advantages.

What advantages the *Heliconidæ* possess to make them so flourishing a group, and consequently the objects of so much mimetic resemblance, it is not easy to discover. There is nothing apparent in their structure or habits which could render them safe from persecution by the numerous insectivorous animals which are constantly on the watch in the same parts of the forest which they inhabit. It is probable they are unpalatable to insect enemies. Some of them (*Lycorea*, *Ituna*) have exsertible glands near the anus, which are protruded when the insects are roughly handled; it is well known that similar organs in other families (*Cambidæ*, *Staphylinidæ*) secrete fetid liquids or gases, and serve as a protection to the species. I have noticed also that recently killed specimens of Danaoid *Heliconidæ*, when set out to dry, were always less subject than other insects to be devoured by vermin. They have all a peculiar smell. I never saw the flocks of slow-flying *Heliconidæ* in the woods persecuted by birds or dragon-flies, to which they would have been easy prey; nor, when at rest on leaves, did they appear to he molested by lizards or the predacious flies of the family *Asilidæ*, which were very often seen pouncing on butterflies of other families. If they owe their flourishing existence to this cause, it would be intelligible why the *Leptalidæ*, whose scanty number of individuals reveals a less protected condition, should be disguised in their dress, and thus share their immunity.

This explanation, however, would not apply to the imitation of Danaoid *Heliconidæ* by other species of the same subfamily. Moreover, there are several genera of other groups (*e. g.*, *Heliconius*, *Papilio*) which contain mimetic species side by side with species that are the objects of mimicry by members of other families. There is no reason to conclude that some of these possess the peculiar means of defense of the Danaoid *Heliconidæ*, whilst their near kindred are deprived of them. It is not unreasonable to suppose that some species are taken by insectivorous animals, whilst others flying in company with them are avoided. I could not, from their excessive scarcity, ascertain on the spot that the *Leptalides* were thus picked out. I noticed, however, that other genera of their family (*Pieridæ*) were much persecuted. We have proof, in the case of sand-wasps, which provision their nests with insects, that a single species is very generally selected out of numbers, even of the same genus, existing in the same locality. I was quite convinced in the case of *Cerceris Binodis* of South America, which destroys numbers of a *Megalostomis* (family *Clythridæ*), that the great rarity of the beetle was owing to its serving as prey to the *Oerceris*. We cannot point out all the conditions of life of each species concerned in these mimetic analogies. All that we can say is, that some species show, by their great abundance in the adult state, that during this period, before they propagate their kind, they enjoy by some means immunity from effective persecution, and that it is therefore an advantage to others not so fortunate, and otherwise unprovided for, if they are so like as to be mistaken for them.

The process by which a mimetic analogy is brought about in nature is a problem which involves that of the origin of all species and all adaptations. What I have previously said regarding the variation of species, and the segregation of local races from variations, the change of species of *Heliconidæ* from one locality to another, and the probable vital necessity of their counterfeits which accompany them keeping to the exact imitation in each locality, has prepared the way to the explanation I have to give. In the cases of local variation of the *Heliconidæ*, there was nothing, as before remarked, very apparent in the conditions of the localities to show why one or more of the varieties should prevail in each over their kindred varieties. There was nothing to show plainly that any cause of the formation of local varieties existed, other than the direct action of physical conditions on the individuals, although this might be seen to be clearly incompetent to explain the occurrence of several varieties of the same species in one locality. We could only conclude, from the way in which the varieties occur in nature, as described in the case of *Mechanitis Polymnia*, that the local conditions favoured the increase of one or more varieties in a district at the expense of the others—the selected ones being different in different districts. What these conditions were, or have been, was not revealed by the facts. With the mimetic species *Leptalis Theonoë* the case is different. We see here a similar segregation of local forms to that of *Mechanitis Polymnia;* but we believe we know the conditions of life of the species, and find that they vary from one locality to another. The existence of the species, in each locality, is seen to depend on its form and colours, or dress, being assimilated to those of the *Ithomiæ* of the same district, which *Ithomiæ* are changed from place to place, such assimilation being apparently its only means of escaping extermination by insectivorous animals. Thus we have here the reason why local races are formed out of the natural variations of a species: the question then remains, how is this brought about?

The explanation of this seems to be quite clear on the theory of natural selection, as recently expounded by Mr. Darwin in the 'Origin of Species.' The local varieties or races cannot be supposed to have been formed by the direct action of physical conditions on the individuals, because, in limited districts where these conditions are the same, the most widely contrasted varieties are found existing together, and it is inexplicable how they could have produced the nice adaptations which these diverse varieties exhibit. All the varieties figured on Pl. LV. figs. 2, 7, 9, and on Pl. LVI. figs. 1, 2, 3, 4, 6, are found at St. Paulo, within a mile of each other, in the same humid forest. Neither can these adapted races, as before remarked, have originated in one generation by *sports* or a single act of variation in each case. It is clear, therefore, that some other active principle must be here at work to draw out, as it were, steadily in certain directions the suitable variations which arise, generation after generation, until forms have resulted which, like our races of *Leptalis Theonoë*, are considerably different from their parent as well as their sister forms. This principle can be no other than natural selection, the selecting agents being insectivorous animals, which gradually destroy those sports or varieties that are not sufficiently like *Ithomiæ* to deceive them. It would seem as though our *Leptalis* naturally produced simple varieties of a nature to resemble *Ithomiæ*; it is not always so, as is proved by many of them figured in the places above quoted. There is some general resemblance, it is true; and this is not purely accidental; for it is quite natural that the parent *Leptalis* should produce offspring varying in the direction of *Ithomiæ*, being itself similar to an *Ithomia*, and having inherited the property of varying in this manner through a long line of ancestors. We cannot ascertain, in this case, whether changed physical conditions have had any effect, quantitative or qualitative, on the variability of the species after migrating to a new district. At any rate, the existing varieties of our *Leptalis* show that the variations of *Leptalis* and *Ithomia* are not quite coincident, and that the agency of natural selection is required to bring the slowly forming race of one to resemble the other. I do not forget that at each step of selection the forms of *Leptalis* must have had sufficient resemblance to an *Ithomia* to lead to their preservation, or, at least, to prevent their complete extinction: as, however, the two analogues so much resemble each other at the commencement of the process, these steps would not be numerous. In many cases of mimetic resemblance, the mimicry is not so exact as in the *Leptalides*. This would show either that the imitator has only inherited its form from remote ancestors who were actively persecuted, the persecution having ceased during the career of its immediate ancestors; or it would show that the persecutor is not keen or rigid in its selection; a moderate degree of resemblance suffices to deceive it, and therefore the process halts at that point. I leave out of consideration all resemblances which can only be accidental, or which are resemblances of affinity.

If a mimetic species varies, some of its varieties must be more and some less faithful imitations of the object mimicked. According, therefore, to the closeness of its persecution by enemies, who seek the imitator, but avoid the imitated, will be its tendency to become an exact counterfeit—the less perfect degrees of resemblance being, generation after generation, eliminated, and only the others left to propagate their kind. The actual state of *Leptalis Theonoë* is not the same in all of its three districts. A few varieties, or sports, are seen at Ega (65° W. long.) and St. Paulo (69° W. long.), namely, those figured Pl. LV. figs. 4, 5, 7, 8, and 9, which have an indeterminate resemblance. On the Cuparí (55° W. long.) the resemblance is perfect (Pl. LV. fig. 1); and this is the only form of the *Leptalis* known in the locality. The varieties figured Pl. LVI. figs. 1, 2, 3, show different degrees of resemblance to *Ithomia Chrysodonia* (fig. 3 *a*); these, therefore, exhibit the selection in process. Thus, although we are unable to watch the process of formation of a new race as it occurs in time, we can see it, as it were, at one glance, by tracing the changes a species is simultaneously undergoing in different parts of the area of its distribution.

The fact of one of the forms of *Leptalis Theonoë*, namely *L. Lysinoë*, mimicking at Ega, not an *Ithomia*, but a flourishing species of another quite distinct family (*Stalachtis Duvalii*), shows that the object of the mimetic tendencies of the species is simply disguise, and that, the simple individual differences in that locality being originally in the direction, not of an *Ithomia*, but of another object equally well answering the purpose, selection operated in the direction of that other object. This point is well illustrated by the species of a small group of longicorn beetles already cited, some of which mimic a piece of bark, and others insects of another family—and by hunting spiders, many of which wear the form of insects, and many that of inanimate objects amongst which they seek their prey.

When the persecution of a variable local form of our *Leptalis* is close or long continued, the indeterminate variations naturally become extinct; nothing then remains in that locality but the one exact counterfeit, whose exactness, it must be added, is henceforward kept up to the mark by the insect pairing necessarily with its exact counterpart, or breeding in and in. This is the condition of *Leptalis Theonoë* (Pl. LV. fig. 1) in its district; and it is the condition of all those numerous species of different orders which now appear fixed and distinct. When (as happens at St. Paulo, where a greater abundance of individuals and species, both of *Ithomia* and *Leptalis*, exists than in the locality of the last-named) many species have been in course of formation out of the varieties of one only, occasional intercrossing may have taken place; this would retard the process of segregation of the species, and, in fact, aid in producing the state of things (varieties and half-formed species) which I have already described as there existing.

In what way our *Leptalis* originally acquired the general form and colours of *Ithomiæ* I must leave undiscussed. We may conclude (if we are to reason at all from existing facts) that, as the antecedent forms of our races of *Leptalis* which are still undergoing change were themselves similar to *Ithomiæ*, the form has been inherited through a long line of ancestors, which have been more or less subjected to similar conditions. The instance of one of our forms leaving the *Ithomiæ* to mimic a species of another family may show us how a new line of mimetic analogy and gradual modification may have been originally opened.

Such, I conceive, is the only way in which the origin of mimetic species can be explained. I believe the case offers a most beautiful proof of the truth of the theory of natural selection. It also shows that a new adaptation, or the formation of a new species, is not effected by great and sudden change, but by numerous small steps of natural variation and selection. Some of the mutual resemblances of the *Heliconidæ* already mentioned seem not to be due to the adaptation of the one to the other, but rather, as they have a real affinity, the genera to which they belong being throughout very similar in colours and markings, and all equally flourishing, to the similar adaptation of all to the same local, probably inorganic, conditions. The selecting agent, which acts in each locality by destroying the variations unsuitable to the locality, would not in these cases be the same as in *Leptalis;* it may act, for anything we know, on the larvæ; in other respects, however, the same law of nature appears, namely, the selection of one or more distinct varieties by the elimination of intermediate gradations. The conditions of life of these creatures are different in each locality where one or more separate local forms prevail, and those conditions are the selecting agents. With regard to the *Leptalides,* I believe we may be said to know these conditions. To exist at all in a given locality, our *Leptalis Theonoë* must wear a certain dress, and those of its varieties which do not come up to the mark are rigidly sacrificed. Our three sets of *Leptalides* may be compared to a variable flowering plant in the hands of a number of floriculturists, whose aims are different, each requiring a different colour of flower, and attaining his end by "roguing" or destroying all variations which depart from the standard.

It may be remarked that a mimetic species need not always be a rare one, although this is very generally the case; it may be highly prolific, or its persecution may be intermitted when the disguise is complete.

The operation of selecting agents, gradually and steadily bringing about the deceptive resemblance of a species to some other definite object, produces the impression of there being some innate principle in species which causes an advance of organization in a special direction. It seems as though the proper variation always arose in the species, and the mimicry were a predestined goal. This suggested the only other explanations that I have heard of, namely, that there may be an innate tendency in the organization to become modified in a given direction—or that the parent insect, being powerfully affected by the desire of concealment from the enemies of its race, may transmit peculiarities to its offspring that help it to become modified, and thus, in the course of many generations, the species becomes gradually assimilated to other forms or objects. On examination, however, these explanations are found to be untenable, and the appearances which suggest them illusory. Those who earnestly desire a rational explanation, must, I think, arrive at the conclusion that these apparently miraculous, but always beautiful and wonderful, mimetic resemblances, and therefore probably every other kind of adaptation in beings, are brought about by agencies similar to those we have here discussed.

Bates, H. W. 1861. "Contributions to an insect fauna of the Amazon Valley." Transactions of the Linnean Society *32; "Lepidoptera:* Heliconidae," *pp. 495–515.*

Vertebrata: Aves: Drepanidae (1903)

Robert C. L. Perkins

DREPANIDIDAE

Remarkable as are some other members of the Hawaiian Avifauna, yet it is upon the Drepanid birds that the interest of the ornithologist will always be centered. The Drepanididae, as here considered, include thirty-five species, belonging to no less than seventeen genera. One genus with one species is restricted to the outlying island of Laysan, as is also a second species not generically peculiar, both being included in these remarks on the family, although with the rest of the Laysan Avifauna they may be excluded from the list of Hawaiian forms. The total number of species here cited is rather less than that given by the latest writers on the Archipelago, owing to the fact that several forms which have been described as distinct appear to be quite unworthy of such rank.

SMALL PROPORTION OF SPECIES AS COMPARED WITH GENERA

If we compare the Drepanid birds with the peculiarly Hawaiian families in other groups of animals, we are at once struck by the very large number of genera accepted, as compared with species. No doubt this is partly due to the very different value attached to characters supposed to be generic by systematic workers in different lines, and also to the huge size of birds as compared with many other creatures, owing to which their characters are obvious on the most casual inspection. If we compare the Drepanididae with such a family as the Proterhinidae in the Beetles, which is also peculiar to the Hawaiian Islands, we do not find the latter susceptible of easy division into well-marked genera as in the birds; indeed, at present the members are all included in a single genus. Yet to the student of both groups it is obvious that the extreme forms of the Proterhinidae exhibit differences of structure as great and varied as are found in the extreme forms of the Drepanididae; in fact the variety of structure is probably greater in the beetles. If, however, we were to reduce the hundred and thirty species of *Proterhinus* to the number of species of the Drepanid birds, and particularly if in doing so we were to eliminate the osculant forms, it is manifest that the condition of the two groups would be strikingly analogous. It is therefore in my opinion clear that, making all allowance for the case with which the one group is studied, and the relatively great difficulty presented by the other, there is a real and great difference between the Drepanididae and the Proterhinidae, and in fact between these birds and most of the other extensive and peculiarly Hawaiian groups of animals, and that the difference is due to the fact that while in the birds there has been a keen competition for existence between the various species and between the individuals of each species, in the Proterhinidae there has been little or none, because the food-supply of the latter, consisting of dead wood, is in a forest covered country almost unlimited. As will be hereafter noticed, there is good reason to believe that the competition between the birds has been much more keen in past times than during the more recent periods of their existence.

ORIGIN OF THE DREPANIDIDAE DOUBTFUL

If we compare the Drepanididae with other families of birds, it is obvious that, considering the few species that exist, they exhibit an unusual diversity of structure. As a proof of this, it is only necessary to mention the fact that competent ornithologists have repeatedly assigned to different families even those forms which without any possible doubt belong to the same. This diversity of structure must have required a vast time for its evolution, and the period at which the ancestral Drepanid immigrated to the islands must have been very remote indeed. Whether all the existing species of this group have been evolved from one original immigrant or from more we cannot say; but the former view is probably more correct, although two ancestral immigrants might be admitted.

That the islands were originally stocked by numerous species which produced the present family is highly improbable, seeing that whole families of birds far better adapted to cross wide extents of ocean were quite unrepresented in the Hawaiian Islands, although we know that some of them thrive exceedingly when imported, and many others would no doubt do so under similar circumstances.

Whence the ancestors of the present Drepanid birds came is, owing to their dubious relationships with outside forms, still an open question; though if it were certain that their closest relationship was, as Dr. Gao has suggested, with the Carbide, little doubt would remain as to their American origin. For the present it is perhaps safer to consider them, with other peculiarly Hawaiian groups, as being of unknown origin.

TWO GROUPS OF HAWAIIAN DREPANID BIRDS EXIST, INDICATING EITHER TWO DISTINCT ORIGINAL IMMIGRANTS OR, MORE PROBABLY, VERY EARLY DIVERGENCE FROM ONE ANCESTOR IN TWO DIRECTIONS.

I have already stated that a dual origin for the present Drepanididae is conceivable, and is indicated by the fact that they fall clearly into two groups. The first of these contains six genera, viz. *Trepans*, *Drepanorhamphus*, *Vestiaria*, *Himatione*, *Palmeria*, and *Ciridops*; the second the remaining eleven.

The genera of the first group are characterized by the truncate apices of the primaries, except in the anomalous *Palmeria*, and by the plumage of the young, which is always partly black or of a dull colour. In the adults white markings are present, either on the wings or on the upper parts of the body. The skin, moreover, is comparatively thick, and sometimes extremely tough and, thick, as cannot fail to be noticed by the collector when using very small charges of powder and shot to procure specimens. The plumage of the sexes is identical or nearly so. Red colours are acquired by the adults of some species in both sections, but in a totally different manner; in the second group it is invariably through a green or olivaceous stage, while green-plumaged forms are never found in the young of the first group. In addition it may be noted that the songs and cries of the members of the first section are of a very different character from those of the second, between most of which there is a striking general resemblance in this particular. Further, all the members of the former which are known to me in life (*Himatione*, *Vestiaria*, *Palmeria*, *Drepanorhamphus*) have a peculiar, noisy flight, so that the sound caused by their wings, when they fly freely, can be heard at a long distance.

In the second group the primaries are never truncate at the apex; the young, moreover, are invariably clothed to a large extent in green or olivaceous plumage; and this colour nearly always persists in the adult female, although it may be totally lost in the adult male. Such is the case: in several species of *Loxops*, the green coloration in this genus being largely permanent in the male of the Kauaian species only (*L. caeruleorostris*). There are almost always well-marked distinctions of colour between the adults of either sex. In a few forms which retain in the adult male and female the green plumage characteristic of immature birds (e.g. *Viridonia* and Chloridops) there is little or no difference in the colour of the sexes; but very rarely is this the case when the adults acquire a special coloration, as in *Loxioides*, in which the head is yellow, though somewhat less brightly coloured in the female.

To those who believe in the great significance of the very different character of the coloration of the young birds in these two groups (whatever change may take place in the adults), as well as of the development of striking sexual characters throughout nearly the whole of one of them, the necessity of distinguishing clearly between them will be apparent.

DEVELOPMENT OF SPECIES IN EACH GROUP ALONG SIMILAR LINES, AND THE REASON FOR THE SAME

When we examine, side by side, a full series of the forms it is obvious at a glance that each group has developed along similar lines. *Himatione* and *Palmeria* of the first are in general structure very like *Chlorodrepanis* and Viridonia of the second; *Drepanis* and *Drepanorhampus* resemble *Hemignathus; Ciridops* may be compared with *Loxops*.

With *Ciridops* in one direction the evolution of forms in the first group ceases, while from *Heterorhynchus* the second proceeds through *Pseudonestor* to a series of thick-billed birds quite unrepresented in the first. Consequently in discussing these remarkably analogous forms the six thick-billed genera will here be excluded. Turning to the habits of the birds of the remaining twelve genera, eleven of those certainly and all probably (the habits of *Ciridops* being little known) contain at least some species accustomed to feed on nectar. At the present time the main supply of this food is derived from the *Metrosideros*—the well-known "Ohia-lehua" of the natives, and the predominant tree in the forests of all the islands. Around the masses of red blossoms of these trees may be seen at the proper season an assemblage of various kinds of birds, the scarlet "Iiwi" (*Vestiaria*) and the green or yellow "Akialoa" (*Hemignathus*)—both with long curved beaks—the crimson "Apapane" (*Himatione*) with moderate straight bill, and the green "Amakihi" (*Chlorodrepanis*) with moderate curved bill. The observer wonders for what purpose such extraordinary developments can have taken place. On the same flowers are numerous bees peculiar to the islands, shortest of all short-tongued bees, with a tongue one millimeter long, yet as well able to feed on the nectar of the "Akialoa" with its tongue of two inches or more. An examination of the *Metrosideros* tree will show that it is a species not peculiar to the islands, although, as above remarked, it forms so huge a part of the whole forest. In its specific characters it is in a remarkably unstable condition, exhibiting many striking variations, as though it were now in process of being differentiated into several species. Many of these variations are of constant occurrence and widely spread; some are deemed worthy even of specific rank.

These facts appeal to me to point to a comparatively recent "immigration" of this tree, and I cannot suppose that it has existed on the islands for the period of time which would have been necessary to produce the exceptionally great variety of structure exhibited by the Drepanididae. Turning to other sources whence the food-supply may have been derived at a period antecedent to the arrival and spread of the "Ohia-lehua", we find very different conditions. All, or practically all, the plants visited by these birds for food had bell-shaped or tubular blossoms, in which the nectar was more or less hard to reach. Of these tubular-flowered plants there are several predominant genera, some of which are themselves restricted to the islands, and belong to various families, comprising hosts of peculiar species. Most striking of all are the arborescent Lobeliaceae, not closely related to forms found in other countries. The multiplicity of these peculiar plants, and their isolation from foreign forms, bears a striking resemblance to that of the Drepanid birds themselves, indicating likewise an extremely ancient occupation of the islands; and as the latter are the pride of the Hawaiian ornithologist, so are the former of the Hawaiian botanist. To these flowers Drepanids of both sections are still partial, and some particularly so, and the development of their extreme forms is not comprehensible without a knowledge of the island flora. That there has been in the past severe competition for food between the various species which have similar habits, and between the individuals of each, cannot be doubted. The number of birds that can exist in a given area is obviously only that which can be supported when the food-supply is at a minimum. At the present day, when the "Ohia" is in bloom over miles of country, the food-supply seems inexhaustible; but between the flowering periods it is limited, and often leads to a decided migration of the birds either from one district to another, or to different elevations in the same district, where, owing to the varying climate, the trees blossom at different seasons. Certainly the arrival of the "Ohia" must have been a powerful agent in the increase of individuals of honey-sucking species; and the competition for food must have been much more keen previously. I can hardly doubt that the primitive Drepanid was a honey-sucker, and that the now purely insectivorous, as well as the thick-billed frugivorous forms, were a later development, although the honey-suckers were no doubt at all times partly insectivorous, as they are at present. With the increase of the insect-fauna there would certainly be a tendency among the honey-sucking forms to become more largely, or even entirely, insectivorous, as in fact has been the case. The examination of a series of species of the Lobeliaceae will show great differences in the length of their flowers; and while in some the nectar can be reached by the moderate tongue of *Chlorodrepanis*, in others it can only be procured by the extremely long-billed and long-tongued forms of Drepanids, and the long-tongued Meliphagine "Oo", the latter also a peculiar and probably very ancient denizen of the islands.

A series of observations made on one of the most superb of the Lobeliaceae showed that it could only be fertilized by these highly specialized birds. In this species the pollen is mature before the stigma is exserted, by which time the pollen has vanished. The latter cannot be wind-borne, because it is shed in a viscid mass on contact, and so is constantly deposited on the bird's forehead, from which it is difficult to remove it. With these considerations in view the cause of the development of the most remarkable forms in each group of birds becomes manifest, and this cause has produced *Hemignathus* in the one, and *Drepanis* in the other, so like one another in general structure, while really but remotely allied. How easily the extraordinary lengthening of the bill, to which the resemblance is mainly due, may have taken place, side by side with the increasing length of the tubular flowers, is apparent from the fact that in some of the birds there is even now individual variation in this respect. It should also be stated that in immature specimens the beak is much shorter, and that in the freshly hatched young of *Chlorodrepanis* it is a short, wide member, instead of having a slender curved form, as in older birds. In the long-billed forms the mandibles are almost invariably shorter in the more "conservative" females, which in my second group retain in the adult the more primitive coloration of the young, though the males assume a totally different dress.

TRANSITION FROM A LARGELY VEGETABLE DIET TO PURELY ANIMAL FOOD

Of the genera *Loxops*, *Oreomyza*, and *Heterorhynchus* the members are mainly insectivorous, but each comprises some species which at times feed on the nectar of flowers. In *Loxops* and *Heterorhynchus* the tubular character of the tongue is fully preserved, yet they very rarely feed from flowers, and some of the species perhaps never do so. Certainly that of the latter genus which is found on Hawaii is purely insectivorous, feeding, after the manner of a woodpecker on beetles and other insects; but the other three allied species are less adapted to such a life, and the Maui form has been known to me to visit blossoms as a very rare occurrence, while the partiality that the extinct species of Oahu had for banana flowers has been noticed by others. That these birds, even when purely insectivorous, still retain the characteristic Drepanid tongue, is clearly due to the fact that it remains a most efficient organ for obtaining insect food—in *Heterorhynchus* for extracting the wood-boring beetles of which it is so fond; and in *Loxops* for securing caterpillars which live in the terminal buds of some forest trees, not to mention other purposes. In *Oreomyza*, on the other hand, the tongue is much degraded from its normal structure, while only two of the species, and those but on the rarest of occasions, have been seen to suck honey, and then only from the shallow "lehua" flowers. The genus is almost entirely insectivorous and feeds chiefly on exposed caterpillars, spiders, and moths.

THE THICK-BILLED SPECIES OF THE SECOND GROUP

There still remain to be considered the thick-billed species of the second division of the Drepanididae, which have no similar forms in the first.

There are seven such forms, distributed in no less than six genera, one of the latter (*Psittacirostra*), with its single unmodified species, ranging over the whole group of forest-bearing islands. One species forming a distinct genus (*Telespiza*) is restricted to the outlying island of Laysan; another, also forming a genus (*Pseudonestor*), is found only on the mountain of Haleakala in Maui; while three genera with four species are confined to the large island of Hawaii, namely, *Rhodacanthis* with two species and *Loxioides* and *Chloridops* each with one. It is now generally conceded that all these forms are only extreme modifications of the more normal Drepanididae. In my published biological notes it is true that I placed this section under the Fringillidae, but I did so merely in deference to the opinions of systematic workers, Messrs Wilson and Evans and Rothschild, and more particularly to those of Dr Gadow, who had availed himself of the opportunity of carefully studying the different forms side by side, whereas at that time I had secured no such facilities. Personally I was convinced that all belonged to one family—whether called Drepanididae, Fringillidae, or otherwise—and always maintained this in my correspondence against general opposition, and that too at a time when Mr. Rothschild himself was setting forth descriptions of the Drepanids under such diverse families as Fringillidae and Meliphagidae! Although biological considerations first suggested to me the common origin of all the present family—honey-suckers and thick-billed birds alike—yet at a very early period of my study of these birds I had excellent reasons apart from such for my belief. Before the body of the first *Pseudonestor* obtained by me was cold I was well aware that its tongue was essentially Drepanine and little modified, and that it indicated a positive connecting-link between the thick- and thin-billed sections, being, in fact, more typically Drepanid than that of the otherwise normal *Oreomyza*. The tongue of *Psittacirostra* likewise was taken from the bird immediately after it fell to show that it was truly Drepanid, although much modified. In a hot country such parts should always be preserved immediately, as after a day's collecting they are liable to dry up and their appearance to become changed. The characters afforded by the nostrils and their opercula in all the important forms, as well as the pattern of colour, had been under my consideration as early as 1894, and it is doubtful whether any other important characters have been advanced since that time.

It is still my belief that the biological or physiological reasons on the strength of which I first concluded that all these birds belonged to one family are of the utmost importance, chief amongst which is the peculiar odour to be noticed in both groups, in the thin-billed and thick-billed forms alike. So far as Hawaiian birds are concerned this odour is absolutely restricted to the Drepanines. Mr Rothschild in his work on Laysan makes the astonishing statement that the Meliphagine Oo has a similar and even more powerful odour; but this is only one of those errors which, for want of due care, the museum naturalist is liable to make in opposing facts ascertained and proved in the field. The explanation is very simple:

the Oo (*Acrulocercus*) freshly killed or alive has no such odour. The specimens supposed to possess it had no doubt been enclosed in boxes with Drepanids, or when collected in the field had been placed in a bag with them, and had thus become impregnated with their odour.

This odour, as I have pointed out in my former notes, cannot be acquired from the food, because it is found in forms of such diverse habits—*e.g.* in *Drepanorhampus* at times when it is feeding solely on the nectar of flowers, in weevil-eating *Heterorhynchus*, in *Psittacirostra* when it is devouring the reel fruit of *Freycinetia*, in *Chloridops* when the sole contents of the crop are the seeds of the bastard sandal. Neither of the Meliphagine birds nor the Flycatchers, when feeding in the same trees and on the same food as Drepanids, possess any such smell. All these facts point to the odour as being an ancestral character in the Drepanididae.

In this connexion it may further be remarked that the song of the thick-billed *Pseudonestor* is practically identical with that of the various species of *Heterorhynchus*, which have always been allowed to be Drepanids, and that *Telespiza*, living isolated on the island of Laysan hundreds of miles distant from its allies, has a song similar to this. I shall not easily forget my astonishment when I first heard it on passing a house in Honolulu, and found on enquiry, not the expected *Heterorhynchus*, but *Telespiza*! Possibly the latter may have other notes, but the fact remains that the song I heard was note for note the same as that of the former species, and I heard it repeatedly.

CAUSE OF FRUGIVOROUS HABITS IN THE THICK-BILLED DREPANIDIDAE

The thick-billed frugivorous Drepanids, like the purely or almost purely insectivorous members of the family, have no doubt assumed their frugivorous habits for the same reason that the latter have become insectivorous, viz., the competition for food, rendered unusually keen from the exceptionally small area of distribution. The development of the beak and the loss of the elaborate sucking-tongue have naturally followed. In this connection it is interesting to note that the rather strong-billed *Chlorodrepanis stejnegeri* of Kauai, so far as I know, stands alone amongst the brush-tongued forms in feeding freely on fruits; for at certain seasons it voraciously devours the berries of the poisonous *Wikstroemia*, in the same manner as *Phaeornis*. Such a species—becoming more and more frugivorous and abandoning flowers for fruits—may be considered as potentially the ancestor of a new series of thick-billed forms; at present it is largely a honey-sucker, largely insectivorous, and on occasion largely frugivorous.

A line may be traced among the thick-billed forms through the purely insectivorous, *Pseudonestor* to the largely frugivorous, but still largely insectivorous, *Psittacirostra*, to end in *Chloridops*, which has become almost entirely frugivorous.

DISTRIBUTION OF GENERA IN THE ISLANDS

The distribution of the genera within the group is very unequal, only five of the eighteen having a range which covers all the islands that are forest-clad. These are *Vestiaria*, *Himatione*, *Chlorodrepanis*, *Oreomyza*, and *Psittacirostra*. Three others, *Hemignathus*, *Heterorhynchus*, and *Loxops*, are found on four islands, a species of each inhabiting Kauai and a second Hawaii, the two extreme forest-bearing islands of the Archipelago. One, *Palmeria*, inhabits only Maui and the neighbouring island of Molokai. *Drepanorhamphus* is peculiar to Molokai, *Pseudonestor* to Maui, *Telespiza* to distant Laysan. Hawaii has no less than six genera peculiar to itself—*Drepanis*, *Ciridops*, *Viridonia*, *Loxioides*, *Rhodacanthis*, and *Chloridops*.

DISTRIBUTION OF SPECIES

One form, *Heterorhynchus lucidus*, is almost certainly extinct, while several others, if not extinct, are so extremely rare as to be very nearly so. It is at once noticeable that the birds may be divided into two very strongly contrasted classes. Thus *Oreomyza* is represented by a distinct species on each of six islands, as are *Hemignathus*, *Heterorhynchus*, and *Loxops* on each of four. On the other hand, *Himatione*, *Vestiaria* and *Psittacirostra* range over the whole main group, each with a single unmodified species. *Chlorodrepanis* occupies an intermediate position with two very distinct forms on Kauai, and another form, sometimes considered divisible into three or more species, ranging over the remaining islands. Of these, however, the distinguishing characters are so slight that it is questionable whether they are worthy even of subspecific rank, and in any case such characters are by no means to be considered equivalent to those which separate the different species of *Oreomyza*. The latter are clearly the results of isolation, one island having been colonized by a species from another, which has subsequently acquired peculiar characters. It might be supposed that the birds in the other class, which show no change on the various islands, are in some way less susceptible to the effects of isolation and change of environment. Probably this is not the case, and the true explanation is to be found in considering the habits of the members of the different genera.

TABLE 1. Table of the distribution of the species of the Drepanidae.

	HAWAII	MAUI	MOLOKAI	LANAI	OAHU	KAUAI	LAYSAN
Drepanis	*pacifica*						
Drephanorhamphus			*funereus*				
Vestiaria	*coccinea*	*coccinea*	*coccinea*	*coccinea*	*coccinea*	*coccinea*	
Palmeria		*dolii*	*dolii*				
Himatione	*sanguinea*	*sanguinea*	*sanguinea*	*sanguinea*	*sanguinea*	*sanguinea*	*freethi*
Ciridops	*anna*						
Hemignathus	*obscurus*			*lanaiensis*	*ellisianus*	*procerus*	
Heterorhynchus	*wilsoni*	*affinis*			*lucidus*	*hanapepe*	
Pseudonestor		*xanthophrys*					
Viridonia	*sagittirostris*						
Chlorodrepanis	*virens*	*virens*	*virens*	*virens*	*virens*	*parva & stejnegeri*	
Loxops	*coccinea*	*ochracea*			*rufa*	*cæruleirostris*	
Oreomyza	*mana*	*newtoni*	*flammea*	*montana*	*maculata*	*bairdi*	
Psittacirostra	*psittacea*	*psittacea*	*psittacea*	*psittacea*	*psittacea*	*psittacea*	
Loxioides	*bailleni*						
Telespiza							*cantans*
Rhodacanthis	*palmeri & flaviceps*						
Chloridops	*kona*						

Himatione, *Vestiaria* and *Psittacirostra* are all birds which take extensive flights, often at a great height in the air, and frequently form small companies in these flights. If we stand on the main ridge of some of the islands the birds may be seen passing high overhead from leeward to windward or vice versa. All freely traverse open country, in passing from one feeding ground to another. Consequently when storms arise they are extremely likely to be carried across the channels between the islands, and no doubt this often happens. The birds of the other class, such as *Hemignathus*, *Oreomyza*, *Heterorhynchus*, etc., do not take these extensive flights, but keep closely to the forest, very rarely—and most of them never—venturing into the open. Very seldom would they be likely to get blown across from one island to another. In short there is little doubt but that individuals of *Vestiaria* and its class are transferred from one island to another sufficiently often to prevent any true isolation, which is not the case with the other class. Who can fail to believe that a *Loxops* or a *Hemignathus* would have prospered on Molokai had they ever reached that island? Although *Himatione* ranges unchanged over six islands, yet after, by some remote chance, reaching the very distant Laysan it has there developed into a distinct form; and the case of the two extreme forms of *Chlorodrepanis* on the rather distant island of Kauai is also greatly in favour of my hypothesis.

That any of the Drepanid birds cross even the narrowest channels between the islands willingly is not to be thought of. In times of storm they are often blown down to the lowlands, sometimes in considerable numbers, in which case they mostly fail to regain the forest and perish after a few days. The only birds that I have myself picked up dead (sometimes in numbers) on the coast after these storms are of the genera *Vestiaria*, *Himatione*, and *Psittacirostra;* in fact, the very forms which by their habits are most liable to be carried away by the wind. Further, it is well known that, after stormy weather, the two former sometimes reach the bare island of Niihau, across the considerable channel which separates it from Kauai, but they cannot live there long on account of its unsuitable nature. No doubt the majority of these unwilling emigrants perish, but it is certain that those blown from a high elevation on one island must not infrequently land in suitable forest-country on one of the others.

There is however another possible explanation of the general distribution over all the islands of the single unmodified species of each of the three genera, *Himatione*, *Vestiaria*, and *Psittacirostra*. The plumage of all these was used by the natives in their feather-work, and the birds themselves were frequently kept alive in captivity to serve as decoys. It might therefore be contended that the bird-catchers established these species on those islands from which they were originally absent. Opposed to this theory are the facts already given, and especially the fact that the red birds are known in time of storms to be occasionally transported from Kauai to Niihau, across a channel not much less formidable than any of the other channels, excepting that between Oahu and Kauai, and wider than some of the others. Moreover had the natives been in the habit of making such transportations one can hardly doubt that they would have established the Mamo (*Drepanis*) on other of the islands, since it could easily be kept alive in captivity. One can only wonder that this was not done, if not in the earlier times, at least at a later period when the whole group was under one sovereignty.

RICHNESS OF THE ISLAND OF HAWAII IN PECULIAR FORMS

The relative richness in birds of Hawaii, with its eleven peculiar species and no less than six peculiar genera, is manifest and interesting, since in other groups of animals with highly peculiar species it is frequently (though not invariably) extremely poor, as compared with the older islands of the group. Probably its large area and very varying climate has favoured the multiplication of peculiar forms, while it must not be forgotten that, owing to its position at the end of the group of islands, it is incapable of sending forth emigrants except in one direction. That this is of importance is rendered more likely from a consideration of the Drepanids of Kauai, at the other end of the group of forest-clad islands. Kauai, it is true, has no peculiar generic forms, although geologically so much more ancient than the large island; but being the most distant of the group, as well as at one extremity of the series of islands, its two species of *Chlorodrepanis* are by far the most isolated, its *Hemignathus* and *Loxops* are similarly circumstanced, while in the Meliphagines its *Acrulocercus* is very different to the other forms, which are closely allied inter se, and in the Turdidae it has the two extreme forms of *Phaeornis*. No doubt its small area and comparatively constant climate tend to render it much less rich in Drepanids than Hawaii.

EXTREME SPECIALIZATION OF MANY FORMS OF DREPANID BIRDS

If, as is natural, we consider the primitive form of Drepanid to have been structurally very similar to such birds as are now comprised in the genera *Himatione* and *Chlorodrepanis*, and side by side with these place such forms as *Loxops*, *Drepanorhamphus*, *Heterorhynchus*, and *Chloridops*, the remarkable specialization of the latter is at once apparent, though we are still able to examine connecting forms. To me this specialization indicates the severe competition that has taken place between the Drepanids in past ages. When a vast portion of the food-supply was derived from the blossoms of flowers, and this source of food, as I have shown, was relatively small to what it became later, change to a purely insectivorous, or largely frugivorous, diet must have been very advantageous to the individuals concerned, and the greater the specialization which resulted in obtaining some particular food (provided that it was sufficiently abundant), the greater the advantage to the species. To the field-naturalist who has examined many specimens of such a form as *Pseudonestor* at various seasons and found that its food consists essentially of the larvae of a group of longicorn beetles peculiarly Hawaiian, and not less remarkable than the Drepanids themselves; who has seen how perfectly modified it is for obtaining these; how perfectly adapted is the bill of such form as *Drepanorhamphus* for obtaining the nectar from the deep tubes of the giant-blossomed Lobeliaceae, inaccessible to other birds; how wonderful is the form of *Heterorhynchus*, which delights in the hard boring weevils, themselves equally noticeable; how powerful are the muscles of the head and beak of *Chloridops*, which can crack the stones of the ripe fruit of the bastard sandal; the extraordinary advantage of this specialization in each form for acquiring a constant supply of food almost or quite inaccessible to its allies, and that too in a country where the small land-area may be supposed to have rendered competition unusually keen, must appeal with the greatest force.

HIGH SPECIALIZATION MAY BECOME A SOURCE OF GREAT DANGER

This high degree of specialization, although of the greatest benefit under stable conditions, with a change of these obviously becomes a source of great danger. Thus, destroy the special food-supply of the birds mentioned above, and there is little doubt but that most of them would very quickly become extinct; for forms so perfectly adapted for special ends are, under ordinary circumstances, but ill-adapted to change their mode of life; and it is amongst such forms that most of the rarest species are found, while a considerable number of them already verge on extinction. It is probable that this state of things has largely been brought about by man, and in particular by the destruction of the lowest forest. Even now, in winter storms, large numbers of birds resort to the lowest skirts of the existing forest, generally at an elevation of 1200–1500 feet; and it is well known that in Cook's time such forms as *Psittacirostra*, *Himatione*, and *Chlorodrepanis* actually came down to the coast in Kealakeakua Bay, though now such flights would mean death to the visitants. Moreover, at these lower altitudes the flowering-season of most plants is different from that in the uplands, and they must have been an important source of food at seasons when it was scarce elsewhere.

Perkins, R. C. L. 1903. In Fauna Hawaiiensis, *D. Sharp, ed. Cambridge University Press, Cambridge, U.K; "Vertebrata: Aves: Drepanidae," pp. 381–395.*

Geography and evolution in the pocket gopher (1927)

Joseph Grinnell

The most universally distributed type of rodent in California is the pocket gopher. It is found thriving at and below sea level, around the southern end of Salton Sea in Imperial County, and above timber line, at 11,500 feet altitude in the vicinity of Mount Whitney; it is found from the arid desert mountain ranges of the Inyo region, such as the Panamint Mountains, to the rainy and foggy coast strip at Humboldt Bay and Crescent City; it is found in the yielding sands of the Colorado River delta at the Mexican line and on the Modoc lava beds at the Oregon line.

This fact of occurrence far and wide might seem to indicate a broad tolerance, tolerance of a number of conditions each varying between wide extremes. How is such an interpretation to be harmonized with the obvious fact that the pocket gopher is an exceedingly specialized type of rodent? Does not specialization ordinarily bring great restriction in habitat? A truism is this statement: The pocket gopher stock has solved successfully the problem of meeting the essential conditions of existence, else its racial line would not have persisted to the present day. Among races of animals the law is evident that only those budding forms persist and continue to evolve that are able to find suitable places, niches for themselves, in the economy of animal existence that are not already preempted and successfully occupied by other forms.

The pocket gophers are rodents restricted to the Western Hemisphere; not only that, they comprise a family (*Geomyidae*) restricted to the continent of North America; furthermore, that family centers in the southern half of the continent. The family *Geomyidae* contains several subdivisions—genera in the parlance of the systematist. The genus *Thomomys*, to which all the gophers of California belong, is still further restricted to that portion of North America lying altogether west of the Mississippi River, and between the 20th and 55th degrees of latitude. As to origin, pocket gophers are of squirrel-like ancestry. But that was in very remote times, geologically speaking: for the particular genus *Thomomys* has been in existence since the Miocene period. Despite this long lapse of time, then, the group of rodents here under consideration has not found its way beyond certain geographic limits, and yet within those limits it is exceedingly abundant and widespread, in other words, successful. What was the place for itself that the nascent ancestral race, just becoming gopher-like, discovered and which its descendants, continually specializing, have found so favorable?

Superficial examination of a garden gopher shows the animal to be remarkably formed throughout for existence underground. Observation of its habits shows that in all probability each individual spends fully 99 percent of its time underground. Its world is limited by the earthen walls of a burrow which the animal is equipped to dig for itself through the soil. In one direction this burrow leads to safety for itself and young, from enemies; at the other end it makes food accessible. Thus the conditions of existence for any vertebrate animal, safe refuges and breeding places, and food of right kind and sufficient amount, are met.

But this discovery of a previously unappropriated means of subsistence, by adoption of the subterranean mode of life, has brought with it deficiencies in certain faculties not bound up with proficiency in digging. To dig, the animal must have short legs and a muscular body, especially anteriorly. The head of a pocket gopher is larger in proportion to its body than is that of any other land mammal in California; there is no obvious neck constriction, and the shoulders are broad. The musculature having to do with the operation of the front feet is massive; and so also are the bones of the skull to which are attached the big muscles which operate the relatively heavy incisor teeth, these being the chief tools with which the gopher loosens the soil as it advances along its underground routes of exploration for food, or digs to greater depths for more secure refuge. Obviously, the acquisition of all these modifications for burrowing has necessitated the loss of that litheness of body and length of limb, which would enable it to move freely over the surface of the ground in search of food or in escape from enemies. The pocket gopher is, indeed, wellnigh lacking in powers of locomotion overland.

Furthermore, the pocket gopher is deficient relatively to other rodents with respect to eyesight, and probably also with respect to hearing. It is almost as helpless outside its burrow as a fish out of water. There may be some compensation in a heightened sense of touch, especially as localized in the nose and surrounding vibrissae and in the tip of the tail. While the animal has little need of being apprised of goings-on outside the walls of its tunnel, it does need to be aware of conditions in front and behind. We find that it moves in its cylinder nearly as well in backward direction as forward.

Since, as seems apparent, the general question of the pocket gopher's occurrence over wide territory must take into account its very special mode of gaining a livelihood, it will be useful in our discussion to inquire further as to its digging proclivities and the structures correlated with these. Comparison of the pocket gopher, as an extreme type of digger, with, say, the California ground squirrel, shows significant differences. While the brain case has, in the two animals, relatively about the same capacity, the skull of a gopher is four times as heavy as that of a ground squirrel, total weights of the two animals being considered. As indicated above, the skull and teeth of the pocket gopher, together with the muscles which operate them, comprise the chief engine of digging. This engine operates in powerful fashion in cutting away the earth, so as to make possible the rapid extension of the gopher's underground system of passageways. The adequate housing of the heavy incisor teeth, and the need of meeting the severe stresses during the action of the muscles which operate the jaws, have resulted in the great thickening and ridging of the bones of the skull. We find that the forefeet are larger than the hind feet, a reversal of the ratio in animals which can run with agility, and the forefeet are provided with long stout curved claws. The forearm and shoulder are heavily muscled, and thus the actions of the jaws and teeth are supplemented, in loosening and particularly in transporting the soil.

So far as is known, no pocket gopher goes into dormancy at any season; none either aestivates or hibernates. The source of food upon which the pocket gopher can depend, year in and year out, and which it can seek in safety, is comprised in the underground stems and rootstalks of various grasses and herbs. These it gets almost altogether by digging its way to them; it gathers food only as it can advance under cover. While it is true that gophers do pull into the temporarily open mouths of burrows, stems and leaves of above-ground plants, these latter, I am led to believe, constitute only a minor fraction of the total annual food supply of the animal. The only dependable food source, continuing throughout the year, is comprised in underground stems and roots. And this is an exceedingly important consideration in our present study; for the general geographic limitation of *Thomomys*; to North America west of the 100th meridian, coincides with the territory where sharp alternation of dry and wet seasons is characteristic of the climate. Linked up with this climatic peculiarity there is undoubtedly, in the Southwest, relatively greater abundance of plants with nutritious roots and thickened underground stems, which tide over the dry season, than in the remainder of North America, where there is no long dry season. In other words, the ancestral pocket gophers of the remote past made the fortunate discovery of an oncoming type of food source correlated with the increasing aridity of what came to be a marked climatic and vegetational province.

Restricting our attention now to *Thomomys* as the genus occurs in California, it will revert to the fact of its well-nigh universal distribution within the State. How can the fact of this wide distribution be harmonized with the restriction in the animal's mode of existence, which we have just pointed out in some detail? Examination of the territory wherever pocket gophers thrive, from one end of the State to the other, does show most emphatically close concordance of occurrence with those very, and special, conditions of suitable food, and of consistency of soil which permits of digging. In other words, these two critical factors are widespread, and wherever they extend pocket gophers have gone.

The hindrances, locally, to the spread of pocket gophers are comprised in, not altitude, not cold, not heat, but in discontinuity of ground wherein the pocket gopher can extend its burrows; in discontinuity of ground in which sufficient food of the kind the pocket gopher can use is available throughout the year; and of course, in impassable bodies or streams of water. In other words, we find operating as outright barriers to their distribution only ground such as lava flows which cannot be penetrated by gophers, or ground which is too dry or too alkaline to support adequate plant growth for the gophers' food throughout the year, or permanent streams or bodies of water which the gopher cannot cross.

In this latter connection, the pocket gopher can thrive, we know, without ever drinking; in many parts of the State the only water it can get for long periods is contained in the plant tissues which it uses for food. On the other hand, the animal can live healthily in soil that is saturated with water. Yet it is forced out of the ground when the land is flooded, as during very heavy rains or when under irrigation. Not only a river itself, but the adjacent bottomland subject to overflow at high water, may thus be effective in limiting the spread of gophers locally. A gopher can swim short distances when forced to; but it does not take to water voluntarily. These facts bear on the problem of geographic differentiation of races now to be discussed.

I have pointed out that, despite the pocket gopher's extreme specializations in structure and habits, despite its restriction to a very narrow range of living conditions, yet the fact that these special living conditions are widespread has permitted the very wide distribution in California, of this type of rodent. Now we come to deal with the observation that while our pocket gopher as a genus exists in every county of California, from below sea level to almost the highest altitudes, from the hottest to the coldest portions of the State, and from the driest to the wettest belts, yet the species represented under all these varying conditions is not the same; the genus is broken up into not less than 33 different races (species or subspecies), no two of them occupying precisely the same territory. And this fact signifies that the varying combinations of conditions resulting from the topographic and climatic diversity in California have made their impress upon the gopher sub-stocks, which are more or less isolated from one another in what may be called differentiation provinces. It is our problem to enquire as to what factors among the present or recently past conditions have resulted in this isolation and consequent differentiation of all these various stocks.

The most conspicuous gopherless areas in California lie in the southeastern desert territory, chiefly on the Mohave Desert. Extended explorations of that arid territory have been made, with the special object of determining the kinds and numbers of rodents and other mammals present. Almost every square mile of those deserts, save on such evaporation floors as those of Searles "Lake" and Panamint and Death Valleys (where there is a heavy deposit of saline substances, and no chance of plant growth), supports a large population of seed-gathering rodents—kangaroo rats, pocket mice, and ground squirrels of certain species. Throughout all the deserts there are, at times, heavy rains, though they may be at intervals of as long as three years; and such rains are followed by luxuriant growths of various herbs. These go to seed and thus give origin to a nutritious type of food, scattered by the winds throughout the drifting sands, to be sought out throughout the long dry intervals by the spermophilous mammals just named. But for the pocket gopher, specialized for gathering, masticating, and digesting roots and stems, and not for seed gathering, the food resources of the desert are, over most of its extent, inadequate. Only here and there, on mountain tops, where rainfall is more copious and of more regular occurrence than in the surrounding territory, and about permanent springs, is there produced the proper type of vegetation for gopher consumption, in permanent supply.

As a result of these special conditions of food supply, the general distribution of gophers on the deserts is conspicuously discontinuous; the animals exist only in colonies here and there, because surrounded by unoccupiable desert; and such colonies are often far isolated from one another. The feeble powers of locomotion of the pocket gophers mean that they are unable to cross the barren intervals, and they are subjected to the same sort of factor, in evolutionary process, as land animals sequestered on islands in the sea. As happens under this circumstance the world over, we find that greater or less degree of inherent, subspecific or even specific, sets of differences characterize the more or less isolated stocks.

As an example of races of pocket gophers which evidently owe their origin to the isolation afforded by discontinuity of food supply, we may cite the form *scapterus* on the Panamint Mountains, which range rises high enough above the general base level of the surrounding desert to enjoy a fairly regular rainfall with a consequent copious growth of biennial or perennial herbs.

Another example is the race *amargosae*, restricted to the immediate vicinity of the permanent springs in the otherwise dry and alkaline valley of the Amargosa "River," at Shoshone, Inyo County. This quite distinct form, in the *perpallidus* "group" of gophers, may appropriately be looked upon as a relict form from earlier times when conditions of moisture much more generally prevailed in the Great Basin territory, and when the dependent fauna and flora were correspondingly widespread. *Amargosae* is not the only mammal at Shoshone dependent, directly or indirectly, on the presence of permanent water; for there is (or was a few years ago) a distinct race of meadow mouse (*Microtus*) occurring around the same springs. Then the springs themselves contain a unique species of fish, residuary of a stock which evidence shows occurred widely in the general region in former times.

No pocket gopher whatsover has been found in the depression of Death Valley, into which the Amargosa "River" empties. The lowest parts of the Valley are too intensely alkaline to support any vegetation at all; and such water as there is around the margins, is either too alkaline, too impermanent, or else too small in amount to have permitted the persistence of gophers there up until the present time.

Even such general areas for the race *mohavensis* are not at all continuously occupied by pocket gophers; and examination of representations from different parts of such a general area shows minor differences characterizing the separated colonies. For example, those animals living in the bottomlands of the Mohave River differ slightly from those inhabiting the somewhat higher tablelands surrounding the Providence Mountains, in extreme eastern San Bernardino County.

The Colorado River is significant in our study, in that it has evidently long served as an impassable obstruction to the passage of pocket gopher populations in either direction. The race *albatus*, occupying the delta and "second-bottom" on the western side of the lower course of the Colorado River, is distinctly different in numerous respects, chiefly relating to the skull, from the form *chrysonotus* of the mesa lands on the eastern or Arizona side of the river. The actual distance apart of the nearest populations of these two species is not more than two miles in places, yet the intervening river and its "first-bottom," of ancient existence and large and permanent volume, has acted effectively in preventing the interbreeding of adjacent stocks. Complete isolation accompanied by even slight difference of environments accomplishes much, granted plenty of time.

A puzzle, at first glance, is offered by the occurrence of the two forms, *albatus* in the delta silts below Salton Sea, and *perpallidus* on the floor of the northwestern end of the same (Colorado) desert. Both races are restricted to fine-textured soil in the vicinity of water or where at least some underground seepage permits the proper growth of salt grass and other plants whose stems or roots are sought by gophers. *Albatus* follows the western distributaries of the Colorado River over the delta, and of late has found wonderfully favorable conditions for itself, with resulting enormous spread and multiplication of its numbers, on the irrigated lands of the Imperial Valley. *Perpallidus* occurs chiefly at the mouths of permanent streams coming down the canyons out of the San Jacinto and Santa Rosa mountains onto the floor of the northwestern end of the Colorado Desert, known locally as the Cahuilla Valley. Why should these two races be as distinct as they are from one another when the floor of the general desert area they occupy is continuous, and only about 150 miles in greatest length?

Not long ago, even measuring in years, the northwestern arm of the Colorado Desert was occupied by "Blake Sea," of which the present Salton Sea is the residuum. More remotely yet, the Gulf of California extended continuously up from its present terminus clear through the Cahuilla Valley; and today the floor of that desert is in many places covered with shells of ocean-inhabiting mollusks, and shore lines at sea level are to be seen along the bases of the mountains which rise abruptly on either hand. The rapidly accumulating silts from the Colorado River filled in the depression opposite its mouth and cut off the basin of Blake Sea; and the arid climate resulted in the disappearance of the waters of that sea by evaporation. But completion of this cutting off of Blake Sea and the evaporation of its waters was of quite recent occurrence. We can, I think, look to the former long and complete separation of the gopher stocks resulting in *perpallidus* and *albatus*, respectively, during the period that the waters of Blake Sea, at sea level, lapped the steeply rising rocks on either side impassable to gophers. Complete isolation was thus afforded for the initial bottomland stocks of gophers at the northwest and to the southeast. With the retraction of the shores of Blake Sea, there is now no barrier between *perpallidus* and *albatus*, save as is comprised in unwatered tracts; and these are getting smaller with the spread of irrigation. It will be interesting to see what happens when and where the two gopher populations meet.

An additional case of isolation by desert conditions is that of *operarius* at Owens Lake, also a member of the *perpallidus* group. *Operarius* is a quite distinct form, so distinct with respect to shape of skull and teeth that individual variation does not bridge over the structural interval between it and its near relative, *perpes*. Hence systematists call it a species, rather than a subspecies, the only criterion for the latter systematic rank being intergradation or blending of its characters with those of another race. *Operarius* is restricted to the vicinity of the permanent springs which occur around the eastern side of Owens Lake in the vicinity of Keeler. There, under conditions of moisture obtaining very locally, there is a luxurious and permanent growth of salt grass upon which almost exclusively this species of gopher depends for food.

An entirely different motif, as one may say, of differentiation is provided on the tops of isolated mountain peaks or ranges. In general, the 33 forms of the genus *Thomomys* in California fall into five "groups," of major systematic significance. These groups probably represent much older periods of differentiation, and the subsidiary forms have budded from each of these major stems.

It will be observed that three of the groups, which may be called the *bottae*, *perpallidus*, and *quadratus* groups, respectively, are essentially lowland groups in that they occupy territory of relatively low altitude; while two, the *monticola* and *alpinus* groups, are high-mountain dwelling. In other words, it would appear that some condition associated with altitude has had critical effect in checking the unlimited spread of certain types upward, and of certain other types downward.

Relatively thoroughgoing trapping of gophers along a typical section, transversely, of the Sierra Nevada, in the Yosemite region, shows a sequence of forms by groups from west to east as follows: *pascalis*, of the San Joaquin Valley floor, and *mewa*, of the somewhat higher foothill, digger pine belt, belong to the *bottae* group; *awahnee*, of the *alpinus* group, occupies middle altitudes on the western slope; *monticola*, of the *monticola* group, occupies the whole upper country from about the 6000 foot level to timber line and through the passes and down onto the upper eastern slopes; and at the eastern base of the Sierra Nevada is *fisheri*, of the *quadratus* group. It would appear from this sequence of forms that temperature does, after all, in some measure, constitute a factor bearing critically upon the successful existence of, and hence determining the geographical limitation of, these several species and subspecies of pocket gophers. It might be supposed that relatively uniform temperatures would everywhere surround an animal staying below-ground; but a fact bearing on this suggestion is that the high-mountain gophers all winter do extensive burrowing through the snow, thereby reaching in safety stems of plants above the ground surface!

Returning now to the subject of isolation of high-mountain types on disconnected mountain masses, interesting examples are afforded on the highest mountains of southern California by members of the *alpinus* group. These are: *alpinus* in the vicinity of Mount Whitney, *neglectus* on Mount San Antonio of the San Gabriel range, *altivallis* on the San Bernardino Mountains, and *jacinteus* on San Jacinto Peak. As a rule, low-level types of gophers intervene in the low passes between these boreal colonies sequestered as they are by some factor involved in altitude. As a further example of montane sequestration, we find at the north, in the *monticola* group, *premaxillaris* set apart on the Yolla Bolly Mountains. But, curiously, we find *mazama*, of the same group, on the Trinity Mountains and on the Siskiyou Mountains, both these representations without any detectable differences between them, separated by the valley of the Klamath River. This intervening valley is occupied by *leucodon*, a member of the *bottae* group.

With respect to differentiation within the lowland groups, we find an obvious association of the areas of differentiation with difference in climatic humidity—rainfall, or perhaps cloudiness. In this connection, the general northwest-southeast trend of the areas of occupancy of the different members of the *bottae* group is significant. Comparison of this map of gopher distribution with a rainfall map of California shows the parallel. Take an east-west section, gopherwise, from the coast at Santa Cruz, and we find *bottae* inhabiting the narrow, most humid, coastal belt; in the interior San Benito or other valleys, of lesser rainfall, we find *angularis*; on the hard-soiled, juniper-clothed ridges of the Diablo range, is *diaboli*; beyond this on the floor of the San Joaquin Valley, but west of the river, is *angularis* again; to the eastward of the San Joaquin river flood-bottom is *pascalis*, chiefly in the bottomlands of the smaller rivers making down from the Sierras; and higher, on the hard-soiled foothills, is *mewa*. *Bottae* is darkest colored of all, *pascalis* is palest colored in this series; *bottae* is largest, but *angularis* and *pascalis* are also large; *diaboli* and *mewa* are small, the latter smallest. It would appear that the effects of varying rainfall, or of cloudiness, or of relative humidity of the air, are registered in varying tone of color. And soil texture affects size. The trend of the long, narrow area of occurrence of each of these races happens to be with that of both rainfall belts and mountain axes. The comparative study of the outlines of the ranges of animals brings clues as to the essential conditions for the special existence of each.

This matter of coloration of gophers presents a rather baffling problem. The paler colored forms are generally associated with arid habitats; the darker colored with humid habitats. *Bottae* of the coast belt as compared with the almost white *albatus* of the Colorado delta presents an extreme amount of difference. Is the factor which has to do with these diverse conditions of coloration, light, or temperature, or is it humidity of the air?

Let us remember that whatever the locality, the pocket gopher stays fully 99 per cent of its time within the underground burrow; and this burrow in desert territory, as well as elsewhere, may run through soil that is nearly or quite saturated with water, though, more often, it must be said, in deserts it extends through soil of relative dryness. In defense of the theory of concealing coloration, it can be urged that the moment of greatest hazard to the animal is the moment when the gopher exposes itself to view at the mouth of its burrow in pushing out earth, even though the total time involved may comprise only a minute a day. It may be that the pallor of the desert-inhabiting gopher, like *canus* or *fisheri*, in a statistical majority of cases, brings success in eluding enemies. Not so consistent with this theory is the fact that there are dark-colored gophers, which inhabit the white sands of river valleys in relatively humid and cloudy belts; also there are pale gophers in arid territories, which live in very dark-colored soil, as in the case of *fisheri* in parts of the Mono Lake district. While in this particular matter of coloration no immediate explanation of the differences between the races is forthcoming, yet I have confidence that not only this, but each and every other character which we find to distinguish races, has its full adaptive justification in the scheme of existence.

Some reader may ask what grounds I have for assuming that long lapse of time is involved in this process of racial differentiation. Why may not the observed differences be induced rather quickly, in one or a very few generations of gophers, as conditions change locally or as the animals move about, say from one kind of soil into another? In reply: for one thing, we find the animals "true to type," in other words, relatively uniform in characters, each within its own distribution area—and this despite the great local differences in conditions. For example, *bottae* from the grassy tops of the Berkeley Hills is quite like *bottae* from the campus at Stanford University, and quite like *bottae* from the wooded slopes of the Santa Cruz Mountains, and also quite like *bottae* from the sand dunes at Seaside near Del Monte.

For another thing, we have in the Museum of Vertebrate Zoology specimens, skins and skulls, preserved by members of the State Geological Survey sixty years and more ago, representative of several of our Californian races. Compared with specimens of the same races collected today I see no appreciable differences. And one other line of evidence pointing to the relative permanence of the species and subspecies with which we are dealing: From the Rancho La Brea asphalt deposits near Los Angeles there have been exhumed an abundance of excellently preserved skulls of *Thomomys* intimately associated with remains of certain mammals, now extinct, of known Pleistocene age. And those gophers show cranial characters identical with not only the species *bottae*, but with the subspecies *pallescens* as it exists in the vicinity of Los Angeles today. In other words, in upwards of 200,000 years which it is thought have passed since those Rancho La Brea gophers lived and died, there have been no changes in cranial features such as the systematist would recognize in separating geographic races existing today in different parts of southern California. This adaptive, evolutionary process is one which involves very long periods of time, therefore, when measured in years.

To summarize: In this essay I have set forth some of the facts in the natural history of the pocket gopher. I have picked out for especial comment those features of the animal, as regards both habits and structure, which seem significant in a consideration of its general distribution. I have also emphasized the fact of the differentiation of the pocket gopher type (genus *Thomomys*) in California, into numerous races—species and subspecies. Furthermore, I have referred to the seeming correlation of area of occupancy in each race with relative uniformity of the conditions of existence for that race. The inferior powers of locomotion of this type of rodent, as compared, say, with the jack rabbit, has brought upon it a condition of extreme provincialism, as it were. That is to say, especially with the contributing agency of more or less impassable barriers here and there, each of the many and diverse "gopher differentiation areas" thus formed in California has each impressed its occupant with its stamp, namely, a peculiar combination of adaptive characters best fitting that gopher to carry on successful existence in that restricted area. One more inference, of far-reaching implications, fairly forces itself upon us—that evolution of habitats (differentiation areas) must have preceded differentiation of the gopher stocks which came eventually under their impress.

Grinnell, J. 1927. "Geography and evolution in the pocket gopher." The University of California Chronicle *28: 247–262.*

Coevolution of mutualism between ants and acacias in Central America (1966)

Daniel H. Janzen

The purpose of this paper is to discuss the coevolution of one of the more thoroughly studied mutualistic systems in the New World tropics: the interdependency between the swollen-thorn acacias and their ant inhabitants. This system has recently been described in detail in respect to one species of plant, *Acacia cornigera* (Mimosoideae; Leguminosae), and one species of ant, *Pseudomyrmex ferruginea* (Pseudomyrmecinae; Formicidae), and shown experimentally to be a case of mutualism (Janzen, 1966a). In this species pair, the ant is dependent upon the acacia for food and domicile, and the acacia is dependent upon the ant for protection from phytophagous insects and neighboring plants. The literature dealing only with the New World tropical acacias (*Acacia* spp.) and their ants (*Pseudomyrmex* spp.) has been re-evaluated by Janzen (1966a) and will not be discussed further in a review sense.

The higher plants that commonly have ant colonies living in them have long been termed myrmecophytes. Ants living in plants range in habit from fortuitous usage of a plant cavity to highly complex interaction systems between the ant and the higher plant. The ant-acacia system represents this latter extreme, and the acacia is by any definition a myrmecophyte. A review paper on the subject of myrmecophytes is in preparation. In the present paper, plants with ants living in them will be called "ant-plants"; the ants will be called "plant-ants."

The "swollen-thorn acacias" are those with 1) enlarged stipular thorns normally tenanted by ants, 2) enlarged foliar nectaries, 3) modified leaflet tips called Beltian bodies (eaten by the ants), and 4) nearly year-round leaf production and maintenance even in areas with a distinct dry season. Swollen-thorn acacias have been shown experimentally to have a virtually obligate dependency on the obligate acacia-ants in lowland eastern Mexico, and are very likely to have a similar relationship in lowland areas throughout Central America (based on field observations). Most members of the genus *Acacia* have no such obligatory dependence on ants, though minor interactions may occur (ant-nectary associations). While the colonies of many species of ants live occasionally in the thorns of swollen-thorn acacias, only those which have colonies solely in swollen-thorn acacias are termed "obligate acacia-ants." In the New World tropics, all of these are in the genus *Pseudomyrmex;* however, most *Pseudomyrmex* do not have an obligate interaction with any living plant, even though they customarily live in hollow branches. The obligate acacia-ants are not specific to anyone species of swollen-thorn acacia, but rather to the swollen-thorn acacia life form.

In the literature, swollen-thorn acacias are referred to by various common names (Janzen, 1966b), the most frequent of which is "bull's-horn acacia"; this name is most appropriate for *Acacia cornigera*. Obligate acacia-ants are commonly referred to as "acacia-ants"; unfortunately this name has also been used for various species, usually in genera other than *Pseudomyrmex*, that occasionally live in unoccupied swollen-thorn acacia thorns.

It should be made abundantly clear that the success or failure of any study of a group of tropical plants and insects weighs heavily on the reliability of the identifications, and the degree of correlation between scientific specific names on the one hand, and actual genetically continuous populations on the other. Much effort to date has been expended in attempts at verification of this correlation by extensive field collection to determine population boundaries, and by close examination of the morphological and behavioral characteristics of the specimens. This portion of the study is still in progress, but it is unlikely that further findings will change the conclusions reached in this paper.

ANT × ACACIA INTERACTION

The basic interaction between *Acacia cornigera* and *Pseudomyrmex ferruginea* in lowland eastern Mexico is representative of a fully developed interdependency. The queen ant finds an unoccupied seedling or sucker shoot of the acacia by flying and running through the vegetation. "Unoccupied" as used here, refers to the absence of workers of obligate acacia-ants outside of the swollen thorns. She either cuts her own entrance hole in a green swollen thorn or uses one cut by a previous worker or queen. She lays her eggs in the thorn and forages out of it for nectar from the foliar nectaries for herself and larvae, and solid food (Beltian bodies) for the larvae. As the colony grows, it occupies all of the thorns of the plant, with up to 25 per cent of the workers normally active at any one time outside of the thorns, but on the acacia. By this time the queen's abdomen has become greatly enlarged (physogastric) and the workers fulfill all of the working, foraging, and protective functions in the colony. About nine months of colony growth are necessary to produce enough old workers to patrol the outside of the acacia in an effective and aggressive manner. By the time the colony contains about 1200 workers, the first winged reproductives are produced, and alate production is thenceforth continuous except under conditions of severe starvation. Colonies may grow to contain as many as 16,000 workers in three years, or 30,000 after an indefinite period. During this colony expansion, the colony often moves out to inhabit neighboring acacias (auxiliary-shoots) within three to 10 meters, but the queen usually stays in the original shoot (queen-shoot). Colonies with over 50 to 150 workers actively patrol the outside of the shoot on a 24-hour basis.

With a few notable exceptions (insects that have evolved mechanisms to feed in the presence of the ants), the workers attack any other insects on the acacia and normally are successful in driving them off by biting and stinging. They also attack (mauling with mandibles) any living foreign plants which touch the swollen-thorn acacia's foliage or grow in an area 10 to 150 cm. in diameter below the acacia (basal circle). Thus the acacia grows in a cylindrical space free of other plants. The efficiency of removal of insects and foreign plant parts from the acacia is a function of worker aggression and patrolling consistency, which is in turn a function of colony size, air temperature, and food availability (foliar nectar and Beltian bodies).

The young acacia grows in young second growth, from a seed or from a stump left by fire, flood, or land clearing. As it grows during the first nine to 24 months without being occupied (assuming that a mature colony has not found it and moved into it as an auxiliary-shoot), it is subject to insect and rodent attacks which greatly reduce the already small population of young acacias. During the early stages of growth, a seedling progressively produces larger thorns, more leaflets with Beltian bodies, and larger foliar nectaries until it has these morphological properties of a mature acacia. Once occupied, the acacia grows very rapidly as an emergent or canopy member during the first six to eight years of undisturbed regenerating second-growth vegetation. Several large seed crops are produced during this time. In undisturbed regenerating vegetation, the acacia shoot rarely lives more than eight to 10 years; in open, repeatedly disturbed sites (pastures, roadsides), it may live at least 20 years. Only after a dry season of over five months' duration is the entire leaf crop dropped without gradual replacement of leaves.

It has been experimentally demonstrated that the swollen-thorn acacia has lost, apparently through evolutionary change, its ability to withstand the phytophagous insect damage and competitive pressure of neighboring plants without the protection of the obligate acacia-ants (Tables 2, 3, 4, and Janzen, 1966a). Unoccupied swollen-thorn acacias show severe defoliation and loss of growing shoot tips. Repeated such losses usually cause death within six to 12 months; moreover, the plant is unable to maintain the high vertical growth rate necessary for a sun loving plant in tropical lowland second growth plant communities. The acacia is then heavily shaded by the surrounding growing plants, and the harm is aggravated by the frequent use of the acacia as a support by vines. Even though the acacia may flower as a damage reaction under these circumstances, this does not result in a seed crop inasmuch as it requires about a year to mature the seeds.

The obligate acacia-ant may thus be regarded as a multipurpose characteristic of the acacia, maintained by swollen thorns, Beltian bodies, enlarged foliar nectaries, and year-round leaf production. Evolution of these features is associated with that of various traits of the ant such as aggressiveness towards foreign organisms, use of Beltian bodies as a protein source, etc. The size of the ant colony is a function of the rate and continual nature of leaf production; the thorns, nectaries, and Beltian bodies are all foliar structures.

In the demonstration of the above dependency, ants were experimentally removed from the acacias by either clipping the thorns, or cutting and burning shoots with their ants, and observing unoccupied growth from the stumps. In the experimental subplots, the unoccupied shoots and new suckers grew for at least seven to nine months without ant colonies and were compared with the occupied growth in the control plots. Tables 2, 3, and 4 contain representative data from these experiments. Fifty subplots were under observation for four to 11 months.

EVOLUTIONARY SYSTEM

The system described for *Acacia cornigera* and *Pseudomyrmex ferruginea* differs somewhat from that of other species pairs. Of importance are variations in the type of plant community, aggressiveness of the ant species, size of the colony, and abundance and presence of insects that attack unoccupied acacias. In drier sites and heavily grazed pastures, unoccupied acacias are in little danger of being shaded (e.g., *Acacia collinsii* in southeastern Nicaragua). It is common for two or more species of obligate acacia-ants to occur at the same site; one usually is more efficient at protecting the acacia than the other (e.g., *Pseudomyrmex belti* is less effective than *Pseudomyrmex spinicola* in central Costa Rica). Some species of obligate acacia-ants will accept newly fecundated queens into the colony and thus increase the colony egg production. The resultant large colonies can effectively patrol several hundred acacias (e.g., *Pseudomyrmex belti* in Nayarit, Mexico) while single-queen colonies can rarely effectively patrol over 20 (e.g., *Pseudomyrmex ferruginea* in Veracruz, Mexico). On a local basis, and even more strikingly over the annual cycle from wet to dry seasons, the presence of insects that severely damage unoccupied acacias is quite variable; these insects must be censused in any experiment designed to verify the presence of a mutualistic interaction. While large vertebrate herbivores may have been of importance in the evolution of the interaction (Brown, 1960), they are not the major present-day selective force in its maintenance. During the end of the dry season, when swollen-thorn acacias may be the only green plants available, small rodents (e.g., *Sig-*

modon hispidus) may do considerable damage to unoccupied acacias; this damage has the same impact as the generally much more common insect damage.

The traits of the ant and the acacia (Tables 1 and 5) can be roughly divided into two groups: (A) those shared with other species of the genera concerned, and therefore presumably present to some degree in the prototype obligate acacia-ant and swollen-thorn acacia before any interaction existed; and (B) those features in which the mutualistic *Acacia* and *Pseudomyrmex* depart from their congeners. Some group A characteristics may be regarded as preadaptations, while group B is richer in novelty. Group B can be further subdivided into those traits essential for the interaction and those that are made possible by it. While some difficulty is encountered in determining what are "normal" traits for the genera *Pseudomyrmex* and *Acacia*, some 70 species of the former and 20 of the latter have been examined in the same areas where the studies of the ant-acacia interactions were made.

SWOLLEN-THORN ACACIAS

(1.) Acacias usually are shrubs or small trees. Swollen-thorn acacias range from low shrubs to trees up to 30 m. tall with 50 cm. DBH (diameter at breast height); they have tough and resilient wood. Such a structure is necessary to support the large weight of woody swollen thorns and the emergent acacia canopy. The canopy of a two-meter-tall acacia may carry one kilogram of swollen thorns; a four-meter plant, three kilograms. The lateral branches are often the stiffest for their age in the second-growth vegetation during the first four to six years of succession. Nevertheless, in some cases the weight of thorns becomes too great and the top of the acacia bends over, decreasing the plant's height by as much as a meter. The heaviest thorns on the tree (type B) are borne on short lateral branches from mature thorn axils on major branches. It appears that further enlargement of the thorns would require strengthening of the lateral branches.

During the first five to 15 years of undisturbed succession, the occupied and maturing swollen-thorn acacia canopy normally maintains a position in the upper part of the general canopy emergent to it; this is accomplished by a very high vertical growth rate (as much as 2.5 cm. per day) accompanied by ant protection of the tender shoot tips and ant removal of encroaching vines. It is of interest that the new green shoot tips of swollen-thorn acacias lack the fibrous material which makes the shoot tips of other acacias very tough and resistant to breakage, and presumably

TABLE 1. *Acacia* traits related to the ant-acacia coevolution.

A. GENERAL FEATURES OF ACACIAS OF IMPORTANCE TO THE INTERACTION	B. SPECIALIZED FEATURES OF SWOLLEN-THORN ACACIAS (COEVOLVED TRAITS)
1. Woody shrub or tree life form	1. Woody but very high growh rate
2. Reproduce from suckers	2. Rapid and year-round sucker production
3. Moderate seedling and sucker mortality	3. Very high unoccupied seedling and sucker mortality
4. Plants of dry areas	4. Plants of moister areas
5. Ecologically widely distributed	5. Ecologically vey widely distributed
6. Leaves shed during dry season	6. Year-round leaf production **
7. Shade-intolerant, sometimes covered by vines	7. Shade-intolerant and free of vines
8. Stipules often persistent	8. Stipules longer persistent, woody with soft pith **
9. Bitter-tasting foliage	9. Bland-tasting foliage
10. Each species with a group of relatively host-specific phytophagous insects, able to feed in the presence of the physical and chemical properties of acacia	10. Each species with a few host-specific phytophagous insects, able to feed in the presence of the ants
11. Foliar nectaries	11. Very enlarged foliar nectaries **
12. Compound unmodified leaves	12. Leaflets with tips modified into Beltian bodies **
13. Flowers insect-pollinated, outcrossing	13. Flowers insect-pollinated, outcrossing
14. Seeds dispersed by water, gravity, and rodents	14. Seeds dispersed by birds
15. Lengthy seed maturation period	15. Lengthy seed maturation period
16. Not dependent upon another species for survival	16. Dependent upon another speices for survival

** Essential to the interaction

resistant to insect feeding; the shoot tips of other acacias grow much more slowly. The caterpillar of *Coxina hadenoides* (Noctuidae) eats shoot tips of swollen-thorn acacias at about three times the rate of other acacias in the area; it chews slower on these tougher structures. In general, high rates of branch elongation tend to be correlated with tender and fragile shoot tips. It is doubtful that the swollen-thorn acacias could maintain their high rate of vertical growth and a tough insect-resistant shoot tip at the same time. In most of the young lowland second-growth plant communities, the other canopy members and emergents are not in the genus *Acacia*.

(2.) The swollen-thorn acacias regenerate rapidly from cut or burned stumps and from lateral roots; under conditions of man-made disturbance, nearly all acacia shoots are younger than their rootstocks. Other species of acacia reproduce readily from root stock, but often fail to produce new suckers from stumps cut during the dry season, until the following rainy season starts; such a pattern would be fatal to the tenant ant colony. Sucker regeneration is important to any woody plant in second growth that is commonly subject to fires during the dry season. When the swollen-thorn acacia shoot is cut or killed by heat from a fire, it must put out new suckers rapidly so that the ant colony does not starve. If the colony dies or moves to a new acacia, the new sucker shoots will lack protection from the ants for from nine to 15 months. The first suckers appear from one to two weeks after the shoot is killed, the ant colony then moves into new thorns. It is notable that the new suckers have fully developed thorns, nectaries, and Beltian bodies (in contrast to seedlings, which gradually develop these features).

The tenacity of the root system is also important in keeping the young seedling (or sucker) alive long enough for a colony to develop from a founding queen (minimum of seven to nine months), or a mature colony to find the unoccupied acacia. This is reflected in the population's structure. There is a very slow rate of input of entirely new plants into a given area, but maturing plants with an ant colony exhibit a very low mortality rate. During any one-generation cycle, the unoccupied shoots do not reach reproductive maturity and incur moderate to heavy mortality.

(3.) Apparently, most acacias have normal seedling and sucker mortality. However unoccupied seedling and sucker swollen-thorn acacias are subject to very high mortality compared to the occupied swollen-thorn acacias. Once a plant has become occupied by a healthy obligate acacia-ant colony, there is a high probability that it will produce a seed crop except in frequently burned or cut areas; in these areas, almost no seed is produced, and the entire reproduction is through stump suckers and sprouts from lateral roots. Seedlings and suckers have the same extreme susceptibility to insect attack that mature plants have; it takes seven to nine months at a minimum to produce an ant colony from a founding queen, and during this time many of the young plants are eaten. Evolution of more rapid colony production could occur through production of Beltian bodies, nectaries, and swollen thorns at the first node of the seedling or sucker. In the seedling, this does not occur until the 10th to 20th node but the seedling probably lacks sufficient energy reserves until this time.

(4.) While the swollen-thorn acacias inhabit wetter habitats than most other members of the genus Acacia, they still appear to be restricted to areas with a distinct dry season. There is no obvious interaction with other members of the community, which makes this restriction necessary. Still, the swollen-thorn acacias maintain a high population density in second-growth plant communities that are in general much wetter (one- to three-month dry season) than those where other acacias are common. This dominance is most immediately associated with their high rate of growth (keeping them in the canopy or above) and their freedom from surrounding vegetation (pruning activities of the obligate acacia-ants), both of which enable them to compete in such rapidly growing vegetation. There appears, however, to be a more direct evolutionary consideration;

the shorter the dry season, the less time the swollen-thorn acacia has a reduced leaf output and the faster the ant colony grows. The faster the ant colony reaches maturity, the quicker the acacia becomes relatively free from surrounding plants and phytophagous insects. Thus it appears that there is selective pressure for the swollen-thorn acacias to inhabit the wettest areas possible. At the dry margins of their ranges, the swollen-thorn acacias are generally found only in the wettest sites (ravines, road shoulders, swamps); it is only in these places that they can hold their leaves long enough to keep the ant colonies alive.

(5.) In general, the genus *Acacia* contains plants of second growth or low primary growth in dry areas. The swollen-thorn acacias occur as apparently reproductive populations under a much wider range of rainfall, temperature, soil, and disturbance absolute values and patterns. They are found at sites with a dry season of one to six months duration, and a mean annual rainfall from 1500 to 4000 mm. They grow in areas with a mean annual temperature of 21° to 30° C., with monthly means varying from about 14° to 32° C., and a cool season of zero to three months duration. Within these areas they are found on almost every imaginable type of soil from old lava to laterite, black swamp mud, and new beach sand. They may be found under almost any pattern of disturbance except that of pastures where all woody plants have been grubbed out by hand, annually plowed fields, and areas burned annually for more than three consecutive years. Swollen-thorn acacias are clearly ruderals—plants of disturbed vegetation—and they nearly disappear when there has been more than 20 years of undisturbed regeneration (no cattle, fire, or cutting). In primary forest of dry areas, they are plants of the small natural disturbance sites; they also grow rarely as suppressed seedlings, with an ant colony, for several years until a hole appears in a low canopy. Then the plant shoots up to become a reproductive individual.

(6.) In any given population of swollen-thorn acacias, a large number of the plants have green leaves on them throughout, or nearly throughout, the year. These acacias are characteristically the last plants in young second growth to drop their leaves in an exceptionally dry year. Other members of the genus Acacia usually drop their leaves and produce no more leaves during the last three-fourths to one-half of the dry season. On swollen-thorn acacias, there is little vertical growth during the dry season, but leaf production is in the form of small auxiliary tufts of leaves and leaves on the flower bearing shoots; these small leaves have Beltian bodies and fully developed nectaries, but are not subtended by swollen thorns. The products of these leaves are essential to keep the ant colony from starving to death or migrating to another acacia that has leaves. The acacia needs an ant colony during the dry season for at least three reasons: 1) to protect the new flowering branches; 2) to protect the woody parts of the plant from rodents and ovipositing wood-boring beetles, and 3) to have a healthy colony in occupation when the rainy season begins. There are two courses of flexibility in the ant's dependency during a severe dry season; a colony can starve for three to five weeks without losing all of its members, and portions of the colony can temporarily move into other acacias occupied by the same colony and that still have leaves.

With respect to completeness and duration of leaf drop during the end of the dry season, there is notable genetic and age variation within a given species of swollen-thorn acacia. Suckers and seedlings less than a year old generally do not drop their leaves at all; this is very significant for the founding queens that need the products of the leaves to stay alive and produce a small colony. Stumps cut or burned during the driest part of the year still put out new suckers immediately, which either enables the old colony to move into them, or a founding queen ant to start a colony promptly. However, seeds do not germinate during the dry season. Some maturing swollen-thorn acacias may hold their leaves and produce new auxiliary tufts as long as a month after others. When a colony occupies several acacias, its components generally consolidate into the one that holds its leaves longest; this plant generally does better than the others when the rainy season begins because at that time it has a higher worker density. The queen is in this acacia, and her acacia generally has more workers; the workers may take several weeks to a month or more to move back into the acacias abandoned during the dry season. Thus there is obvious selection against leaf drop. That the plants show any periodicity in leaf drop is probably a reflection of a serious physiological disadvantage in holding the leaves during a severe dry season.

(7.) The entire genus *Acacia* in the Central American tropics is shade intolerant; the plants do not achieve reproductive maturity as long as the acacia canopy is in full shade. Leguminosae in general are not plants of shaded understories, but rather are members of the canopy and second growth. However, occupied swollen-thorn acacias are capable of surviving as shade-stunted seedlings for several years (with thin branches, very small thorns, thin nectar, and few Beltian bodies). Lack of shade tolerance is of great importance to the interaction; when unoccupied swollen-thorn acacias are damaged by insects or covered by vines, they are submerged below the general canopy. The immediate reduction in growth due to shading greatly reduces the acacia's ability to recover from the insect damage, and consequently to recover from the submerged position even if it becomes occupied. Thus it is important that seedlings and suckers in newly cleared sites become occupied quickly; within a year the vegetation is often too heavy or too high for the shaded seedling to grow up through.

In general, other species of acacia are swamped by vines and overtopped by the surrounding upright vegetation when they attempt to grow in the dense second-growth communities where swollen-thorn acacias are so common. In more open communities, such as heavily browsed pastures, unoccupied swollen-thorn acacias are in little danger of being shaded. However, insect damage in these sites is very severe. Emergent unoccupied acacias are especially liable to insect damage and use by vines as standards. Associated with the gap in the vegetation around occupied swollen-thorn acacias, the acacias have leafy branches much closer to the ground than do other general canopy member species in the area. While not due to any direct trait of the occupied swollen-thorn acacia, the acacias are characteristically free of vines and do not have foreign vegetation projecting into the canopy; this is due to the obligate acacia-ants' pruning activities. Unoccupied swollen-thorn acacias make excellent standards for vines, with their stiff upright form, thorny and strong lateral branches, compound leaves, and emergent position. The pruning abilities of the ants have allowed the swollen-thorn acacias to move into the lush second-growth communities where direct competition by vines is a major factor.

(8.) A large number of the New World acacias have woody stipules of various lengths with sharp tips. As their name implies, the persistent woody stipular spines of swollen-thorn acacias are enlarged in diameter and often in length. If the thorns were not persistent, they would be of little use to the ant as domatia; they stay on the acacia for two to four years but after one to two years they are not accepted as domatia by the ants. The total volume of a thorn ranges from 0.1 to 7.0 cc.; usually a swollen-thorn acacia canopy contains 100 to 800 cc. of internal thorn volume. The entire brood and a large part of the workers in the colony are found in the swollen thorns of the acacia(s) that it occupies. The ants cut an entrance hole through one wall of the thorn near one apex while the thorn is still green but fully expanded. The soft and sweet parenchyma is excavated and dropped from the acacia. The outside of the thorn is waterproof for at least two years; the inside wall is water-absorbent and as such is probably functional in regulating the moisture content of the thorns. Especially during the dry season when insects are scarce, the hard-walled thorns serve to protect the ants from birds; the thorns with thinner walls are sometimes opened and if the queen is not in a heavy-walled thorn, she may be lost with resultant death of the colony.

The evolution of spines as a possible deterrent to browsing vertebrates has received sporadic attention in the literature (see Brown, 1960). It is generally the case that spiny plants, such as many acacias, are often found in dry areas where green plant matter is scarce during parts of the year. Presumably these spiny plants are not absolutely free from browsing pressure, but the spines probably serve to place the plant farther down the vertebrate's host preference list.

It is very easy to postulate that all of the modifications of the swollen thorns from a "normal" stipular spine were evolved to maintain a larger and more stable obligate acacia-ant colony. This ant colony should then in turn favor the genotype that has better thorns. In the several other species of acacia, which have fairly large thorns, obligate acacia-ant queens occasionally try to excavate a cavity; the internal pith is too hard and the thorn wall, even when green, is too tough. It is of obvious advantage to the acacia to have the entire ant colony located in its canopy; in order for this to occur, there must be a substantial volume of domatia on the acacia. The prototype acacias have some thorns that are adequate as domatia, but there are not nearly enough to support an ant colony large enough to patrol the tree. The colony in the swollen-thorn acacia canopy is fully exposed to sunlight, bird predation, rain, and desiccation; the thorn walls must be waterproof and somewhat resistant to sudden increases in temperature. When fires are hot enough to kill the shoot, they often do not kill the ant colony, and later it moves into the new sucker shoots from the rootstock. While the swollen thorns have retained their sharp tips throughout this evolution, they do not adequately cover the long leaves (much longer and larger than those of most other acacias) and browsing mammals (cattle, deer) easily pluck the large leaves from unoccupied branches without coming into contact with the thorns.

(9.) A striking feature of swollen-thorn acacias is that their foliage does not have the extremely bitter taste, to the author and other humans, that characterizes other members of the genus Acacia (at least 12 species sampled in Central America). While the responsible compound in the other acacias has not been identified, such a mild taste in the swollen-thorn acacias is indicative of a lack of some so-called "secondary" plant substance. It is postulated that this bland chemical nature of the swollen-thorn acacia explains in great part why the unoccupied acacia is so heavily used as a food plant by a large variety of insects that normally feed only on other acacias (see Janzen, 1966a, for a list of these insects). Each of these insects is relatively specific to some other species of acacia, yet the unoccupied swollen-thorn acacia is very much acceptable to all of them. It also appears that though the compounds which normally serve to restrict the insect to other plants are lacking in the swollen-thorn acacia, it still has enough of the chemical properties of the entire genus to be recognized as a host plant by many phytophagous insects.

If it is the case that the swollen-thorn acacias really lack some chemical properties (which are replaced in function by the obligate acacia-ants), then it seems reasonable to postulate that 1) the swollen-thorn acacias have lost these properties because the ants take their place, and 2) these compounds are not mere "incidental" by-products but instead are relatively "expensive" materials which were evolved directly in response to feeding pressure from phytophagous insects. At the present day, it seems unnecessary to implicate browsing vertebrates as a selective force in the production of such compounds in other acacias (due to the ubiquity and importance of browsing insects), but it is definitely the case that unoccupied swollen-thorn acacia foliage is much preferred over that of other acacias by cattle, brocket deer, and small rodents. With the exception of the small rodents, which may kill many small unoccupied shoots during the dry season, modern browsing vertebrates appear to have much less impact on second-growth vegetation than do insects, in the lowland tropical second-growth plant communities under consideration (Janzen, 1966a).

(10.) Most acacia species have a small group of insects that feed specifically on each of them, and a variably larger group of somewhat more general feeders which occasionally attack them. Often the usual feeder on one or two species is an occasional feeder on a variety of other species in the subfamily (Mimosoideae). In addition there are the few generalized feeders (mostly Orthoptera and Lepidoptera) in the community that feed on a wide variety of plant families. This complex of insects and acacias has evolved to a continually shifting equilibrium whereby the insects eat enough of the specific plant parts to supply the necessary energy flow into their populations, yet do not eat so much that they have an immediate

TABLE 2. Representative mean length increments of occupied suckers from stumps (subplot A-2) and unoccupied suckers from stumps (subplot A-1) cut on May 25, 1964 (Temascal, Oaxaca, Mexico). Ants removed from subplot A-1 by burning the ant-infested canopies, once removed from the subplot (*Acacia cornigera* × *Pseudomyrmex ferruginea*, Jansen 1966a).

		Occupied stumps/total		*Length*		*Increment*	
Sub-plot	TIME INTERVAL	BEGIN	END	AVE (CM)	SD (CM)	AVE (CM)	SD (CM)
A-1	25 May –	0/42	0/42	0.00	0.0	6.23	5.6
A-2	16 June	29/29	26/29	0.00	0.0	30.96	17.1
A-1	16 June –	0/42	1/42	6.23	5.6	10.23	9.8
A-2	3 August	26/29	29/29	30.96	17.1	72.86	37.4

drastic effect on the host acacia population. The bruchid-acacia seed interaction, discussed in the last section, is an example of such a system. The obligate acacia-ants, inasmuch as they are phytophagous and specific to the swollen-thorn acacias, represent a special case.

As do the other acacias, the swollen-thorn acacias have a small group of phytophagous insects that feed on them. These insects have various morphological and behavioral traits which allow them to feed in the presence of the ants (e.g., 1) impenetrable cuticle coupled with no avoidance reaction to ants—*Pelidnota punctulata*, Scarabaeidae, 2) totally ignored by the ants—*Syssphinx mexicana*, Saturniidae, and 3) throw off the attacking ants—*Coxina hadenoides*, Noctuidae). It can be assumed that the prototype swollen-thorn acacia had a group of insects feeding on it. As the ant × acacia interaction gradually developed, these insects were very likely forced off the acacia. However, this leaves the swollen-thorn acacia as a large food source for any insect that can evolve a mechanism to feed in the presence of the obligate acacia-ant, an event that has happened in at least nine different species of insects now found only on swollen-thorn acacias (e.g., *Syssphinx mexicana*, Saturniidae; *Pelidnota punctulata*, Scarabaeidae; *Coxina hadenoides*, Noctuidae; *Rosema dentifera*, Notodontidae). There are at least two insects specific to the small, naturally unoccupied seedling and sucker population (*Mozena tomentosa*, Coreidae, and *Aristotelia corallina*, Gelechiidae). They, along with rodents in the dry season, are responsible for much of the mortality of these small unoccupied swollen-thorn acacias.

For swollen-thorn acacias, the obligate acacia-ants may be regarded as replacing in some sense the physical and chemical protective properties of other acacias, and the ants are a multipurpose property at that, since they attempt to remove a very large number of species of insects. That this system is so amenable to study is due to the facility with which the ants can be removed without doing direct damage to the acacia; such things as secondary plant substances and some micromorphological properties cannot be so readily removed or altered, but their significance to the plants not associated with ants appears to be of the same nature as the obligate acacia-ants to the swollen-thorn acacia.

(11.) Raised, nectar-producing glands (nectaries) on the upper side of the leaf petiole and/or rachis are almost universally present on Central American acacias. They are visited by a large assortment of predaceous and parasitic Hymenoptera, beetles, cockroaches, and flies. The presence of the predaceous and parasitic members of this assortment, and especially the ants, is significant in that they may start their predating activities at the site of the nectar, and then radiate outward; thus it is that the acacia may be the plant that is searched for phytophagous insects first and most thoroughly. It is clear that this removal of phytophagous insects, and especially those that feed on shoot tips, is likely to be an advantage to the acacia. Even if this is not the basis for the original evolution of the nectaries, they have such a function at present and the most extreme examples are in the swollen-thorn acacias with their obligate acacia ant colonies, which receive virtually all their sugars from the nectaries. It should be noted that the evolution of foliar nectaries should be no more difficult morphologically than floral nectaries, since flowers are believed to be in general modified branches and leaves. In acacias in general, the development of the nectaries has probably evolved to the point where the energy necessary for their production and maintenance about balances with the protective value of the predaceous and parasitic insects.

The foliar nectaries of swollen-thorn acacias are much larger and have a much heavier nectar flow than those of other acacias examined. This nectar is about the consistency of Karo syrup, and a two-meter acacia produces about 1 cc. per day (40 mg. of glucose and fructose combined). This sugar is virtually the entire sugar source of the obligate acacia-ant colony; it may be slightly augmented by some sugars from the sweet parenchyma in the new green swollen thorns. It is postulated that the evolutionary elaboration of the foliar nectaries was in large part responsible for the gradually increasing attraction of the prototype *Pseudomyrmex* to the prototype swollen-thorn acacia. The acacia provides a large and reliable sugar source to the ant species, and this leads into the more complex interaction. The nectar production is a critical part of the coevolutionary aspects of the ant × acacia interaction system; more nectar means more ants and more ants mean more nectar through increased foraging and/or patrolling on the acacia foliage, which in turn leads to more leaf production, which thus leads to more nectar production.

(12.) Acacias frequently have twice-compound leaves with 500 to 2000 leaflets. Only the swollen-thorn acacias, or acacias presumably genetically related to them, have the tips of these leaflets modified into Beltian bodies (named after Thomas Belt, the first man to describe the interaction—see Belt, 1874). Beltian bodies are constricted leaflets, pinnae, and rachis ends, and the midvein runs nearly to their tip; they are not glands.

The large thin-walled cells in Beltian bodies are apparently full of cytoplasm and other nutrients of a proteinaceous and fatty nature. They also include vitamins; German cockroaches grow on them nearly as well as they do on a yeast diet. The ants harvest them shortly after they are fully expanded on the new leaves, cut them up, and feed them to the larvae; the adult ants (workers) may receive some nutrients from the fluids released during cutting. The nectar and Beltian bodies constitute virtually all of the food of the obligate acacia-ant colony; only rarely is a small insect caught and parts of it used as food for the larvae.

The Beltian bodies are the most unique botanical part of the interaction; they represent no preadaptation by the swollen-thorn acacia and there are no analogous structures on other legumes. The only clue to their origin lies in an undescribed species of eastern Costa Rican acacia, which may be regarded as a prototype; some plants of this species have leaflets with a small, clear apical portion. If any one of the six *Pseudomyrmex* species that lives in the occasional large thorns of this acacia and uses its nectaries were to evolve the trait of harvesting these clear leaflet tips to augment their insect food, especially during the early part of the dry season when food becomes very scarce, the system would be started. Such a food supplement would increase the size of the colony and therefore increase the foraging on the surface of the acacia. This should lead to more leaf and flowering branch production, which would in turn lead to more ants, plus selection for increased nutritional quality of the new structure, leading eventually to a fully developed Beltian body. In addition, continuity and density of reproductives from a *Pseudomyrmex* colony is directly proportional to the size of the ant colony; thus selection for use of a new food such as this will be favored.

(13.) It appears that all of the Central American acacias are insect pollinated, and there is no reason to believe that they do not regularly outcross. Moderate outcrossing is of importance to an evolutionary interaction system that is to be sensitive to slight selective differences favoring one genotype over another, and lead to gradual evolution of behavioral and morphological systems favorable to the interaction. The habit of flowering during the end of the dry season, when obligate acacia-ant activities are at their lowest, probably lowers ant interference with the visiting bees. However, most of the bees (Megachilidae, Halictidae, Trigona, Ceratina)

TABLE 3. Total wet weight of suckers regenerated, and leaf crop from occupied stumps (sub-plot H-3) and unoccupied stumps (sub-plot H-2); stumps cut on October 17, 1963, regeneration harvested August 5, 1964 (Temascal, Oaxaca, Mexico). Ants removed as in Table 1 (*Acacia cornigera* × *Pseudomyrmex ferruginea*, Janzen 1966a).

Sub-plot	NUMBER OF STUMPS	TOTAL WET WEIGHT, GMS.	TOTAL NUMBER LEAVES	TOTAL NUMBER SWOLLEN THORNS
H-2	66	2,900	3,460	2,596
H-3	72	41,750	7,785	7,483

are quite persistent and when chased from one flower, they immediately go to another; this of course promotes outcrossing.

(14.) The swollen-thorn acacias are unusual among Central American acacias in that they have bird-dispersed seeds; the seeds are imbedded in a sweet pulp in an indehiscent or dehiscent pod. This pulp is bright yellow in the dehiscent species. The ants generally attack the visiting birds, but the birds move rapidly enough so that they are not sufficiently bothered to leave. Orioles (*Icterus* spp.) eat large quantities of obligate acacia-ants along with the fruit, but rarely eat the seeds. The most effective distributors are birds of second-growth vegetation (e.g., *Psilorhinus mexicanus*), which then tend to drop the seeds in the same types of habitat where they picked them up. There is high seed mortality from bruchid beetles while the pods are still on the swollen-thorn acacias; it is the birds' activity in eating some seeds before the bruchids oviposit on them that is the primary factor that saves enough seeds to maintain the population. Both in man-made and natural disturbance sites, a relatively high and continuous viable seed input is necessary to produce enough seedlings to get a very few acacias to the stage where they are occupied by obligate acacia-ants; in view of the heavy weight of acacia seeds, birds seem to be the only wide-ranging candidates for such a dispersal agent. Other acacias depend on gravity, water, and rodents for dispersal, and are often very slow to move into newly cleared areas.

(15.) It appears to be a general rule among the woody Central American acacias that the seeds maturing during one dry season are from the flowers pollinated during the preceding dry season. Therefore a swollen-thorn acacia must live six to nine months after it produces flowers in order to contribute seed to the population; flower production as a damage reaction (a common phenomenon) is of little use if the damage is fatal within this time.

The delay in seed production may be due to the very large seed crop which is produced; after flowering, the acacia may need a rainy season's energy production to mature the crop. The large number of large seeds produced by acacias is apparently associated with the 90 percent or higher seed destruction by bruchid beetles; it appears that acacias have been singularly unsuccessful in evolving means to reduce this beetle damage. The obligate acacia-ants are of little use in this connection because the ovipositing female bruchids are very persistent. In the newly created but extensive habitat of man's pastures and roadsides, it appears that there is strong selective pressure favoring the shortening of the time for seed maturation; very often the sucker shoots from stumps do not grow long enough for seed production (though they often flower) before they are cut off or burned. If the above hypothesized reason for a year-long seed maturation period is correct, then a reduction in the necessary size of the seed crop, either through increased protection from bruchids or increased seedling survival, should be selected for; the obligate acacia-ants are an obvious instrument through which this selection may take its effect.

(16.) The swollen-thorn acacias are virtually dependent on the obligate acacia-ants for survival in nature. The experimental work with unoccupied swollen-thorn acacias indicates that if the obligate acacia-ants were abruptly exterminated, the acacia population would be drastically reduced to the point of extinction in nearly all locations (e.g., see Tables 2, 3, and 4); the same can be said of the obligate acacia-ants, were the swollen-thorn acacias to be abruptly removed. Other members of the genus *Acacia* do not depend so heavily on other higher organisms; they are not even noted for specific pollinators.

As a by-product of this study, it has become obvious that at the present time natural insect populations can maintain sufficient feeding pressure on a plant population to cause and maintain large morphological and physiological evolutionary deviations from the other members of the genus. To a less obvious but significant degree, it can likewise be said that other plants in the community can provide such selective pressure as well. The ant-acacia system happens to be easy to study because the obligate acacia-ants are easily removed; however, there is reason to believe that if one could remove the secondary plant substances and alter the micromorphology of many plants, one would often be able to show a similar immediate dramatic effect of the insects and other plants in the community on the plant species concerned. In a sense, the obligate acacia-ants are the secondary plant substances and micromorphology of the swollen-thorn acacias, and they are the means through which the swollen-thorn acacias

TABLE 4. Mortality of unoccupied stumps (I-1, I-2) and occupied stumps (I-3), between October 18, 1963 and August 6, 1964 (Temescal, Oaxaca, Mexico). Ants removed as in Table 1 (*Acacia cornigera* × *Pseudomyrmex ferruginea*, Janzen, 1966a).

Subplot	*I-1*		*I-2*		*I-3*	
Date	NUMBER ALIVE	NUMBER DEAD	NUMBER ALIVE	NUMBER DEAD	NUMBER ALIVE	NUMBER DEAD
Oct. 18	38	0	31	0	39	0
Mar. 13	22	16	21	10	34	5
Jun. 10	18	20	17	14	28	11
Aug. 6	17	21	13	18	28	11

TABLE 5. *Pseudomyrmex* traits related to the ant-acacia coevolution (worker traits unless otherwise indicated)

A. GENERAL FEATURES OF *PSEUDOMYRMEX* OF IMPORTANCE TO THE INTERACTION	B. SPECIALIZED OF OBLOGATE ACACIA-ANTS (COEVOLVED TRAITS)
1. Fast and agile runners, not aggressive	1. Very fast and agile runners, aggressive
2. Good vision	2. Good vision
3. Independent foragers	3. Independent foragers
4. Smooth sting, barbed sting sheatch not inserted	4. Smooth sting, barbed sting sheath often inserted
5. Lick substrate, from buccal pellet	5. Lick substrate, from buccal pellet
6. Prey items retrieved entire	6. Prey items retrieved entire
7. Ignore living vegetation	7. Maul living vegetation contacting the swollen-thorn acacia
8. Workers without morphological castes	8. Workers without morphological castes
9. Arboreal colony	9. Arboreal colony
10. Highly mobile colony	10. Highly mobile colony
11. Larvae resistant to mortality by starvation	11. Larvae resistant to mortality by starvation
12. One queen per colony	12. Sometimes more than one queen per colony
13. Colonies small	13. Colonies large
14. Diurnal activity outside nest	14. 24-hour activity outside nest
15. Few workers per unit plant surface	15. Many workers active on small plant surface area
16. Discontinuous food sources and unpredictable new nest site	16. Continuous food source and predictable new nest sites
17. Founding queens forage far for food	17. Founding queens forage short distances for food
18. Not dependent on another species	18. Dependent on another species group

interact with a large number of the other organisms in the community. Despite the generally accepted concept that, with the exception of browsing mammals, the physiognomy of plant communities is a direct and total reaction by the plants to physical parts of the environment (fire, water, soil, temperature, wind, etc.), consideration of the ant × acacia interaction system leads to the hypothesis that evolutionary abilities of the entire group of organisms in the community is responsible for the structure and properties of the plants in the community, especially at the microstructure level, operating within the framework of the physical environment.

ACACIA-ANTS

(1, 2, 3, 4, 5, 6.) The entire ant genus *Pseudomyrmex* is characterized by fast and agile worker ants, and the queens are no exception. The evolution of such speed and agility was undoubtedly very important in the movement of the first members of the genus into an exposed arboreal habitat, where visual predators are extremely abundant and a fall means getting lost. The behavior of a single worker *Pseudomyrmex* is much like that of a hunted squirrel, dodging behind objects and to the far side of branches. The ants have large compound eyes that cover one-third to two-thirds of the side of the head, and they often visually locate prey, enemies, and nest markers. Odor trails are rarely used. *Pseudomyrmex* workers forage independently, and bring in entire or fragmented insects, plant nectars, and coccid nectars. They do not solicit the cooperation of other workers in prey capture or retrieval. Two workers contest a prey object (tug-of-war) until one gets it and then returns to the nest. Foraging workers are constantly picking up small bits of organic matter from the foliage and these are incorporated into a buccal pellet which is fed to the larvae; this is undoubtedly an important food supplement. This ability to collect and consolidate tiny food items from surfaces from which larger items often fall has likely been of importance in the success of *Pseudomyrmex* as an arboreal forager. Usually only the barbless sting is inserted in prey insects; the barbed sting sheath is apparently only rarely inserted. *Pseudomyrmex* workers are well known for their bad sting among people who work with vegetation. Judging from the inordinately large number of apparent Batesian mimics of *Pseudomyrmex*, the sting is well respected by visual predators. The well-developed sting is probably of greater importance as an individual worker defense mechanism than in predation; thus it is not surprising that it is not barbed. Prey items are generally retrieved entire, and are rarely more than twice the size of the worker. Common items are small lepidopterous larvae, insect eggs, flies, and small beetles, plus unidentifiable insect fragments. While removal of these insects from the tree where the colony is located (in a hollow twig) has some effect, the mere presence of the ants is of equal or greater importance. Phytophagous insects generally display strong avoidance reactions when approached by ants.

The workers of obligate acacia-ants do not differ greatly in the above six characteristics of *Pseudomyrmex*, but they are in addition often extremely aggressive to foreign objects. There are few insects that can outrun or dodge away from an aggressive obligate acacia-ant (Table 6); those few that can do so live in association with the interaction (see Table 1, B-10). Intruders are often located through vision (day) and contact (24 hours). Workers can see small moving insects at distances up to four centimeters, and will even attack shadows from a flashlight at night. Attacking workers often move so fast that they miss their object and are distracted by some other moving worker. When contacted, the object is bitten and stung. This is generally sufficient to cause an insect to leave. However, the excited worker liberates an alarm odor which causes other workers to run much faster and increase their frequency of turning. The net result is a rapid accumulation of workers on the intruder. There is no mutual cooperation among the workers in this attack. Only very small insects are held

TABLE 6. Incidence of phytophagous insects on shoots of *Acacia cornigera* occupied and unoccupied by *Pseudomyrmex ferruginea*, between June 13 and July 29, 1964 (Temascal, Oaxaca, Mexico, first part of the rainy season, Janzen, 1966a.)

	OCCUPIED	UNOCCUPIED
Daylight		
No. of shoots examined	1,241	1,109
Percentage of shoots with insects	2.7	38.5
Mean no. of insects per shoot	0.039	0.881
Mean no. of insects per shoot known to feed on *A. cornigera*	0.036	0.806
Nighttime		
No. of shoots examined	847	793
Percentage of shoots with insects	12.9	58.8
Mean no. of insects per shoot	0.226	2.707
Mean no. of insects per shoot known to feed on *A. cornigera*	0.220	2.665

until dead, and only rarely these are dismembered and some pieces used for food for the larvae; most pieces are dropped off the tree. Mammals are attacked with equal vigor, and the barbed sting sheath is often inserted to hold the ant in place. To humans, the obligate acacia-ants have a much more painful sting than do other *Pseudomyrmex*.

More than 99 per cent of the colony's solid food comes from the Beltian bodies that are harvested as single units and taken back to the thorns. Obligate acacia-ants lick the surface of the acacia and make buccal pellets. Not only are all fungal spores, etc., removed, but a sticky material is removed from the newly expanding leaves which allows normal expansion. Necrotic areas on the surface of the plant are chewed off by the workers. Odor trails exist between two different shoots occupied by the same colony and the surface of the acacia very likely has a colony odor; however, it appears that workers use visual orientation on the surface of the acacia as well.

Several points bear consideration in the original evolution of worker aggressiveness. The queens are not aggressive. The prototype acacia that provided the *Pseudomyrmex* colony with a progressively larger and more reliable food and domatia, not only had more workers active on its surface but it had older workers since the members of the colony had to forage less distance from the thorns and subject themselves less frequently to predation and getting lost. Secondly, it is reasonable to expect that there were several species of *Pseudomyrmex* and other ants living in the thorns of the prototype acacia, and the one that became an obligate acacia-ant was very likely initially either the most aggressive or had the genetic ability to rapidly become aggressive with little selective pressure.

(7.) While attack of plants has not been noted among the *Pseudomyrmex* that are not associated with living plants, obligate acacia-ants habitually bite and/or sting any foreign vegetation that contacts the canopy of the swollen-thorn acacia. Seedling acacias are killed unless they have swollen thorns and nectaries. In addition, the ants kill the vegetation attempting to grow in a circular area below the swollen-thorn acacia. This circle is of varying size, depending on colony age, size, and genotype. The net result of this activity is that the swollen-thorn acacia often grows in a cylindrical space that is virtually free of foreign vegetation and the various influences that it can have (shading, mechanical interference, insect pathways, fuel for fires—Janzen, 1966d). Dense stands of swollen-thorn acacias are completely free of other green vegetation.

It is postulated that the reaction to foreign plants is an outgrowth of the aggressiveness to other insects and to strange objects in general. Some nutrients may be obtained from the sap of the chewed plants but since (1) they are attacked even when there is a surplus of nectar and Beltian bodies, (2) all species are attacked, and (3) workers become aggressive to plants after they start attacking insects (aging process), it is unlikely that this is the present-day cause. The selective force for increased aggressiveness against plants is a feedback in exactly the same manner as is selection for aggression towards animals by the obligate acacia-ants.

(8.) The workers of *Pseudomyrmex* display monomorphic allometry (terminology of Wilson, 1953) accompanied by size variation up to about one and a third the length of the smallest worker in the colony. The size of the workers increases until the first reproductives are produced and then levels off.

The obligate acacia-ants apparently have temporal division of labor. As a worker ages, it moves through a progression of duties in the colony, the last one being patrolling the surface of the swollen-thorn acacia. It is these oldest workers that often are first to encounter foreign organisms. A small but old colony is sometimes more effective in patrolling a swollen-thorn acacia than a large young colony. The presence of workers old enough to be very aggressive may be due to the considerably lower death rate of workers when they are concentrated in the canopy of a swollen-thorn acacia (in contrast to being scattered through the vegetation on foraging activities).

Provided that food is available, the reproductive castes are produced in small numbers on a year-round basis. For the species living in hollow twigs, there is rarely enough food for this. However, obligate acacia-ants generally have enough food throughout the dry season, except at the dry margins of distribution. This continuous production of new queens insures that seedlings and sucker shoots have founding queens in them as soon as a thorn is present on the shoot. More queens mean more ant colonies. The upper limit is apparently set through competition for thorns, and seedling and sucker production by the swollen-thorn acacias.

(9, 10.) The genus *Pseudomyrmex* lives almost entirely in and on living and dead plant structures above the ground. The adaptive radiation into the arboreal habitat is reflected in much of the ants' behavior and morphology (fast, agile, good vision, good sting, buccal pellet formation, slender body, mobile colony, dispersed colony units, etc.). The nest is in hollow twigs and branches with a narrow inside diameter; when vertical, the larvae are either wedged in or hung by dorsal hooked hairs from the walls. The larva has a ventral pouch to hold food particles while feeding. The colony is often distributed among several hollow branches. When one begins to rot, that portion of the colony is easily moved to some other site previously selected by the workers. They commonly hollow out the pith area of a dead stem. These dead and hollow twigs rarely last over a year without rotting or falling, and thus the colony moves frequently. Except for destruction of the unit containing the queen, destruction of one of the colony portions does not destroy the colony. The thorns of prototype acacias, other plants, and unoccupied swollen-thorn acacias are commonly used by *Pseudomyrmex* that are not associated with living plants. They use these structures like hollow twigs, either cutting their own entrance holes, using natural cracks, or using the exit holes of microlepidoptera or other insects that matured in the thorn parenchyma.

The evolution of nest site preference from hollow twigs to swollen thorns is not a large change for obligate acacia-ants, but extremely important for the interaction. It concentrates the ant colony in the canopy of the acacia, and means that it is most efficient for the ant colony to get its nutrients from the acacia canopy. Even at the lowest level of interaction, the colony is partially protecting the acacia just in its own defense reactions.

The swollen thorn is an excellent domatium (stronger, more predictable in occurrence, waterproof, more permanent, usually bird-proof, etc.) but portions of the colony continue to move as the thorns dehisce. By continually moving into the new green thorns, and gradually phasing out the old thorns, the colony moves upward with the rising acacia canopy. The queen has to move one to three times a year to stay in the biggest new thorn. The mobility of the entire colony is important in areas where the acacias are occasionally cut or scorched, or when the acacia dies of old age. The colony then has to move out of the old colony and into new stump suckers or find a new unoccupied seedling. Very often a large colony is much older than the shoot that it occupies, and because of its size and age offers maximal protection to the shoot.

(11.) *Pseudomyrmex* larvae can live isolated in glass tubes for two to four weeks without any food or water. This resistance to starvation is important to the species that live in hollow dead branches; the food supply for these ants fluctuates greatly with the seasons. During the dry season, colony size and the number of colony subdivisions is greatly reduced. Even during the rainy season, larvae may have to wait several days between food items.

A strong obligate acacia-ant colony can survive three to five weeks without food and apparently very little cannibalism takes place. The very shrunken larvae readily accept pieces of Beltian body after starving for three weeks. This means that a swollen-thorn acacia can drop its leaves for nearly a month during the dry season without losing its ant colony by starvation. However, the colony may move if another leafy acacia is available. Secondly, the colony can survive in a cut canopy until new suckers are produced by the cut stump. Finally, if some insect happens to eat off all the new growing shoot tips (e.g., *Coxina hadenoides*), the colony can easily starve until more new shoot tips are produced (two to five days).

(12, 13.) The majority of *Pseudomyrmex* species have one egg-laying queen per colony, and the colony contains 100 to 2000 workers when reproductively mature. Determining colony size of the species living in hollow branches is difficult due to its subdivision into several different branches. The queen remains in one branch and the eggs or first-instar larvae are carried to the other parts of the colony. In single-queen colonies, the workers do not tolerate the entrance of new founding queens. This insures that there is a continual production of new colonies, rather than established colonies absorbing the newly produced queens; this avoids large density fluctuations from mortality of single large colonies, and insures colony dispersal among the vegetation. In view of the difficulty of finding food, when not associated with a living plant, it appears that one queen can easily produce enough larvae to use all the food brought in.

Obligate acacia-ants have single-queen colonies with two to 30 thousand workers in at least four species. Even with abundant food, the colonies do not get much larger. Where conditions highly favor acacia reproduction, the colony may become too dispersed among the many shoots and growth rates are reduced. In at least three species, the colonies will accept new founding queens. This results in a high rate of egg production and a colony with several million workers (several thousand queens) spread over as much as a half acre of pure swollen-thorn acacia vegetation. In areas where the acacias grow well for most of the year, and phytophagous insects are abundant, there is probably strong selective pressure for maximal egg production and acceptance of founding queens by the workers of established colonies. Such a tremendous number of new queens is produced by multiple-queen colonies that there is little danger of the entire queen crop being absorbed by established colonies and abandonment of new seedlings.

(14.) Obligate acacia-ants patrol the surface of the swollen-thorn acacia both day and night. Other *Pseudomyrmex* workers are active outside the nest only during the middle six to 10 hours of the day. Nocturnal activity of obligate acacia-ants is a necessity since much of the damage done to unoccupied shoots is at night. There are distinct rhythms in the density of workers on the surface of the acacia and the peaks coincide with the times of maximum insect activity (dusk and noon) and of nectar flow (dawn).

(15.) Almost no workers are seen on dead branches containing *Pseudamyrmex* nests; on the other hand, obligate acacia-ants are very active on the surface of the acacia. As many as 25 per cent may normally be outside of the thorns, and up to 50 per cent may leave the thorns under severe disturbance conditions. This difference is directly related to the great dispersion of foraging workers of other *Pseudomyrmex* throughout the vegetation, and the concentration of the activities of the obligate acacia-ants within the acacia canopy; this concentration is possible since all the food and nest sites of the ants are found there. There is a small dispersion of obligate acacia-ants away from the acacia, and it is these ants which find the various unoccupied swollen-thorn acacias that the entire colony later occupies.

It is this concentration of the workers in the acacia canopy, through provision of food (at least nectar) and thorns, which was probably the major step by the acacia in the initiation of the interaction. This, coupled with increased aggressiveness by the ants, is the basic start to an effective ant × acacia interaction. When the workers are concentrated at one site, they are much less prone to predation since they protect each other by multiple attacks on invaders (lizards, birds, spiders, etc.). Secondly, with a large and predictable food and domatia source at hand, the workers do not subtract grossly from the colony energy economy by "spending their time" patrolling the surface of the acacia.

(16.) The obligate acacia-ants have their food and nest sites dependably provided by the acacia. In view of the fluctuating nature of food density, and extreme competition for food that other *Pseudomyrmex* experience among themselves and with other ants, the obligate acacia-ant colony could not grow large enough to patrol the acacia effectively if it had to depend on its food-gathering abilities off the acacias. Thus the interaction has required a switch to dependency on a plant product—Beltian bodies. The major change is the behavior of harvesting Beltian bodies. In this connection it should be noted that ants in the genus *Solenopsis* harvest them as well (no connection with the interaction).

The predictable production of many new thorns each year to house the enlarging ant colony is of utmost importance to the system. Unless the obligate acacia-ants nested in the ground, it would be virtually impossible to find enough hollow space near the acacia to house the ant colony.

(17.) Once a founding queen obligate acacia-ant finds an unoccupied acacia, and becomes established in an untenanted thorn, she is assured of a close food source. She has to move only a few centimeters to obtain nectar for herself and to obtain nectar and Beltian bodies for her larvae. Founding queens not associated with living plants forage for food in the vegetation in open competition with other ants. Examinations of bird and lizard intestines, and spider webs, show that there is heavy predation on searching and founding queens. The advantages of having good nest sites and nectar close at hand for founding queens were undoubtedly important in increasing the density of *Pseudomyrmex* colonies on the prototype swollen-thorn acacias.

Severe competition for the thorns on unoccupied acacia seedlings or suckers is common in man-made disturbance sites. The queens interfere with each other when cutting entrance holes and searching the acacia. As many as 100 queens may be found on a single small seedling and these ants spend much time dodging away from each other. Once a queen is inside a thorn, she may still lose it and her brood to any other searching queen which gets into the thorn while she is out collecting food. In areas where many queens are being produced by large mature colonies, seedlings and suckers may go for many months without a new colony being successfully started in them. This may serve as a population regulation mechanism since increased searching queen density lowers both rate of new colony establishment and swollen-thorn acacia seedling and sucker survival. There is very high mortality of the searching queens on such sites; the system is apparently an artifact of times when swollen-thorn acacias were more widely dispersed (natural disturbance sites) and large numbers of searching queens were necessary.

(18.) The obligate acacia-ants are completely dependent, as a colony in nature, on the swollen-thorn acacias. In the laboratory, glass tubes and honey can substitute for thorns and nectar, but a substitute has not been found for Beltian bodies. The obligate acacia-ants have committed themselves to the fate of the swollen-thorn acacias, provided that something happens to the swollen-thorn acacias so rapidly that the obligate acacia-ants cannot evolve out of the interaction system. However, the obligate acacia-ants are in the rather unique position of being able to evolve characteristics that can clearly affect, in a positive direction, the survival and growth of the swollen-thorn acacia. In a less clear manner, such mutual interference or assistance is probably present in the majority of interspecies interactions; however, it is much less easy to demonstrate since most species leave themselves several alternatives in their interactions with other species. In this connection it is important to note that the obligate acacia-ant is a multipurpose trait of the swollen-thorn acacia (and vice versa) and is involved in its interaction with many other species in the environment. It therefore has a large chance to influence the outcome of these interactions.

A further feedback system is that the presence of an ant colony old enough to produce reproductives indicates that there is a swollen-thorn acacia present that is old enough, or will be old enough, to produce seeds or suckers; that there is a swollen-thorn acacia old enough to produce seeds or suckers is a very strong indication that there is an obligate acacia-ant colony present that is large enough to produce reproductives.

DISCUSSION

The interaction system between obligate acacia-ants and swollen-thorn acacias is clearly an example of mutualism between a higher plant and an animal. The present deductive discussion of the evolution of the traits of the ant and the acacia leads without reasonable doubt to the conclusion that the system evolved through mutually interactive evolution, or coevolution, rather than by chance coincidence in time and space of two highly specialized yet mutually beneficial organisms. A great part of the necessary traits are beneficial to the ants and acacias only when interacting. With the exception of a few other *Pseudomyrmex* that apparently have obligatory interactions with other genera of plants, at least 10 traits of the ant and 14 traits of the acacia are possessed by only those members of the genera *Pseudomyrmex* and *Acacia* that are obligatorily associated with each other. These traits run the gamut from purely morphological (e.g., swollen thorns) to purely behavioral and physiological (e.g., aggressiveness, year-round leaf production); their putative prototypes can be found among other members of the genera *Pseudomyrmex* and *Acacia*, some of which may be presently in the process of coevolving an ant × acacia interaction system. Neither organism evolved solely to match the properties of the other, but rather there must have been gradual evolution by both the ant and the acacia toward the present-day interaction.

If the coevolution of the ant × acacia interaction is viewed as a system between two species (or life-forms), rather than a unique case of ant × acacia coevolution, certain basic properties of interaction systems can be discussed.

1. The coevolution of an ant × acacia mutualistic interaction is the product of an evolutionary feedback system par excellence. There have been a multiplicity of changes in the swollen-thorn acacia which resulted directly in an increase in density or "healthiness" of the occupant ants and indirectly in selection for the acacia genotype which possessed the changes. There have likewise been changes in the ant which resulted directly in an increase in the reproductive capacity or "healthiness" of the acacia being occupied and indirectly in selection for the *Pseudomyrmex* genotype which exhibited the changes. This system requires two basic components. One of the organisms must supply the other with a requisite (sensu lato, and the receiving population must have a mechanism to favor the donating genotype of the donor population. In addition, the requisite is usually a quantitatively defined variable, and the donor can increase its value. For example, the obligate acacia-ant can supply a swollen-thorn acacia's requisite of a means to deter most of the phytophagous insects in the plant community, and can increase this flow from a trickle (taking an occasional insect as prey) to a massive output (keeping the tree virtually free of insects, except for those that can evolve into the system). The ant genotype that does this will be favored by an increased supply of the requisites it receives from the acacia. The swollen-thorn acacia can supply an obligate acacia-ant's requisite of a sugar source, and can increase this flow from a trickle (the tiny output of the tiny nectaries found on many legumes and other plants that lack an obligate relationship with ants) to a massive output. The acacia genotype that does this will be favored by an increased supply of the requisites it receives from the ant.

This system involves selection for a genotype that differs only slightly from the other members of the species' population at any point in the system's evolution. Further, such a system need not involve any long-term changes in total density of either population, although it does appear that the ant × acacia coevolution has increased both the density and ranges of the species involved in manmade disturbance sites. The cumulative result of this selective system is a set of populations which are greatly affected if a trait necessary to the interaction is suddenly removed (e.g., remove the ant's aggressiveness or the acacia's Beltian bodies).

The system is self-reinforcing, operating through the medium of the other member of the species pair. Systems of this type are very common in nature, though they usually involve one-way, rather than two-way, energy flow. The evolution of the traits of predator and prey is such a case. Here, success in either of their interactive operations (prey capture by predator, or prey escape by prey) favors the genotype that was successful, causing selection against the unsuccessful genotype in both populations. This in turn promotes selection for an improvement of the successful genotype. For example, if a predator catches a prey organism, it usually promotes the prey genotype that was not captured. However, this latter prey genotype is presumably harder to capture, and the predator must then evolve further capturing abilities. Such a system is balanced by the evolution of the prey in the same manner, changing hosts or predators, inefficiency, etc. In a mutualistic system such as the ant × acacia interaction, both organisms are simultaneously successful and thus both genotypes responsible for the success are favored. It should be noted that both members of the mutualistic pair have predator-prey (sensu lato) evolutionary systems with other organisms outside of the ant × acacia interaction system.

2. The ant × acacia system contains an excellent example of convergence or parallelism, and if the entire Pseudomyrmecinae and other ant-plants are considered, the case is even stronger. Based on morphological similarity, the obligate acacia-ants are composed of two groups. *Pseudomyrmex nigropilosa* is the only member of one group, and appears to be a species which is "evolving into" the ant × acacia interaction system. It is morphologically very similar to a large group of *Pseudomyrmex* that are not associated with living plants. Since it offers no protection to the swollen-thorn acacia that it occupies it must be regarded as parasitic on the other obligate acacia-ant species which keep the swollen-thorn acacia population alive. The other obligate acacia-ants are so similar morphologically and behaviorally that they are very likely evolved from a common ancestor which was an obligate acacia-ant. There is no reason to believe that the conditions of living in association with a swollen-thorn acacia are so exacting that one would expect the *Pseudomyrmex* becoming involved in the interaction to converge as strongly in their morphology as are these obligate acacia-ants (see Janzen, 1966c).

The swollen-thorn acacias, if flower and fruit diversity are to be trusted as indicators of lineages, represent at least five independent evolutions into the system: *Acacia cornigera* and *Acacia sphaerocephala*; *Acacia hindsii* and *Acacia collinsii*; *Acacia melanoceras*; and two undescribed species.

A progression for the production of these interactive species is hypothesized as follows. The coevolution of an obligate interaction between one species of *Pseudomyrmex* and one species of *Acacia* occurred somewhere in the New World tropics. The simplest further complication lies in what apparently occurred to produce *A. cornigera* and *A. sphaerocephala*, two extremely similar species occupied by *Pseudomyrmex ferruginea* in eastern lowland Mexico. *A. cornigera* is a species of wetter areas and *A. sphaerocephala* a species of drier areas. At some time in the past, they apparently represented two populations of an ancestral species (or perhaps *A. cornigera* or *A. sphaerocephala*) which became genetically isolated and at present are just beginning to reinvade each other's ranges with very little introgression. A similar situation appears to have occurred with *A. collinsii* and *A. hindsii*. Once this occurred, other acacia lineages must have entered the system. Once the interaction system is present at the same locality as another acacia with the preadaptations to become involved, such involvement is facilitated by three means. (1) The obligate acacia-ant will readily occupy another acacia which develops swollen thorns, nectaries, and Beltian bodies in sufficient numbers to support the ant colony. If such resources are insufficient in quantity on one plant but there is a nearby swollen-thorn acacia, the colony will live in both species of plant at the same time. (2) The swollen-thorn acacias frequently form interspecific hybrids with other acacias; the traits for enlarged nectaries, swollen thorns, and Beltian bodies are readily transferred through introgression and the hybrids apparently produce viable seed (Janzen, 1966b). (3) The trait of extreme susceptibility to insect attack is also transmitted and this makes it imperative that the hybrids, to survive, be thoroughly occupied. The entire system described above appears to be occurring at the present time between *A. cornigera* and *A. chiapensis* in southwestern Veracruz, Mexico.

To become thoroughly occupied, the acacia species must reinforce the possession of nectaries, swollen thorns, and Beltian bodies by evolution on its own, or by very thorough incorporation of the donated traits into its own genotype. The population of acacias which has newly received the swollen-thorn acacia traits will likely have a different total distribution from the donor swollen-thorn acacia; thus, the obligate acacia-ant, in following the incorporation of traits into its new host plant's population, will undoubtedly find itself under new environmental conditions. If it can tolerate these, and the acacia can maintain the swollen-thorn acacia traits over at least part of its range, the acacia will likely find itself being a swollen-thorn acacia, in spite of itself, so to speak. It should be noted that this will, in all likelihood, result in such by-product effects as changing the acacia's range to those areas where the interaction can survive, and extinction or speciation of the portions of the population living in areas where the interaction cannot occur. Under such circumstances, there is a chance that the obligate acacia-ant species will be cut off from its parent population, leading to further speciation among the ants.

3. There is a growing body of literature which strongly suggests that the chemical and microphysical properties of the vegetative parts of plants may be evolved in partial or entire response to the use of the plants as energy sources and substrates by insects (see Southwood, 1961; Beck, 1965; Stark, 1965; and Ehrlich and Raven, 1965; and included references). It appears that the swollen-thorn acacias have substituted the abilities of the obligate acacia-ants for many of the physical and chemical traits that other acacias use in their defense; this is strong evidence that these properties exist. In fact, the swollen-thorn acacia gains properties from the ants that other members of the genus Acacia do not possess. The demonstration of these two statements was made possible in a clear manner only through the ease with which the obligate acacia-ants could be removed from the acacia. To perform similar experimental demonstrations of the significance of apparently meaningless chemical and physical properties of plants will require removal of the secondary plant substances or alteration of the micromorphology without directly damaging the plant. This is exceedingly difficult, and becomes even more so if one tries to remove as many factors as the obligate acacia-ant provides.

The ant × acacia system is representative of plant × insect interactions in that the swollen-thorn acacia has a small group of insects which feed on it in spite of its normally deterrent properties (obligate acacia-ants). These insects do not do enough damage to remove the acacia from the habitat although they take enough of the acacia to keep their own populations in existence. The actual numerical values of such a set of mutually sensitive energy flows are primarily a function of their position in evolutionary time, the evolutionary abilities of the organisms in the community, and their energy requirements at any given time of year. The energy flow values change as the plants and insects evolve in and out of interactions. Further, when the ants are removed, not all the insects in the community feed on the unoccupied swollen-thorn acacia; such attack is generally only by those that feed on closely related plant species or genera, or those that normally show long and broad host plant preference lists. It is the net effect of all of these insects coming from other host plants to feed on the unoccupied swollen-thorn acacia that mounts sufficient continuous feeding pressure to kill the acacia. Or, stated in another way, the behavior of the ants places the occupied swollen-thorn acacia low on the host lists of all but a very few insects; when the ants are removed, this suddenly moves the swollen-thorn acacia very near the top of the host preference lists of many insects that normally feed on closely related plants, or have broad food tolerances. The insects that feed in the presence of the ants are rarely found on unoccupied swollen-thorn acacias; such plants have generally been so badly damaged that the intact foliage preferred by these insects is absent.

In addition to affecting the use of the swollen-thorn acacia by insects, the obligate acacia-ants also affect its interactions with other plants, especially vines. Other plants in the lush second-growth community keep their canopies sufficiently free of vines by various structural and behavioral traits. These traits are singularly difficult to recognize and demonstrate. They apparently include such things as drooping leaves with very smooth upper surfaces, off which vine tendrils easily slide while the supple and smooth woody stems move in the breeze. This results in the general vine layer being slightly below the rapidly rising canopy rather than over it. However, at times the vine mat actually does cover and heavily shade the woody vegetation. The swollen-thorn acacias serve as excellent vine supports when unoccupied. Their stiff lateral branches adorned with thorns and compound leaves make excellent attachment points for vines and are quickly covered and bound into the vine mat. On the occupied swollen-thorn acacia, the vine leaders and tendrils are killed by the obligate acacia-ants when they come in contact with the acacia. The effect can be very dramatic when the swollen-thorn acacia grows up through a solid vine mat that is shading all the neighboring plants. The pruning of lateral branch tips from neighboring shrubs, when they contact the swollen-thorn acacia, is another function of the obligate acacia-ants which aids the acacia in its competition with other plants (reduces shading). A by-product of the pruning activities of the obligate acacia-ants is the deterrent effect of the clear area around the acacia on certain types of fires (Janzen, 1966d).

4. The ant × acacia interaction system is possible because of various preadaptations. Both ant and acacia display many morphological and behavioral traits which are necessary to the interaction as it now exists and which apparently were possessed before the species became more interrelated than are the many other species of *Pseudomyrmex* and *Acacia* at present. Further, nearly all of the traits of the ant and the acacia which are important to the interaction can be seen to be the results of gradual reduction or magnification of some pre-interaction property. The three clear exceptions are the Beltian bodies, their use by the obligate acacia-ants, and the attack on foreign plants by the ants.

That the Pseudomyrmecinae are exceptionally well preadapted to the evolution of such ant × plant interactions is suggested by the existence of at least five apparently independently evolved pseudomyrmecine × plant mutualistic interactions (*Pseudomyrmex* × *Acacia*, *Pseudomyrmex* × *Triplaris*, *Pseudomyrmex* × *Tachigalia*, *Pachysima* × *Barteria*, and *Viticicola* × *Vitex*) and only a few doubtful cases in all the other ant subfamilies. It must be stressed that only the first of the five has been experimentally demonstrated, but the descriptions in the literature on the others hint strongly that they too are mutualistic. There remain a large number of ants in the Pseudomyrmecinae and other subfamilies that receive some of their requisites directly from plants, and though less obvious, it appears that there may be many plants besides the classic ant-plants which receive some protection from ants and other predaceous Hymenoptera (Janzen, 1966e).

It appears that in most wet tropical lowland communities there is ample "room" for the presence of more ant × plant interactions. In view of the many cases where it is extremely difficult to demonstrate clear mutualism, it appears that the active evolution of such interactions is now taking place. An ant species which enters into such an interaction generally removes itself from the apparently highly competitive interactions with other foraging ants and other insects, and places itself in a generally more isolated and more immediately reliable fraction of the habitat. It may at the same time make room for expansion of existing generalized ants or the entry of a new species. By this method species density of the ant community may increase. However, species density of the insects that feed on the newly occupied plant will have to decrease, or else they will have to shift their population structures. Thus it is that one cannot predict increases or decreases in total species density through the evolution of ant × plant interactions, but in general the complexity of the community increases and this may be associated with an increase in species density.

5. As can be seen from the concluding paragraph of the last section, a study of this type has a bearing on the problems of high species density in the tropics. It was earlier noted that as one approaches areas with a progressively longer dry season, the ant × acacia system (and thus its members) drops out because the acacia cannot hold its leaves long enough to keep the ant colony alive, and the unoccupied shoot does not survive to maturity due to insect damage. Though less thoroughly studied, the same system appears to be operating as one moves from warm to cool areas (altitudinal or latitudinal); in this case, the ant is not sufficiently active in cool weather to deter the more cool weather adapted phytophagous insects and vertebrate browsers, and thus the acacia receives more damage than it can tolerate and the ant colony starves to death due to lack of leaf products (slower acacia growth as well in cool weather).

As one moves gradually out of warm and moist areas, the complexity and number of plant-insect interspecies interactions appears to decrease per unit area; this is apparently most strongly associated with the gradual reduction in standing biomass produced over the year. As the fluctuations in the physical environment increase and their predictability decreases, and as the standing biomass decreases, the conditions for the evolution of both numbers and quality of interspecies interactions decreases. This is apparently in part due to failure of species because they cannot evolve further to tolerate the physical aspects of the environment. There is, however, another and less obvious reason: the more genetic environment (organisms) there is to evolve with and to match, the more precise can be the evolved interaction, and the more efficiently the given quantity of biomass can be partitioned among the various species concerned. In such a system, the more uniform and predictable the physical environment, the more precise can be the interaction among species. While so much more diverse than the physical environment, the biotic environment (total genetic background) is nevertheless much more predictable in many respects and thus the more biotic environment present, the more complex interactions that can be evolved; this is of course circular in the sense that the more complex biotic environment fosters a more predictable one, and vice versa. Upper limits to such complexity are set by such things as the size of the organisms, the energy requirements of their minimum-sized populations, their own predictability, etc.

Thus, as one goes from the "tropics" to areas with proportionately less and less biotic environment, and more severe unpredictable physical environments, the members of the ant × acacia interaction system gradually disappear. In certain areas there are as many as four species of obligate acacia-ants and three swollen-thorn acacias at the same site (Janzen, 1966b, 1966c). It is characteristic of these distributions that as one moves toward the margins of distribution of swollen-thorn acacias and obligate acacia-ants, all the species but one ant and acacia eventually drop out, and finally these last ones disappear. The point at hand appears to be that the non-uniformity and fluctuations in the physical environment take their toll of species as one moves out of the wet and warm tropics by the breakdown of interaction systems as well as by being inimical to the individual species' requirements.

ACKNOWLEDGMENTS

Throughout this study, I have been aided and inspired by the comments and assistance of E. O. Wilson (Harvard University), W. L. Brown, Jr. (Cornell University), V. E. Rudd (U. S. National Herbarium), C. D. Michener (University of Kansas), and H. V. Daly and R. F. Smith (University of California, Berkeley). I am likewise grateful to the Evolutionists' Club at The University of Kansas, for which the first draft was prepared.

The following people have read the manuscript and offered helpful criticisms, but I do not mean to imply that they necessarily agree with all of the conclusions reached in the final draft: C. D. Michener, W. L. Brown, Jr., G. L. Stebbins, and V. E. Rudd.

LITERATURE CITED

Beck, S. D. 1965. Resistance of plants to insects. Ann. Rev. Ent., 10: 207–232.

Belt, T. 1874. The naturalist in Nicaragua. E. Bumpas, London, 403 pp.

Brown, W. L., Jr. 1960. Ants, acacias and browsing mammals. Ecology, 41: 587–592.

Ehrlich, P. R., and P. H. Raven. 1965. Butterflies and plants: a study in coevolution. Evolution, 18: 586–608.

Janzen, D. H. 1966a. The interaction of the bull's-horn Acacia (Acacia cornigera L.) with one of its ant inhabitants (Pseudomyrmex ferruginea F. Smith) in eastern Mexico. Univ. California Publ. Ent., in press.

———. 1966b. The swollen-thorn acacias of Central America. MS.

———. 1966c. The obligate acacia-ants of Central America. MS.

———. 1966d. Fire, vegetation structure, and the ant × acacia interaction in Central America. MS.

———. 1966e. The ecological significance of myrmecophytes. MS.

Rudd, V. E. 1964. Nomenclatural problems in the Acacia cornigera complex. Madrono, 17: 198–201.

Southwood, T. R. E. 1961. The evolution of the insect-host tree relationship—a new approach. XI. Inter. Congo Ent. Wien 1960, 1: 651–655.

Stark, R W. 1965. Recent trends in forest entomology. Ann. Rev. Ent., 10: 303–324.

Wilson, E. O. 1953. The origin and evolution of polymorphism in ants. Quart. Rev. Biol., 28: 136–156.

Janzen, D. H. 1966. "Coevolution of mutualism between ants and acacias in Central America." Evolution *20:249–275. Reprinted with permission of Wiley-Blackwell.*

Contributors

GAGE H. DAYTON
University of California, Santa Cruz
Natural Reserve System
Physical and Biological Sciences
Santa Cruz, CA 95064

PAUL K. DAYTON
Scripps Institution of Oceanography
University of California, San Diego
La Jolla, CA 92037

MICHAEL H. GRAHAM
Moss Landing Marine Laboratories
Moss Landing, CA 95039

PETER R. GRANT
Department of Ecology and Evolutionary Biology
Princeton University
Princeton, NJ 08544-2016

HARRY W. GREENE
Cornell University
Department of Ecology and Evolutionary Biology
Ithaca, NY 14853

NANCY KNOWLTON
Scripps Institution of Oceanography
University of California, San Diego
La Jolla, CA 92093

SHAHID NAEEM
Department of Ecology, Evolution, and Environmental Biology
Columbia University
New York, NY 10027

ROBERT T. PAINE
University of Washington
Seattle, WA 98195-1800

JOAN PARKER
Moss Landing Marine Laboratories
Moss Landing, CA 95039